BUILDING CONSTRUCTION HANDBOOK

BUILDING CONSTRUCTION HANDBOOK

Eighth edition

R. Chudley

and

R. Greeno

AMSTERDAM • BOSTON • HEIDELBERG • LONDON • NEW YORK • OXFORD
PARIS • SAN DIEGO • SAN FRANCISCO • SINGAPORE • SYDNEY • TOKYO

Butterworth-Heinemann is an imprint of Elsevier

ELSEVIER

Butterworth-Heinemann is an imprint of Elsevier
The Boulevard, Langford Lane, Oxford OX5 1GB, UK
30 Corporate Drive, Suite 400, Burlington, MA 01803, USA

Eighth edition 2010

Notices

Knowledge and best practice in this field are constantly changing. As new research and
experience broaden our understanding, changes in research methods, professional practices, or
medical treatment may become necessary.

Practitioners and researchers must always rely on their own experience and knowledge in
evaluating and using any information, methods, compounds, or experiments described herein. In
using such information or methods they should be mindful of their own safety and the safety of
others, including parties for whom they have a professional responsibility.

To the fullest extent of the law, neither the Publisher nor the authors, contributors, or editors,
assume any liability for any injury and/or damage to persons or property as a matter of products
liability, negligence or otherwise, or from any use or operation of any methods, products,
instructions, or ideas contained in the material herein.

British Library Cataloguing in Publication Data
A catalogue record for this book is available from the British Library

Library of Congress Cataloging-in-Publication Data Control Number:
A catalog record for this book is available from the Library of Congress

ISBN: 978-1-85617-805-1

For information on all Butterworth-Heinemann publications
visit our website at elsevierdirect.com

Typeset by MPS Limited, a Macmillan Company
Printed and bound in Great Britain

10 11   11 10 9 8 7 6 5 4 3 2

Working together to grow
libraries in developing countries

www.elsevier.com | www.bookaid.org | www.sabre.org

ELSEVIER   BOOK AID International   Sabre Foundation

CONTENTS

Preface to eighth edition xi

Part One General

Built environment   2
The structure   5
Primary and secondary elements   12
Component parts and functions   15
Construction activities   19
Construction documents   20
Construction drawings   21
Building survey   28
HIPs/Energy Performance Certificates   32
Method statement and programming   33
Weights and densities of building materials   35
Imposed floor loads   37
Drawings – notations   38
Planning application   42
Modular coordination   47
Construction regulations   49
CDM regulations   50
Safety signs and symbols   51
Building Regulations   53
Code for Sustainable Homes   62
British Standards   63
European Standards   64
Product and practice accreditation   66
CPI System of Coding   67
CI/SfB system of coding   68

Part Two Site Works

Site survey   70
Site investigations   71
Soil investigation   74
Soil assessment and testing   81
Site layout considerations   88
Site security   91
Site lighting and electrical supply   94
Site office accommodation   98

Contents

Materials storage   101
Materials testing   106
Dry and wet rot   121
Protection orders for trees and structures   123
Locating public utility services   124
Setting out   125
Levels and angles   129
Road construction   132
Tubular scaffolding and scaffolding systems   140
Shoring systems   153
Demolition   162

Part Three   Builders Plant

General considerations   168
Bulldozers   171
Scrapers   172
Graders   173
Tractor shovels   174
Excavators   175
Transport vehicles   180
Hoists   183
Rubble chutes and skips   185
Cranes   186
Concreting plant   198

Part Four   Substructure

Foundations – function, materials and sizing   206
Foundation beds   215
Short bored pile foundations   221
Foundation types and selection   223
Piled foundations   228
Retaining walls   248
Gabions and mattresses   262
Basement construction   269
Waterproofing basements   272
Excavations   278
Concrete production   284
Cofferdams   290
Caissons   292
Underpinning   294
Ground water control   303
Soil stabilisation and improvement   313
Reclamation of waste land   318
Contaminated sub-soil treatment   319

Part Five   Superstructure – 1

Choice of materials   322
Brick and block walls   323
Cavity walls   338
Damp-proof courses and membranes   344
Gas resistant membranes   351
Calculated brickwork   353
Mortars   356
Arches and openings   359
Windows   366
Glass and glazing   379
Doors   391
Crosswall construction   400
Framed construction   404
Rendering to external walls   408
Cladding to external walls   410
Roofs – basic forms   417
Pitched roofs   420
Double lap tiling   437
Single lap tiling   439
Slating   441
Flat roofs   447
Dormer windows   456
Green roofs   465
Thermal insulation   467
'U' values   472
Thermal bridging   488
Access for the disabled   492

Part Six   Superstructure – 2

Reinforced concrete slabs   496
Reinforced concrete framed structures   500
Reinforcement types   510
Structural concrete, fire protection   513
Formwork   516
Precast concrete frames   521
Prestressed concrete   525
Structural steelwork sections   532
Structural steelwork connections   537
Structural fire protection   542
Portal frames   549
Composite timber beams   557
Multi-storey structures   560
Roof sheet coverings   564

Contents

Long span roofs   569
Shell roof construction   579
Membrane roofs   588
Rooflights   590
Panel walls   594
Rainscreen cladding   600
Structural glazing   602
Curtain walling   603
Concrete claddings   607
Concrete surface finishes   614
Concrete surface defects   616

Part Seven   Internal Construction and Finishes

Internal elements   618
Internal walls   619
Construction joints   624
Internal walls, fire protection   626
Party/separating walls   628
Partitions   629
Strut design   631
Plasters and plastering   636
Dry lining techniques   639
Plasterboard   642
Wall tiling   645
Domestic floors and finishes   647
Large cast in-situ ground floors   654
Concrete floor screeds   656
Timber suspended floors   658
Lateral restraint   661
Timber beam design   664
Timber floors, fire protection   667
Reinforced concrete suspended floors   668
Precast concrete floors   673
Raised access floors   678
Sound insulation   679
Timber, concrete and metal stairs   685
Internal doors   716
Doorsets   718
Fire resisting doors   719
Plasterboard ceilings   725
Suspended ceilings   726
Paints and painting   730
Joinery production   734
Composite boarding   739
Plastics in building   741

Part Eight Domestic Services

Drainage effluents 746
Subsoil drainage 747
Surface water removal 749
Road drainage 752
Rainwater installations 754
Drainage systems 758
Drainage-pipe sizes and gradients 766
Water supply 767
Cold water installations 769
Hot water installations 771
Flow controls 774
Cisterns and cylinders 775
Pipework joints 777
Sanitary fittings 778
Single and ventilated stack systems 781
Hot water heating systems 784
Electrical supply and installation 788
Gas supply and gas fires 797
Open fireplaces and flues 801
Telephone installations 811
Electronic communications installations 812

Index 813

PREFACE TO EIGHTH EDITION

This edition retains the predominantly illustrative format of earlier editions, presenting the principles of building construction with comprehensive guidance to procedures with numerous examples of formulated and empirical design. Summary notes are supplemented with references to further reading where appropriate.

The content applies to both current and established UK construction practice. This includes the building and maintenance of housing and other low-rise structures and the more advanced techniques applied to medium and high-rise commercial and large industrial buildings. Many examples from previous editions are kept as important references and benchmarks for newer applications. These have evolved in response to material developments and in consideration for environmental issues, not least with regard to energy conservation measures and sustainable building.

The UK's housing stock of about 25 million dwellings includes approximately 2 million units built in the past decade. Therefore, the aftercare of older buildings is an important part of the construction industry's economy. In order to represent this important sector of maintenance, refurbishment, renovation and remedial work, many established practices are included in the *Handbook*.

Modern construction processes and associated technology are incorporated in this new edition, however the content is not extensive, nor is it intended to be prescriptive. Building design and subsequent construction techniques are varied and diverse depending on availability of materials and skills. This *Handbook* provided guidance to achieving these objectives, but sufficient publishing space cannot cover every possibility. Therefore, the reader is encouraged to supplement their study with site observation and practice, with further reading of professional journals, legislative papers and manufacturer's catalogues.

Roger Greeno 2010

1 GENERAL

BUILT ENVIRONMENT

THE STRUCTURE

PRIMARY AND SECONDARY ELEMENTS

CONSTRUCTION ACTIVITIES

CONSTRUCTION DOCUMENTS

CONSTRUCTION DRAWINGS

BUILDING SURVEY

HIPs/EPCs

MATERIAL WEIGHTS AND DENSITIES

IMPOSED FLOOR LOADS

PLANNING APPLICATION

MODULAR COORDINATION

CONSTRUCTION REGULATIONS

CDM REGULATIONS

SAFETY SIGNS AND SYMBOLS

BUILDING REGULATIONS

CODE FOR SUSTAINABLE HOMES

BRITISH STANDARDS

EUROPEAN STANDARDS

CPI SYSTEM OF CODING

CI/SFB SYSTEM OF CODING

Environment = surroundings which can be natural, man-made or a combination of these.

Built Environment = created by man with or without the aid of the natural environment.

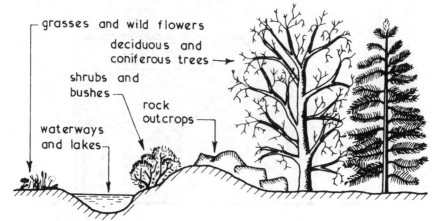

grasses and wild flowers

deciduous and coniferous trees →

shrubs and bushes

rock outcrops

waterways and lakes

ELEMENTS of the NATURAL ENVIRONMENT

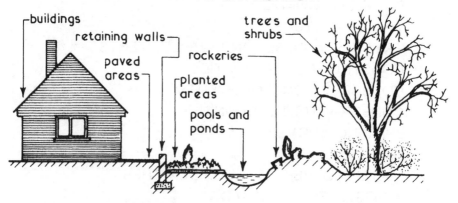

buildings

retaining walls

trees and shrubs

paved areas

rockeries

planted areas

pools and ponds

ELEMENTS of the BUILT ENVIRONMENT (EXTERNAL)

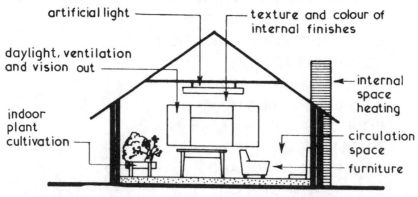

artificial light

texture and colour of internal finishes

daylight, ventilation and vision out

internal space heating

indoor plant cultivation

circulation space

furniture

ELEMENTS of the BUILT ENVIRONMENT (INTERNAL)

Environmental Considerations

1. Planning requirements.
2. Building Regulations.
3. Land restrictions by vendor or lessor.
4. Availability of services.
5. Local amenities including transport.
6. Subsoil conditions.
7. Levels and topography of land.
8. Adjoining buildings or land.
9. Use of building.
10. Daylight and view aspects.

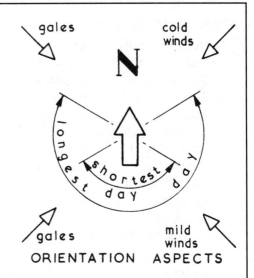

ORIENTATION   ASPECTS

Examples:~

HOUSES

SCHOOLS

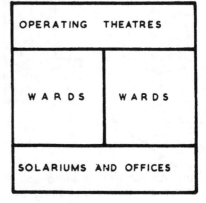

FACTORIES

HOSPITALS

Physical considerations

1. Natural contours of land.
2. Natural vegetation and trees.
3. Size of land and/or proposed building.
4. Shape of land and/or proposed building.
5. Approach and access roads and footpaths.
6. Services available.
7. Natural waterways, lakes and ponds.
8. Restrictions such as rights of way; tree preservation and ancient buildings.
9. Climatic conditions created by surrounding properties, land or activities.
10. Proposed future developments.

Examples:~

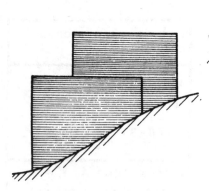

Split level construction to form economic shape.

Shape determined by existing trees.

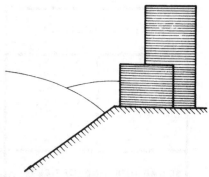

Plateau or high ground solution giving dry site conditions on sloping sites.

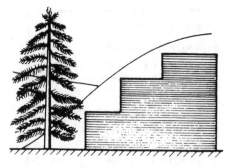

Stepped elevation or similar treatment to blend with the natural environment.

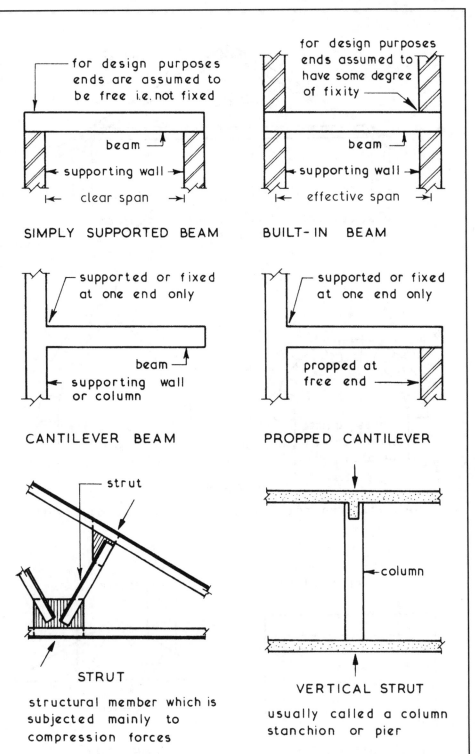

for design purposes ends are assumed to be free i.e. not fixed

beam

supporting wall

clear span

SIMPLY SUPPORTED BEAM

for design purposes ends assumed to have some degree of fixity

beam

supporting wall

effective span

BUILT-IN BEAM

supported or fixed at one end only

beam

supporting wall or column

CANTILEVER BEAM

supported or fixed at one end only

propped at free end

PROPPED CANTILEVER

strut

STRUT

structural member which is subjected mainly to compression forces

column

VERTICAL STRUT

usually called a column stanchion or pier

5

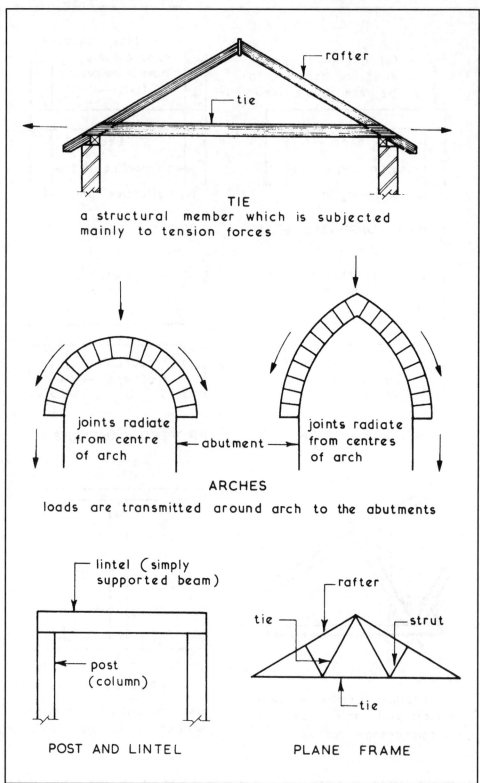

TIE
a structural member which is subjected mainly to tension forces

joints radiate from centre of arch

←— abutment —→

joints radiate from centres of arch

ARCHES
loads are transmitted around arch to the abutments

lintel (simply supported beam)

post (column)

POST AND LINTEL

rafter

tie

strut

tie

PLANE FRAME

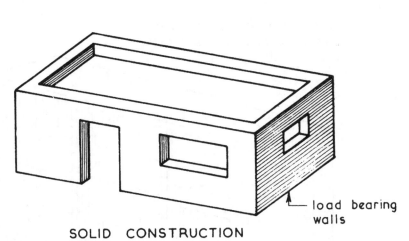

SOLID CONSTRUCTION

structurally limited confined usually to buildings of low height and short spans

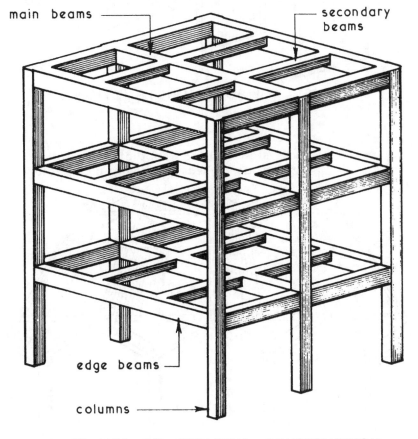

FRAMED OR SKELETAL CONSTRUCTION

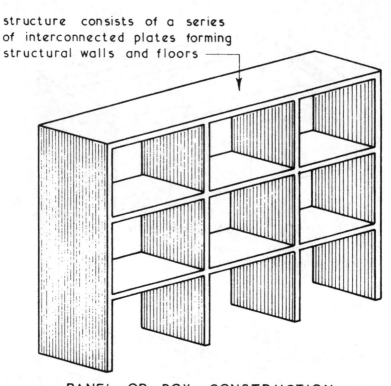

structure consists of a series
of interconnected plates forming
structural walls and floors

PANEL OR BOX CONSTRUCTION

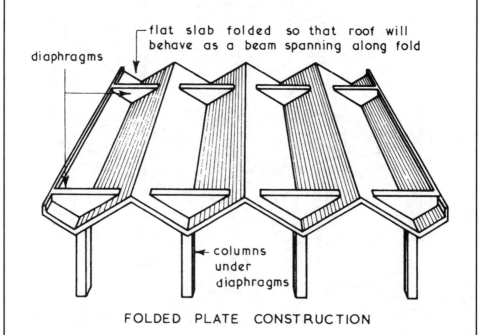

flat slab folded so that roof will
behave as a beam spanning along fold

diaphragms

columns
under
diaphragms

FOLDED PLATE CONSTRUCTION

Shell Roofs ~ these are formed by a structural curved skin covering a given plan shape and area.

Examples ~

double curvature shell formed by rotating a plain curved shape about a vertical axis

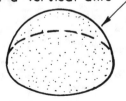

DOME OR ROTATIONAL SHELL

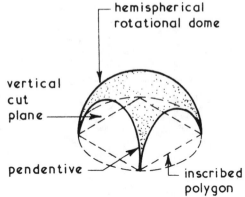

hemispherical rotational dome

vertical cut plane

pendentive

inscribed polygon

PENDENTIVE DOME

formed by a curved line moving over another curved line

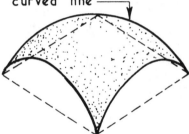

TRANSLATIONAL DOME

cut cylinder giving a single curvature shell

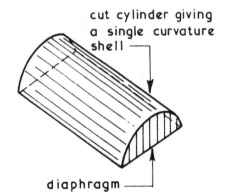

diaphragm

BARREL VAULT

double curvature shells

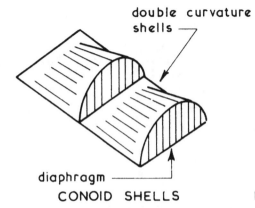

diaphragm

CONOID SHELLS

double curvature saddle shaped shell

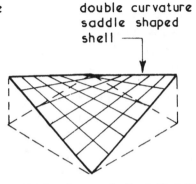

HYPERBOLIC PARABOLOID

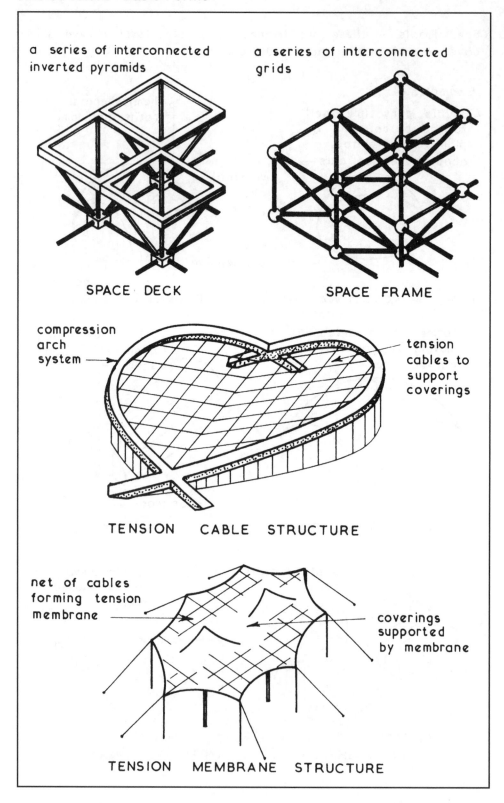

a series of interconnected inverted pyramids

a series of interconnected grids

SPACE DECK

SPACE FRAME

compression arch system

tension cables to support coverings

TENSION CABLE STRUCTURE

net of cables forming tension membrane

coverings supported by membrane

TENSION MEMBRANE STRUCTURE

Substructure ~ can be defined as all structure below the superstructure which in general terms is considered to include all structure below ground level but including the ground floor bed.

Typical Examples~

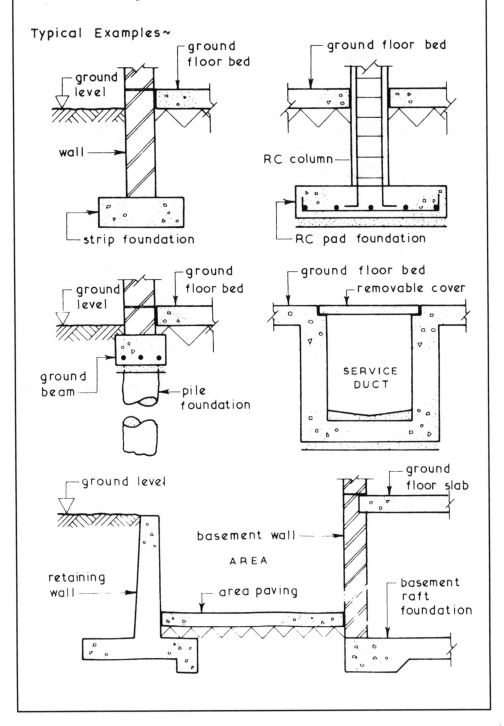

Superstructure ~ can be defined as all structure above substructure both internally and externally.

Primary Elements ~ basically components of the building carcass above the substructure excluding secondary elements, finishes, services and fittings.

Typical Examples~

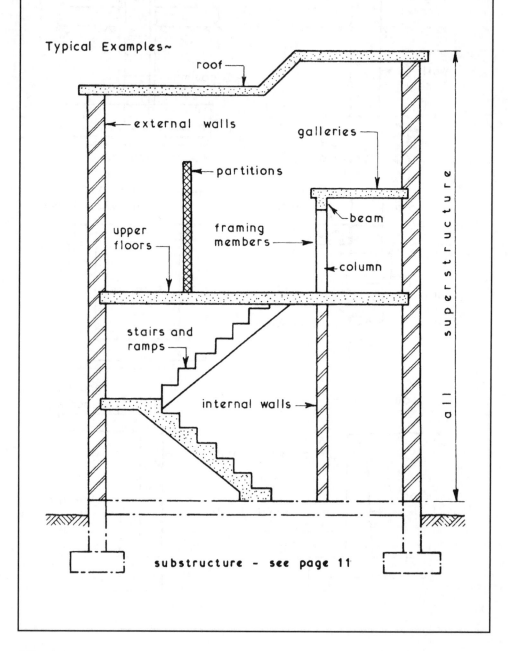

substructure - see page 11

Secondary Elements ~ completion of the structure including completion around and within openings in primary elements.

Typical Examples~

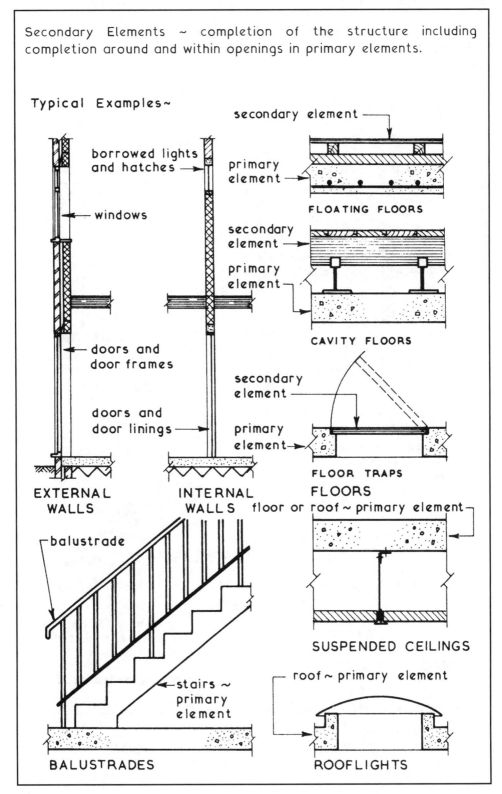

borrowed lights and hatches →

windows

doors and door frames

doors and door linings →

EXTERNAL WALLS

INTERNAL WALLS

secondary element

primary element

FLOATING FLOORS

secondary element →

primary element →

CAVITY FLOORS

secondary element

primary element→

FLOOR TRAPS

FLOORS

floor or roof ~ primary element

SUSPENDED CEILINGS

balustrade

stairs ~ primary element

BALUSTRADES

roof ~ primary element

ROOFLIGHTS

Finish ~ the final surface which can be self finished as with a trowelled concrete surface or an applied finish such as floor tiles.

Typical Examples~

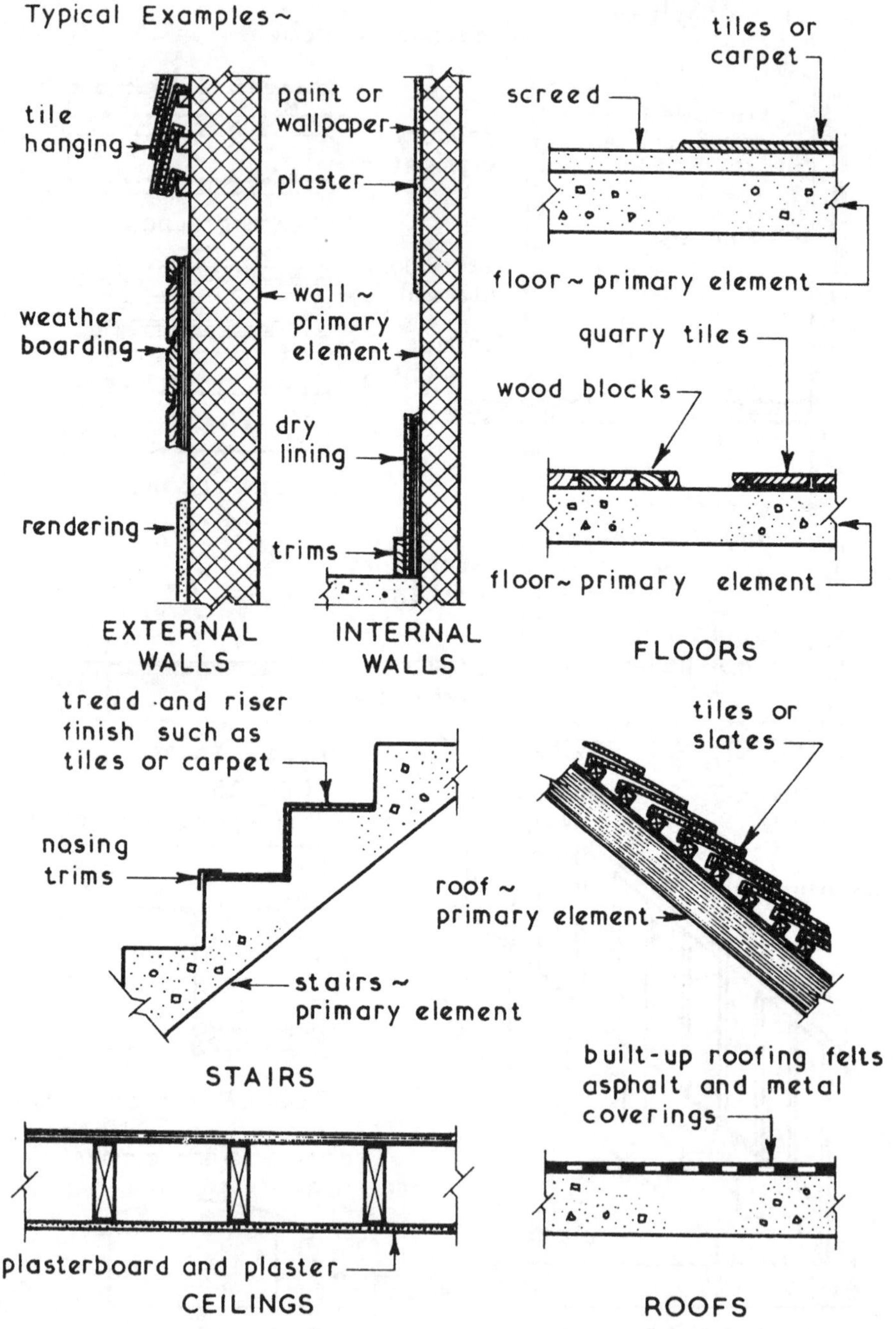

tiles or carpet

screed

tile hanging

paint or wallpaper

plaster

weather boarding

wall ~ primary element

floor ~ primary element

dry lining

quarry tiles

wood blocks

rendering

trims

floor~ primary element

EXTERNAL WALLS

INTERNAL WALLS

FLOORS

tread and riser finish such as tiles or carpet

tiles or slates

nosing trims

roof ~ primary element

stairs ~ primary element

STAIRS

built-up roofing felts asphalt and metal coverings

plasterboard and plaster

CEILINGS

ROOFS

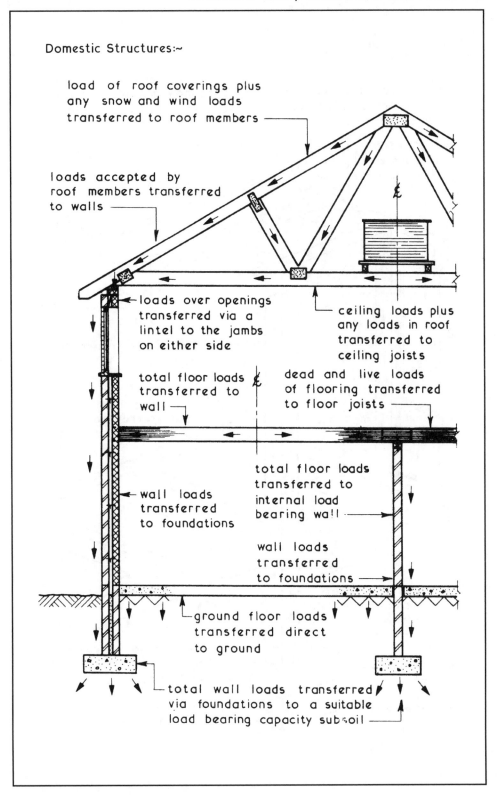

Domestic Structures:~

load of roof coverings plus any snow and wind loads transferred to roof members

loads accepted by roof members transferred to walls

loads over openings transferred via a lintel to the jambs on either side

ceiling loads plus any loads in roof transferred to ceiling joists

total floor loads transferred to wall

dead and live loads of flooring transferred to floor joists

wall loads transferred to foundations

total floor loads transferred to internal load bearing wall

wall loads transferred to foundations

ground floor loads transferred direct to ground

total wall loads transferred via foundations to a suitable load bearing capacity subsoil

Framed Structures:~

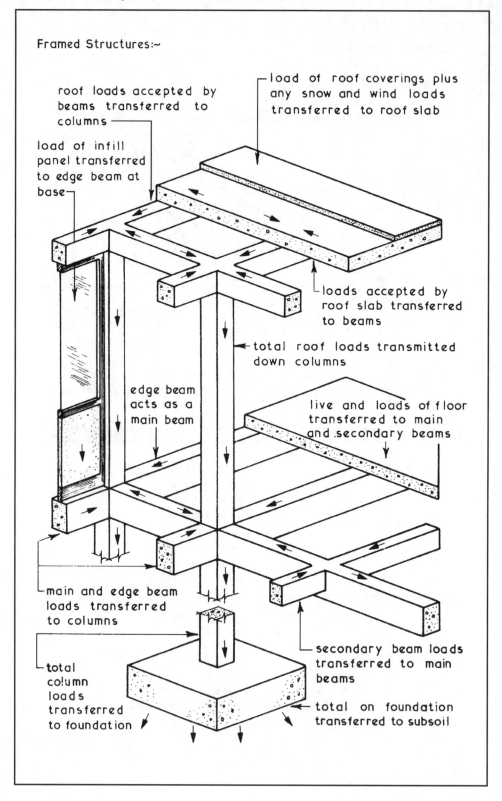

roof loads accepted by beams transferred to columns

load of roof coverings plus any snow and wind loads transferred to roof slab

load of infill panel transferred to edge beam at base

loads accepted by roof slab transferred to beams

total roof loads transmitted down columns

edge beam acts as a main beam

live and loads of floor transferred to main and secondary beams

main and edge beam loads transferred to columns

secondary beam loads transferred to main beams

total column loads transferred to foundation

total on foundation transferred to subsoil

External Envelope ~ consists of the materials and components which form the external shell or enclosure of a building. These may be load bearing or non-load bearing according to the structural form of the building.

Primary Functions:~

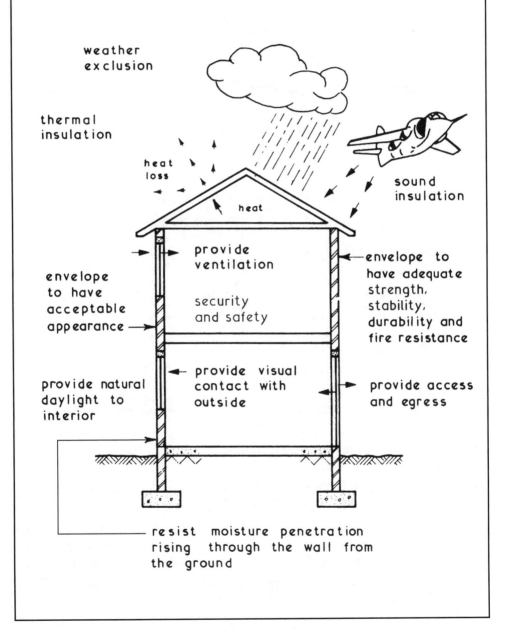

weather exclusion

thermal insulation

heat loss

heat

sound insulation

provide ventilation

security and safety

envelope to have acceptable appearance

envelope to have adequate strength, stability, durability and fire resistance

provide natural daylight to interior

provide visual contact with outside

provide access and egress

resist moisture penetration rising through the wall from the ground

Dwelling houses ~

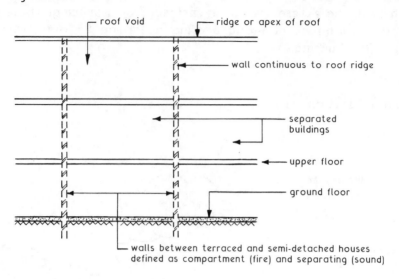

- roof void
- ridge or apex of roof
- wall continuous to roof ridge
- separated buildings
- upper floor
- ground floor

walls between terraced and semi-detached houses defined as compartment (fire) and separating (sound)

Flats ~

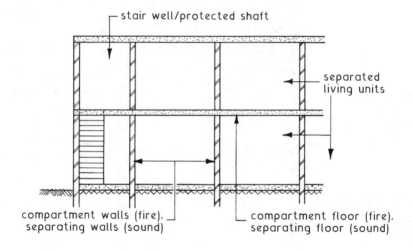

- stair well/protected shaft
- separated living units

compartment walls (fire), separating walls (sound)

compartment floor (fire), separating floor (sound)

Note: Floors within a maisonette are not required to be "compartment".

For non-residential buildings, compartment size is limited by floor area depending on the building function (purpose group) and height.

Compartment ~ a building or part of a building with walls and floors constructed to contain fire and to prevent it spreading to another part of the same building or to an adjoining building.

Separating floor/wall ~ element of sound resisting construction between individual living units.

A Building or Construction Site can be considered as a temporary factory employing the necessary resources to successfully fulfil a contract.

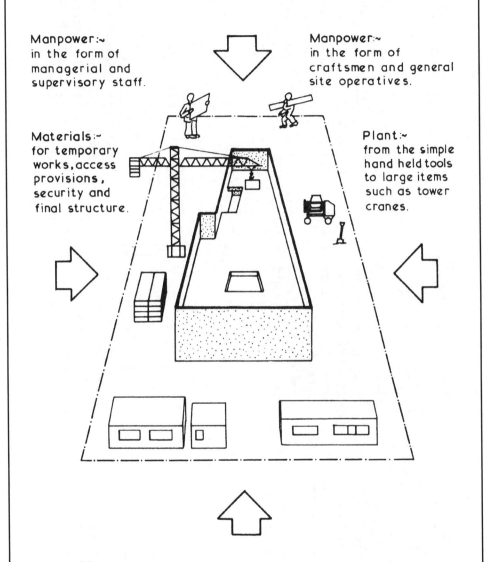

Manpower:~
in the form of managerial and supervisory staff.

Manpower:~
in the form of craftsmen and general site operatives.

Materials:~
for temporary works, access provisions, security and final structure.

Plant:~
from the simple hand held tools to large items such as tower cranes.

Money:~
in the form of capital investment from the building owner to pay for the land, design team fees and a building contractor who uses his money to buy materials, buy or hire plant and hire labour to enable the project to be realised.

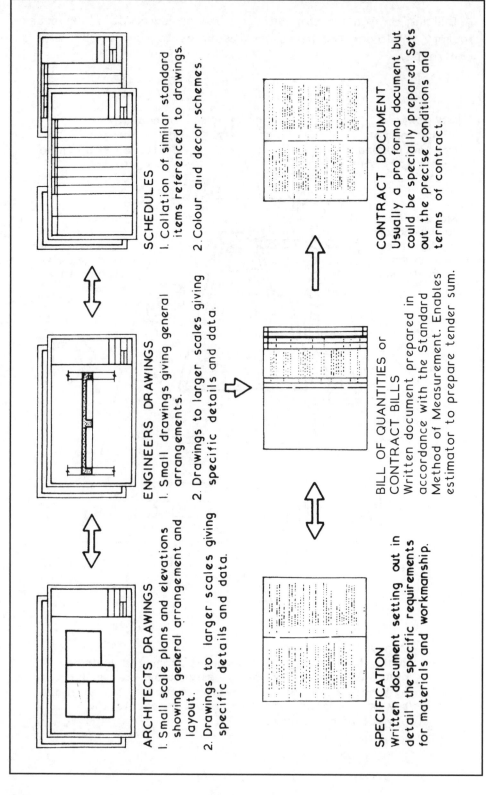

ARCHITECTS DRAWINGS
1. Small scale plans and elevations showing general arrangement and layout.
2. Drawings to larger scales giving specific details and data.

ENGINEERS DRAWINGS
1. Small drawings giving general arrangements.
2. Drawings to larger scales giving specific details and data.

SCHEDULES
1. Collation of similar standard items referenced to drawings.
2. Colour and decor schemes.

SPECIFICATION
Written document setting out in detail the specific requirements for materials and workmanship.

BILL OF QUANTITIES or CONTRACT BILLS
Written document prepared in accordance with the Standard Method of Measurement. Enables estimator to prepare tender sum.

CONTRACT DOCUMENT
Usually a pro forma document but could be specially prepared. Sets out the precise conditions and terms of contract.

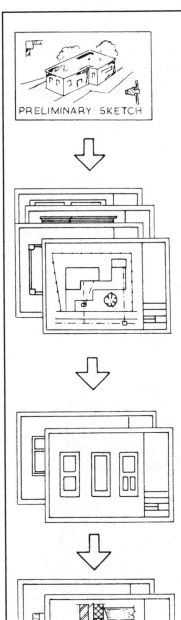

PRELIMINARY SKETCH

Location Drawings ~

Site Plans – used to locate site, buildings, define site levels, indicate services to buildings, identify parts of site such as roads, footpaths and boundaries and to give setting out dimensions for the site and buildings as a whole. Suitable scale not less than 1:2500

Floor Plans – used to identify and set out parts of the building such as rooms, corridors, doors, windows, etc., Suitable scale not less than 1:100

Elevations – used to show external appearance of all faces and to identify doors and windows. Suitable scale not less than 1:100

Sections – used to provide vertical views through the building to show method of construction. Suitable scale not less than 1:50

Component Drawings ~

used to identify and supply data for components to be supplied by a manufacturer or for components not completely covered by assembly drawings. Suitable scale range 1:100 to 1:1

Assembly Drawings ~

used to show how items fit together or are assembled to form elements. Suitable scale range 1:20 to 1:5

All drawings should be fully annotated, fully dimensioned and cross referenced.

Ref. BS EN ISO 7519: Technical drawings. Construction drawings. General principles of presentation for general arrangement and assembly drawings.

Sketch ~ this can be defined as a draft or rough outline of an idea, it can be a means of depicting a three-dimensional form in a two-dimensional guise. Sketches can be produced free-hand or using rules and set squares to give basic guide lines.

All sketches should be clear, show all the necessary detail and above all be in the correct proportions.

Sketches can be drawn by observing a solid object or they can be produced from conventional orthographic views but in all cases can usually be successfully drawn by starting with an outline 'box' format giving length, width and height proportions and then building up the sketch within the outline box.

Example~ Square Based Chimney Pot.

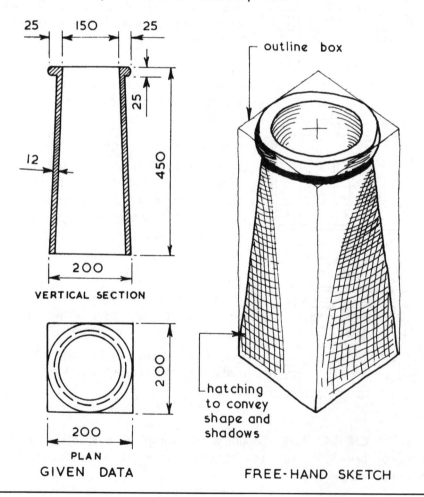

VERTICAL SECTION

PLAN

GIVEN DATA

outline box

hatching
to convey
shape and
shadows

FREE-HAND SKETCH

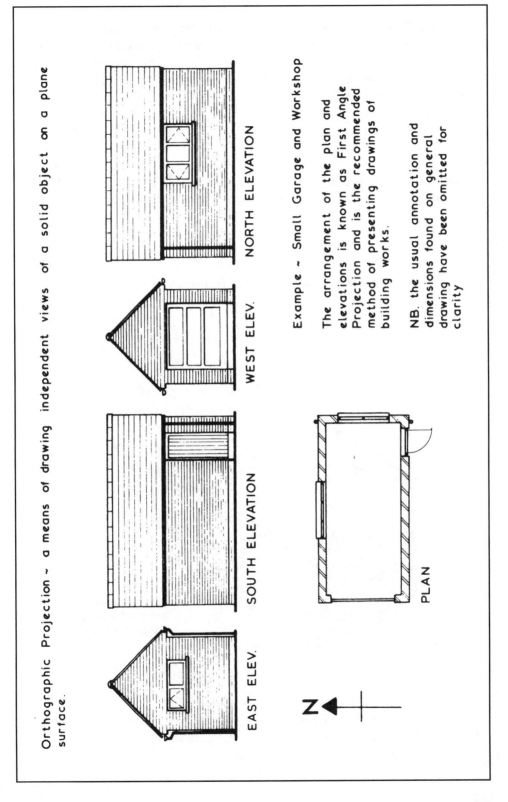

Orthographic Projection ~ a means of drawing independent views of a solid object on a plane surface.

NORTH ELEVATION

WEST ELEV.

SOUTH ELEVATION

EAST ELEV.

PLAN

N

Example ~ Small Garage and Workshop

The arrangement of the plan and elevations is known as First Angle Projection and is the recommended method of presenting drawings of building works.

NB. the usual annotation and dimensions found on general drawing have been omitted for clarity

Isometric Projections ~ a pictorial projection of a solid object on a plane surface drawn so that all vertical lines remain vertical and of true scale length, all horizontal lines are drawn at an angle of 30° and are of true scale length therefore scale measurements can be taken on the vertical and 30° lines but cannot be taken on any other inclined line.

A similar drawing can be produced using an angle of 45° for all horizontal lines and is called an Axonometric Projection

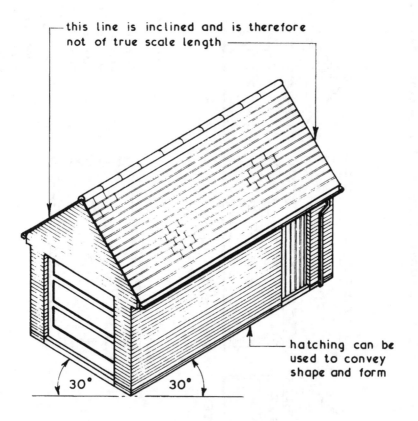

this line is inclined and is therefore not of true scale length

hatching can be used to convey shape and form

30°   30°

ISOMETRIC PROJECTION SHOWING SOUTH AND WEST ELEVATIONS OF SMALL GARAGE AND WORKSHOP ILLUSTRATED ON PAGE 23

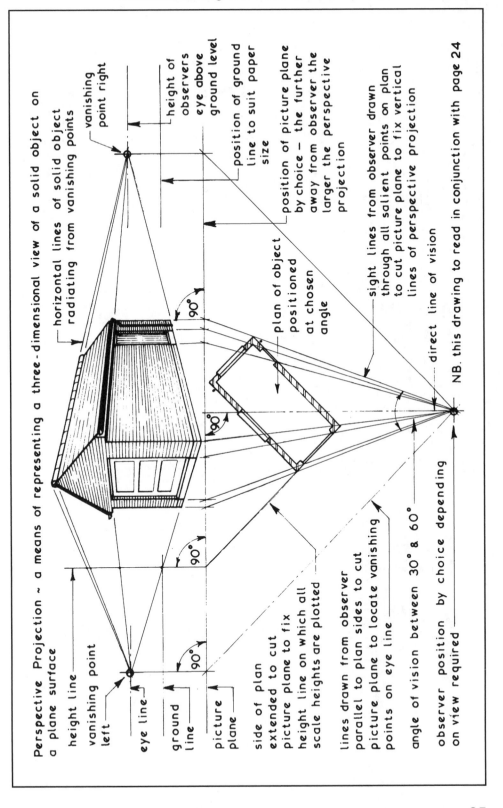

Perspective Projection ~ a means of representing a three-dimensional view of a solid object on a plane surface

height line

vanishing point left

eye line

ground line

picture plane

side of plan extended to cut picture plane to fix height line on which all scale heights are plotted

lines drawn from observer parallel to plan sides to cut picture plane to locate vanishing points on eye line

angle of vision between 30° & 60°

observer position by choice depending on view required

horizontal lines of solid object radiating from vanishing points

vanishing point right

height of observers eye above ground level

position of ground line to suit paper size

position of picture plane by choice — the further away from observer the larger the perspective projection

plan of object positioned at chosen angle

sight lines from observer drawn through all salient points on plan to cut picture plane to fix vertical lines of perspective projection

direct line of vision

NB. this drawing to read in conjunction with page 24

90°

90°

90°

90°

25

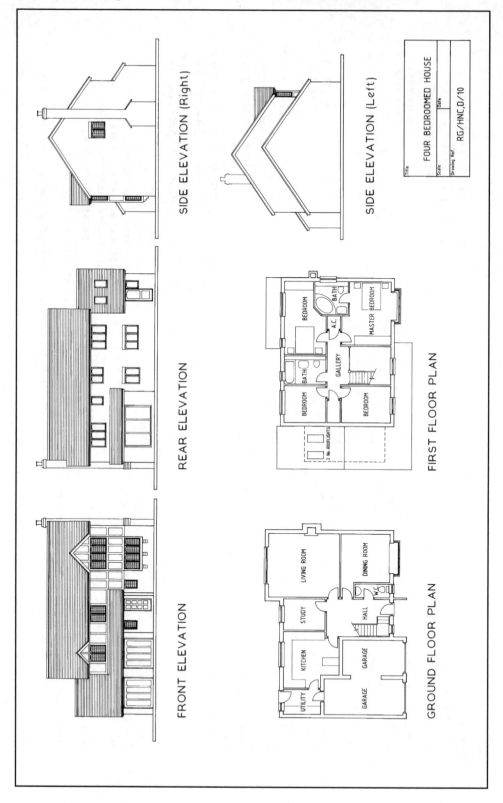

SIDE ELEVATION (Right)

SIDE ELEVATION (Left)

FOUR BEDROOMED HOUSE

Title

Scale Date

Drawing Ref. RG/HNC,D/10

REAR ELEVATION

FIRST FLOOR PLAN

BEDROOM

BATH

A.C.

MASTER BEDROOM

BEDROOM

BATH

GALLERY

BEDROOM

BEDROOM

2 No ROOFLIGHTS

FRONT ELEVATION

GROUND FLOOR PLAN

LIVING ROOM

DINING ROOM

STUDY

W.C.

HALL

KITCHEN

GARAGE

UTILITY

GARAGE

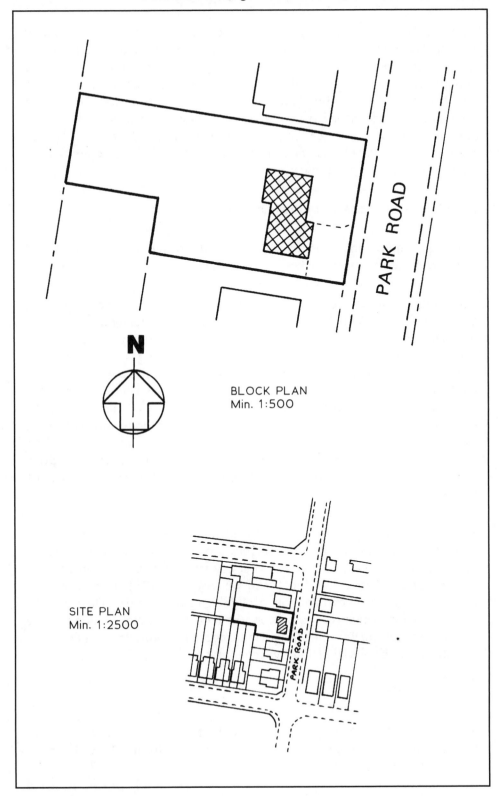

N

BLOCK PLAN
Min. 1:500

SITE PLAN
Min. 1:2500

PARK ROAD

Construction Defects – correct application of materials produced to the recommendations of British, European and International Standards authorities, in accordance with local building regulations, by-laws and the rules of building guarantee companies, i.e. National House Building Council (NHBC) and MD Insurance Services, should ensure a sound and functional structure. However, these controls can be seriously undermined if the human factor of quality workmanship is not fulfilled. The following guidance is designed to promote quality controls:

BS 8000: Workmanship on building sites.

Building Regulations, Approved Document to support Regulation 7 – materials and workmanship.

No matter how good the materials, the workmanship and supervision, the unforeseen may still affect a building. This may materialise several years after construction. Some examples of these latent defects include: woodworm emerging from untreated timber, electrolytic decomposition of dissimilar metals inadvertently in contact, and chemical decomposition of concrete. Generally, the older a building the more opportunity there is for its components and systems to have deteriorated and malfunctioned. Hence the need for regular inspection and maintenance. The profession of facilities management has evolved for this purpose and is represented by the British Institute of Facilities Management (BIFM).

Property values, repairs and replacements are of sufficient magnitude for potential purchasers to engage the professional services of a building surveyor. Surveyors are usually members of the Royal Institution of Chartered Surveyors (RICS). The extent of survey can vary, depending on a client's requirements. This may be no more than a market valuation to secure financial backing, to a full structural survey incorporating specialist reports on electrical installations, drains, heating systems, etc.

Further reading:

BRE Digest No. 268 – Common defects in low-rise traditional housing. Available from Building Research Establishment Bookshop – www.brebookshop.com.

Established Procedure – the interested purchaser engages a building surveyor.

UK Government Requirements – the seller to provide a property/ home information pack (HIP) which can include `A survey report on the condition of the property, including requirements for urgent or significant repairs ...'.

Survey document preliminaries:

* Title and address of property

* Client's name, address and contacts

* Survey date and time

* Property status – freehold, leasehold or commonhold

* Occupancy – occupied or vacant. If vacant, source of keys

* Extent of survey, e.g. full structural + services reports

* Specialists in attendance, e.g. electrician, heating engineer, etc.

* Age of property (approx. if very dated or no records)

* Disposition of rooms, i.e. number of bedrooms, etc.

* Floor plans and elevations if available

* Elevation (flooding potential) and orientation (solar effect)

* Estate/garden area and disposition if appropriate

* Means of access – roads, pedestrian only, rights of way

Survey tools and equipment:

* Drawings + estate agent's particulars if available

* Notebook and pencil/pen

* Binoculars and a camera with flash facility

* Tape measure, spirit level and plumb line

* Other useful tools, to include small hammer, torch, screwdriver and manhole lifting irons

* Moisture meter

* Ladders – eaves access and loft access

* Sealable bags for taking samples, e.g. wood rot, asbestos, etc.

Estate and garden:

* Location and establishment of boundaries
* Fences, gates and hedges – material, condition and suitability
* Trees – type and height, proximity to building
* Pathways and drives – material and condition
* Outbuildings – garages, sheds, greenhouses, barns, etc.
* Proximity of water courses

Roof:

* Tile type, treatment at ridge, hips, verge and valleys
* Age of covering, repairs, replacements, renewals, general condition, defects and growths
* Eaves finish, type and condition
* Gutters – material, size, condition, evidence of leakage
* Rainwater downpipes as above
* Chimney – dpcs, flashings, flaunching, pointing, signs of movement
* Flat roofs – materials, repairs, abutments, flashings and drainage

Walls:

* Materials – type of brick, rendering, cladding, etc., condition and evidence of repairs
* Solid or cavity construction, if cavity extent of insulation and type
* Pointing of masonry, painting of rendering and cladding
* Air brick location, function and suitability
* Dpc, material and condition, position relative to ground level
* Windows and doors, material, signs of rot or damage, original or replacement, frame seal
* Settlement – signs of cracking, distortion of window and door frames – specialist report

Drainage:

A building surveyor may provide a general report on the condition of the drainage and sanitation installation. However, a full test for leakage and determination of self-cleansing and flow conditions to include fibre-optic scope examination is undertaken as a specialist survey.

Roof space:
* Access to all parts, construction type – traditional or trussed
* Evidence of moisture due to condensation – ventilation at eaves, ridge, etc.
* Evidence of water penetration – chimney flashings, abutments and valleys
* Insulation – type and quantity
* Party wall in semi-detached and terraced dwellings – suitability as fire barrier
* Plumbing – adequacy of storage cistern, insulation, overflow function

Floors:
* Construction – timber, pre-cast or cast in-situ concrete? Finish condition?
* Timber ground floor – evidence of dampness, rot, woodworm, ventilation, dpcs
* Timber upper floor stability, ie. wall fixing, strutting, joist size, woodworm, span and loading

Stairs:
* Type of construction and method of fixing – built in-situ or preformed
* Soffit, re. fire protection (plasterboard?)
* Balustrading – suitability and stability
* Safety – adequate screening, balusters, handrail, pitch angle, open tread, tread wear

Finishes:
* General décor, i.e. paint and wallpaper condition – damaged, faded
* Woodwork/joinery – condition, defects, damage, paintwork
* Plaster – ceiling (plasterboard or lath and plaster?) – condition and stability
* Plaster – walls – render and plaster or plasterboard, damage and quality of finish
* Staining – plumbing leaks (ceiling), moisture penetration (wall openings), rising damp
* Fittings and ironmongery – adequacy and function, weather exclusion and security

Supplementary enquiries should determine the extent of additional building work, particularly since the planning threshold of 1948. Check for planning approvals, permitted development and Building Regulation approvals, exemptions and completion certificates.

Services – apart from a cursory inspection to ascertain location and suitability of system controls, these areas are highly specialised and should be surveyed by those appropriately qualified.

Home Information Packs ~ otherwise known as HIPS or "seller's packs". A HIP is provided as supplementary data to the estate agent's sales particulars by home sellers when marketing a house. The packs place emphasis on an energy use assessment and contain some contract preliminaries such as evidence of ownership. Property developers are required to provide a HIP as part of their sales literature. Preparation is by a surveyor, specifically trained in energy performance assessment.

Compulsory Content ~
• Index
• Energy performance certificate
• Sales statement
• Standard searches, e.g. LA enquiries, planning consents, drainage arrangements, utilities providers
• Evidence of title (ownership)
• Leasehold and commonhold details (generally flats and maisonettes)
• Property information questionnaire, to include flood risk, gas and electricity safety, service charges, structural damage and parking arrangements

Optional Content ~
• Home condition report (general survey)
• Legal summary – terms of sale
• Home use and contents form (fixtures and fittings)
• Guarantees and warrantees
• Other relevant information, e.g. access over ancillary land

Energy Performance Certificate (EPC) ~ provides a rating between A and G. A is the highest possible grade for energy efficiency and lowest impact on environmental damage in terms of CO_2 emissions. The certificate is similar to the EU energy label (see page 480 as applied to windows) and it relates to SAP numerical ratings (see page 477). The certificate is an asset rating based on a building's performance relating to its age, location/exposure, size, appliance efficiency e.g. boiler, glazing type, construction, insulation and general condition.

EPC rating (SAP rating) ~

A (92–100)   B (81–91)   C (69–80)   D (55–68)
E (39–54)   F (21–38)   G (1–20)

Ref. The Home Information Pack Regulations 2006.

A method statement precedes preparation of the project programme and contains the detail necessary for construction of each element of a building. It is prepared from information contained in the contract documents – see page 20. It also functions as a brief for site staff and operatives in sequencing activities, indicating resource requirements and determining the duration of each element of construction. It complements construction programming by providing detailed analysis of each activity.

A typical example for foundation excavation could take the following format:

Activity	Quantity	Method	Output/hour	Labour	Plant	Days
Strip site for excavation	300 m²	Exc. to reduced level over construction area – JCB-4CX face shovel/ loader. Topsoil retained on site.	50 m²/hr	Exc. driver +2 labourers	JCB-4CX backhoe/ loader	0·75
Excavate for foundations	60 m³	Excavate foundation trench to required depth – JCB-4CX backhoe. Surplus spoil removed from site.	15 m³/hr	Exc. driver +2 labourers. Truck driver.	JCB-4CX backhoe/ loader. Tipper truck.	0·50

PROJECT	TWO STOREY OFFICE AND WORKSHOP																					CONTRACT No. 1234															

MONTH/YEAR																																								

DATE: W/E ● ← pin

No.	Activity	Week No.	1	2	3	4	5	6	7	8	9	10	11	12	13	14	15	16	17	18	19	20	21	22	23	24	25	26	27	28	29	30	31	32	33	34	35	36	37
1	Set up site																																							
2	Level site and fill																																							
3	Excavate founds																																							
4	Conc. foundations																																							
5	Brickwork < dpc																																							
6	Ground floor																																							
7	Drainage																																							
8	Scaffold																																							
9	Brickwork > dpc																																							
10	1st. floor carcass																																							
11	Roof framing																																							
12	Roof tiling																																							
13	1st. floor deck																																							
14	Partitions																																							
15	1st. fix joiner																																							
16	1st. fix services																																							
17	Glazing																																							
18	Plaster & screed																																							
19	2nd. fix joiner																																							
20	2nd. fix services																																							
21	Paint & dec.																																							
22	Floor finishes																																							
23	Fittings & fixtures																																							
24	Clean & make good																																							
25	Roads & landscape																																							
26	Clear site																																							
27	Commissioning																																							

← activity duration

← string line

← progress to date

planned completion →

Material	Weight (kg/m^2)
BRICKS, BLOCKS and PAVING –	
Clay brickwork – 102.5 mm	
low density	205
medium density	221
high density	238
Calcium silicate brickwork – 102.5 mm	205
Concrete blockwork, aerated	78
.... lightweight aggregate	129
Concrete flagstones (50 mm)	115
Glass blocks (100 mm thick) 150 × 150	98
..........................200 × 200	83
ROOFING –	
Slates – see page 443	
Thatching (300 mm thick)	40·00
Tiles – plain clay	63·50
.. – plain concrete	93·00
.. single lap, concrete	49·00
Tile battens (50 × 25) and felt underlay	7·70
Bituminous felt underlay	1·00
Bituminous felt, sanded topcoat	2·70
3 layers bituminous felt	4·80
HD/PE breather membrane underlay	0·20
SHEET MATERIALS –	
Aluminium (0·9 mm)	2·50
Copper (0·9 mm)	4·88
Cork board (standard) per 25 mm thickness	4·33
.... (compressed)	9·65
Hardboard (3·2 mm)	3·40
Glass (3 mm)	7·30
Lead (1·25 mm)	14·17
.. .. (3 mm)	34·02
Particle board/chipboard (12 mm)	9·26
.. .. (22 mm)	16·82
Planking, softwood strip flooring (ex 25 mm)	11·20
.... hardwood	16·10
Plasterboard (9·5 mm)	8·30
.. .. (12·5 mm)	11·00
.. .. (19 mm)	17·00
Plywood per 25 mm	15·00
PVC floor tiling (2·5 mm)	3·90
Strawboard (25 mm)	9·80

Typical Weights of Building Materials and Densities

Material	Weight (kg/m^2)
Weatherboarding (20 mm)	7·68
Woodwool (25 mm)	14·50
INSULATION	
Glass fibre thermal (100 mm)	2·00
.. .. acoustic	4·00
APPLIED MATERIALS -	
Asphalte (18 mm)	42
Plaster, 2 coat work	22
STRUCTURAL TIMBER -	
Rafters and Joists (100 × 50 @ 400 c/c)	5·87
Floor joists (225 × 50 @ 400 c/c)	14·93

Densities -

Material	Approx. Density (kg/m^3)
Cement	1440
Concrete (aerated)	640
.. .. (broken brick)	2000
.. .. (natural aggregates)	2300
.. .. (no-fines)	1760
.. .. (reinforced)	2400
Metals -	
Aluminium	2770
Copper	8730
Lead	11325
Steel	7849
Timber (softwood/pine)	480 (average)
.. .. .. (hardwood, e.g. maple, teak, oak)	720 .. .. ..
Water	1000

Ref. BS 648: Schedule of Weights of Building Materials.

Structural design of floors will be satisfied for most situations by using the minimum figures given for uniformly distributed loading (UDL). These figures provide for static loading and for the dynamics of occupancy. The minimum figures given for concentrated or point loading can be used where these produce greater stresses.

Application	UDL (kN/m^2)	Concentrated (kN)
Dwellings ~		
Communal areas	1.5	1.4
Bedrooms	1.5	1.8
Bathroom/WC	2.0	1.8
Balconies (use by 1 family)	1.5	1.4
Commercial/Industrial ~		
Hotel/motel bedrooms	2.0	1.8
Communal kitchen	3.0	4.5
Offices and general work areas	2.5	2.7
Kitchens/laundries/ laboratories	3.0	4.5
Factories and workshops	5.0	4.5
Balconies – guest houses	3.0	1.5/m run at outer edge
Balconies – communal areas in flats	3.0	1.5/m run at outer edge
Balconies – hotels/motels	4.0	1.5/m run at outer edge
Warehousing/Storage ~		
General use for static items	2.0	1.8
Reading areas/libraries	4.0	4.5
General use, stacked items	2.4/m height	7.0
Filing areas	5.0	4.5
Paper storage	4.0/m height	9.0
Plant rooms	7.5	4.5
Book storage	2.4/m height (min. 6.5)	7.0

See also:

BS 6399-1: Loading for buildings. Code of practice for dead and imposed loads.

BS 6399-2: Loading for buildings. Code of practice for wind loads.

BS 6399-3: Loading for buildings. Code of practice for imposed roof loads.

Drawings ~ these are the principal means of communication between the designer, the builder and other parties to a contract.

Drawings should therefore be clear, accurate, contain all the necessary information and be capable of being easily read.

Design practices have their own established symbols and notations for graphical communication. Some of which are shown on this and the next three pages. Other guidance can be found in BS EN ISOs 4157 and 7519.

Typical Examples~

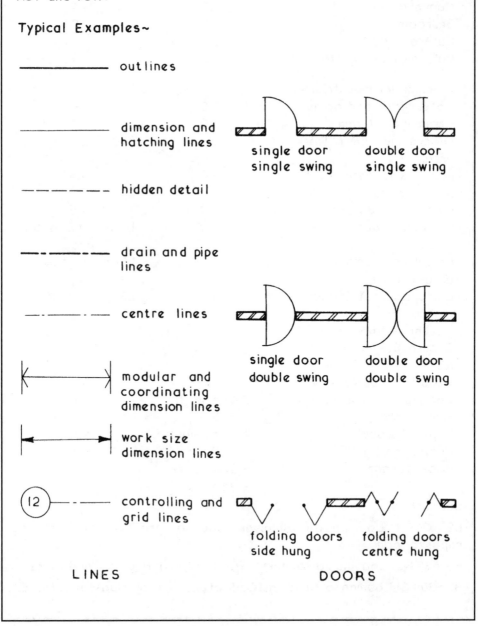

——————————— outlines

——————————— dimension and hatching lines

single door   double door
single swing   single swing

— — — — — — hidden detail

—·———··— drain and pipe lines

—·———··— centre lines

single door   double door
double swing   double swing

modular and coordinating dimension lines

work size dimension lines

(12)—·——— controlling and grid lines

folding doors   folding doors
side hung   centre hung

LINES   DOORS

Hatchings ~ the main objective is to differentiate between the materials being used thus enabling rapid recognition and location. Whichever hatchings are chosen they must be used consistently throughout the whole set of drawings. In large areas it is not always necessary to hatch the whole area.

Symbols ~ these are graphical representations and should wherever possible be drawn to scale but above all they must be consistent for the whole set of drawings and clearly drawn.

Typical Examples~

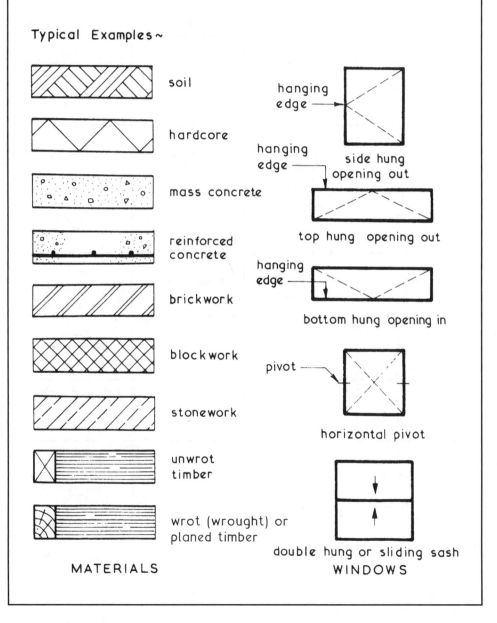

MATERIALS

WINDOWS

Name	Symbol	Name	Symbol
Rainwater pipe	◯ RWP	Distribution board	▢
Gully	▢ G	Electricity meter	⊡
Inspection chambers	IC —▢— soil or foul / IC —⊖— surface water	Switched socket outlet	▷
Boiler	▢ B	Switch	●
Sink	▢ S	Two way switch	●
Bath	▭	Pendant switch	●
Wash basin	W B	Filament lamp	◯
Shower unit	▢ S	Fluorescent lamp	◯
Urinal	⌂ stall ⌓ bowl	Bed	▭
Water closet	▭ ◯	Table and chairs	▦

TYPICAL COMPONENT, FITMENT AND ELECTRICAL SYMBOLS

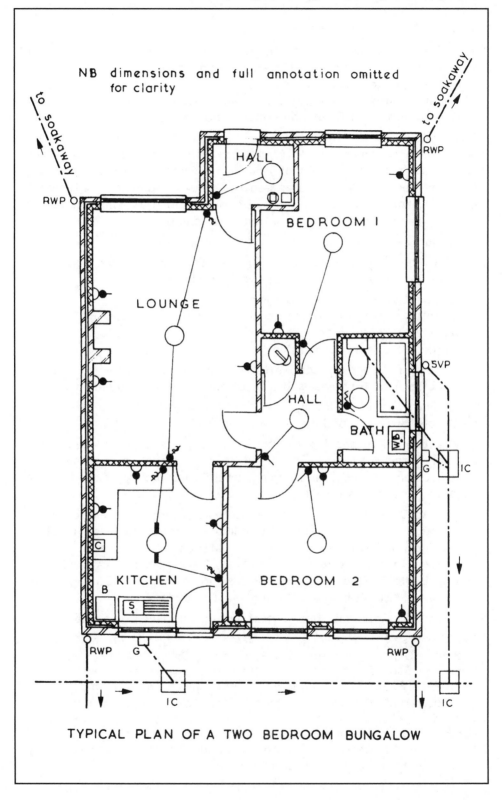

NB dimensions and full annotation omitted for clarity

to soakaway

to soakaway

RWP

RWP

HALL

BEDROOM 1

LOUNGE

SVP

HALL

BATH

wb

G

IC

C

B

S

KITCHEN

BEDROOM 2

RWP

G

RWP

IC

IC

TYPICAL PLAN OF A TWO BEDROOM BUNGALOW

Principal legislation: ~

The Town & Country Planning Act 1990 - Effects control over volume of development, appearance and layout of buildings. The Public Health Acts 1936 to 1961 - Limits development with regard to emission of noise, pollution and public nuisance. The Highways Act 1980 - Determines layout and construction of roads and pavements.

The Building Act 1984 - Effects the Building Regulations 2000, which enforce minimum material and design standards. The Civic Amenities Act 1967 - Establishes conservation areas, providing local authorities with greater control of development. The Town & Country Amenities Act 1974 - Local authorities empowered to prevent demolition of buildings and tree felling.

Procedure: ~

Outline Planning Application - This is necessary for permission to develop a proposed site. The application should contain:

An application form describing the work.
A site plan showing adjacent roads and buildings (1:2500).
A block plan showing the plot, access and siting (1:500).
A certificate of land ownership.

Detail or Full Planning Application - This follows outline permission and is also used for proposed alterations to existing buildings.

It should contain: details of the proposal, to include trees, materials, drainage and any demolition.

Site and block plans (as above). A certificate of land ownership. Building drawings showing elevations, sections, plans, material specifications, access, landscaping, boundaries and relationship with adjacent properties (1:100).

Permitted Developments - House extensions may be exempt formal application. Conditions vary depending on house position relative to its plot and whether detached or attached. Ref. The Town and Country Planning (General Permitted Development) (Amendment) (No.2) (England) Order, 2008. Porches are exempt if <3m^2 external floor area, <3m in height >2m from the boundary.

Note: All developments are subject to Building Regulation approval.

Certificates of ownership - Article 7 of the Town & Country Planning (General Development Procedure) Order 1995:
Cert. A - States the applicant is sole site freeholder.
Cert. B - States the applicant is part freeholder or prospective purchaser and all owners of the site know of the application.
Cert. C - As Cert. B, but the applicant is only able to ascertain some of the other land owners.
Cert. D - As Cert. B, but the applicant cannot ascertain any owners of the site other than him/herself.

PLANNING APPLICATION

	APPLICATION No

Use this form to apply for Planning Permission for:-
- an Extension
- a Loft Conversion
- a New or Altered Access
- a High Wall or Fence
- a Garage or Outbuilding
- a Satellite Dish

Please return:-
- 6 copies of the Form
- 6 copies of the Plans
- a Certificate under Article 7
- the correct fee

DATE RECEIVED

1. NAME AND ADDRESS OF APPLICANT

Post Code _____

Tel. No. _____

2. NAME AND ADDRESS OF AGENT (If Used)

Post Code _____

Tel. No. _____

3. ADDRESS OF PROPERTY TO BE ALTERED OR EXTENDED

4. OWNERSHIP
Please indicate applicants interest in the property and complete the appropriate Certificate under Article 7.

Freeholder ☐   Other ☐

Leaseholder ☐

Purchaser ☐

5. BRIEF DESCRIPTION OF WORKS (include any demolition work)

6. DESCRIPTION OF EXTERNAL MATERIALS

7. ACCESS AND PARKING

Will your proposal affect?   Please tick appropriate boxes

Vehicular Access   Yes ☐   No ☐
A Public Right of Way   Yes ☐   No ☐
Existing Parking   Yes ☐   No ☐

8. DRAINAGE

a. Please indicate method of Surface Water Disposal

_____

b. Please indicate method of Foul Water Disposal
Please tick one box

Mains Sewer ☐   Septic Tank ☐

Cesspit ☐   Other ☐

9. TREES
Does the proposal involve the felling of any trees?
Please tick box   Yes ☐   No ☐
If yes, please show details on plans

10. PLEASE SIGN AND DATE THIS FORM BEFORE SUBMITTING
I/We hereby apply for Full Planning Permission for the development described above and shown on the accompanying plans.

Signed _____   Date _____

Date
On behalf of (if agent) _____

43

Planning Application—New Build (1)

Use this form to apply for **Planning Permission for:-**

Outline Permission

Full Permission

Approval of Reserved Matters

Renewal of Temporary Permission

Change of Use

Please return:-

* 6 copies of the Form

* 6 copies of the Plans

* a Certificate under
   Article 7

* the correct fee

DATE RECEIVED

DATE VALID

1. NAME AND ADDRESS OF APPLICANT

_____

_____

_____

_____

Post Code _____

Day Tel. No. _____ Fax No. _____

Email: _____

2. NAME AND ADDRESS OF AGENT (If Used)

_____

_____

_____

_____

Post Code _____

Tel. No. _____ Fax No. _____

Email: _____

3. ADDRESS OR LOCATION OF LAND TO WHICH APPLICATION RELATES

_____

_____

_____

State Site Area _____ Hectares

This must be shown edged in Red on the site plan

4. OWNERSHIP

Please indicate applicants interest in the property and complete the appropriate Certificate under Article 7.

Freeholder ☐   Other ☐

Leaseholder ☐   Purchaser ☐

Any adjoining land owned or controlled and not part of application must be edged Blue on the site plan

5. WHAT ARE YOU APPLYING FOR?   Please tick one box and then answer relevant questions.

☐ **Outline Planning Permission** Which of the following are to be considered?

☐ Siting ☐ Design ☐ Appearance ☐ Access ☐ Landscaping

☐ **Full Planning Permission/Change of use**

☐ **Approval of Reserved Matters following Outline Permission.**

O/P No. _____ Date_____ No. of Condition this application refers to: _____

☐ **Continuance of Use without complying with a condition of previous permission**

P/P No. _____ Date_____ No. of Condition this application relates to: _____

☐ **Permission for Retention of works.**

Date of Use of land or when buildings or works were constructed: _____ Length of temporary permission: _____

Is the use temporary or permanent? _____ No. of previous temporary permission if applicable: _____

6. BRIEF DESCRIPTION OF PROPOSED DEVELOPMENT

Please indicate the purpose for which the land or buildings are to be used. _____

_____

_____

7. NEW RESIDENTIAL DEVELOPMENTS. Please answer the following if appropriate:

What type of building is proposed? _____

No. of dwellings: _____ No. of storeys: _____ No. of Habitable rooms: _____

No. of Garages: _____ No. of Parking Spaces: _____ Total Grass Area of all buildings: _____

How will surface water be disposed of? _____

How will foul sewage be dealt with? _____

8. ACCESS

Does the proposed development involve any of the following? Please tick the appropriate boxes.

New access to a highway ☐ Pedestrian ☐ Vehicular

Alteration of an existing highway ☐ Pedestrian ☐ Vehicular

The felling of any trees ☐ Yes ☐ No

If you answer Yes to any of the above, they should be clearly indicated on all plans submitted.

9. BUILDING DETAIL

Please give details of all external materials to be used, if you are submitting them at this stage for approval.

_____

_____

List any samples that are being submitted for consideration. _____

_____

10. LISTED BUILDINGS OR CONSERVATION AREA

Are any Listed buildings to be demolished or altered? ☐ Yes ☐ No

If Yes, then Listed Building Consent will be required and a separate application should be submitted.

Are any non-listed buildings within a Conservation Area to be demolished? ☐ Yes ☐ No

If Yes, then Conservation Area consent will be required to demolish. Again, a separate application should be submitted.

11. NOTES

A special Planning Application Form should be completed for all applications involving Industrial, Warehousing, Storage, or Shopping development.

An appropriate Certificate must accompany this application unless you are seeking approval to Reserved Matters. A separate application for Building Regulation approval is also required.

Separate applications may also be required if the proposals relate to a Listed Building or non-listed building within a Conservation Area.

12. PLEASE SIGN AND DATE THIS FORM BEFORE SUBMITTING

I/We hereby apply for Planning Permission for the development described above and shown on the accompanying plans.

Signed _____

TOWN AND COUNTRY PLANNING ACT

TOWN AND COUNTRY PLANNING (General Development Procedure) ORDER
Certificates under Article 7 of the Order

CERTIFICATE A   **For Freehold Owner (or his/her Agent)**

I hereby certify that:-

1.   No person other than the applicant was an owner of any part of the land to which the application relates at the beginning of the period of 21 days before the date of the accompanying application.

2.   ***Either (i)** None of the land to which the application relates constitutes or forms part of an agricultural holding:

*or (ii) *(I have) (the applicant has) given the requisite notice to every person other than *(myself) (himself) (herself) who, 21 days before the date of the application, was a tenant of any agricultural holding any part of which was comprised in the land to which the application relates, viz:-

Name and Address of Tenant..

....

.... Signed .... Date........................

Date of Service of Notice.. *On Behalf of ....

CERTIFICATE B   **For Part Freehold Owner or Prospective Purchaser (or his/her Agent) able to ascertain all the owners of the land**

I hereby certify that:-

1.   *(I have) (the applicant has) given the requisite notice to all persons other than (myself) (the applicant) who, 21 days before the date of the accompanying application were owners of any part of the land to which the application relates, viz:-

Name and Address of Owner ....

....

.... Date of Service of Notice ....

2.   ***Either (i)** None of the land to which the application relates constitutes or forms part of an agricultural holding;

*or (ii) *(I have) (the applicant has) given the requisite notice to every person other than *(myself) (himself) (herself) who, 21 days before the date of the application, was a tenant of any agricultural holding any part of which was comprised in the land to which the application relates, viz:-

Name and Address of Tenant..

....

.... Signed .... Date........................

Modular Coordination ~ a module can be defined as a basic dimension which could for example form the basis of a planning grid in terms of multiples and submultiples of the standard module.

Typical Modular Coordinated Planning Grid ~

Let M = the standard module

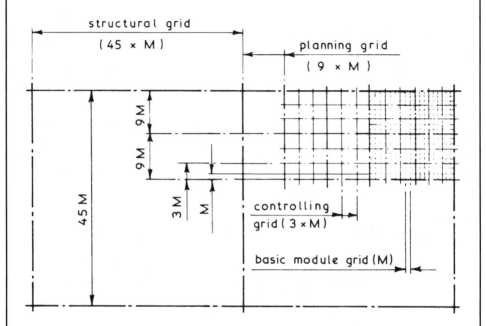

Structural Grid ~ used to locate structural components such as beams and columns.

Planning Grid ~ based on any convenient modular multiple for regulating space requirements such as rooms.

Controlling Grid ~ based on any convenient modular multiple for location of internal walls, partitions etc.

Basic Module Grid ~ used for detail location of components and fittings.

All the above grids, being based on a basic module, are contained one within the other and are therefore interrelated. These grids can be used in both the horizontal and vertical planes thus forming a three dimensional grid system. If a first preference numerical value is given to M dimensional coordination is established – see next page.

Ref. BS 6750: Specification for modular coordination in building.

Dimensional Coordination ~ the practical aims of this concept are to:-

1. Size components so as to avoid the wasteful process of cutting and fitting on site.
2. Obtain maximum economy in the production of components.
3. Reduce the need for the manufacture of special sizes.
4. Increase the effective choice of components by the promotion of interchangeability.

BS 6750 specifies the increments of size for coordinating dimensions of building components thus:-

Preference	1st	2nd	3rd	4th
Size (mm)	300	100	50	25

the 3rd and 4th preferences having a maximum of 300mm

Dimensional Grids – the modular grid network as shown on page 47 defines the space into which dimensionally coordinated components must fit. An important factor is that the component must always be undersized to allow for the joint which is sized by the obtainable degree of tolerance and site assembly:-

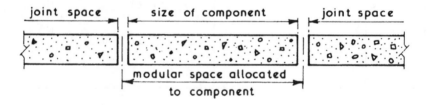

Controlling Lines, Zones and Controlling Dimensions – these terms can best be defined by example:-

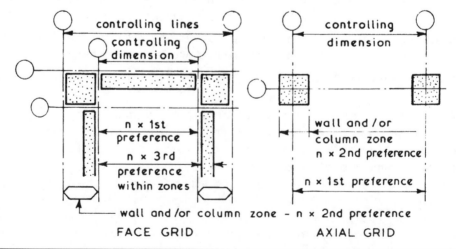

FACE GRID   AXIAL GRID

Construction Regulations ~ these are Statutory Instruments made under the Factories Acts of 1937 and 1961 and come under the umbrella of the Health and Safety at Work etc., Act 1974. They set out the minimum legal requirements for construction works and relate primarily to the health, safety and welfare of the work force. The requirements contained within these documents must therefore be taken into account when planning construction operations and during the actual construction period. Reference should be made to the relevant document for specific requirements but the broad areas covered can be shown thus:-

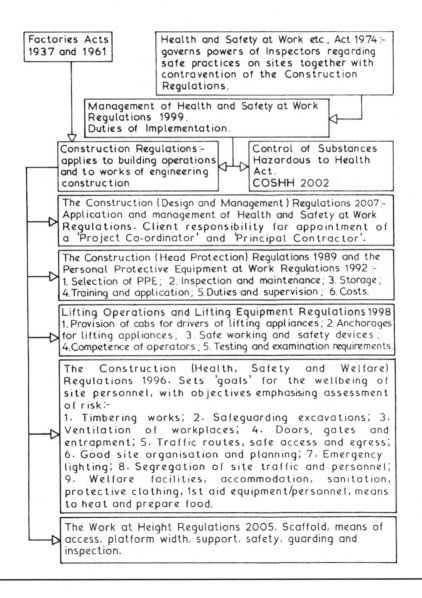

Objective – To create an all-party integrated and planned approach to health and safety throughout the duration of a construction project.

Administering Body – The Health and Safety Executive (HSE).

Scope – The CDM Regulations are intended to embrace all aspects of construction, with the exception of very minor works.

Responsibilities – The CDM Regulations apportion responsibility to everyone involved in a project to cooperate with others and for health and safety issues to all parties involved in the construction process, i.e. client, designer, project coordinator and principal contractor.

Client – Appoints a project coordinator and the principal contractor. Provides the project coordinator with information on health and safety matters and ensures that the principal contractor has prepared an acceptable construction phase plan for the conduct of work. Ensures adequate provision for welfare and that a health and safety file is available.

Designer – Establishes that the client is aware of their duties. Considers the design implications with regard to health and safety issues, including an assessment of any perceived risks. Coordinates the work of the project coordinator and other members of the design team.

Project Coordinator – Ensures that:
* a pre-tender, construction phase plan is prepared.
* the HSE are informed of the work.
* designers are liaising and conforming with their health and safety obligations.
* a health and safety file is prepared.
* contractors are of adequate competence with regard to health and safety matters and advises the client and principal contractor accordingly.

Principal Contractor – Develops a construction phase plan, collates relevant information and maintains it as the work proceeds. Administers day-to-day health and safety issues. Co-operates with the project coordinator, designers and site operatives preparing risk assessments as required.

Note: The CDM Regulations include requirements defined under The Construction (Health, Safety and Welfare) Regulations.

Health and Safety at Work etc. Act 1974

| The Health and Safety (Safety Signs and Signals) Regulations 1996 | Management of Health and Safety at Work Regulations 1999 (Management Regulations) |

Under these regulations, employers are required to provide and maintain health and safety signs conforming to European Directive 92/58 EEC:

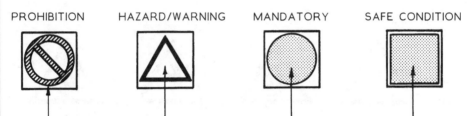

PROHIBITION — Red

HAZARD/WARNING — Yellow

MANDATORY — Blue

SAFE CONDITION — Green

In addition, employers obligations include the need to provide:

Risk Assessment – provide and maintain safety signs where there is a risk to health and safety, e.g. obstacles. Train staff to comprehend safety signs.

Pictorial Symbols – pictograms alone are acceptable but supplementary text, e.g. FIRE EXIT, is recommended.

Fire/Emergency Escape Signs – A green square or rectangular symbol.

Positioning of signs – primarily for location of fire exits, fire equipment, alarms, assembly points, etc. Not to be located where they could be obscured.

Marking of Hazardous Areas – to identify designated areas for storing dangerous substances: Dangerous Substances (Notification and Marking of Sites) Regulations 1990. Yellow triangular symbol.

Pipeline Identification – pipes conveying dangerous substances to be labelled with a pictogram on a coloured background conforming to BS 1710: Specification for identity of pipelines and services and BS 4800: Schedule of paint colours for building purposes. Non-dangerous substances should also be labelled for easy identification.

Typical Examples on Building Sites ~

PROHIBITION (Red)

| Authorised personnel only | Children must not play on this site | Smoking prohibited | Access not permitted |

HAZARD/WARNING (Yellow)

| Dangerous substance | Flammable liquid | Danger of electric shock | Compressed gas |

MANDATORY (Blue)

| Safety helmets must be worn | Protective footwear must be worn | Use ear protectors | Protective clothing must be worn |

SAFE CONDITIONS (Green)

 FIRE EXIT

| Emergency escapes | Treatment area | Safe area |

Ref. BS 5499-1: Graphical symbols and signs. Safety signs, including fire safety signs. Specification for geometric shapes, colours and layout.

The Building Regulations ~ this is a Statutory Instrument which sets out the minimum performance standards for the design and construction of buildings and where applicable to the extension of buildings. The regulations are supported by other documents which generally give guidance on how to achieve the required performance standards. The relationship of these and other documents is set out below:-

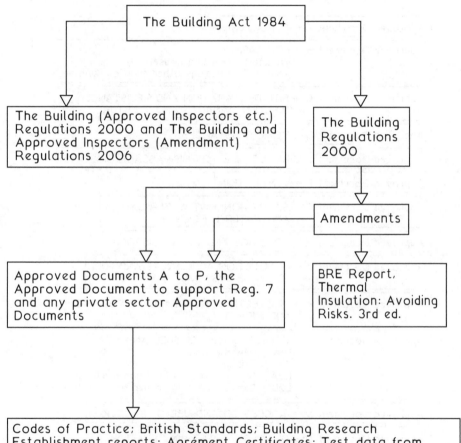

NB. The Building Regulations apply to England and Wales but not to Scotland and Northern Ireland which have separate systems of control.

53

Approved Documents ~ these publications support the Building Regulations. They are prepared by the Department for Communities and Local Government approved by the Secretary of State and issued by The Stationery Office. The Approved Documents (ADs) have been compiled to give practical guidance to comply with the performance standards set out in the various regulations. They are not mandatory but show compliance with the requirements of the Building Regulations. If other solutions are used to satisfy the requirements of the Regulations, proving compliance rests with the applicant or designer.

Approved Document A — STRUCTURE

Approved Document B — FIRE SAFETY
Volume 1 — Dwelling houses
Volume 2 — Buildings other than dwelling houses

Approved Document C — SITE PREPARATION AND RESISTANCE TO CONTAMINANTS AND MOISTURE

Approved Document D — TOXIC SUBSTANCES

Approved Document E — RESISTANCE TO THE PASSAGE OF SOUND

Approved Document F — VENTILATION

Approved Document G — SANITATION, HOT WATER SAFETY AND WATER EFFICIENCY

Approved Document H — DRAINAGE AND WASTE DISPOSAL

Approved Document J — COMBUSTION APPLIANCES AND FUEL STORAGE SYSTEMS

Approved Document K — PROTECTION FROM FALLING, COLLISION AND IMPACT

Approved Document L — CONSERVATION OF FUEL AND POWER
L1A — New dwellings
L1B — Existing dwellings
L2A — New buildings other than dwellings
L2B — Existing buildings other than dwellings

Approved Document M — ACCESS TO AND USE OF BUILDINGS

Approved Document N — GLAZING — SAFETY IN RELATION TO IMPACT, OPENING AND CLEANING

Approved Document P — ELECTRICAL SAFETY

Approved Document to support Regulation 7
MATERIALS AND WORKMANSHIP

BASEMENTS FOR DWELLINGS — A government approved private sector AD published by The Basement Information Centre

Example in the Use of Approved Documents

Problem:- the sizing of suspended upper floor joists to be spaced at 400 mm centres with a clear span of 3·600 m for use in a two storey domestic dwelling.

Building Regulation A1:- states that the building shall be constructed so that the combined dead, imposed and wind loads are sustained and transmitted by it to the ground –

(a) safely, and
(b) without causing such deflection or deformation of any part of the building, or such movement of the ground, as will impair the stability of any part of another building.

Approved Document A:- guidance on sizing floor joists can be found in 'Span Tables for Solid Timber Members in Dwellings', published by the Timber Research And Development Association (TRADA), and BS5268-2: Structural use of timber. Code of practice for permissible stress design, materials and workmanship.

Solution :-

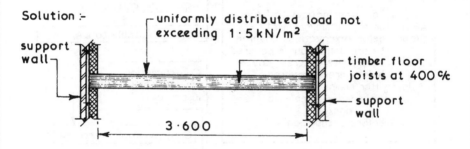

Dead load (kN/m²) supported by joist excluding mass of joist:-

Floor finish – carpet	– 0·03	weights of materials from BS648
Flooring – 20 mm thick particle board	– 0·15	
Ceiling – 9·5 mm thick plasterboard	0·08	
Ceiling finish – 3 mm thick plaster	– 0·04	
total dead load –	0·30 kN/m³	

Dead loading is therefore in the 0·25 to 0·50 kN/m² band
From table on page 633 suitable joist sizes are:- 38 × 200, 50 × 175, 63 × 175 and 75 × 150.

Final choice of section to be used will depend upon cost; availability; practical considerations and/or personal preference.

Building Control ~ unless the applicant has opted for control by a private approved inspector under The Building (Approved Inspectors etc.) Regulations 2000 the control of building works in the context of the Building Regulations is vested in the Local Authority. There are two systems of control namely the Building Notice and the Deposit of Plans. The sequence of systems is shown below:-

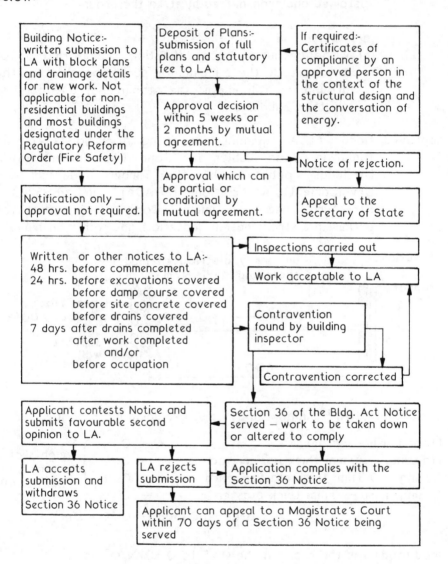

NB. In some stages of the above sequence statutory fees are payable as set out in The Building (Local Authority Charges) Regulations 1998.

Building Regulations Approval ~ required if ``Building Work'' as defined in Regulation 3 of the Building Regulations is proposed. This includes:

- Construction or extension of a building.
- Alterations to an existing building that would bring into effect any of the complying regulations.
- Installing replacement windows where the installer is not known to the local Building Control Authority as being a ``competent'' registered installer, e.g. FENSA (FENestration Self Assessment) scheme.
- Alteration or installation of building services and fittings that bring into effect any of the complying regulations.
- Installation of cavity wall insulation.
- Underpinning of a building's foundations.
- Change of purpose or use of a building.

``Competent'' persons are appropriately qualified and experienced to the satisfaction of a relevant scheme organiser. For example, Capita Group's ``Gas Safe Register'' of engineers for gas installation and maintenance services. They can ``self certify'' that their work complies with Building Regulations, thereby removing the need for further inspection.

Local Authority Building Control ~ the established procedure as set out diagrammatically on the preceding page with an application form of the type shown on page 61 and accompanying documents as indicated on the next page.

Private Sector Building Control ~ an alternative, where suitably qualified and experienced inspectors approved by the local authority undertake the application approval and site inspections. An ``Initial Notice'' from the client and their appointed inspector is lodged with the local authority.

Whichever building control procedure is adopted, the methodology is the same, i.e. Deposit of Plans or Building Notice (see page 59).

Refs. The Building (Approved Inspectors, etc.) Regulations.
　　　The Association of Consultant Approved Inspectors.

Local Authority Building Control ~ as described in the previous two pages. A public service administered by borough and unitary councils through their building control departments.

Approved Inspectors ~ a private sector building control alternative as outlined on the preceding page. Approved inspectors may be suitably qualified individuals or corporate bodies employing suitably qualified people, e.g. National House Building Council (NHBC Ltd.) and MD Insurance Services Ltd.

Borough councils can contract out the building control process to approved inspectors. Validation and site examinations follow the established format shown on page 56, with approved inspectors substituting for LA.

Both NHBC and MD Insurance publish their own construction rules and standards that supplement the Building Regulations. These form the basis for their own independent quality control procedures whereby their Inspectors will undertake stage and periodic examinations of work in progress to ensure that these standards are adhered to. The objective is to provide new home buyers with a quality assured product warranted against structural defects (10–15 years), provided the house builder has satisfied certain standards for registration. Therefore, the buyer should be provided with a completion certificate indicating Building Regulations approval and a warranty against defects.

Robust Details ~ Building Regulations A.D. E – Resistance to the passage of sound; requires that the separating walls, floors and stairs in new dwellings are sufficiently resistant to airborne and impact sound transmission. Sound measurement tests defined in the associated BSs specified in the Approved Document must be undertaken by an approved inspector/building control official before completion.

An alternative or a means for exemption of pre-completion testing is for the builder to notify the building inspector that sound insulation construction details are registered and specified to those approved by Robust Details Ltd. This is a not-for-profit company established by the house building industry to produce guidance manuals containing details of acceptable sound resistant construction practice.

Deposit of Plans or Full Plans Application ~

- Application form describing the proposed work.
- Location plan, scale not less than 1:2500.
- Block plan, scale not less than 1:1250 showing north point, lines of drains (existing and proposed) and size and species of trees within 30 m.
- Plans, sections and elevations, scale not less than 1:50 (1:100 may be acceptable for elevations).
- Materials specification.
- Structural calculations where appropriate, e.g. load bearing beams.
- Fee depending on a valuation of work.

The appointed inspector examines the application and subject to any necessary amendments, an approval is issued. This procedure ensures that work on site is conducted in accordance with the approved plans. Also, where the work is being financed by a loan, the lender will often insist the work is only to a Full Plans approval.

Building Notice ~

- A simplified application form.
- Block plan as described above.
- Construction details, materials specification and structural calculations if considered necessary by the inspector.
- Fee depending on a valuation of work.

This procedure is only really appropriate for minor work such as extensions to existing small buildings such as houses. Building control/inspection occurs as each element of the work proceeds. Any Building Regulation contravention will have to be removed or altered to attain an acceptable standard.

Regularisation ~

- Application form.
- Structural calculations if relevant.
- A proportionally higher fee.

Applies to unauthorised work undertaken since Nov. 1985. In effect a retrospective application that will involve a detailed inspection of the work. Rectification may be necessary before approval is granted.

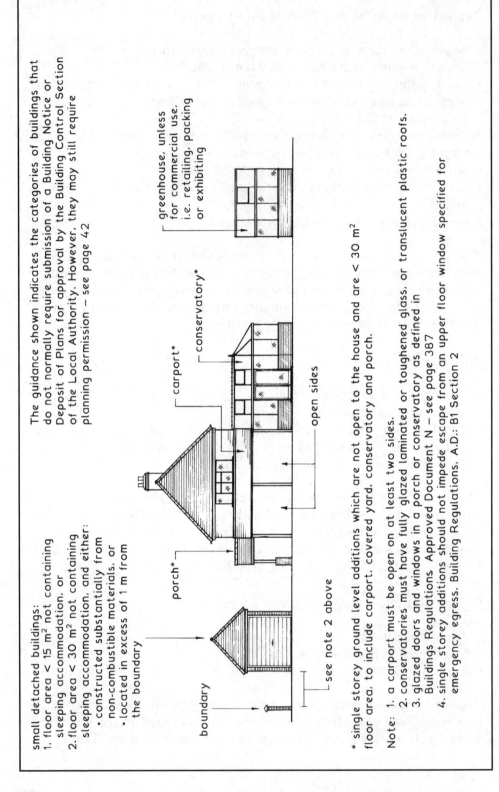

small detached buildings:
1. floor area < 15 m² not containing sleeping accommodation, or
2. floor area < 30 m² not containing sleeping accommodation, and either:
 • constructed substantially from non-combustible materials, or
 • located in excess of 1 m from the boundary

The guidance shown indicates the categories of buildings that do not normally require submission of a Building Notice or Deposit of Plans for approval by the Building Control Section of the Local Authority. However, they may still require planning permission – see page 42

greenhouse, unless for commercial use, i.e. retailing, packing or exhibiting

conservatory*

carport*

porch*

boundary

see note 2 above

open sides

* single storey ground level additions which are not open to the house and are < 30 m² floor area, to include carport, covered yard, conservatory and porch.

Note: 1. a carport must be open on at least two sides.
 2. conservatories must have fully glazed laminated or toughened glass, or translucent plastic roofs.
 3. glazed doors and windows in a porch or conservatory as defined in Buildings Regulations Approved Document N – see page 387
 4. single storey additions should not impede escape from an upper floor window specified for emergency egress. Building Regulations. A.D.: B1 Section 2

BUILDING REGULATIONS APPLICATION

APPLICATION No

Use this form to give notice of intention to erect, extend, or alter a building, install fittings or make a material change of use of the building.

Unless specified differently overleaf, Please return:-
- 2 copies of the Form
- 4 copies of the Plans
- the correct fee

DATE RECEIVED

1. NAME AND ADDRESS OF APPLICANT
Applicant will be invoiced on commencement of work.

Post Code _____

Tel. No. _____

2. NAME AND ADDRESS OF AGENT (If Used)

Post Code _____

Tel. No. _____

3. ADDRESS OR LOCATION OF PROPOSED WORK

4. DESCRIPTION OF PROPOSED WORKS

5. IF NEW BUILDING OR EXTENSION PLEASE STATE PROPOSED USE

6. IF EXISTING BUILDING PLEASE STATE PRESENT USE

7. DRAINAGE

Please state means of:-

Water Supply _____

Foul Water Disposal _____

Storm Water Disposal _____

8. CONDITIONS

Do you consent to the Plans being passed subject to conditions where appropriate?   Yes ☐   No ☐

Do you agree to an extension of time if this is required by the Council?   Yes ☐   No ☐

9. COMPLETION CERTIFICATE

Do you wish the Council to issue a Completion Certificate upon satisfactory completion of the work?

Yes ☐   No ☐

10. REGULATORY REFORM ORDER (Fire Safety) 2005

Is the building intended for any other purpose than occupation as a domestic living unit by one family group?

Yes ☐   No ☐

11. FEE
Please state estimated cost of the work (at current market value) £......................... Amount of Fee submitted £.......................

Has Planning Permission been sought?   Yes ☐   No ☐   If Yes, please give Application No _____

12. PLEASE SIGN AND DATE THIS FORM BEFORE SUBMITTING

I/We hereby give notice of intention to carry out the work set out above and deposit the attached drawings and documents in accordance with the requirements of Regulations 11 (1) (b). Also enclosed is the appropriate Plan Fee and I understand that a further Fee will be payable when the first inspection of work on site is made by the Local Authority.

Signed _____ Date _____ On behalf of (if agent) _____

Published ~ 2006 by the Department for Communities and Local Government (DCLG) in response to the damaging effects of climate change. The code promotes awareness and need for new energy conservation initiatives in the design of new dwellings.

Objective ~ to significantly reduce the 27% of UK CO_2 emissions that are produced by 25 million homes. This is to be a gradual process, with the target of reducing CO_2 emissions from all UK sources by 60% by 2050.

Sustainability ~ measured in terms of a quality standard designed to provide new homes with a factor of environmental performance. This measure is applied primarily to categories of thermal energy, use of water, material resources, surface water run-off and management of waste.

Measurement ~ a 'green' star rating that indicates environmental performance ranging from one to six stars. Shown below is the star rating criteria applied specifically to use of thermal energy. A home with a six star rating is also regarded as a zero carbon home.

Proposed Progression ~

Percentage Improvement compared with AD L 2006	Year	Star rating
10	–	1
18	–	2
25	2010	3
44	2013	4
100	2016	5 and 6

Zero Carbon Home ~ zero net emissions of CO_2 from all energy use in the home. This incorporates insulation of the building fabric, heating equipment, hot water systems, cooling, washing appliances, lighting and other electrical/electronic facilities. Net zero emissions can be measured by comparing the carbon emissions produced in consuming on- or off-site fossil fuel energy use in the home, with the amount of on-site renewable energy produced. Means for producing low or zero carbon energy include micro combined heat and power units, photovoltaic (solar) panels, wind generators and ground energy heat pumps, (see Building Services Handbook).

British Standards ~ these are publications issued by the British Standards Institution which give recommended minimum standards for materials, components, design and construction practices. These recommendations are not legally enforceable but some of the Building Regulations refer directly to specific British Standards and accept them as deemed to satisfy provisions. All materials and components complying with a particular British Standards are marked with the British Standards kitemark thus:- ♡ together with the appropriate BS number.

This symbol assures the user that the product so marked has been produced and tested in accordance with the recommendations set out in that specific standard. Full details of BS products and services can be obtained from, Customer Services, BSI, 389 Chiswick High Road, London, W4 4AL. Standards applicable to building may be purchased individually or in modules, GBM 48, 49 and 50; Construction in General, Building Materials and Components and Building Installations and Finishing, respectively. British Standards are constantly under review and are amended, revised and rewritten as necessary, therefore a check should always be made to ensure that any standard being used is the current issue. There are over 1500 British Standards which are directly related to the construction industry and these are prepared in four formats:-

1. British Standards – these give recommendations for the minimum standard of quality and testing for materials and components. Each standard number is prefixed BS.

2. Codes of Practice – these give recommendations for good practice relative to design, manufacture, construction, installation and maintenance with the main objectives of safety, quality, economy and fitness for the intended purpose. Each code of practice number is prefixed CP or BS.

3. Draft for Development – these are issued instead of a British Standard or Code of Practice when there is insufficient data or information to make firm or positive recommendations. Each draft number is prefixed DD. Sometimes given a BS number and suffixed DC, ie. Draft for public Comment.

4. Published Document – these are publications which cannot be placed into any one of the above categories. Each published document is numbered and prefixed PD.

European Standards – since joining the European Union (EU), trade and tariff barriers have been lifted. This has opened up the market for manufacturers of construction-related products, from all EU and European Economic Area (EEA) member states. Before 2004, the EU was composed of 15 countries: Austria, Belgium, Denmark, Finland, France, Germany, Greece, Ireland, Italy, Luxemburg, Netherlands, Portugal, Spain, Sweden and the United Kingdom. It now includes Bulgaria, Cyprus, the Czech Republic, Estonia, Hungary, Latvia, Lithuania, Malta, Poland, Romania, Slovakia and Slovenia. The EEA extends to: Iceland, Liechtenstein and Norway. Nevertheless, the wider market is not so easily satisfied, as regional variations exist. This can create difficulties where product dimensions and performance standards differ. For example, thermal insulation standards for masonry walls in Mediterranean regions need not be the same as those in the UK. Also, preferred dimensions differ across Europe in items such as bricks, timber, tiles and pipes.

European Standards are prepared under the auspices of Comité Européen de Normalisation (CEN), of which the BSI is a member. European Standards that the BSI have not recognised or adopted, are prefixed EN. These are EuroNorms and will need revision for national acceptance.

For the time being, British Standards will continue and where similarity with other countries' standards and ENs can be identified, they will run side by side until harmonisation is complete and approved by CEN.

e.g. BS EN 295, complements the previous national standard:
    BS 65 – Vitrefied clay pipes . . . . . for drains and sewers.

European Pre-standards are similar to BS Drafts for Development. These are known as ENVs.

Some products which satisfy the European requirements for safety, durability and energy efficiency, carry the CE mark. This is not to be assumed a mark of performance and is not intended to show equivalence to the BS kitemark. However, the BSI is recognised as a Notified Body by the EU and as such is authorised to provide testing and certification in support of the CE mark.

International Standards – these are prepared by the International Organisation for Standardisation and are prefixed ISO. Many are compatible with and complement BSs, e.g. the ISO 9000 Quality Management series and BS 5750: Quality systems.

For manufacturers' products to be compatible and uniformly acceptable in the European market, there exists a process for harmonising technical specifications. These specifications are known as harmonised European product standards (hENs), produced and administered by the Comité Européen de Normalisation (CEN). European Technical Approvals (ETAs) are also acceptable where issued by the European Organisation for Technical Approvals (EOTA). These standards are not a harmonisation of regulations. Whether or not the technical specification satisfies regional and national regulations is for local determination. However, for commercial purposes a technical specification should cover the performance characteristics required by regulations established by any member state in the European Economic Area (EEA).

CPD harmonises:
* methods and criteria for testing
* methods for declaring product performance
* methods and measures of conformity assessment

UK attestation accredited bodies include: BBA, BRE and BSI.

CE mark – a marking or labelling for conforming products. A 'passport' permitting a product to be legally marketed in any EEA. It is not a quality mark, e.g. BS Kitemark, but where appropriate this may appear with the CE marking.

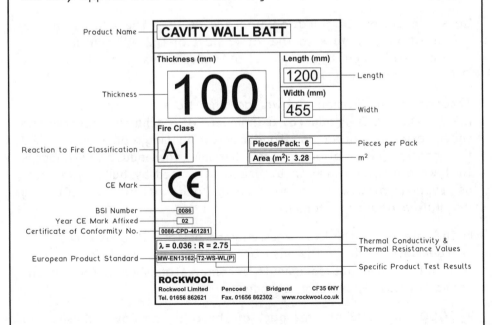

CE marking – reproduced with kind permission of Rockwool Ltd.

Building Research Establishment ~ The BRE was founded as a UK Government agency in 1921 and was known until the early 1970s as the Building Research Station.

In addition to UK Government funding, some financial support is now provided by the European Union. Additional funding is derived from a variety of sources, including commercial services for private industry and from publications. The latter includes the BRE's well known regular issue of research information products, i.e. Digests, Information Papers, Good Building Guides and Good Repair Guides.

UK Government support is principally through the Department for Business Enterprise and Regulatory Reform (BERR) and the Department for Communities and Local Government (DCLG). The DCLG works with the BRE in formulating specific aspects of the Approved Documents to the Building Regulations. Commissioned research is funded by BRE Trust.

The BRE incorporates and works with other specialised research and material testing organisations, e.g. see LPCB, below. It is accredited under the United Kingdom Accreditation Service (UKAS) as a testing laboratory authorised to issue approvals and certifications such as CE product marking (see pages 64 and 65). Certification of products, materials and applications is effected through BRE Certification Ltd.

Loss Prevention Certification Board (LPCB) ~ The origins of this organisation date back to the latter part of the 19th century, when it was established by a group of building insurers as the Fire Offices' Committee (FOC).

Through a subdivision known as the Loss Prevention Council (LPC), the FOC produced a number of technical papers and specifications relating to standards of building construction and fire control installations. These became the industry standards that were, and continue to be, frequently used by building insurers as supplementary to local byelaws and latterly the Building Regulation Approved Documents.

In the late 1980s the LPC was renamed as the LPCB as a result of reorganisation within the insurance profession. At this time the former LPC guidance documents became established in the current format of Loss Prevention Standards.

In 2000 the LCPB became part of the BRE and now publishes its Standards under BRE Certification Ltd.

CPI System of Coding ~ the Co-ordinated Project Information initiative originated in the 1970s in response to the need to establish a common arrangement of document and language communication, across the varied trades and professions of the construction industry.

However, it has only been effective in recent years with the publication of the Standard Method of Measurement 7th edition (SMM 7), the National Building Specification (NBS) and the Drawings Code. (Note: The NBS is also produced in CI/SfB format.)

The arrangement in all documents is a coordination of alphabetic sections, corresponding to elements of work, the purpose being to avoid mistakes, omissions and other errors which have in the past occurred between drawings, specification and bill of quantities descriptions.

The coding is a combination of letters and numbers, spanning 3 levels:-

Level 1 has 24 headings from A to Z (omitting I and O). Each heading relates to part of the construction process, such as groundwork (D), Joinery (L), surface finishes (M), etc.

Level 2 is a sub-heading, which in turn is sub-grouped numerically into different categories. So for example, Surface Finishes is sub-headed; Plaster, Screeds, Painting, etc. These sub-headings are then extended further, thus Plaster becomes; Plastered/Rendered Coatings, Insulated Finishes, Sprayed Coatings etc.

Level 3 is the work section sub-grouped from level 2, to include a summary of inclusions and omissions.

As an example, an item of work coded M21 signifies:-

        M – Surface finishes

        2 – Plastered coatings

        1 – Insulation with rendered finish

The coding may be used to:-
(a) simplify specification writing
(b) reduce annotation on drawings
(c) rationalise traditional taking-off methods

CI/SfB System ~ this is a coded filing system for the classification and storing of building information and data. It was created in Sweden under the title of Samarbetskommittën för Byggnadsfrågor and was introduced into this country in 1961 by the RIBA. In 1968 the CI (Construction Index) was added to the system which is used nationally and recognised throughout the construction industry. The system consists of 5 sections called tables which are subdivided by a series of letters or numbers and these are listed in the CI/SfB index book to which reference should always be made in the first instance to enable an item to be correctly filed or retrieved.

Table 0 – Physical Environment

This table contains ten sections 0 to 9 and deals mainly with the end product (i.e. the type of building.) Each section can be further subdivided (e.g. 21, 22, et seq.) as required.

Table 1 – Elements

This table contains ten sections numbered (--) to (9-) and covers all parts of the structure such as walls, floors and services. Each section can be further subdivided (e.g. 31, 32 et seq.) as required.

Table 2 – Construction Form

This table contains twenty five sections lettered A to Z (O being omitted) and covers construction forms such as excavation work, blockwork, cast in-situ work etc., and is not subdivided but used in conjunction with Table 3.

Table 3 – Materials

This table contains twenty five sections lettered a to z (l being omitted) and covers the actual materials used in the construction form such as metal, timber, glass etc., and can be subdivided (e.g. n1, n2 et seq.) as required.

Table 4 – Activities and Requirements

This table contains twenty five sections lettered (A) to (Z), (O being omitted) and covers anything which results from the building process such as shape, heat, sound, etc. Each section can be further subdivided ((M1), (M2) et seq.) as required.

2 SITE WORKS

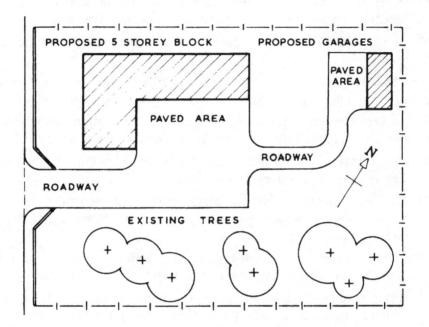

SITE INVESTIGATIONS
SOIL INVESTIGATION
SOIL ASSESSMENT AND TESTING
SITE LAYOUT CONSIDERATIONS
SITE SECURITY
SITE LIGHTING AND ELECTRICAL SUPPLY
SITE OFFICE ACCOMMODATION
MATERIALS STORAGE
MATERIALS TESTING-BRICKS AND CONCRETE
MATERIALS TESTING-SOFTWOOD CHARACTERISTICS
TIMBER DECAY AND TREATMENT
SETTING OUT
LEVELS AND ANGLES
ROAD CONSTRUCTION
TUBULAR SCAFFOLDING AND SCAFFOLDING SYSTEMS
SHORING SYSTEMS
DEMOLITION

Site Analysis – prior to purchasing a building site it is essential to conduct a thorough survey to ascertain whether the site characteristics suit the development concept. The following guidance forms a basic checklist:

* Refer to Ordnance Survey maps to determine adjacent features, location, roads, facilities, footpaths and rights of way.

* Conduct a measurement survey to establish site dimensions and levels.

* Observe surface characteristics, i.e. trees, steep slopes, existing buildings, rock outcrops, wells.

* Inquire of local authority whether preservation orders affect the site and if it forms part of a conservation area.

* Investigate subsoil. Use trial holes and borings to determine soil quality and water table level.

* Consider flood potential, possibilities for drainage of water table, capping of springs, filling of ponds, diversion of streams and rivers.

* Consult local utilities providers for underground and overhead services, proximity to site and whether they cross the site.

* Note suspicious factors such as filled ground, cracks in the ground, subsidence due to mining and any cracks in existing buildings.

* Regard neighbourhood scale and character of buildings with respect to proposed new development.

* Decide on best location for building (if space permits) with regard to 'cut and fill', land slope, exposure to sun and prevailing conditions, practical use and access.

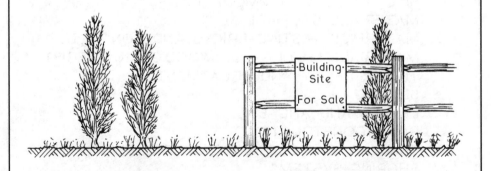

See also, desk and field studies on page 72

Site Investigation For New Works ~ the basic objective of this form of site investigation is to collect systematically and record all the necessary data which will be needed or will help in the design and construction processes of the proposed work. The collected data should be presented in the form of fully annotated and dimensioned plans and sections. Anything on adjacent sites which may affect the proposed works or conversely anything appertaining to the proposed works which may affect an adjacent site should also be recorded.

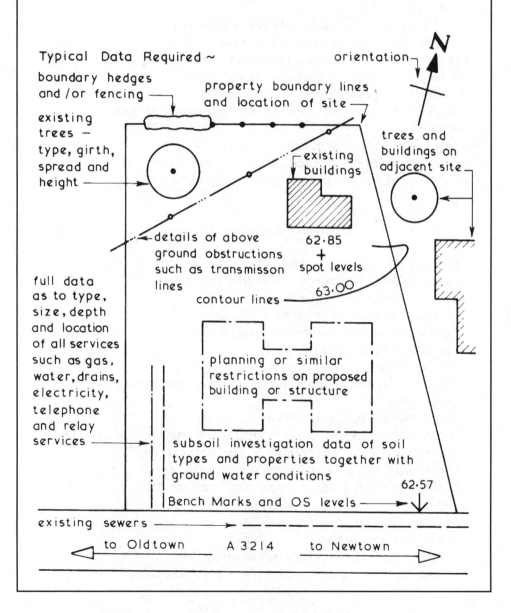

Typical Data Required ~

boundary hedges and /or fencing

existing trees — type, girth, spread and height

property boundary lines and location of site

orientation

N

trees and buildings on adjacent site

existing buildings

details of above ground obstructions such as transmisson lines

62·85
+
spot levels

contour lines
63·00

full data as to type, size, depth and location of all services such as gas, water, drains, electricity, telephone and relay services

planning or similar restrictions on proposed building or structure

subsoil investigation data of soil types and properties together with ground water conditions

62·57

Bench Marks and OS levels

existing sewers

to Oldtown   A 3214   to Newtown

71

Procedures ~

1. Desk study
2. Field study or walk-over survey
3. Laboratory analysis (see pages 81-82 and 85-87)

Desk Study ~ collection of known data, to include:

- Ordnance Survey maps - historical and modern, note grid reference.
- Geological maps - subsoil types, radon risk.
- Site history - green-field/brown-field.
- Previous planning applications/approvals.
- Current planning applications in the area.
- Development restrictions - conservation orders.
- Utilities - location of services on and near the site.
- Aerial photographs.
- Ecology factors - protected wildlife.
- Local knowledge - anecdotal information/rights of way.
- Proximity of local land fill sites - methane risk.

Field Study ~ intrusive visual and physical activity to:

- Establish site characteristics from the desk study.
- Assess potential hazards to health and safety.
- Appraise surface conditions:
  * Trees - preservation orders.
  * Topography and geomorphological mapping.
- Appraise ground conditions:
  * Water table.
  * Flood potential - local water courses and springs.
  * Soil types.
  * Contamination - vegetation die-back.
  * Engineering risks - ground subsidence, mining, old fuel tanks.
  * Financial risks - potential for the unforeseen.
- Take subsoil samples and conduct in-situ tests.
- Consider the need for subsoil exploration, trial pits and bore holes.
- Appraise existing structures:
  * Potential for re-use/refurbishment.
  * Archaeological value/preservation orders.
  * Demolition - costs, health issues e.g. asbestos.

Purpose ~ primarily to obtain subsoil samples for identification, classification and ascertaining the subsoil's characteristics and properties. Trial pits and augered holes may also be used to establish the presence of any geological faults and the upper or lower limits of the water table.

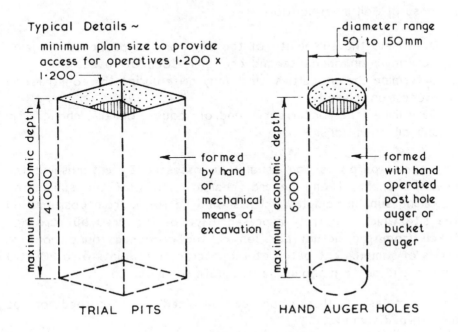

Typical Details ~

minimum plan size to provide access for operatives 1·200 x 1·200

diameter range 50 to 150 mm

maximum economic depth 4·000

formed by hand or mechanical means of excavation

maximum economic depth 6·000

formed with hand operated post hole auger or bucket auger

TRIAL PITS

HAND AUGER HOLES

General use ~

dry ground which requires little or no temporary support to sides of excavation.

Subsidiary use~
to expose and/or locate underground services.

Advantages ~
subsoil can be visually examined in-situ – both disturbed and undisturbed samples can be obtained.

General use ~

dry ground but liner tubes could be used if required to extract subsoil samples at a depth beyond the economic limit of trial holes.

Advantages ~
generally a cheaper and simpler method of obtaining subsoil samples than the trial pit method.

Trial pits and holes should be sited so that the subsoil samples will be representative but not interfering with works.

Site Investigation ~ this is an all embracing term covering every aspect of the site under investigation.

Soil Investigation ~ specifically related to the subsoil beneath the site under investigation and could be part of or separate from the site investigation.

Purpose of Soil Investigation ~

1. Determine the suitability of the site for the proposed project.
2. Determine an adequate and economic foundation design.
3. Determine the difficulties which may arise during the construction process and period.
4. Determine the occurrence and/or cause of all changes in subsoil conditions.

The above purposes can usually be assessed by establishing the physical, chemical and general characteristics of the subsoil by obtaining subsoil samples which should be taken from positions on the site which are truly representative of the area but are not taken from the actual position of the proposed foundations. A series of samples extracted at the intersection points of a 20·000 square grid pattern should be adequate for most cases.

Soil Samples ~ these can be obtained as disturbed or as undisturbed samples.

Disturbed Soil Samples ~ these are soil samples obtained from bore holes and trial pits. The method of extraction disturbs the natural structure of the subsoil but such samples are suitable for visual grading, establishing the moisture content and some laboratory tests. Disturbed soil samples should be stored in labelled airtight jars.

Undisturbed Soil Samples ~ these are soil samples obtained using coring tools which preserve the natural structure and properties of the subsoil. The extracted undisturbed soil samples are labelled and laid in wooden boxes for dispatch to a laboratory for testing. This method of obtaining soil samples is suitable for rock and clay subsoils but difficulties can be experienced in trying to obtain undisturbed soil samples in other types of subsoil.

The test results of soil samples are usually shown on a drawing which gives the location of each sample and the test results in the form of a hatched legend or section.

Depth of Soil Investigation ~ before determining the actual method of obtaining the required subsoil samples the depth to which the soil investigation should be carried out must be established. This is usually based on the following factors –

1. Proposed foundation type.
2. Pressure bulb of proposed foundation.
3. Relationship of proposed foundation to other foundations.

Typical Examples ~

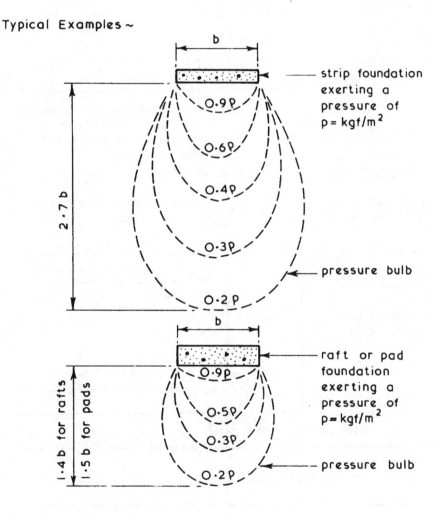

Pressure bulbs of less than 20% of original loading at foundation level can be ignored – this applies to all foundation types.

For further examples see next page.

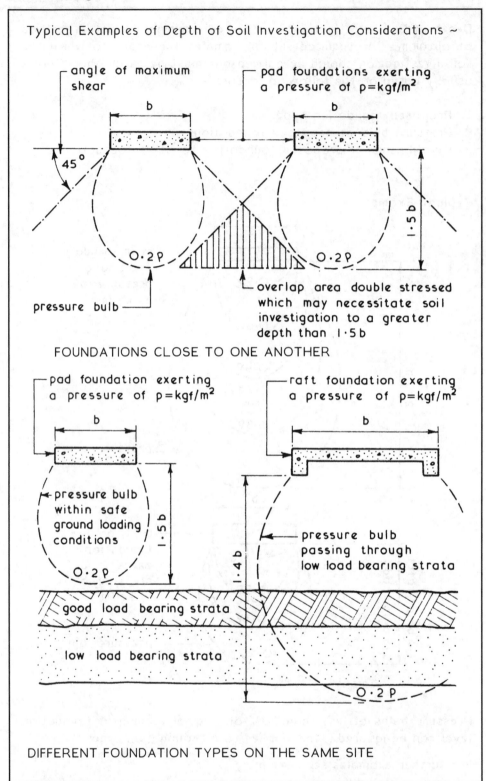

Typical Examples of Depth of Soil Investigation Considerations ~

angle of maximum shear

pad foundations exerting a pressure of $p = kgf/m^2$

b

b

45°

1·5 b

O·2p

O·2p

overlap area double stressed which may necessitate soil investigation to a greater depth than 1·5b

pressure bulb

FOUNDATIONS CLOSE TO ONE ANOTHER

pad foundation exerting a pressure of $p = kgf/m^2$

raft foundation exerting a pressure of $p = kgf/m^2$

b

b

pressure bulb within safe ground loading conditions

1·5 b

1·4 b

pressure bulb passing through low load bearing strata

O·2p

good load bearing strata

low load bearing strata

O·2p

DIFFERENT FOUNDATION TYPES ON THE SAME SITE

Soil Investigation Methods ~ method chosen will depend on several factors -

1. Size of contract.
2. Type of proposed foundation.
3. Type of sample required.
4. Type of subsoils which may be encountered.

As a general guide the most suitable methods in terms of investigation depth are -

1. Foundations up to 3·000 deep - trial pits.
2. Foundations up to 30·000 deep - borings.
3. Foundations over 30·000 deep - deep borings and in-situ examinations from tunnels and/or deep pits.

Typical Trail Pit Details ~

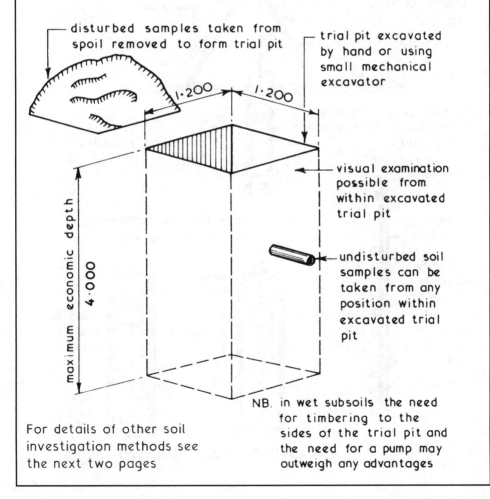

disturbed samples taken from spoil removed to form trial pit

trial pit excavated by hand or using small mechanical excavator

1·200   1·200

maximum economic depth
4·000

visual examination possible from within excavated trial pit

undisturbed soil samples can be taken from any position within excavated trial pit

For details of other soil investigation methods see the next two pages

NB. in wet subsoils the need for timbering to the sides of the trial pit and the need for a pump may outweigh any advantages

Boring Methods to Obtain Disturbed Soil Samples ~

1. Hand or Mechanical Auger ~ suitable for depths up to 3·000 using a 150 or 200mm diameter flight auger.
2. Mechanical Auger ~ suitable for depths over 3·000 using a flight or Cheshire auger ~ a liner or casing is required for most granular soils and may be required for other types of subsoil.
3. Sampling Shells ~ suitable for shallow to medium depth borings in all subsoils except rock.

Typical Details ~

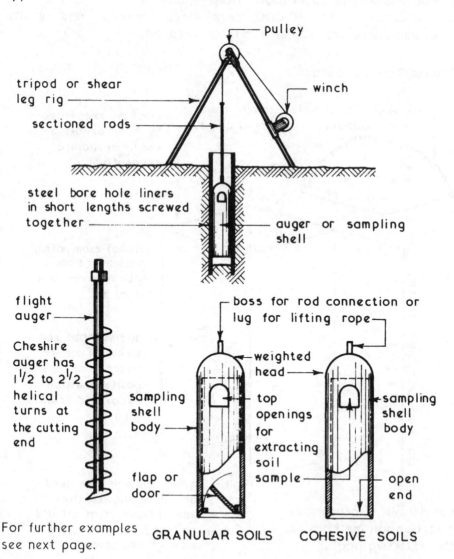

pulley

tripod or shear leg rig

winch

sectioned rods

steel bore hole liners in short lengths screwed together

auger or sampling shell

flight auger

Cheshire auger has 1½ to 2½ helical turns at the cutting end

boss for rod connection or lug for lifting rope

weighted head

sampling shell body

top openings for extracting soil sample

sampling shell body

flap or door

open end

GRANULAR SOILS   COHESIVE SOILS

For further examples see next page.

Wash Boring ~ this is a method of removing loosened soil from a bore hole using a strong jet of water or bentonite which is a controlled mixture of fullers earth and water. The jetting tube is worked up and down inside the bore hole, the jetting liquid disintegrates the subsoil which is carried in suspension up the annular space to a settling tank. The settled subsoil particles can be dried for testing and classification. This method has the advantage of producing subsoil samples which have not been disturbed by the impact of sampling shells however it is not suitable for large gravel subsoils or subsoils which contain boulders.

Typical Wash Boring Arrangement ~

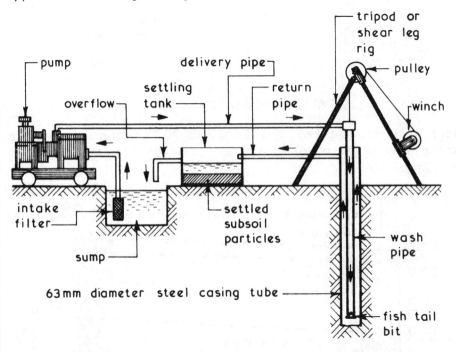

Mud-rotary Drilling ~ this is a method which can be used for rock investigations where bentonite is pumped in a continuous flow down hollow drilling rods to a rotating bit. The cutting bit is kept in contact with the bore face and the debris is carried up the annular space by the circulating fluid. Core samples can be obtained using coring tools.

Core Drilling ~ water or compressed air is jetted down the bore hole through a hollow tube and returns via the annular space. Coring tools extract continuous cores of rock samples which are sent in wooden boxes for laboratory testing.

Bore Hole Data ~ the information obtained from trial pits or bore holes can be recorded on a pro forma sheet or on a drawing showing the position and data from each trial pit or bore hole thus:-

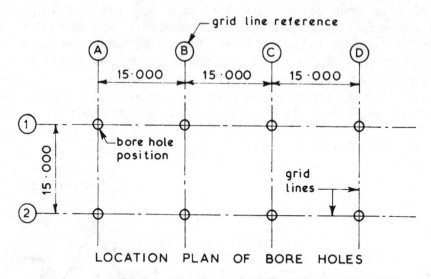

grid line reference

15·000   15·000   15·000

bore hole position

grid lines

15·000

LOCATION PLAN OF BORE HOLES

Bore holes can be taken on a 15·000 to 20·000 grid covering the whole site or in isolated positions relevant to the proposed foundation(s).

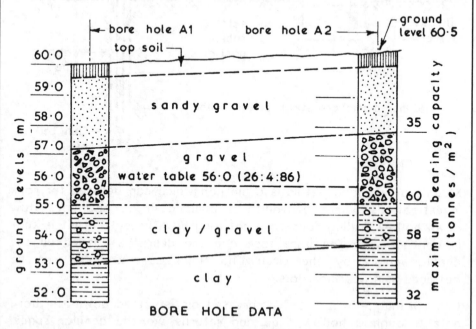

bore hole A1   bore hole A2   ground level 60·5

top soil

60·0

sandy gravel

ground levels (m)

59·0
58·0   35
57·0
gravel
56·0   water table 56·0 (26:4:86)
55·0   60
clay / gravel
54·0   58
53·0
clay
52·0   32

maximum bearing capacity (tonnes/m²)

BORE HOLE DATA

As a general guide the cost of site and soil investigations should not exceed 1% of estimated project costs.

Soil Assessment ~ prior to designing the foundations for a building or structure the properties of the subsoil(s) must be assessed. These processes can also be carried out to confirm the suitability of the proposed foundations. Soil assessment can include classification, grading, tests to establish shear strength and consolidation. The full range of methods for testing soils is given in BS 1377: Methods of test for soils for civil engineering purposes.

Classification ~ soils may be classified in many ways such as geological origin, physical properties, chemical composition and particle size. It has been found that the particle size and physical properties of a soil are closely linked and are therefore of particular importance and interest to a designer.

Particle Size Distribution ~ this is the percentages of the various particle sizes present in a soil sample as determined by sieving or sedimentation. BS 1377 divides particle sizes into groups as follows:-

Gravel particles – over 2mm
Sand particles   – between 2mm and 0·06mm
Silt particles   – between 0·06mm and 0·002mm
Clay particles   – less than 0·002mm

The sand and silt classifications can be further divided thus:-

CLAY	SILT			SAND			GRAVEL
	fine	medium	coarse	fine	medium	coarse	
0·002	0·006	0·02	0·06	0·2	0·6	2	

The results of a sieve analysis can be plotted as a grading curve thus:-

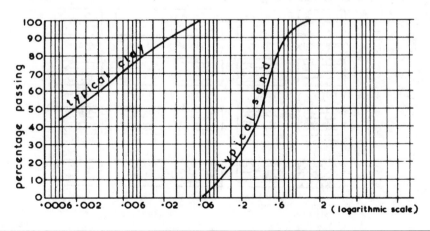

81

Triangular Chart ~ this provides a general classification of soils composed predominantly from clay, sand and silt. Each side of the triangle represents a percentage of material component. Following laboratory analysis, a sample's properties can be graphically plotted on the chart and classed accordingly.

e.g. Sand – 70%. Clay – 10% and Silt – 20% = Sandy Loam.

Note:

Silt is very fine particles of sand, easily suspended in water.
Loam is very fine particles of clay, easily dissolved in water.

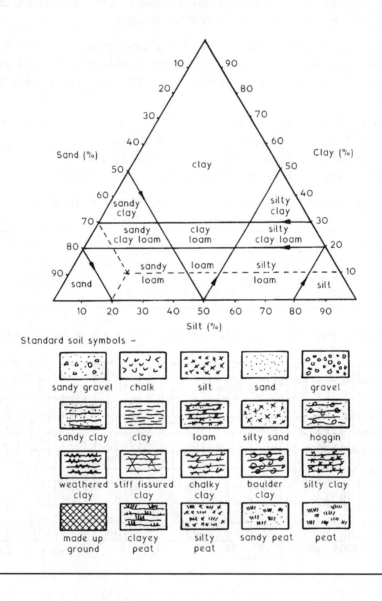

Site Soil Tests ~ these tests are designed to evaluate the density or shear strength of soils and are very valuable since they do not disturb the soil under test. Three such tests are the standard penetration test, the vane test and the unconfined compression test all of which are fully described in BS 1377; Methods of test for soils for civil engineering purposes.

Standard Penetration Test ~ this test measures the resistance of a soil to the penetration of a split spoon or split barrel sampler driven into the bottom of a bore hole. The sampler is driven into the soil to a depth of 150 mm by a falling standard weight of 65 kg falling through a distance of 760 mm. The sampler is then driven into the soil a further 300 mm and the number of blows counted up to a maximum of 50 blows. This test establishes the relative density of the soil.

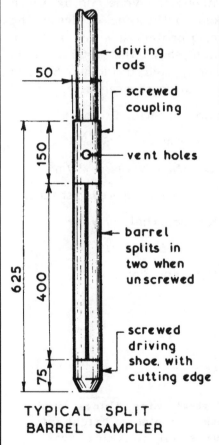

TYPICAL SPLIT BARREL SAMPLER

- driving rods
- screwed coupling
- vent holes
- barrel splits in two when unscrewed
- screwed driving shoe, with cutting edge

50 · 150 · 625 · 400 · 75

TYPICAL RESULTS
Non-cohesive soils:-

No. of Blows	Relative Density
0 to 4	very loose
4 to 10	loose
10 to 30	medium
30 to 50	dense
50+	very dense

Cohesive soils:-

No. of Blows	Relative Density
0 to 2	very soft
2 to 4	soft
4 to 8	medium
8 to 15	stiff
15 to 30	very stiff
30+	hard

The results of this test in terms of number of blows and amounts of penetration will need expert interpretation.

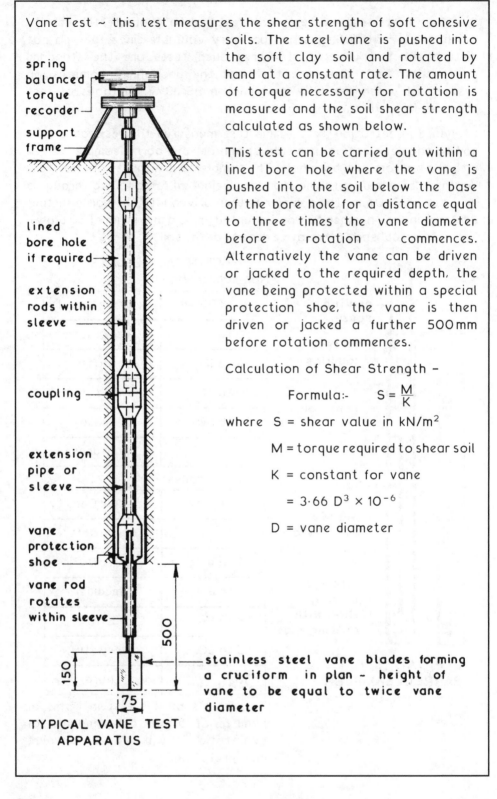

Vane Test ~ this test measures the shear strength of soft cohesive soils. The steel vane is pushed into the soft clay soil and rotated by hand at a constant rate. The amount of torque necessary for rotation is measured and the soil shear strength calculated as shown below.

This test can be carried out within a lined bore hole where the vane is pushed into the soil below the base of the bore hole for a distance equal to three times the vane diameter before rotation commences. Alternatively the vane can be driven or jacked to the required depth, the vane being protected within a special protection shoe, the vane is then driven or jacked a further 500mm before rotation commences.

Calculation of Shear Strength –

$$\text{Formula:-} \quad S = \frac{M}{K}$$

where S = shear value in kN/m^2

M = torque required to shear soil

K = constant for vane

$= 3.66 \, D^3 \times 10^{-6}$

D = vane diameter

spring balanced torque recorder

support frame

lined bore hole if required

extension rods within sleeve

coupling

extension pipe or sleeve

vane protection shoe

vane rod rotates within sleeve

500

150

75

stainless steel vane blades forming a cruciform in plan – height of vane to be equal to twice vane diameter

TYPICAL VANE TEST APPARATUS

Unconfined Compression Test ~ this test can be used to establish the shear strength of a non-fissured cohesive soil sample using portable apparatus either on site or in a laboratory. The 75 mm long × 38 mm diameter soil sample is placed in the apparatus and loaded in compression until failure occurs by shearing or lateral bulging. For accurate reading of the trace on the recording chart a transparent viewfoil is placed over the trace on the chart.

Typical Apparatus Details~

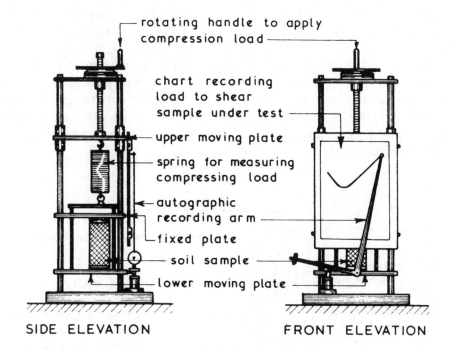

rotating handle to apply compression load

chart recording load to shear sample under test

upper moving plate

spring for measuring compressing load

autographic recording arm

fixed plate

soil sample

lower moving plate

SIDE ELEVATION   FRONT ELEVATION

Typical Results ~ showing compression strengths of clays:-

Very soft clay   – less than 25 kN/m²

Soft clay   – 25 to 50 kN/m²

Medium clay   – 50 to 100 kN/m²

Stiff clay   – 100 to 200 kN/m²

Very stiff clay   – 200 to 400 kN/m²

Hard clay   – more than 400 kN/m²

NB. The shear strength of clay soils is only half of the compression strength values given above.

Laboratory Testing ~ tests for identifying and classifying soils with regard to moisture content, liquid limit, plastic limit, particle size distribution and bulk density are given in BS 1377.

Bulk Density ~ this is the mass per unit volume which includes mass of air or water in the voids and is essential information required for the design of retaining structures where the weight of the retained earth is an important factor.

Shear Strength ~ this soil property can be used to establish its bearing capacity and also the pressure being exerted on the supports in an excavation. The most popular method to establish the shear strength of cohesive soils is the Triaxial Compression Test. In principle this test consists of subjecting a cylindrical sample of undisturbed soil (75mm long × 38mm diameter) to a lateral hydraulic pressure in addition to a vertical load. Three tests are carried out on three samples (all cut from the same large sample) each being subjected to a higher hydraulic pressure before axial loading is applied. The results are plotted in the form of Mohr's circles.

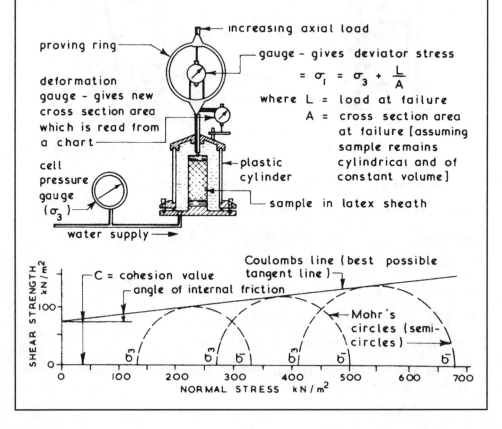

Shear Strength ~ this can be defined as the resistance offered by a soil to the sliding of one particle over another. A simple method of establishing this property is the Shear Box Test in which the apparatus consists of two bottomless boxes which are filled with the soil sample to be tested. A horizontal shearing force (S) is applied against a vertical load (W) causing the soil sample to shear along a line between the two boxes.

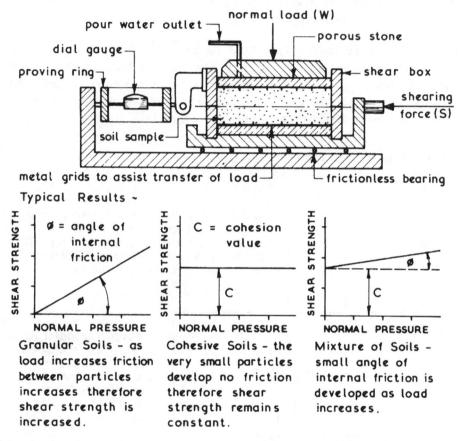

Typical Results ~

$\emptyset$ = angle of internal friction

C = cohesion value

Granular Soils - as load increases friction between particles increases therefore shear strength is increased.

Cohesive Soils - the very small particles develop no friction therefore shear strength remains constant.

Mixture of Soils - small angle of internal friction is developed as load increases.

Consolidation of Soil ~ this property is very important in calculating the movement of a soil under a foundation. The laboratory testing apparatus is called an Oedometer.

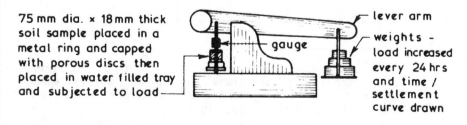

75 mm dia. × 18 mm thick soil sample placed in a metal ring and capped with porous discs then placed in water filled tray and subjected to load

lever arm

weights - load increased every 24 hrs and time / settlement curve drawn

General Considerations ~ before any specific considerations and decisions can be made regarding site layout a general appreciation should be obtained by conducting a thorough site investigation at the pre-tender stage and examining in detail the drawings, specification and Bill of Quantities to formulate proposals of how the contract will be carried out if the tender is successful. This will involve a preliminary assessment of plant, materials and manpower requirements plotted against the proposed time scale in the form of a bar chart (see page 34).

Access Considerations ~ this must be considered for both on- and off-site access. Routes to and from the site must be checked as to the suitability for transporting all the requirements for the proposed works. Access on site for deliveries and general circulation must also be carefully considered.

Typical Site Access Considerations ~

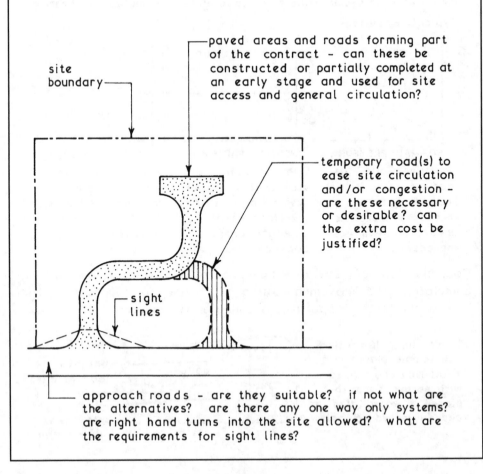

site boundary

paved areas and roads forming part of the contract - can these be constructed or partially completed at an early stage and used for site access and general circulation?

temporary road(s) to ease site circulation and /or congestion - are these necessary or desirable? can the extra cost be justified?

sight lines

approach roads - are they suitable? if not what are the alternatives? are there any one way only systems? are right hand turns into the site allowed? what are the requirements for sight lines?

Storage Considerations ~ amount and types of material to be stored, security and weather protection requirements, allocation of adequate areas for storing materials and allocating adequate working space around storage areas as required, siting of storage areas to reduce double handling to a minimum without impeding the general site circulation and/or works in progress.

Accommodation Considerations ~ number and type of site staff anticipated, calculate size and select units of accommodation and check to ensure compliance with the minimum requirements of the Construction (Health, Safety and Welfare) Regulations 1996, select siting for offices to give easy and quick access for visitors but at the same time giving a reasonable view of the site, select siting for messroom and toilets to reduce walking time to a minimum without impeding the general site circulation and/or works in progress.

Temporary Services Considerations ~ what, when and where are they required? Possibility of having permanent services installed at an early stage and making temporary connections for site use during the construction period, coordination with the various service undertakings is essential.

Plant Considerations ~ what plant, when and where is it required? static or mobile plant? If static select the most appropriate position and provide any necessary hard standing, if mobile check on circulation routes for optimum efficiency and suitability, provision of space and hard standing for on-site plant maintenance if required.

Fencing and Hoarding Considerations ~ what is mandatory and what is desirable? Local vandalism record, type or types of fence and/or hoarding required, possibility of using fencing which is part of the contract by erecting this at an early stage in the contract.

Safety and Health Considerations ~ check to ensure that all the above conclusions from the considerations comply with the minimum requirements set out in the various Construction Regulations and in the Health and Safety at Work etc., Act 1974.

For a typical site layout example see next page.

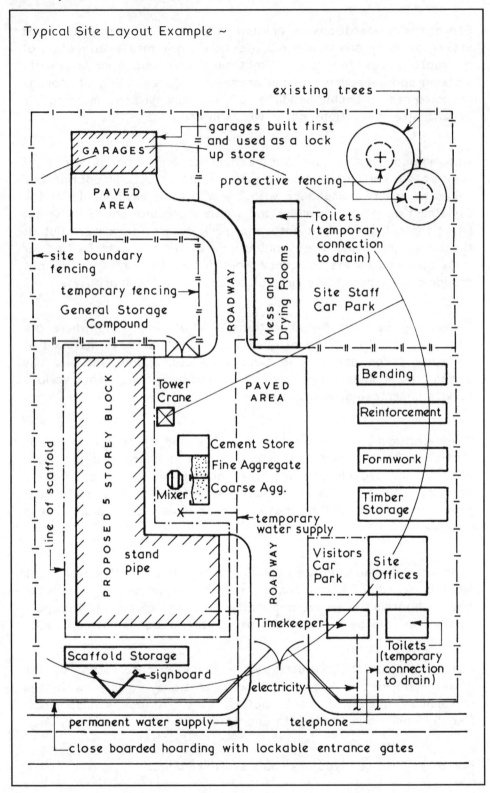

Typical Site Layout Example ~

Site Security ~ the primary objectives of site security are –

1. Security against theft.
2. Security from vandals.
3. Protection from innocent trespassers.

The need for and type of security required will vary from site to site according to the neighbourhood, local vandalism record and the value of goods stored on site. Perimeter fencing, internal site protection and night security may all be necessary.

Typical Site Security Provisions ~

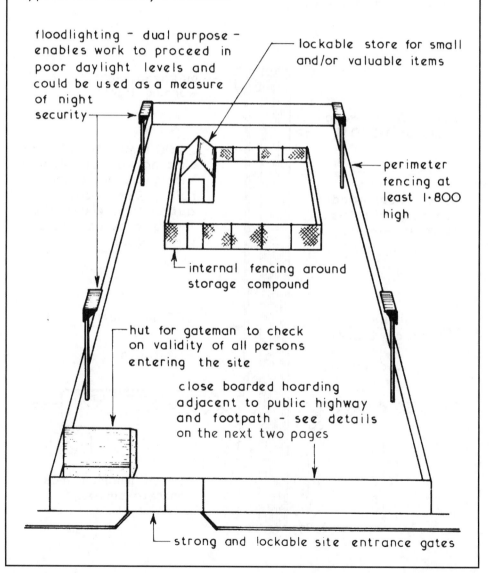

floodlighting – dual purpose – enables work to proceed in poor daylight levels and could be used as a measure of night security

lockable store for small and/or valuable items

perimeter fencing at least 1·800 high

internal fencing around storage compound

hut for gateman to check on validity of all persons entering the site

close boarded hoarding adjacent to public highway and footpath – see details on the next two pages

strong and lockable site entrance gates

Hoardings ~ under the Highways Act 1980 a close boarded fence hoarding must be erected prior to the commencement of building operations if such operations are adjacent to a public footpath or highway. The hoarding needs to be adequately constructed to provide protection for the public, resist impact damage, resist anticipated wind pressures and adequately lit at night. Before a hoarding can be erected a licence or permit must be obtained from the local authority who will usually require 10 to 20 days notice. The licence will set out the minimum local authority requirements for hoardings and define the time limit period of the licence.

Typical Hoarding Details ~

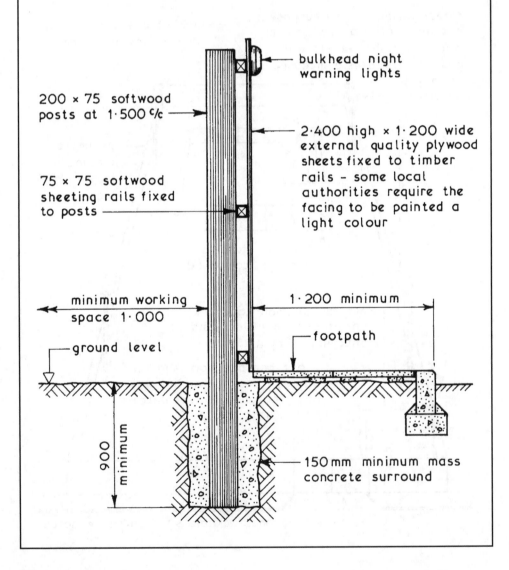

200 × 75 softwood posts at 1·500 ℅

bulkhead night warning lights

2·400 high × 1·200 wide external quality plywood sheets fixed to timber rails - some local authorities require the facing to be painted a light colour

75 × 75 softwood sheeting rails fixed to posts

minimum working space 1·000

1·200 minimum

footpath

ground level

900 minimum

150 mm minimum mass concrete surround

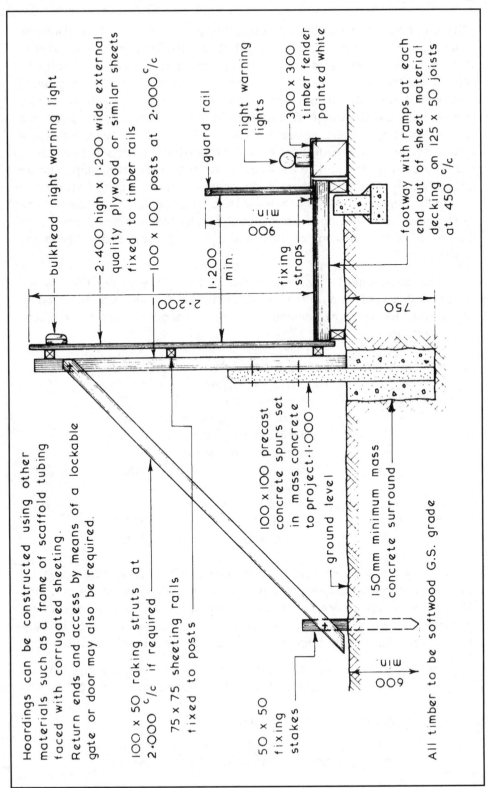

Hoardings can be constructed using other materials such as a frame of scaffold tubing faced with corrugated sheeting.
Return ends and access by means of a lockable gate or door may also be required.

bulkhead night warning light

2·400 high x 1·200 wide external quality plywood or similar sheets fixed to timber rails

100 x 100 posts at 2·000 c/c

guard rail

night warning lights

300 x 300 timber fender painted white

footway with ramps at each end out of sheet material decking on 125 x 50 joists at 450 c/c

900 min

1·200 min.

fixing straps

2·200

750

100 x 50 raking struts at 2·000 c/c if required

75 x 75 sheeting rails fixed to posts

100 x 100 precast concrete spurs set in mass concrete to project 1·000

ground level

150mm minimum mass concrete surround

50 x 50 fixing stakes

600 min

All timber to be softwood G.S. grade

Site Lighting ~ this can be used effectively to enable work to continue during periods of inadequate daylight. It can also be used as a deterrent to would-be trespassers. Site lighting can be employed externally to illuminate the storage and circulation areas and internally for general movement and for specific work tasks. The types of lamp available range from simple tungsten filament lamps to tungsten halogen and discharge lamps. The arrangement of site lighting can be static where the lamps are fixed to support poles or mounted on items of fixed plant such as scaffolding and tower cranes. Alternatively the lamps can be sited locally where the work is in progress by being mounted on a movable support or hand held with a trailing lead. Whenever the position of site lighting is such that it can be manhandled it should be run on a reduced voltage of 110 V single phase as opposed to the mains voltage of 230 V.

To plan an adequate system of site lighting the types of activity must be defined and given an illumination target value which is quoted in lux (lx). Recommended minimum target values for building activities are:-

External lighting   –   general circulation $\left.\right\}$ 10 lx
materials handling

Internal lighting   –   general circulation   5 lx
general working areas   15 lx
concreting activities   50 lx
carpentry and joinery $\left.\right\}$ 100 lx
bricklaying
plastering
painting and decorating $\left.\right\}$
site offices   200 lx
drawing board positions   300 lx

Such target values do not take into account deterioration, dirt or abnormal conditions therefore it is usual to plan for at least twice the recommended target values. Generally the manufacturers will provide guidance as to the best arrangement to use in any particular situation but lamp requirements can be calculated thus:-

$$\text{Total lumens required} = \frac{\text{area to be illluminated (m}^2\text{)} \times \text{target value (lx)}}{\text{utilisation factor 0·23 [dispersive lights 0·27]}}$$

After choosing lamp type to be used:-

$$\text{Number of lamps required} = \frac{\text{total lumens required}}{\text{lumen output of chosen lamp}}$$

Typical Site Lighting Arrangement:-

Area lighting using high mounted lamps ~

area of illumination

tungsten
halogen
lamps
mounted
on posts
or mast
supports

limit of effective throw
4 × height

maximum
0·6 × height

maximum spacing 1·5 × height

maximum
0·6 × height

lamp

post

height

ground
level

Typical minimum heights for tungsten
halogen lamps :-

500 watts - 7·500 metres
1000 watts - 9·000 metres
2000 watts - 15·000 metres

Area lighting using overhead dispersive lights suspended from
a grid or from the structure ~

edge of illuminated
area

lamp fittings to be resistant
to corrosion, rust and rain

lamps
at a
height
of H
above
floor
level

grid
lines

0·75 H
max.

maximum spacing 1·5 × H

0·75 H
max.

Typical minimum heights for dispersive lamps:

Fluorescent 40 to 125 W – 2·500 m; Tungsten filament 300 W – 3·000 m

Walkway and Local Lighting ~ to illuminate the general circulation routes bulkhead and/or festoon lighting could be used either on a standard mains voltage of 230 V or on a reduced voltage of 110 V. For local lighting at the place of work hand lamps with trailing leads or lamp fittings on stands can be used and positioned to give the maximum amount of illumination without unacceptable shadow cast.

Typical Walkway and Local Lighting Fittings ~

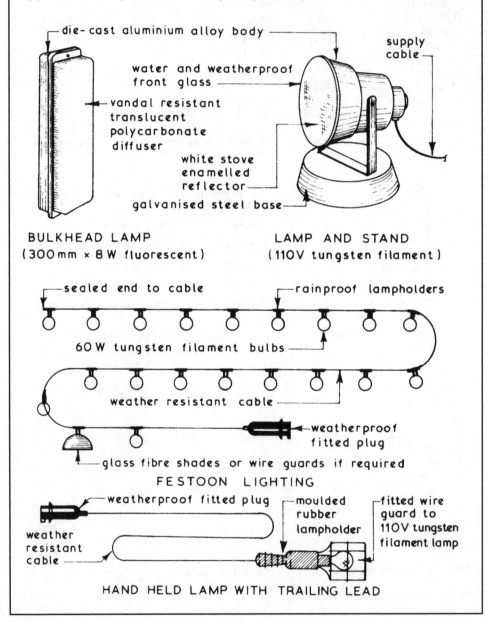

die-cast aluminium alloy body

supply cable

water and weatherproof front glass

vandal resistant translucent polycarbonate diffuser

white stove enamelled reflector

galvanised steel base

BULKHEAD LAMP
(300 mm × 8 W fluorescent)

LAMP AND STAND
(110V tungsten filament)

sealed end to cable

rainproof lampholders

60 W tungsten filament bulbs

weather resistant cable

weatherproof fitted plug

glass fibre shades or wire guards if required

FESTOON LIGHTING

weatherproof fitted plug

moulded rubber lampholder

fitted wire guard to 110V tungsten filament lamp

weather resistant cable

HAND HELD LAMP WITH TRAILING LEAD

Electrical Supply to Building Sites ~ a supply of electricity is usually required at an early stage in the contract to provide light and power to the units of accommodation. As the work progresses power could also be required for site lighting, hand held power tools and large items of plant. The supply of electricity to a building site is the subject of a contract between the contractor and the local area electricity company who will want to know the date when supply is required; site address together with a block plan of the site; final load demand of proposed building and an estimate of the maximum load demand in kilowatts for the construction period. The latter can be estimated by allowing $10\,W/m^2$ of the total floor area of the proposed building plus an allowance for high load equipment such as cranes. The installation should be undertaken by a competent electrical contractor to ensure that it complies with all the statutory rules and regulations for the supply of electricity to building sites.

Typical Supply and Distribution Equipment ~

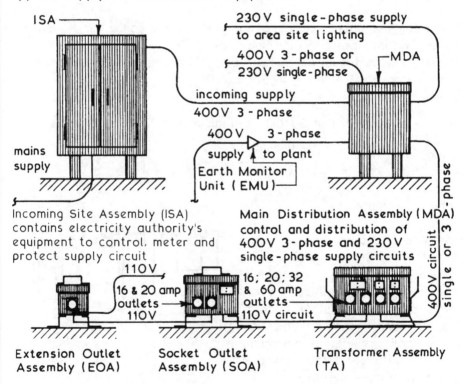

Incoming Site Assembly (ISA) contains electricity authority's equipment to control, meter and protect supply circuit

Main Distribution Assembly (MDA) control and distribution of 400V 3-phase and 230V single-phase supply circuits

Extension Outlet Assembly (EOA)

Socket Outlet Assembly (SOA)

Transformer Assembly (TA)

The units must be strong, durable and resistant to rain penetration with adequate weather seals to all access panels and doors. All plug and socket outlets should be colour coded :- 400V - red; 230V - blue; 110V - yellow.

Office Accommodation ~ the arrangements for office accommodation to be provided on site is a matter of choice for each individual contractor. Generally separate offices would be provided for site agent, clerk of works, administrative staff, site surveyors and sales staff.

The minimum requirements of such accommodation is governed by the Offices, Shops and Railway Premises Act 1963 unless they are ~

1. Mobile units in use for not more then 6 months.
2. Fixed units in use for not more than 6 weeks.
3. Any type of unit in use for not more than 21 man hours per week.
4. Office for exclusive use of self employed person.
5. Office used by family only staff.

Sizing Example ~

Office for site agent and assistant plus an allowance for 3 visitors.
Assume an internal average height of 2·400.
Allow 3·7 m² minimum per person and 11·5 m³ minimum per person.
Minimum area = 5 × 3·7 = 18·5 m²
Minimum volume = 5 × 11·5 = 57·5 m³

Assume office width of 3·000 then minimum length required is
$= \dfrac{57\cdot5}{3 \times 2\cdot4} = \dfrac{57\cdot5}{7\cdot2} = 7\cdot986$ say 8·000
Area check 3 × 8 = 24 m² which is > 18·5 m² ∴ satisfactory

Typical Example ~

Portable cabin with four adjustable steel legs with attachments for stacking. Panelling of galvanised steel sheet and rigid insulation core. Plasterboard inner lining to walls and ceiling. Pyro-shield windows with steel shutters and a high security steel door.

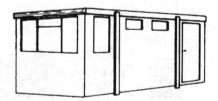

Ref. Fire prevention on construction sites – the joint code of practice on protection from fire of construction sites and buildings undergoing renovation. Published by Construction Confederation and The Fire Protection Association.

The requirements for health and wellbeing of persons on construction sites are enforced by the Health and Safety Executive, through the Health and Safety at Work etc. Act 1974 and the Construction (Health, Safety and Welfare) Regulations 1996. The following minimum requirements apply and the numbers of persons on site were established by the Construction Regulations of 1966.

Provision	Requirement	No of persons employed on site
FIRST AID	Box to be distinctively marked and in charge of responsible person.	5 to 50 – first aid boxes 50 + first aid box and a person trained in first aid
AMBULANCES	Stretcher(s) in charge of responsible person	25 + notify ambulance authority of site details within 24 hours of employing more than 25 persons
FIRST AID ROOM	Used only for rest or treatment and in charge of trained person	If more than 250 persons employed on site each employer of more than 40 persons to provide a first aid room
SHELTER AND ACCOMMODATION FOR CLOTHING	All persons on site to have shelter and a place for changing, drying and depositing clothes. Separate facilities for male and female staff.	Up to 5 where possible a means of warming themselves and drying wet clothes 5 + adequate means of warming themselves and drying wet clothing
REST ROOM	Drinking water, means of boiling water, preparing and eating meals for all persons on site. Arrangements to protect non-smokers from tobacco smoke.	10 + facilities for heating food if hot meals are not available on site
WASHING FACILITIES	Washing facilities to be provided for all persons on site for more than 4 hours. Ventilated and lit. Separate facilities for male and female staff.	20 to 100 if work is to last more than 6 weeks – hot and cold or warm water, soap and towel. 100 + work lasting more than 12 months – 4 wash places + 1 for every 35 persons over 100
SANITARY FACILITIES	To be maintained, lit, ventilated and kept clean. Separate facilities for male and female staff	Up to 100 – 1 convenience for every 25 persons 100 + convenience for every 35 persons

Site Storage ~ materials stored on site prior to being used or fixed may require protection for security reasons or against the adverse effects which can be caused by exposure to the elements.

Small and Valuable Items ~ these should be kept in a secure and lockable store. Similar items should be stored together in a rack or bin system and only issued against an authorised requisition.

Large or Bulk Storage Items ~ for security protection these items can be stored within a lockable fenced compound. The form of fencing chosen may give visual security by being of an open nature but these are generally easier to climb than the close boarded type of fence which lacks the visual security property.

Typical Storage Compound Fencing ~

Close boarded fences can be constructed on the same methods used for hoardings - see pages 92 & 93.

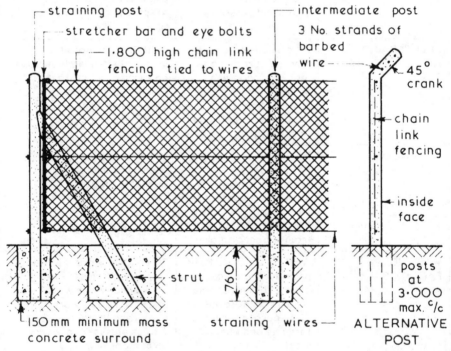

CHAIN LINK FENCING WITH PRECAST CONCRETE POSTS

Alternative Fence Types ~ woven wire fence, strained wire fence, cleft chestnut pale fence, wooden palisade fence, wooden post and rail fence and metal fences - see BS 1722: Fences, for details.

Storage of Materials ~ this can be defined as the provision of adequate space, protection and control for building materials and components held on site during the construction process. The actual requirements for specific items should be familiar to students who have completed studies in construction technology at an introductory level but the need for storage and control of materials held on site can be analysed further:-

1. Physical Properties – size, shape, weight and mode of delivery will assist in determining the safe handling and stacking method(s) to be employed on site, which in turn will enable handling and storage costs to be estimated.

2. Organisation – this is the planning process of ensuring that all the materials required are delivered to site at the correct time, in sufficient quantity, of the right quality, the means of unloading is available and that adequate space for storage or stacking has been allocated.

3. Protection – building materials and components can be classified as durable or non-durable, the latter will usually require some form of weather protection to prevent deterioration whilst in store.

4. Security – many building materials have a high resale and/or usage value to persons other than those for whom they were ordered and unless site security is adequate material losses can become unacceptable.

5. Costs – to achieve an economic balance of how much expenditure can be allocated to site storage facilities the following should be taken into account:-

   a. Storage areas, fencing, racks, bins, etc.
   b. Protection requirements.
   c. Handling, transporting and stacking requirements.
   d. Salaries and wages of staff involved in storage of materials and components.
   e. Heating and/or lighting if required.
   f. Allowance for losses due to wastage, deterioration, vandalism and theft.
   g. Facilities to be provided for subcontractors.

6. Control – checking quality and quantity of materials at delivery and during storage period, recording delivery and issue of materials and monitoring stock holdings.

Site Storage Space ~ the location and size(s) of space to be allocated for any particular material should be planned by calculating the area(s) required and by taking into account all the relevant factors before selecting the most appropriate position on site in terms of handling, storage and convenience. Failure to carry out this simple planning exercise can result in chaos on site or having on site more materials than there is storage space available.

Calculation of Storage Space Requirements ~ each site will present its own problems since a certain amount of site space must be allocated to the units of accommodation, car parking, circulation and working areas, therefore the amount of space available for materials storage may be limited. The size of the materials or component being ordered must be known together with the proposed method of storage and this may vary between different sites of similar building activities. There are therefore no standard solutions for allocating site storage space and each site must be considered separately to suit its own requirements.

Typical Examples ~

Bricks – quantity = 15,200 to be delivered in strapped packs of 380 bricks per pack each being 1100 mm wide × 670 mm long × 850 mm high. Unloading and stacking to be by forklift truck to form 2 rows 2 packs high.

Area required :- number of packs per row = $\dfrac{15,200}{380 \times 2}$ = 20

length of row = 10 × 670 = 6·700
width of row = 2 × 1100 = 2·200

allowance for forklift approach in front of stack = 5·000 ∴ minimum brick storage area = 6·700 long × 7·200 wide

Timber – to be stored in open sided top covered racks constructed of standard scaffold tubes. Maximum length of timber ordered = 5·600. Allow for rack to accept at least 4 No. 300 mm wide timbers placed side by side then minimum width required = 4 × 300 = 1·200
Minimum plan area for timber storage rack = 5·600 × 1·200
Allow for end loading of rack equal to length of rack
∴ minimum timber storage area = 11·200 long × 1·200 wide
Height of rack to be not more than 3 × width = 3·600

Areas for other materials stored on site can be calculated using the basic principles contained in the examples above.

Site Allocation for Materials Storage ~ the area and type of storage required can be determined as shown on pages 100 to 102, but the allocation of an actual position on site will depend on:-

1. Space available after areas for units of accommodation have been allocated.
2. Access facilities on site for delivery, vehicles.
3. Relationship of storage area(s) to activity area(s) – the distance between them needs to be kept as short as possible to reduce transportation needs in terms of time and costs to the minimum. Alternatively storage areas and work areas need to be sited within the reach of any static transport plant such as a tower crane.
4. Security – needs to be considered in the context of site operations, vandalism and theft.
5. Stock holding policy – too little storage could result in delays awaiting for materials to be delivered, too much storage can be expensive in terms of weather and security protection requirements apart from the capital used to purchase the materials stored on site.

Typical Example ~

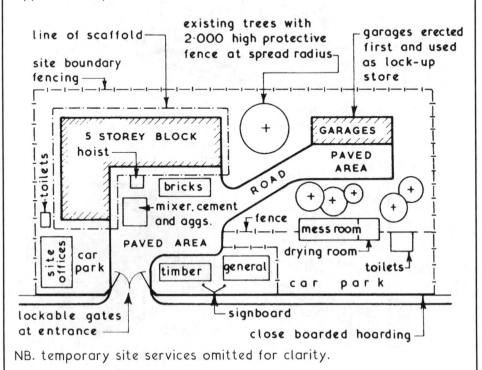

NB. temporary site services omitted for clarity.

Bricks ~ may be supplied loose or strapped in unit loads and stored on timber pallets

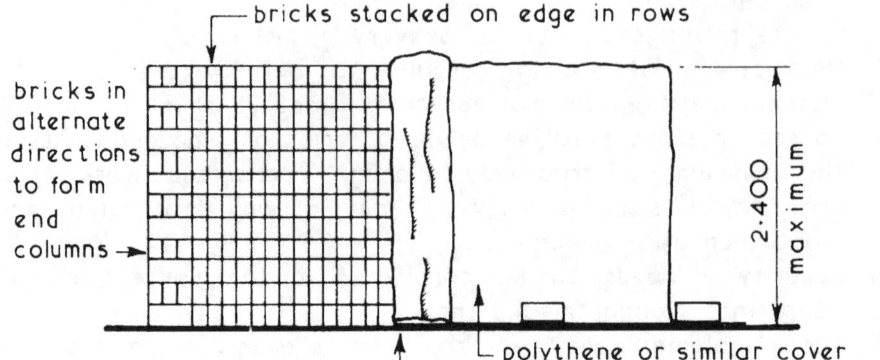

bricks stacked on edge in rows

bricks in alternate directions to form end columns

2·400 maximum

level well drained ground

polythene or similar cover weighted at bottom to protect bricks against atmospheric pollution and/or inclement weather

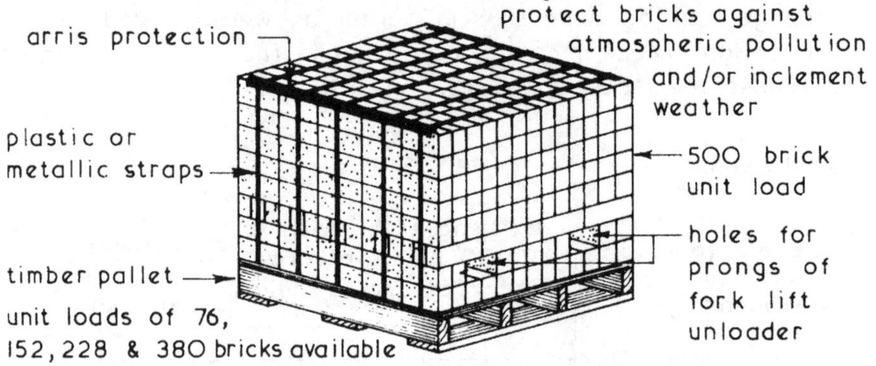

arris protection

plastic or metallic straps

500 brick unit load

holes for prongs of fork lift unloader

timber pallet

unit loads of 76, 152, 228 & 380 bricks available

Blocks ~ may be supplied loose or in unit loads on timber pallets

blocks stacked in 'columns'

Roofing Tiles ~ may be supplied loose, in plastic wrapped packs or in unit loads on timber pallets

protective cover

8 courses maximum

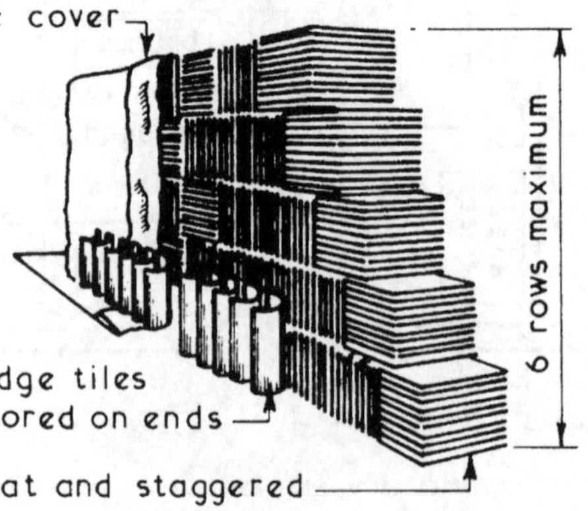

6 rows maximum

ridge tiles stored on ends

end laid flat and staggered

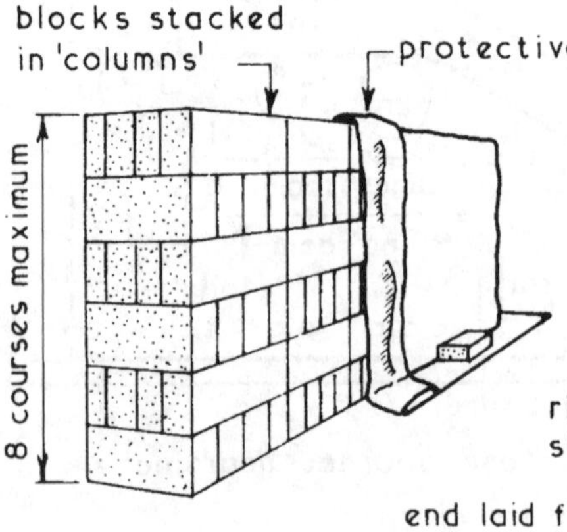

Drainage Pipes ~ supplied loose or strapped together on timber pallets

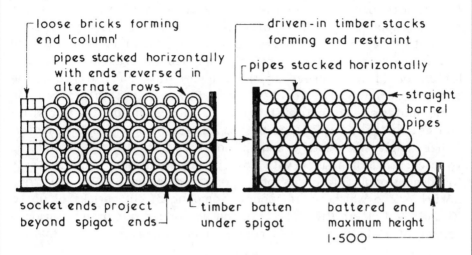

loose bricks forming end 'column'

pipes stacked horizontally with ends reversed in alternate rows

driven-in timber stacks forming end restraint

pipes stacked horizontally

straight barrel pipes

socket ends project beyond spigot ends

timber batten under spigot

battered end maximum height 1·500

Gullies etc., should be stored upside down and supported to remain level

Baths ~ stacked or nested vertically or horizontally on timber battens

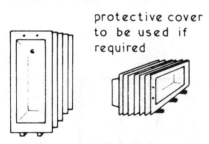

protective cover to be used if required

Timber and Joinery Items ~ should be stored horizontally and covered but with provison for free air flow

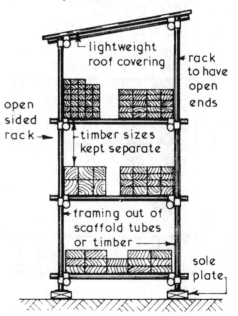

lightweight roof covering

rack to have open ends

open sided rack

timber sizes kept separate

framing out of scaffold tubes or timber

sole plate

Basins ~ stored similar to baths but not more than four high if nested one on top of another

Corrugated and Similar Sheet Materials ~ stored flat on a level surface and covered with a protective polythene or similar sheet material

Cement, Sand and Aggregates ~ for supply and storage details see pages 285 & 289.

Site Tests ~ the majority of materials and components arriving on site will conform to the minimum recommendations of the appropriate British Standard and therefore the only tests which need be applied are those of checking quantity received against amount stated on the delivery note, ensuring quality is as ordered and a visual inspection to reject damaged or broken goods. The latter should be recorded on the delivery note and entered in the site records. Certain site tests can however be carried out on some materials to establish specific data such as the moisture content of timber which can be read direct from a moisture meter. Other simple site tests are given in the various British Standards to ascertain compliance with the recommendations, such as tests for dimensional tolerances and changes given in BS EN 771-1 and BS EN 772-16 which cover random sampling of clay bricks of up to 10 units. An alternative site test can be carried out by measuring a sample of 24 bricks taken at random from a delivered load thus:-

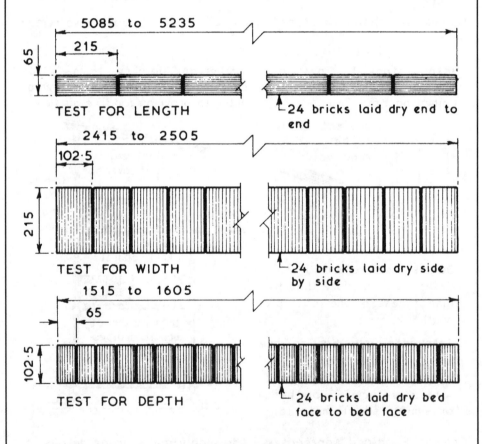

TEST FOR LENGTH   24 bricks laid dry end to end

TEST FOR WIDTH   24 bricks laid dry side by side

TEST FOR DEPTH   24 bricks laid dry bed face to bed face

Refs. BS EN 772-16: Methods of test for masonry units.
BS EN 771-1: Specification for masonry units.

Site Test ~ apart from the test outlined on page 83 site tests on materials which are to be combined to form another material such as concrete can also be tested to establish certain properties which if not known could affect the consistency and/or quality of the final material.

Typical Example ~ Testing Sand for Bulking

This data is required when batching concrete by volume – test made at commencement of mixing and if change in weather

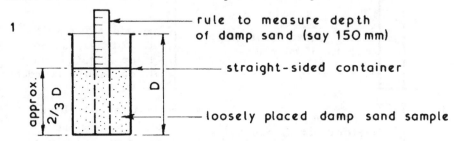

1 — rule to measure depth of damp sand (say 150 mm)

— straight-sided container

approx. 2/3 D

D

— loosely placed damp sand sample

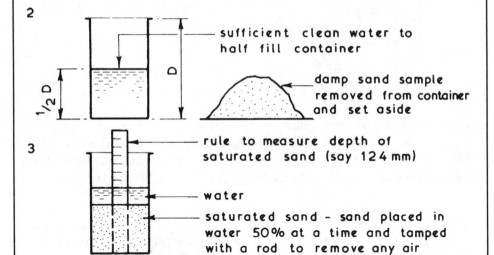

2 — sufficient clean water to half fill container

1/2 D

D

damp sand sample removed from container and set aside

3 — rule to measure depth of saturated sand (say 124 mm)

— water

— saturated sand – sand placed in water 50% at a time and tamped with a rod to remove any air

4   Calculation :-

$$\text{bulking} = \frac{\text{difference in height between damp \& saturated sand}}{\text{depth of saturated sand}}$$

$$\% \text{ bulking} = \frac{150 - 124}{124} \times 100 = \frac{26}{124} \times 100 = 20 \cdot 96774 \%$$

Therefore volume of sand should be increased by 21% over that quoted in the specification

NB. a given weight of saturated sand will occupy the same space as when dry but more space when damp

Silt Test for Sand ~ the object of this test is to ascertain the cleanliness of sand by establishing the percentage of silt present in a natural sand since too much silt will weaken the concrete

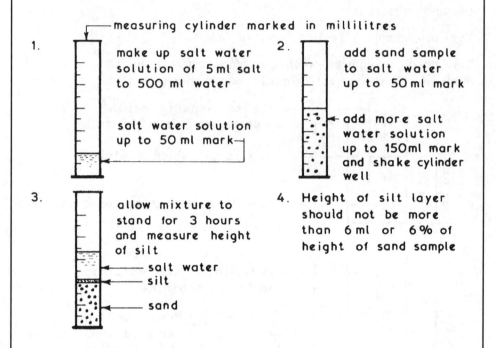

measuring cylinder marked in millilitres

1. make up salt water solution of 5 ml salt to 500 ml water

salt water solution up to 50 ml mark

2. add sand sample to salt water up to 50 ml mark

add more salt water solution up to 150 ml mark and shake cylinder well

3. allow mixture to stand for 3 hours and measure height of silt

salt water
silt
sand

4. Height of silt layer should not be more than 6 ml or 6% of height of sand sample

Obtaining Samples for Laboratory Testing ~ these tests may be required for checking aggregate grading by means of a sieve test, checking quality or checking for organic impurities but whatever the reason the sample must be truly representative of the whole:-

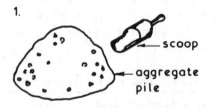

1. scoop

aggregate pile

samples extracted by means of a scoop from at least ten different positions in the pile

sample required :-
fine aggregate - 50 kg
coarse aggregate - 200 kg

2. well mixed sample divided into four equal parts - opposite quarters are discarded - remainder of sample remixed and quartered - whole process is repeated until required size of sample is left.
samples required :-
fine aggregate ⇥ 4 mm - 3 kg
coarse aggregate ⇥ 10 mm - 6 kg
⇥ 20 mm - 25 kg
⇥ 40 mm - 50 kg

Ref. BS EN 12620: Aggregates for concrete.

Concrete requires monitoring by means of tests to ensure that subsequent mixes are of the same consistency and this can be carried out on site by means of the slump test and in a laboratory by crushing test cubes to check that the cured concrete has obtained the required designed strength.

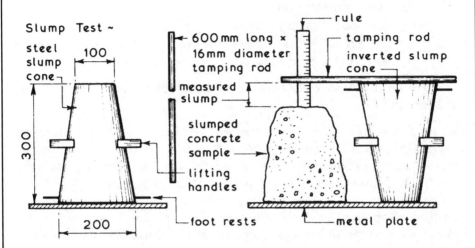

Slump Test ~

steel slump cone

100

300

200

rule

600 mm long × 16 mm diameter tamping rod

measured slump

slumped concrete sample

lifting handles

foot rests

tamping rod inverted slump cone

metal plate

The slump cone is filled to a quarter depth and tamped 25 times – filling and tamping is repeated three more times until the cone is full and the top smoothed off. The cone is removed and the slump measured, for consistent mixes the slump should remain the same for all samples tested. Usual specification 50 mm or 75 mm slump.

Test Cubes – these are required for laboratory strength tests~

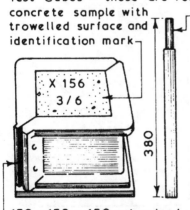

concrete sample with trowelled surface and identification mark

X 156
3 / 6

380

150 × 150 × 150 standard steel test cube mould thinly coated inside with mould oil

25 × 25 mm square end tamping bar

1. Sample taken from discharge outlet of mixer or from point of placing using random selection by means of a scoop.

2. Mould filled in three equal layers each layer well tamped with at least 35 strokes from the tamping bar.

3. Sample left in mould for 24 hours and covered with a damp sack or similar at a temperature of 4·4 to 21°C

4. Remove sample from mould and store in water at temperature of 10 to 21°C until required for testing

Refs. BS EN 12350-2 (Slump) and BS EN 12390-1 (Cubes)

Non destructive testing of concrete. Also known as in-place or in-situ tests.

Changes over time and in different exposures can be monitored.

References: BS 6089: Guide to assessment of concrete strength in existing structures;
BS 1881: Testing concrete.
BS EN 13791: Assessment of in-situ compressive strength in structures and pre-cast concrete components.

Provides information on: strength in-situ, voids, flaws, cracks and deterioration.

Rebound hammer test – attributed to Ernst Schmidt after he devised the impact hammer in 1948. It works on the principle of an elastic mass rebounding off a hard surface. Varying surface densities will affect impact and propagation of stress waves. These can be recorded on a numerical scale known as rebound numbers. It has limited application to smooth surfaces of concrete only. False results may occur where there are local variations in the concrete, such as a large piece of aggregate immediately below the impact surface. Rebound numbers can be graphically plotted to correspond with compressive strength.

Ref: BS EN 12504-2: Testing concrete in structures.

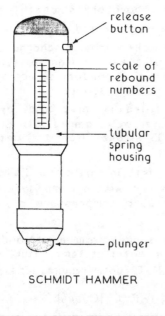

SCHMIDT HAMMER

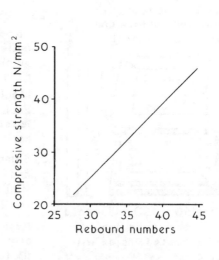

CONVERSION CHART
(illustration only)

Penetration or Windsor probe test ~ there are various interpretations of this test. It is a measure of the penetration of a steel alloy rod, fired by a predetermined amount of energy into concrete. In principle, the depth of penetration is inversely proportional to the concrete compressive strength. Several recordings are necessary to obtain a fair assessment and some can be discarded particularly where the probe cannot penetrate some dense aggregates. The advantage over the rebound hammer is provision of test results at a greater depth (up to 50 mm).

Pull out test ~ this is not entirely non destructive as there will be some surface damage, albeit easily repaired. A number of circular bars of steel with enlarged ends are cast into the concrete as work proceeds. This requires careful planning and location of bars with corresponding voids provided in the formwork. At the appropriate time, the bar and a piece of concrete are pulled out by tension jack. Although the concrete fails in tension and shear, the pull out force can be correlated to the compressive strength of the concrete.

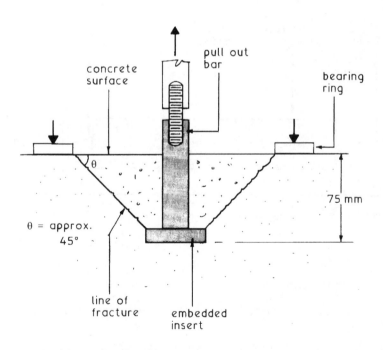

Ref: BS 1881-207: Testing concrete. Recommendations for the assessment of concrete strength by near-to-surface tests.

111

Vibration test ~ a number of electronic tests have been devised, which include measurement of ultrasonic pulse velocity through concrete. This applies the principle of recording a pulse at predetermined frequencies over a given distance. The apparatus includes transducers in contact with the concrete, pulse generator, amplifier, and time measurement to digital display circuit. For converting the data to concrete compressive strength, see BS EN 12504-4: Testing concrete. Determination of ultrasonic pulse velocity.

A variation, using resonant frequency, measures vibrations produced at one end of a concrete sample against a receiver or pick up at the other. The driving unit or exciter is activated by a variable frequency oscillator to generate vibrations varying in resonance, depending on the concrete quality. The calculation of compressive strength by conversion of amplified vibration data is by formulae found in BS 1881-209: Testing concrete. Recommendations for the measurement of dynamic modulus of elasticity.

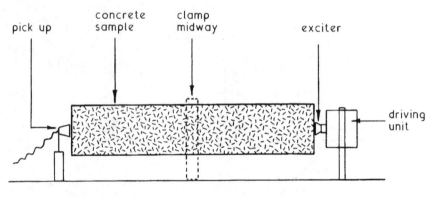

RESONANT FREQUENCY TEST

Other relevant standards:-

BS 1881-122: Testing concrete. Method for determination of water absorption.
BS 1881-124: Testing concrete. Methods for analysis of hardened concrete.
BS EN 12390-7: Testing hardened concrete. Density of hardened concrete.

The quality of softwood timber for structural use depends very much on the environment in which it is grown and the species selected. Timber can be visually strength graded, but this is unlikely to occur at the construction site except for a general examination for obvious handling defects and damage during transit. Site inspection will be to determine that the grading authority's markings on the timber comply with that specified for the application.

Format of strength grade markings on softwood timber for structural uses ~

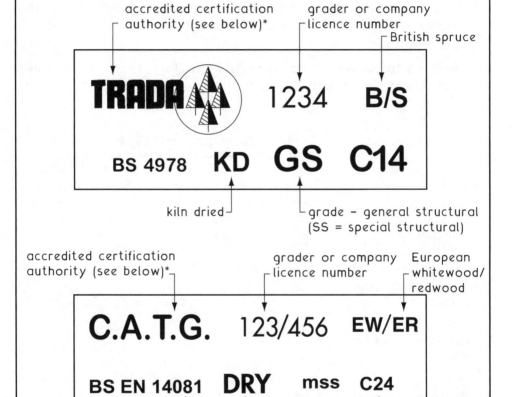

*Accredited certification authorities include
BM TRADA Certification Ltd. and Certification And Timber Grading Ltd.

Refs. BS 4978: Visual strength grading of softwood. Specification.
   BS EN 14081: Timber structures. Strength graded structural
   timber with rectangular cross section. (In 4 parts).

Grading ~ either visually or by computerised machine. Individual rectangular timber sections are assessed against permissible defect limitations and grade marked accordingly.

UK grading standard ~ BS 4978.
European grading standard ~ BS EN 14081 (4 parts).

The two principal grades apart from rejects are, GS (general structural) and SS (special structural) preceded with an M if graded by machine.

Additional specification is to BS EN 338: Structural timber. Strength classes. This standard provides softwood strength classifications from C14 to C40 as well as a separate classification of hardwoods.

A guide to softwood grades with strength classes for timber from the UK, Europe and North America ~

Source/species	Strength class (BS EN 338)						
	C14	C16	C18	C22	C24	C27	C30
UK:							
British pine	GS			SS			
British spruce	GS		SS				
Douglas Fir	GS		SS				
Larch	GS				SS		
Ireland:							
Sitka and Norway spruce	GS		SS				
Europe:							
Redwood or white-wood		GS			SS		
USA:							
Western whitewood	GS		SS				
Southern pine		GS			SS		
USA/Canada:							
Spruce/pine/fir or hemlock		GS			SS		
Douglas fir and larch		GS			SS		
Canada:							
Western red cedar	GS		SS				
Sitka spruce	GS		SS				

BS EN 338: Structural softwood classifications and typical strength properties ~

BS EN 338 strength class	Bending parallel to grain (N/mm²)	Tension parallel to grain (N/mm²)	Compression parallel to grain (N/mm²)	Compression perpendicular to grain (N/mm²)	Shear parallel to grain (N/mm²)	Modulus of Mean (N/mm²)	Elasticity Minimum (N/mm²)	Characteristic density (kg/m³)	Average density (kg/m³)
C14	4.1	2.5	5.2	2.1	0.60	6800	4600	290	350
C16	5.3	3.2	6.8	2.2	0.67	8800	5800	310	370
C18	5.8	3.5	7.1	2.2	0.67	9100	6000	320	380
C22	6.8	4.1	7.5	2.3	0.71	9700	6500	340	410
C24	7.5	4.5	7.9	2.4	0.71	10800	7200	350	420
TR26	10.0	6.0	8.2	2.5	1.10	11000	7400	370	450
C27	10.0	6.0	8.2	2.5	1.10	12300	8200	370	450
C30	11.0	6.6	8.6	2.7	1.20	12300	8200	380	460
C35	12.0	7.2	8.7	2.9	1.30	13400	9000	400	480
C40	13.0	7.8	8.7	3.0	1.40	14500	10000	420	500

Notes: 1. Strength class TR26 is specifically for the manufacture of trussed rafters.

2. Characteristic density values are given specifically for the design of joints. Average density is appropriate for calculation of dead load.

3. For a worked example of a softwood timber joist/beam design using data for strength classification C24 (e.g. SS graded European redwood) see pages 664 and 665.

Visual strength grading ~ "process by which a piece of timber can be sorted, by means of visual inspection, into a grade to which characteristic values of strength, stiffness and density may be allocated". Definition from BS EN 14081-1.

Characteristics:

Knots ~ branch growth from or through the main section of timber weakening the overall structural strength. Measured by comparing the sum of the projected cross sectional knot area with the cross sectional area of the piece of timber. This is known as the knot area ratio (KAR). Knots close to the edge of section have greater structural significance therefore this area is represented as a margin condition at the top and bottom quarter of a section. A margin condition exists when more than half the top or bottom quarter of a section is occupied by knots.

MKAR = Margin knot area ratio.
TKAR = Total knot area ratio.

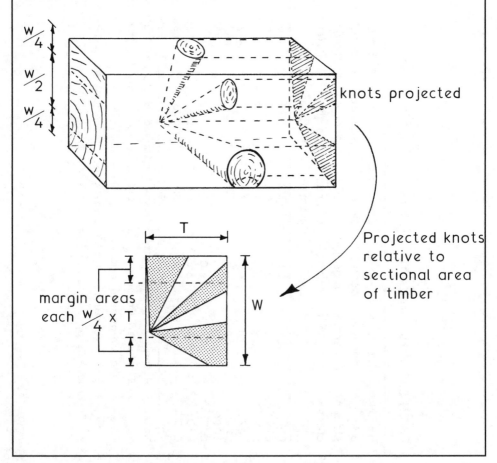

knots projected

Projected knots relative to sectional area of timber

margin areas each $\frac{W}{4}$ x T

Fissures and resin pockets ~ defects in growth. Fissures, also known as shakes, are usually caused by separation of annual growth rings. Fissures and resin pockets must be limited in structural timber as they reduce resistance to shear and bending parallel to the grain.

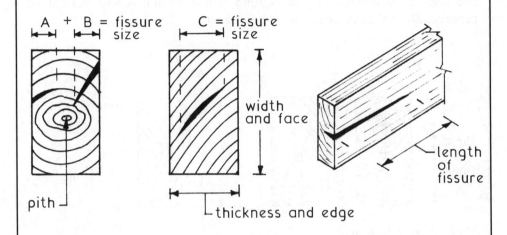

Slope of grain ~ an irregularity in growth or where the log is not cut parallel to the grain. If excessive this will produce a weakness in shear. Measurement is by scoring a line along the grain of the timber surface and comparing this with the parallel sides of the section.

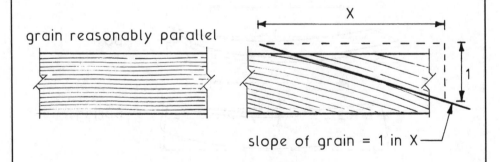

Insect damage ~ no active allowed. Wood-worm holes acceptable if only nominal. Wood wasp holes not permitted.

Sapstain ~ acceptable.

Wane or waney edge ~ occurs on timber cut close to the outer surface of the log producing incomplete corners. Measurement. is parallel to the edge or face of section and it is expressed as a fraction of the surface dimension.

Growth rate ~ measurement is applied to the annual growth ring separation averaged over a line 75 mm long. If pith is present the line should commence 25 mm beyond and if 75 mm is impractical to achieve, the longest possible line is taken.

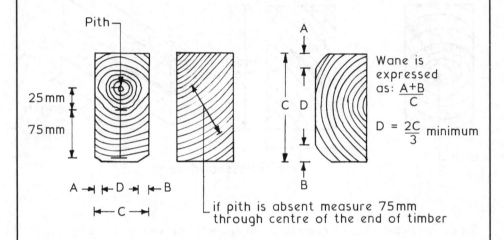

Wane is expressed as: $\frac{A+B}{C}$

$D = \frac{2C}{3}$ minimum

if pith is absent measure 75 mm through centre of the end of timber

Distortion ~ measurement over the length and width of section to determine the amount of bow, spring and twist.

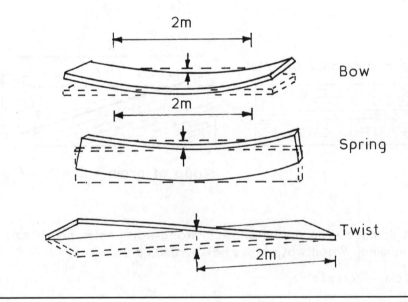

Bow

Spring

Twist

Characteristics and tolerances of GS and SS graded timber ~

Criteria	GS	SS
KAR: No margin condition	MKAR ≤1/2 TKAR ≤1/2	MKAR ≤1/2 TKAR ≤1/3
	Or:	
KAR: Margin condition	MKAR >1/2 TKAR ≤1/3	MKAR >1/2 TKAR ≤1/5
Fissures and resin pockets: Not through thickness Through thickness	Defects ≤1/2 timber thickness <1.5 m or 1/2 timber length take lesser <1.0 m or 1/4 timber length take lesser If at ends fissure length maximum 2 × timber width	Defects ≤1/2 timber thickness <1.0 m or 1/4 timber length take lesser <0.5 m or 1/4 timber length take lesser If at ends fissure length < width of timber section
Slope of grain:	Maximum 1 in 6	Maximum 1 in 10
Wane:	Maximum 1/3 of the full edge and face of the section – length not limited	
Resin pockets: Not through thickness Through thickness	Unlimited if shorter than width of section otherwise as for fissures Unlimited if shorter than 1/2 width of section otherwise as for fissures	
Growth rate of annual rings:	Average width or growth <10 mm	Average width or growth <6 mm
Distortion: –bow –spring –twist	<20 mm over 2 m <12 mm over 2 m <2 mm per 25 mm width over 2 m	<10 mm over 2 m <8 mm over 2 m <1 mm per 25 mm width over 2 m

Structural softwood cross sectional size has established terminology such as, sawn, basic and unwrought as produced by conversion of the log into commercial dimensions, e.g. 100 × 50 mm and 225 × 75 mm (4″ × 2″ and 9″ × 3″ respectively, as the nearest imperial sizes).

Timber is converted in imperial and metric sizes depending on its source in the world. Thereafter, standardisation can be undertaken by machine planing the surfaces to produce uniformly compatible and practically convenient dimensions, i.e. 225 mm is not the same as 9″. Planed timber has been variously described as, nominal, regularised and wrought, e.g. 100 × 50 mm sawn becomes 97 × 47 mm when planed and is otherwise known as ex. 100 × 50 mm, where ex means out of.

Guidance in BS EN 336 requires the sizes of timber from a supplier to be redefined as 'Target Sizes' within the following tolerances:

T1 ~ Thickness and width$\leq$100 mm, −1 to +3 mm.

　　　Thickness and width>100 mm, −2 to +4 mm.

T2 ~ Thickness and width$\leq$100 mm, −1 to +1 mm.

　　　Thickness and width>100 mm, −1.5 to +1.5 mm.

T1 applies to sawn timber, e.g. 100 × 75 mm.

T2 applies to planed timber, e.g. 97 × 72 mm.

Further example ~ a section of timber required to be 195 mm planed × 50 mm sawn is specified as: 195 (T2) × 50 (T1).

Target sizes for sawn softwood (T1) ~

50, 63, 75, 100, 125, 150, 175, 200, 225, 250 and 300 mm.

Target sizes for planed/machined softwood (T2) ~

47, 60, 72, 97, 120, 145, 170, 195, 220, 245 and 295 mm.

Ref. BS EN 336: Structural timber. Sizes, permitted deviations.

Damp conditions can be the source of many different types of wood-decaying fungi. The principal agencies of decay are –

* Dry rot (Serpula lacrymans or merulius lacrymans), and
* Wet rot (Coniophora cerabella)

Dry rot – this is the most difficult to control as its root system can penetrate damp and porous plaster, brickwork and concrete. It can also remain dormant until damp conditions encourage its growth, even though the original source of dampness is removed.

Appearance – white fungal threads which attract dampness from the air or adjacent materials. The threads develop strands bearing spores or seeds which drift with air movements to settle and germinate on timber having a moisture content exceeding about 25%. Fruiting bodies of a grey or red flat profile may also identify dry rot.

Typical surface appearance of dry rot –

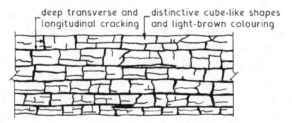

deep transverse and longitudinal cracking — distinctive cube-like shapes and light-brown colouring

Wet rot – this is limited in its development and must have moisture continually present, e.g. a permanent leaking pipe or a faulty dpc. Growth pattern is similar to dry rot, but spores will not germinate in dry timber.

Appearance – fungal threads of black or dark brown colour. Fruiting bodies may be olive-green or dark brown and these are often the first sign of decay.

Typical surface appearance of wet rot –

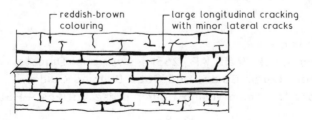

reddish-brown colouring — large longitudinal cracking with minor lateral cracks

Causes –
* Defective construction, e.g. broken roof tiles; no damp-proof course.
* Installation of wet timber during construction, e.g. framing sealed behind plasterboard linings; wet joists under floor decking.
* Lack of ventilation, e.g. blocked air bricks to suspended timber ground floor; condensation in unventilated roof spaces.
* Defective water services, e.g. undetected leaks on internal pipework; blocked or broken rainwater pipes and guttering.

General treatment –
* Remove source of dampness.
* Allow affected area to dry.
* Remove and burn all affected timber and sound timber within 500 mm of fungal attack.
* Remove contaminated plaster and rake out adjacent mortar joints to masonry.

Note: This is normally sufficient treatment where wet rot is identified. However, where dry rot is apparent the following additional treatment is necessary:
* Sterilise surface of concrete and masonry.
  Heat with a blow torch until the surface is too hot to touch. Apply a proprietary fungicide† generously to warm surface. Irrigate badly affected masonry and floors, i.e. provide 12 mm diameter bore holes at about 500 mm spacing and flood or pressure inject with fungicide.

† 20:1 dilution of water and sodium pentachlorophenate, sodium orthophenylphate or mercuric chloride. Product manufacturers' safety in handling and use measures must be observed when applying these chemicals.

Replacement work should ensure that new timbers are pressure impregnated with a preservative. Cement and sand mixes for rendering, plastering and screeds should contain a zinc oxychloride fungicide.

Further reading –
BRE: Timber pack (ref. AP 265) – various Digests, Information Papers, Good Repair Guides and Good Building Guides.
In-situ timber treatment using timber preservatives – HSE Books.

Ref: Bldg. Regs. Approved Document C, Site preparation and resistance to contaminants and moisture.

Trees ~ these are part of our national heritage and are also the source of timber – to maintain this source a control over tree felling has been established under the Forestry Act 1967 which places the control responsibility on the Forestry Commission. Local planning authorities also have powers under the Town and Country Planning Act 1990 and the Town and Country Amenities Act 1974 to protect trees by making tree preservation orders. Contravention of such an order can lead to a substantial fine and a compulsion to replace any protected tree which has been removed or destroyed. Trees on building sites which are covered by a tree preservation order should be protected by a suitable fence.

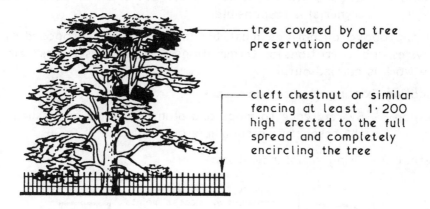

tree covered by a tree preservation order

cleft chestnut or similar fencing at least 1·200 high erected to the full spread and completely encircling the tree

Trees, shrubs, bushes and tree roots which are to be removed from site can usually be grubbed out using hand held tools such as saws, picks and spades. Where whole trees are to be removed for relocation special labour and equipment is required to ensure that the roots, root earth ball and bark are not damaged.

Structures ~ buildings which are considered to be of historic or architectural interest can be protected under the Planning Acts provisions. The Department for Communities and Local Government lists buildings according to age, architectural, historical and/or intrinsic value. It is an offence to demolish or alter a listed building without first obtaining 'listed building consent' from the local planning authority. Contravention is punishable by a fine and/or imprisonment. It is also an offence to demolish a listed building without giving notice to the Royal Commission on Historical Monuments, this is to enable them to note and record details of the building.

Services which may be encountered on construction sites and the authority responsible are:-

Water – Local Water Company

Electricity – transmission ~ RWE npower, BNFL and E-on.

distribution ~ Area Electricity Companies in England and Wales. Scottish Power and Scottish Hydro-Electric, EDF Energy.

Gas – Local gas or energy service providers, e.g. British Gas.

Telephones – National Telecommunications Companys, e.g. BT, C&W, etc.

Drainage – Local Authority unless a private drain or sewer when owner(s) is responsible.

All the above authorities must be notified of any proposed new services and alterations or terminations to existing services before any work is carried out.

Locating Existing Services on Site ~

Method 1 – By reference to maps and plans prepared and issued by the respective responsible authority.

Method 2 – Using visual indicators ~

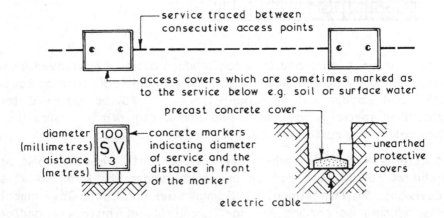

Method 3 – Detection specialist contractor employed to trace all forms of underground services using electronic subsurface survey equipment.

Once located, position and type of service can be plotted on a map or plan, marked with special paint on hard surfaces and marked with wood pegs with indentification data on earth surfaces.

Setting Out the Building Outline ~ this task is usually undertaken once the site has been cleared of any debris or obstructions and any reduced level excavation work is finished. It is usually the responsibility of the contractor to set out the building(s) using the information provided by the designer or architect. Accurate setting out is of paramount importance and should therefore only be carried out by competent persons and all their work thoroughly checked, preferably by different personnel and by a different method.

The first task in setting out the building is to establish a base line to which all the setting out can be related. The base line very often coincides with the building line which is a line, whose position on site is given by the local authority in front of which no development is permitted.

Typical Setting Out Example ~

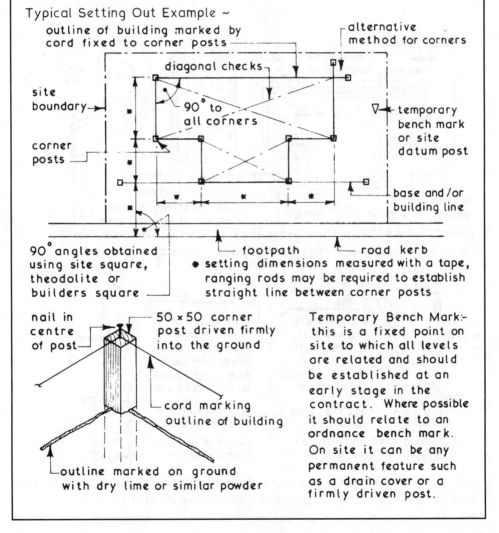

outline of building marked by cord fixed to corner posts

alternative method for corners

diagonal checks

site boundary →

90° to all corners

temporary bench mark or site datum post

corner posts

base and/or building line

90° angles obtained using site square, theodolite or builders square

footpath ⌐ road kerb

* setting dimensions measured with a tape, ranging rods may be required to establish straight line between corner posts

nail in centre of post

50 × 50 corner post driven firmly into the ground

cord marking outline of building

outline marked on ground with dry lime or similar powder

Temporary Bench Mark:- this is a fixed point on site to which all levels are related and should be established at an early stage in the contract. Where possible it should relate to an ordnance bench mark. On site it can be any permanent feature such as a drain cover or a firmly driven post.

Setting Out Trenches ~ the objective of this task is twofold. Firstly it must establish the excavation size, shape and direction and secondly it must establish the width and position of the walls. The outline of building will have been set out and using this outline profile boards can be set up to control the position, width and possibly the depth of the proposed trenches. Profile boards should be set up at least 2·000 clear of trench positions so they do not obstruct the excavation work. The level of the profile crossboard should be related to the site datum and fixed at a convenient height above ground level if a traveller is to be used to control the depth of the trench. Alternatively the trench depth can be controlled using a level and staff related to site datum. The trench width can be marked on the profile with either nails or sawcuts and with a painted band if required for identification.

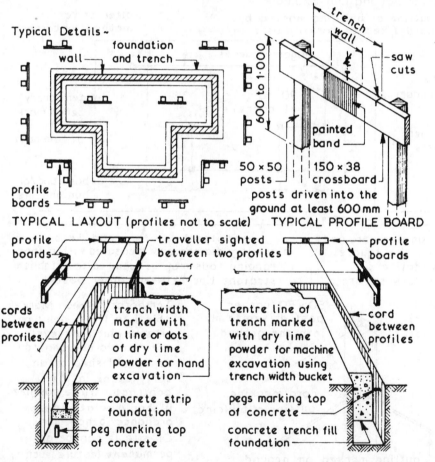

Typical Details ~

wall
foundation and trench
profile boards

TYPICAL LAYOUT (profiles not to scale)

trench
wall
saw cuts
600 to 1·000
painted band
50 × 50 posts
150 × 38 crossboard
posts driven into the ground at least 600mm

TYPICAL PROFILE BOARD

profile boards
traveller sighted between two profiles
profile boards

cords between profiles
trench width marked with a line or dots of dry lime powder for hand excavation
centre line of trench marked with dry lime powder for machine excavation using trench width bucket
cord between profiles

concrete strip foundation
peg marking top of concrete
pegs marking top of concrete
concrete trench fill foundation

NB. Corners of walls transferred from intersecting cord lines to mortar spots on concrete foundations using a spirit level

Setting Out a Framed Building ~ framed buildings are usually related to a grid, the intersections of the grid lines being the centre point of an isolated or pad foundation. The grid is usually set out from a base line which does not always form part of the grid. Setting out dimensions for locating the grid can either be given on a drawing or they will have to be accurately scaled off a general layout plan. The grid is established using a theodolite and marking the grid line intersections with stout pegs. Once the grid has been set out offset pegs or profiles can be fixed clear of any subsequent excavation work. Control of excavation depth can be by means of a traveller sighted between sight rails or by level and staff related to site datum.

Typical Details ~

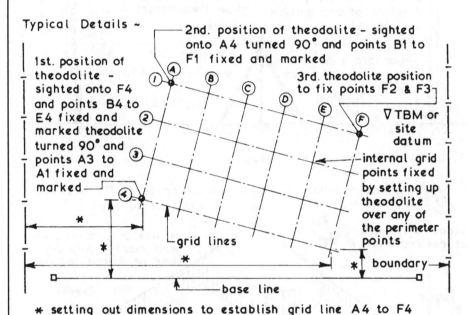

* setting out dimensions to establish grid line A4 to F4

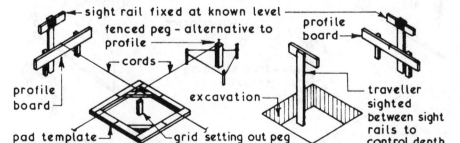

1. Pad template positioned with cords between profiles and pad outline marked with dry lime or similar powder.

2. Pad pits excavated using traveller sighted between sight rails fixed at a level related to site datum.

127

Setting Out Reduced Level Excavations ~ the overall outline of the reduced level area can be set out using a theodolite, ranging rods, tape and pegs working from a base line. To control the depth of excavation, sight rails are set up at a convenient height and at positions which will enable a traveller to be used.

Typical Details ~

1. Setting up sight rails :-

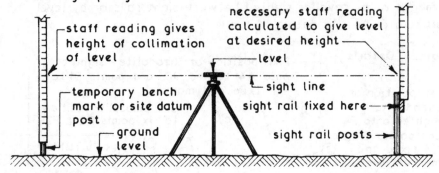

staff reading gives height of collimation of level

necessary staff reading calculated to give level at desired height

level

sight line

temporary bench mark or site datum post

sight rail fixed here

sight rail posts

ground level

2. Controlling excavation depth :-

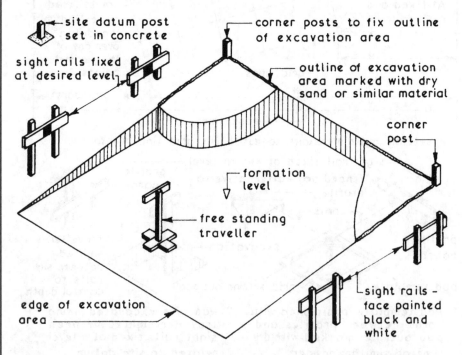

site datum post set in concrete

corner posts to fix outline of excavation area

sight rails fixed at desired level

outline of excavation area marked with dry sand or similar material

corner post

formation level

free standing traveller

edge of excavation area

sight rails - face painted black and white

height of traveller = desired level of sight rail - formation level

Levelling ~ the process of establishing height dimensions, relative to a fixed point or datum. Datum is mean sea level, which varies between different countries. For UK purposes this is established at Newlyn in Cornwall, from tide data recorded between May 1915 and April 1921. Relative levels defined by benchmarks are located throughout the country. The most common, identified as carved arrows, can be found cut into walls of stable structures. Reference to Ordnance Survey maps of an area will indicate benchmark positions and their height above sea level, hence the name Ordnance Datum (OD).

On site it is usual to measure levels from a temporary benchmark (TBM), i.e. a manhole cover or other permanent fixture, as an OD may be some distance away.

Instruments consist of a level (tilting or automatic) and a staff. A tilting level is basically a telescope mounted on a tripod for stability. Correcting screws establish accuracy in the horizontal plane by air bubble in a vial and focus is by adjustable lens. Cross hairs of horizontal and vertical lines indicate image sharpness on an extending staff of 3, 4 or 5m length. Staff graduations are in 10mm intervals, with estimates taken to the nearest millimetre. An automatic level is much simpler to use, eliminating the need for manual adjustment. It is approximately levelled by centre bulb bubble. A compensator within the telescope effects fine adjustment.

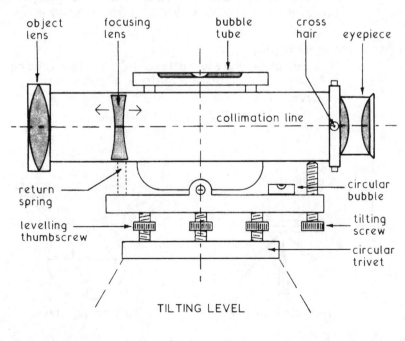

object lens   focusing lens   bubble tube   cross hair   eyepiece

collimation line

return spring

circular bubble

levelling thumbscrew

tilting screw

circular trivet

TILTING LEVEL

Application ~ methods to determine differences in ground levels for calculation of site excavation volumes and costs.

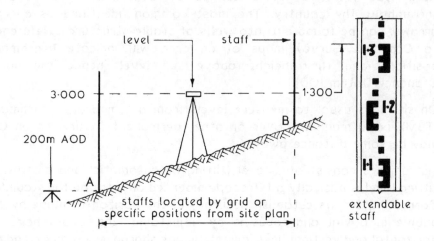

level   staff

3·000   1·300

200m AOD

B

A

staffs located by grid or specific positions from site plan

extendable staff

Rise and fall:
      Staff reading at A = 3·00 m, B = 1·30 m
      Ground level at A = 200 m above ordnance datum (AOD)
      Therefore level at B = 200 m + rise (– fall if declining)
      So level at B = 200 + (3·00 – 1·30) = 201·7 m

Height of collimation (HC):
      HC at A = Reduced level (RL) + staff reading
            = 200 m + 3·00 m = 203 m AOD
      Level at B = HC at A – staff reading at B
            = 203 – 1·30 = 201·7 m

height above mean sea level at Newlyn

cross hairs

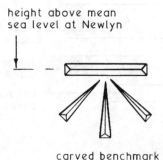

carved benchmark

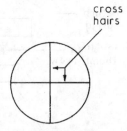

view through level

Theodolite – a tripod mounted instrument designed to measure angles in the horizontal or vertical plane.

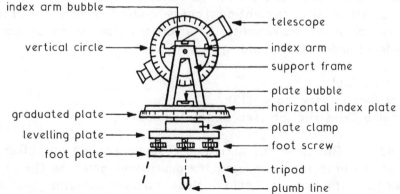

The theodolite in principle

Measurement – a telescope provides for focal location between instrument and subject. Position of the scope is defined by an index of angles. The scale and presentation of angles varies from traditional micrometer readings to computer compatible crystal displays. Angles are measured in degrees, minutes and seconds, e.g. 165° 53′ 30″.

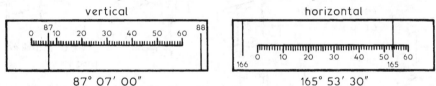

Direct reading micrometer scale

Application – at least two sightings are taken and the readings averaged. After the first sighting, the horizontal plate is rotated through 180° and the scope also rotated 180° through the vertical to return the instrument to its original alignment for the second reading. This process will move the vertical circle from right face to left face, or vice-versa. It is important to note the readings against the facing – see below.

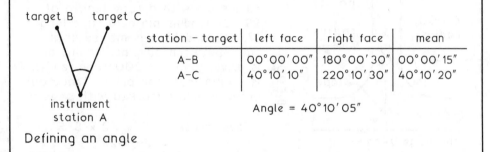

station – target	left face	right face	mean
A–B	00°00′00″	180°00′30″	00°00′15″
A–C	40°10′10″	220°10′30″	40°10′20″

Angle = 40°10′05″

Defining an angle

Road Construction ~ within the context of building operations roadworks usually consist of the construction of small estate roads, access roads and driveways together with temporary roads laid to define site circulation routes and/or provide a suitable surface for plant movements. The construction of roads can be considered under three headings:-

1. Setting out.
2. Earthworks (see page 133).
3. Paving Construction (see pages 133-135).

Setting Out Roads ~ this activity is usually carried out after the topsoil has been removed using the dimensions given on the layout drawing(s). The layout could include straight lengths junctions, hammer heads, turning bays and intersecting curves.

Straight Road Lengths - these are usually set out from centre lines which have been established by traditional means

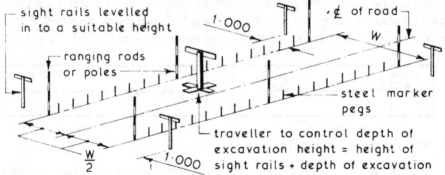

NB. curve road lengths set out in a similar manner

Junctions and Hammer Heads -

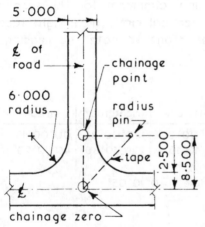

Centre lines fixed by traditional methods. Tape hooked over pin at chainage zero and passed around chainage point pin at 8.500 then returned to chainage zero via the radius pin with a tape length of 29.021. Radius pin held tape length 17.000 and tape is moved until tight between all pins. Radius pin is driven and a 6.000 tape length is swung from the pin to trace out curve which is marked with pegs or pins

$$\text{Tape length} = 17 + \sqrt{2} \times 8.5$$
$$= 29.021$$

Earthworks ~ this will involve the removal of topsoil together with any vegetation, scraping and grading the required area down to formation level plus the formation of any cuttings or embankments. Suitable plant for these operations would be tractor shovels fitted with a 4 in 1 bucket (page 174); graders (page 173) and bulldozers (page 171). The soil immediately below the formation level is called the subgrade whose strength will generally decrease as its moisture content rises therefore if it is to be left exposed for any length of time protection may be required. Subgrade protection may take the form of a covering of medium gauge plastic sheeting with 300mm laps or alternatively a covering of sprayed bituminous binder with a sand topping applied at a rate of 1 litre per m². To preserve the strength and durability of the subgrade it may be necessary to install cut off subsoil drains alongside the proposed road (see Road Drainage on page 752).

Paving Construction ~ once the subgrade has been prepared and any drainage or other buried services installed the construction of the paving can be undertaken. Paved surfaces can be either flexible or rigid in format. Flexible or bound surfaces are formed of materials applied in layers directly over the subgrade whereas rigid pavings consist of a concrete slab resting on a granular base (see pages 134 & 135).

Typical Flexible Paving Details ~

surfacing = base layer + wearing course

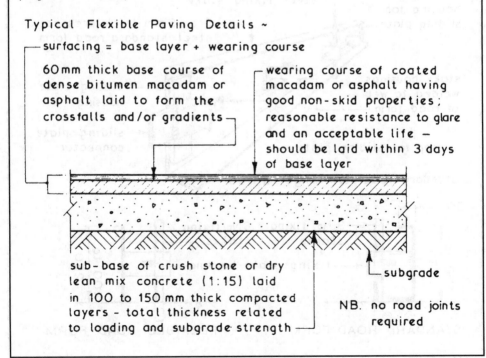

60mm thick base course of dense bitumen macadam or asphalt laid to form the crossfalls and/or gradients

wearing course of coated macadam or asphalt having good non-skid properties; reasonable resistance to glare and an acceptable life — should be laid within 3 days of base layer

sub-base of crush stone or dry lean mix concrete (1:15) laid in 100 to 150 mm thick compacted layers - total thickness related to loading and subgrade strength

subgrade

NB. no road joints required

Rigid Pavings ~ these consist of a reinforced or unreinforced in-situ concrete slab laid over a base course of crushed stone or similar material which has been blinded to receive a polythene sheet slip membrane. The primary objective of this membrane is to prevent grout loss from the in-situ slab.

Typical Rigid Paving Details ~

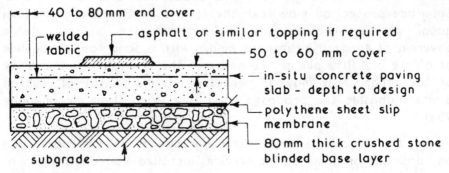

40 to 80mm end cover

welded fabric

asphalt or similar topping if required

50 to 60 mm cover

in-situ concrete paving slab - depth to design

polythene sheet slip membrane

80mm thick crushed stone blinded base layer

subgrade

The paving can be laid between metal road forms or timber edge formwork. Alternatively the kerb stones could be laid first to act as permanent formwork.

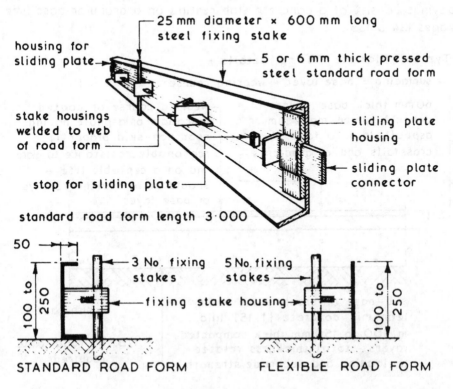

25 mm diameter × 600 mm long steel fixing stake

housing for sliding plate

5 or 6 mm thick pressed steel standard road form

stake housings welded to web of road form

sliding plate housing

stop for sliding plate

sliding plate connector

standard road form length 3·000

50

3 No. fixing stakes

5 No. fixing stakes

100 to 250

fixing stake housing

100 to 250

STANDARD ROAD FORM

FLEXIBLE ROAD FORM

Joints in Rigid Pavings ~ longitudinal and transverse joints are required in rigid pavings to:-

1. Limit size of slab.
2. Limit stresses due to subgrade restraint.
3. Provide for expansion and contraction movements.

The main joints used are classified as expansion, contraction or longitudinal, the latter being the same in detail as the contraction joint differing only in direction. The spacing of road joints is determined by:-

1. Slab thickness.
2. Whether slab is reinforced or unreinforced.
3. Anticipated traffic load and flow rate.
4. Temperature at which concrete is laid.

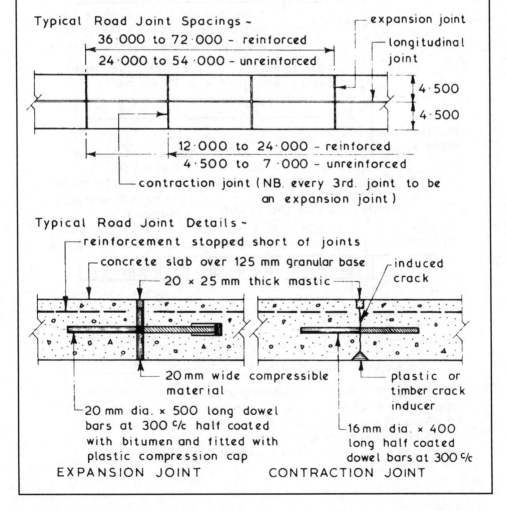

Typical Road Joint Spacings ~

36·000 to 72·000 - reinforced
24·000 to 54·000 - unreinforced

expansion joint

longitudinal joint

4·500
4·500

12·000 to 24·000 - reinforced
4·500 to 7·000 - unreinforced

contraction joint (NB. every 3rd. joint to be an expansion joint)

Typical Road Joint Details ~

reinforcement stopped short of joints
concrete slab over 125 mm granular base
20 × 25 mm thick mastic
induced crack

20 mm wide compressible material
20 mm dia. × 500 long dowel bars at 300 c/c half coated with bitumen and fitted with plastic compression cap

plastic or timber crack inducer
16 mm dia. × 400 long half coated dowel bars at 300 c/c

EXPANSION JOINT   CONTRACTION JOINT

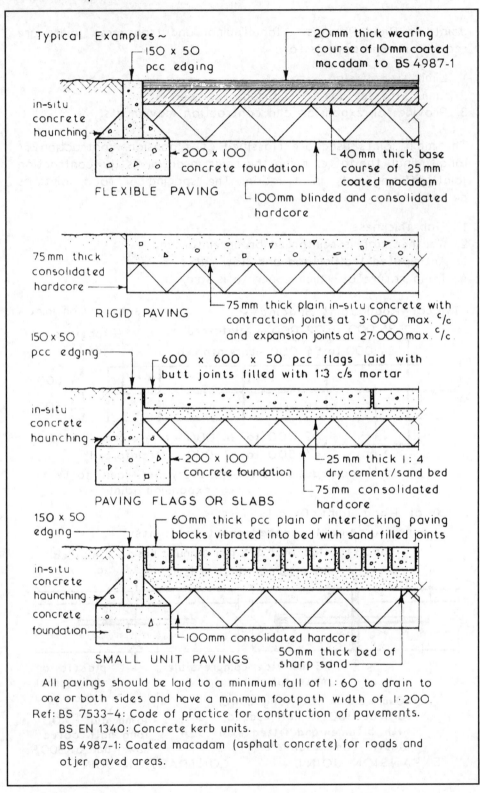

Typical Examples ~

150 x 50 pcc edging

20mm thick wearing course of 10mm coated macadam to BS 4987-1

in-situ concrete haunching

200 x 100 concrete foundation

40mm thick base course of 25mm coated macadam

FLEXIBLE PAVING

100mm blinded and consolidated hardcore

75mm thick consolidated hardcore

RIGID PAVING

75mm thick plain in-situ concrete with contraction joints at 3·000 max. ^{c}/c and expansion joints at 27·000 max. ^{c}/c

150 x 50 pcc edging

600 x 600 x 50 pcc flags laid with butt joints filled with 1:3 c/s mortar

in-situ concrete haunching

200 x 100 concrete foundation

25mm thick 1:4 dry cement/sand bed

PAVING FLAGS OR SLABS

75mm consolidated hardcore

150 x 50 edging

60mm thick pcc plain or interlocking paving blocks vibrated into bed with sand filled joints

in-situ concrete haunching

concrete foundation

100mm consolidated hardcore

50mm thick bed of sharp sand

SMALL UNIT PAVINGS

All pavings should be laid to a minimum fall of 1:60 to drain to one or both sides and have a minimum footpath width of 1·200.
Ref: BS 7533–4: Code of practice for construction of pavements.
    BS EN 1340: Concrete kerb units.
    BS 4987-1: Coated macadam (asphalt concrete) for roads and otjer paved areas.

Available sections ~ manufactured in 915mm lengths from silver/grey aggregate concrete.

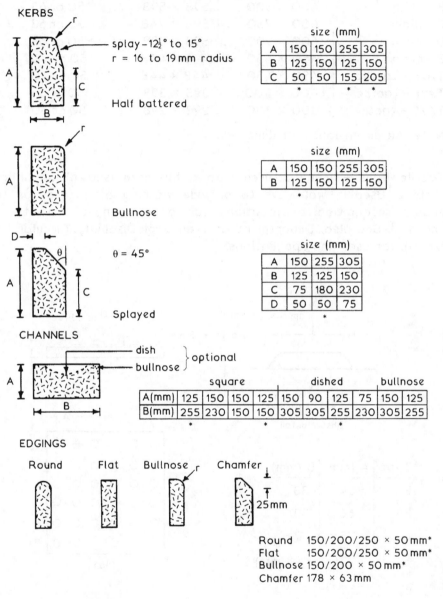

KERBS

splay – $12\frac{1}{2}°$ to $15°$
r = 16 to 19 mm radius

Half battered

size (mm)				
A	150	150	255	305
B	125	150	125	150
C	50	50	155	205
	*		*	*

Bullnose

size (mm)				
A	150	150	255	305
B	125	150	125	150
	*			

$θ = 45°$

Splayed

size (mm)			
A	150	255	305
B	125	125	150
C	75	180	230
D	50	50	75
		*	

CHANNELS

dish
bullnose } optional

	square				dished				bullnose	
A(mm)	125	150	150	125	150	90	125	75	150	125
B(mm)	255	230	150	150	305	305	255	230	305	255
		*				*			*	

EDGINGS

Round   Flat   Bullnose   Chamfer

25mm

Round   150/200/250 × 50 mm*
Flat   150/200/250 × 50 mm*
Bullnose 150/200 × 50 mm*
Chamfer 178 × 63 mm

*denotes BS sections

Further components such as drop/tapered kerbs are available for vehicle accesses. Quadrants and angles provide for directional change.

Concrete paving flags – BS dimensions:

Type	Size (nominal)	Size (work)	Thickness (T)
A – plain	600 × 450	598 × 448	50 or 63
B – plain	600 × 600	598 × 598	50 or 63
C – plain	600 × 750	598 × 748	50 or 63
D – plain	600 × 900	598 × 898	50 or 63
E – plain	450 × 450	448 × 448	50 or 70
TA/E – tactile	450 × 450	448 × 448	50 or 70
TA/F – tactile	400 × 400	398 × 398	50 or 65
TA/G – tactile	300 × 300	298 × 298	50 or 60

Note: All dimensions in millimetres.

Tactile flags – manufactured with a blistered (shown) or ribbed surface. Used in walkways to provide warning of hazards or to enable recognition of locations for people whose visibility is impaired. See also, Department of Transport Disability Circular DU 1/86[1], for uses and applications.

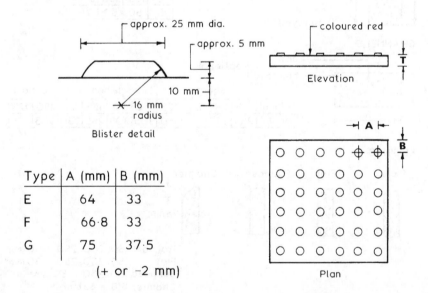

Type	A (mm)	B (mm)
E	64	33
F	66·8	33
G	75	37·5

(+ or −2 mm)

Ref. BS EN 1339: Concrete paving flags. Requirements and test methods.

Landscaping ~ in the context of building works this would involve reinstatement of the site as a preparation to the landscaping in the form of lawns, paths, pavings, flower and shrub beds and tree planting. The actual planning, lawn laying and planting activities are normally undertaken by a landscape subcontractor. The main contractor's work would involve clearing away all waste and unwanted materials, breaking up and levelling surface areas, removing all unwanted vegetation, preparing the subsoil for and spreading topsoil to a depth of at least 150 mm.

Services ~ the actual position and laying of services is the responsibility of the various service boards and undertakings. The best method is to use the common trench approach, avoid as far as practicable laying services under the highway.

Typical Common Trench Details ~

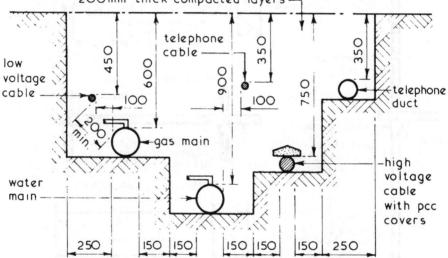

Road Signs ~ these can range from markings painted on roads to define traffic lanes, rights of way and warnings of hazards to signs mounted above the road level to give information, warning or directives, the latter being obligatory. See also, pages 51 and 52.

Typical Examples ~

INFORMATION   WARNING - ROAD WORKS   DIRECTIVE - NO LEFT TURN

Scaffolds ~ these are temporary working platforms erected around the perimeter of a building or structure to provide a safe working place at a convenient height. They are usually required when the working height or level is 1·500 or more above the ground level. All scaffolds must comply with the minimum requirements and objectives of the Work at Height Regulations 2005.

Component Parts of a Tubular Scaffold ~

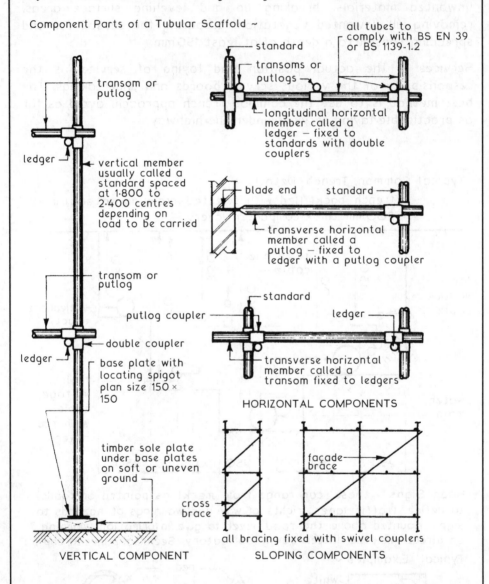

all tubes to comply with BS EN 39 or BS 1139-1.2

standard

transoms or putlogs

longitudinal horizontal member called a ledger — fixed to standards with double couplers

transom or putlog

ledger

vertical member usually called a standard spaced at 1·800 to 2·400 centres depending on load to be carried

blade end   standard

transverse horizontal member called a putlog — fixed to ledger with a putlog coupler

transom or putlog

putlog coupler

standard

ledger

double coupler

ledger

base plate with locating spigot plan size 150 × 150

transverse horizontal member called a transom fixed to ledgers

HORIZONTAL COMPONENTS

timber sole plate under base plates on soft or uneven ground

façade brace

cross brace

all bracing fixed with swivel couplers

VERTICAL COMPONENT   SLOPING COMPONENTS

Refs. BS EN 39: Loose steel tubes for tube and coupler scaffolds.
BS 1139-1.2: Metal scaffolding. Tubes. Specification for aluminium tube.

Putlog Scaffolds ~ these are scaffolds which have an outer row of standards joined together by ledgers which in turn support the transverse putlogs which are built into the bed joints or perpends as the work proceeds, they are therefore only suitable for new work in bricks or blocks.

Typical Details ~

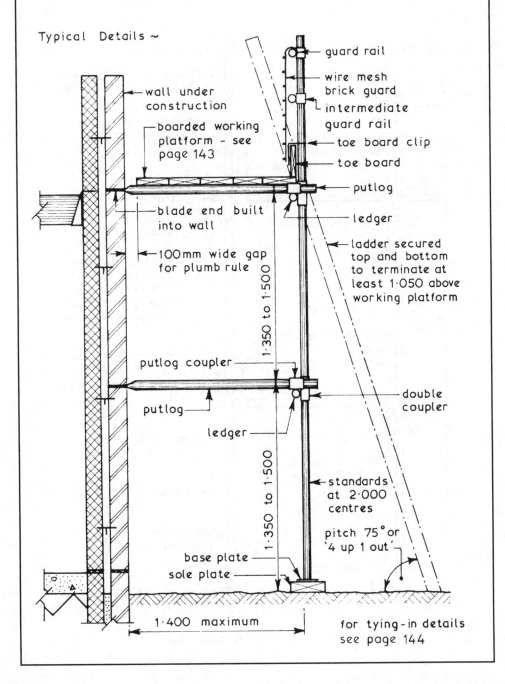

guard rail

wire mesh brick guard

intermediate guard rail

wall under construction

boarded working platform - see page 143

toe board clip

toe board

putlog

blade end built into wall

ledger

ladder secured top and bottom to terminate at least 1·050 above working platform

100mm wide gap for plumb rule

1·350 to 1·500

putlog coupler

putlog

double coupler

ledger

1·350 to 1·500

standards at 2·000 centres

pitch 75° or '4 up 1 out'

base plate

sole plate

1·400 maximum

for tying-in details see page 144

141

Independent Scaffolds ~ these are scaffolds which have two rows of standards each row joined together with ledgers which in turn support the transverse transoms. The scaffold is erected clear of the existing or proposed building but is tied to the building or structure at suitable intervals – see page 144

Typical Details ~

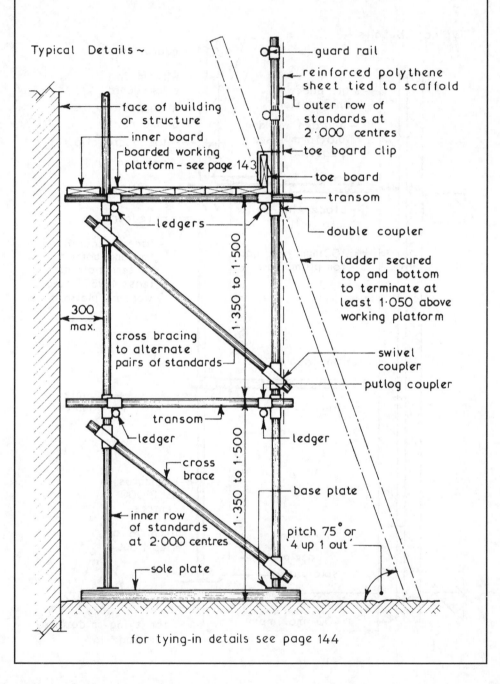

guard rail

reinforced polythene sheet tied to scaffold

outer row of standards at 2·000 centres

toe board clip

toe board

transom

double coupler

face of building or structure

inner board

boarded working platform – see page 143

ledgers

ladder secured top and bottom to terminate at least 1·050 above working platform

1·350 to 1·500

300 max.

cross bracing to alternate pairs of standards

swivel coupler

putlog coupler

transom

ledger

ledger

cross brace

base plate

1·350 to 1·500

inner row of standards at 2·000 centres

pitch 75° or '4 up 1 out'

sole plate

for tying-in details see page 144

Working Platforms ~ these are close boarded or plated level surfaces at a height at which work is being carried out and they must provide a safe working place of sufficient strength to support the imposed loads of operatives and/or materials. All working platforms above the ground level must be fitted with a toe board and a guard rail.

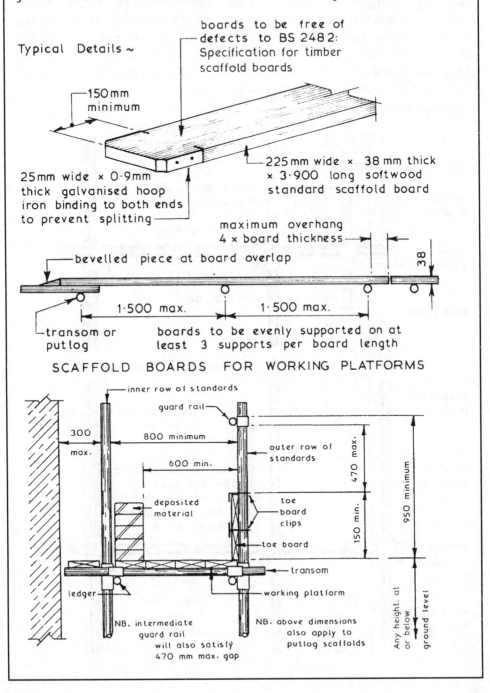

Typical Details ~

boards to be free of defects to BS 2482: Specification for timber scaffold boards

150 mm minimum

25 mm wide × 0·9 mm thick galvanised hoop iron binding to both ends to prevent splitting

225 mm wide × 38 mm thick × 3·900 long softwood standard scaffold board

maximum overhang 4 × board thickness

38

bevelled piece at board overlap

1·500 max.   1·500 max.

transom or putlog

boards to be evenly supported on at least 3 supports per board length

SCAFFOLD BOARDS FOR WORKING PLATFORMS

inner row of standards

guard rail

300 max.

800 minimum

600 min.

outer row of standards

470 max.

950 minimum

150 min.

deposited material

toe board clips

toe board

transom

ledger

working platform

NB. intermediate guard rail will also satisfy 470 mm max. gap

NB. above dimensions also apply to putlog scaffolds

Any height. at or below ground level

143

Tying-in ~ all putlog and independent scaffolds should be tied securely to the building or structure at alternate lift heights vertically and at not more than 6·000 centres horizontally. Putlogs should not be classified as ties.

Suitable tying-in methods include connecting to tubes fitted between sides of window openings or to internal tubes fitted across window openings, the former method should not be used for more than 50% of the total number of ties. If there is an insufficient number of window openings for the required number of ties external rakers should be used.

Typical Details ~

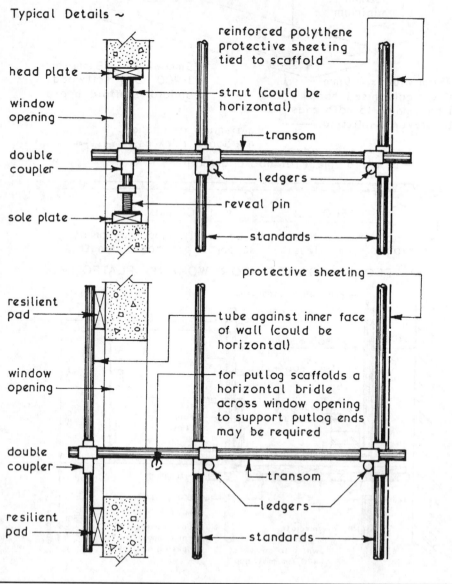

Mobile Scaffolds ~ otherwise known as mobile tower scaffolds. They can be assembled from pre-formed framing components or from standard scaffold tube and fittings. Used mainly for property maintenance. Must not be moved whilst occupied by persons or equipment.

Typical detail ~

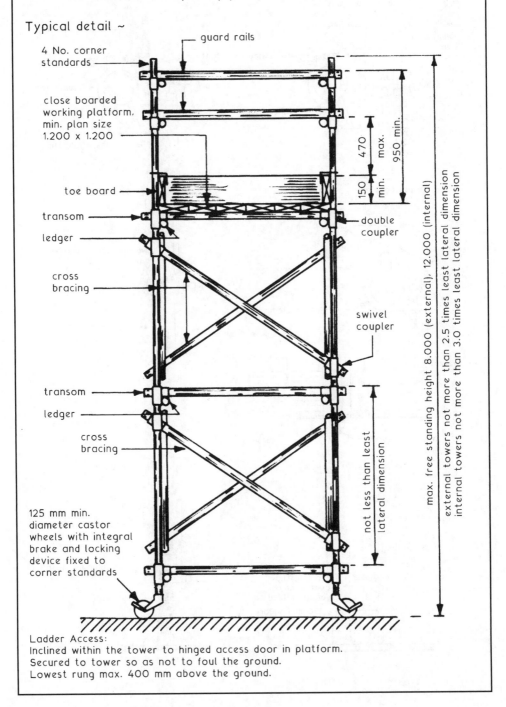

4 No. corner standards

guard rails

close boarded working platform, min. plan size 1.200 x 1.200

toe board

transom

ledger

cross bracing

transom

ledger

cross bracing

double coupler

swivel coupler

470 max.

150 min.

950 min.

not less than least lateral dimension

max. free standing height 8.000 (external), 12.000 (internal)

external towers not more than 2.5 times least lateral dimension

internal towers not more than 3.0 times least lateral dimension

125 mm min. diameter castor wheels with integral brake and locking device fixed to corner standards

Ladder Access:
Inclined within the tower to hinged access door in platform.
Secured to tower so as not to foul the ground.
Lowest rung max. 400 mm above the ground.

Some basic fittings ~

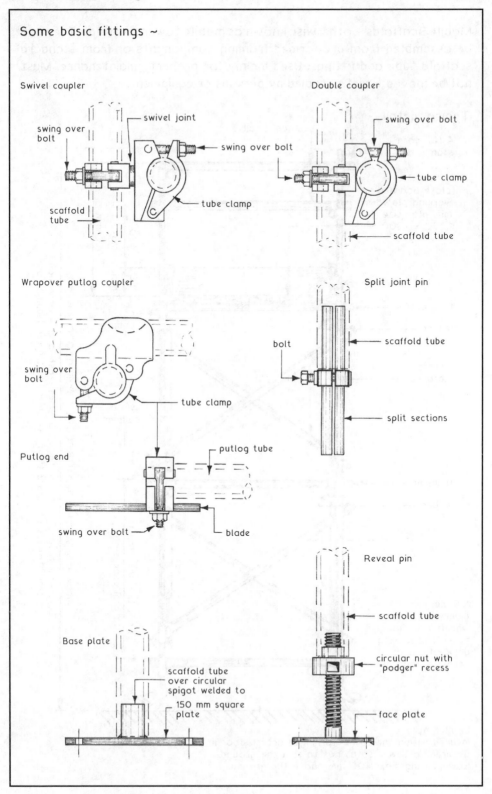

Swivel coupler

swing over bolt

swivel joint

swing over bolt

tube clamp

scaffold tube

Double coupler

swing over bolt

tube clamp

scaffold tube

Wrapover putlog coupler

swing over bolt

tube clamp

Putlog end

putlog tube

swing over bolt

blade

Split joint pin

bolt

scaffold tube

split sections

Reveal pin

scaffold tube

circular nut with "podger" recess

face plate

Base plate

scaffold tube over circular spigot welded to 150 mm square plate

Patent Scaffolding ~ these are systems based on an independent scaffold format in which the members are connected together using an integral locking device instead of conventional clips and couplers used with traditional tubular scaffolding. They have the advantages of being easy to assemble and take down using semi-skilled labour and should automatically comply with the requirements set out in the Work at Height Regulations 2005. Generally cross bracing is not required with these systems but façade bracing can be fitted if necessary. Although simple in concept patent systems of scaffolding can lack the flexibility of traditional tubular scaffolds in complex layout situations.

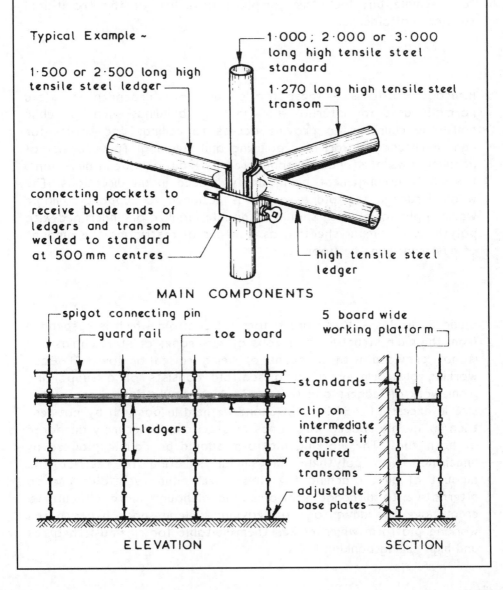

Typical Example ~

1·500 or 2·500 long high tensile steel ledger

1·000 ; 2·000 or 3·000 long high tensile steel standard

1·270 long high tensile steel transom

connecting pockets to receive blade ends of ledgers and transom welded to standard at 500 mm centres

high tensile steel ledger

MAIN COMPONENTS

spigot connecting pin
guard rail
toe board
5 board wide working platform

ledgers

standards

clip on intermediate transoms if required

transoms

adjustable base plates

ELEVATION

SECTION

Scaffolding Systems ~ these are temporary stagings to provide safe access to and egress from a working platform. The traditional putlog and independent scaffolds have been covered on pages 140 to 144 inclusive. The minimum legal requirements contained in the Construction (Health Safety and Welfare) Regulations 1996 applicable to traditional scaffolds apply equally to special scaffolds. Special scaffolds are designed to fulfil a specific function or to provide access to areas where it is not possible and or economic to use traditional formats. They can be constructed from standard tubes or patent systems, the latter complying with most regulation requirements are easy and quick to assemble but lack the complete flexibility of the traditional tubular scaffolds.

Birdcage Scaffolds ~ these are a form of independent scaffold normally used for internal work in large buildings such as public halls and churches to provide access to ceilings and soffits for light maintenance work like painting and cleaning. They consist of parallel rows of standards connected by leaders in both directions, the whole arrangement being firmly braced in all directions. The whole birdcage scaffold assembly is designed to support a single working platform which should be double planked or underlined with polythene or similar sheeting as a means of restricting the amount of dust reaching the floor level.

Slung Scaffolds ~ these are a form of scaffold which is suspended from the main structure by means of wire ropes or steel chains and is not provided with a means of being raised or lowered. Each working platform of a slung scaffold consists of a supporting framework of ledgers and transoms which should not create a plan size in excess of 2·500 × 2·500 and be held in position by not less than six evenly spaced wire ropes or steel chains securely anchored at both ends. The working platform should be double planked or underlined with polythene or similar sheeting to restrict the amount of dust reaching the floor level. Slung scaffolds are an alternative to birdcage scaffolds and although more difficult to erect have the advantage of leaving a clear space beneath the working platform which makes them suitable for cinemas, theatres and high ceiling banking halls.

Suspended Scaffolds ~ these consist of a working platform in the form of a cradle which is suspended from cantilever beams or outriggers from the roof of a tall building to give access to the façade for carrying out light maintenance work and cleaning activities. The cradles can have manual or power control and be in single units or grouped together to form a continuous working platform. If grouped together they are connected to one another at their abutment ends with hinges to form a gap of not more than 25 mm wide. Many high rise buildings have a permanent cradle system installed at roof level and this is recommended for all buildings over 30·000 high.

Typical Example ~

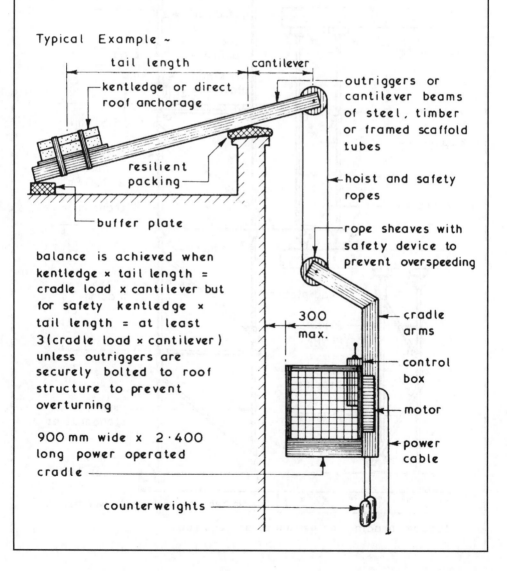

tail length   cantilever

kentledge or direct roof anchorage

outriggers or cantilever beams of steel, timber or framed scaffold tubes

resilient packing

hoist and safety ropes

buffer plate

rope sheaves with safety device to prevent overspeeding

balance is achieved when kentledge × tail length = cradle load × cantilever but for safety kentledge × tail length = at least 3(cradle load × cantilever) unless outriggers are securely bolted to roof structure to prevent overturning

300 max.

cradle arms

control box

motor

900 mm wide × 2·400 long power operated cradle

power cable

counterweights

149

Cantilever Scaffolds ~ these are a form of independent tied scaffold erected on cantilever beams and used where it is impracticable, undesirable or uneconomic to use a traditional scaffold raised from ground level. The assembly of a cantilever scaffold requires special skills and should therefore always be carried out by trained and experienced personnel.

Typical Example ~

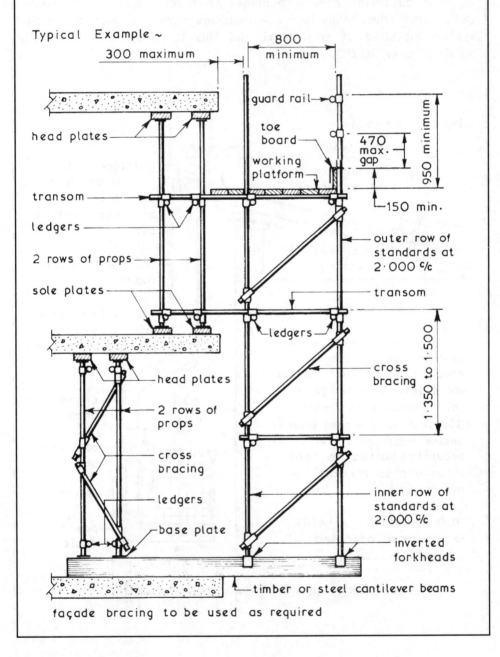

façade bracing to be used as required

Truss-out Scaffold ~ this is a form of independent tied scaffold used where it is impracticable, undesirable or uneconomic to build a scaffold from ground level. The supporting scaffold structure is known as the truss-out. The assembly of this form of scaffold requires special skills and should therefore be carried out by trained and experienced personnel.

Typical Example ~

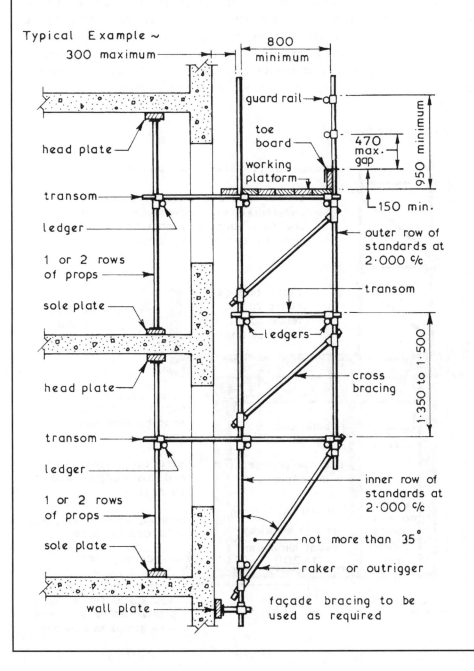

300 maximum

800 minimum

head plate

guard rail

toe board

working platform

470 max. gap

950 minimum

150 min.

transom

ledger

1 or 2 rows of props

sole plate

head plate

transom

ledgers

outer row of standards at 2·000 c/c

transom

cross bracing

1·350 to 1·500

transom

ledger

1 or 2 rows of props

sole plate

inner row of standards at 2·000 c/c

not more than 35°

raker or outrigger

wall plate

façade bracing to be used as required

Gantries ~ these are elevated platforms used when the building being maintained or under construction is adjacent to a public footpath. A gantry over a footpath can be used for storage of materials, housing units of accommodation and supporting an independent scaffold. Local authority permission will be required before a gantry can be erected and they have the power to set out the conditions regarding minimum sizes to be used for public walkways and lighting requirements. It may also be necessary to comply with police restrictions regarding the loading and unloading of vehicles at the gantry position. A gantry can be constructed of any suitable structural material and may need to be structurally designed to meet all the necessary safety requirements.

Typical Example ~

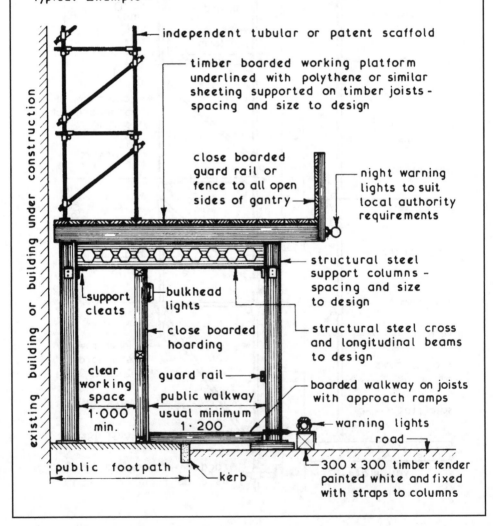

independent tubular or patent scaffold

timber boarded working platform underlined with polythene or similar sheeting supported on timber joists - spacing and size to design

close boarded guard rail or fence to all open sides of gantry

night warning lights to suit local authority requirements

support cleats

bulkhead lights

structural steel support columns - spacing and size to design

close boarded hoarding

structural steel cross and longitudinal beams to design

clear working space 1·000 min.

guard rail

public walkway usual minimum 1·200

boarded walkway on joists with approach ramps

warning lights

road

existing building or building under construction

public footpath

kerb

300 × 300 timber fender painted white and fixed with straps to columns

152

Shoring ~ this is a form of temporary support which can be given to existing buildings with the primary function of providing the necessary precautions to avoid damage to any person from collapse of structure as required by the Construction (Health, Safety and Welfare) Regulations 1996.

Shoring Systems ~ there are three basic systems of shoring which can be used separately or in combination with one another to provide the support(s) and these are namely:-

1. Dead Shoring - used primarily to carry vertical loadings.
2. Raking Shoring - used to support a combination of vertical and horizontal loadings.
3. Flying Shoring - an alternative to raking shoring to give a clear working space at ground level.

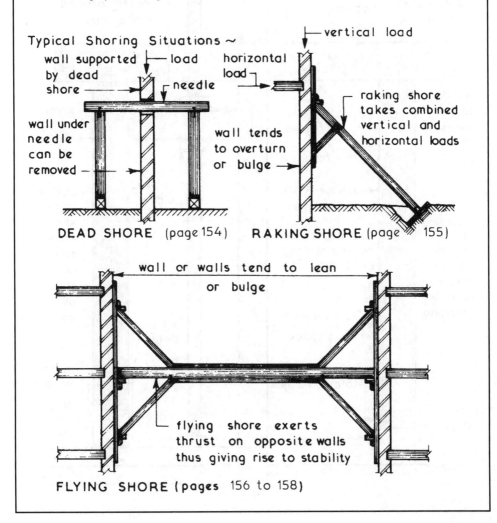

Typical Shoring Situations ~

DEAD SHORE (page 154)   **RAKING SHORE** (page 155)

FLYING SHORE (pages 156 to 158)

Dead Shores ~ these shores should be placed at approximately 2·000 c/c and positioned under the piers between the windows, any windows in the vicinity of the shores being strutted to prevent distortion of the openings. A survey should be carried out to establish the location of any underground services so that they can be protected as necessary. The sizes shown in the detail below are typical, actual sizes should be obtained from tables or calculated from first principles. Any suitable structural material such as steel can be substituted for the timber members shown.

Typical Detail ~

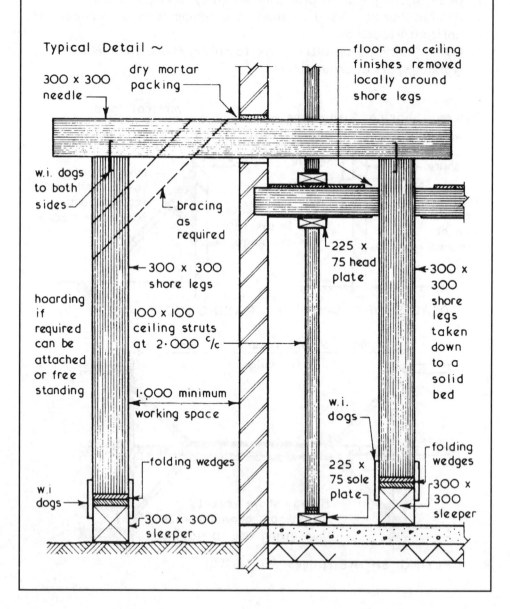

300 x 300 needle

dry mortar packing

floor and ceiling finishes removed locally around shore legs

w.i. dogs to both sides

bracing as required

300 x 300 shore legs

225 x 75 head plate

300 x 300 shore legs taken down to a solid bed

hoarding if required can be attached or free standing

100 x 100 ceiling struts at 2·000 c/c

1·000 minimum working space

w.i. dogs

w.i dogs

folding wedges

225 x 75 sole plate

folding wedges

300 x 300 sleeper

300 x 300 sleeper

Raking Shoring ~ these are placed at 3·000 to 4·500 c/c and can be of single, double, triple or multiple raker format. Suitable materials are timber, structural steel and framed tubular scaffolding.

Typical Multiple Raking Shore Detail ~

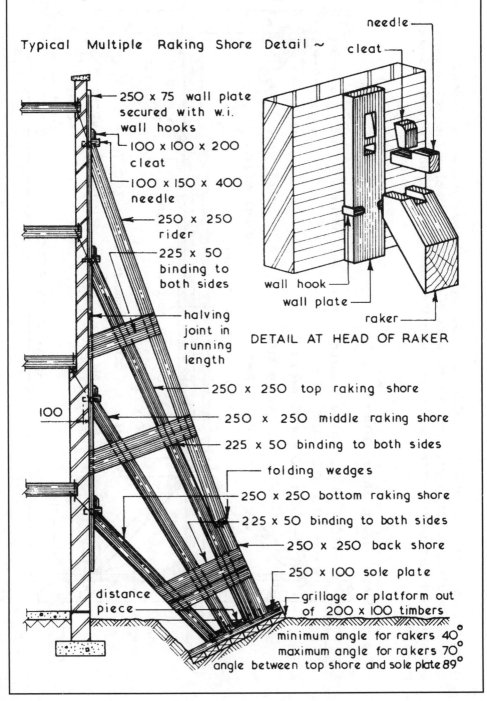

- 250 x 75 wall plate secured with w.i. wall hooks
- 100 x 100 x 200 cleat
- 100 x 150 x 400 needle
- 250 x 250 rider
- 225 x 50 binding to both sides
- halving joint in running length
- 250 x 250 top raking shore
- 250 x 250 middle raking shore
- 225 x 50 binding to both sides
- folding wedges
- 250 x 250 bottom raking shore
- 225 x 50 binding to both sides
- 250 x 250 back shore
- 250 x 100 sole plate
- grillage or platform out of 200 x 100 timbers

100

distance piece

needle
cleat
wall hook
wall plate
raker

DETAIL AT HEAD OF RAKER

minimum angle for rakers 40°
maximum angle for rakers 70°
angle between top shore and sole plate 89°

155

Flying Shores ~ these are placed at 3.000 to 4.500 c/c and can be of a single or double format. They are designed, detailed and constructed to the same basic principles as that shown for raking shores on page 155. Unsymmetrical arrangements are possible providing the basic principles for flying shores are applied – see page 158.

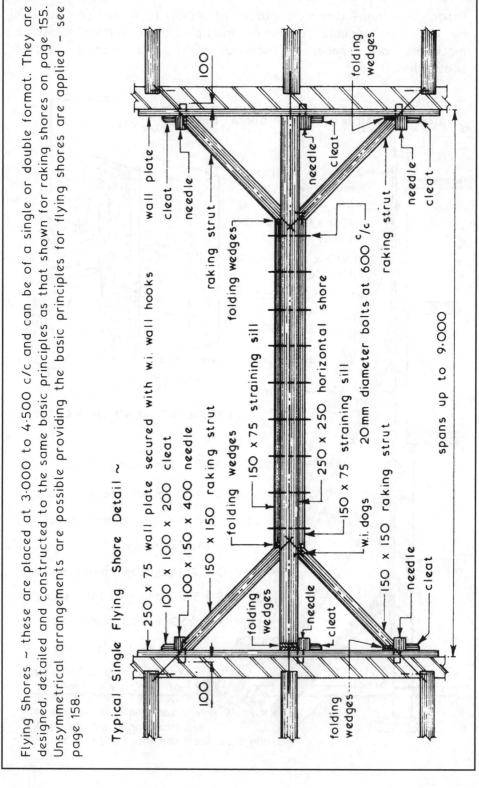

Typical Single Flying Shore Detail ~

- wall plate
- cleat
- needle
- raking strut
- folding wedges
- needle
- cleat
- raking strut
- needle
- cleat
- folding wedges
- 100
- 250 × 75 wall plate secured with w.i. wall hooks
- 100 × 100 × 200 cleat
- 100 × 150 × 400 needle
- 150 × 150 raking strut
- folding wedges
- 150 × 75 straining sill
- 250 × 250 horizontal shore
- 150 × 75 straining sill
- 20mm diameter bolts at 600 c/c
- w.i. dogs
- 150 × 150 raking strut
- needle
- cleat
- folding wedges
- needle
- cleat
- folding wedges
- 100
- spans up to 9.000

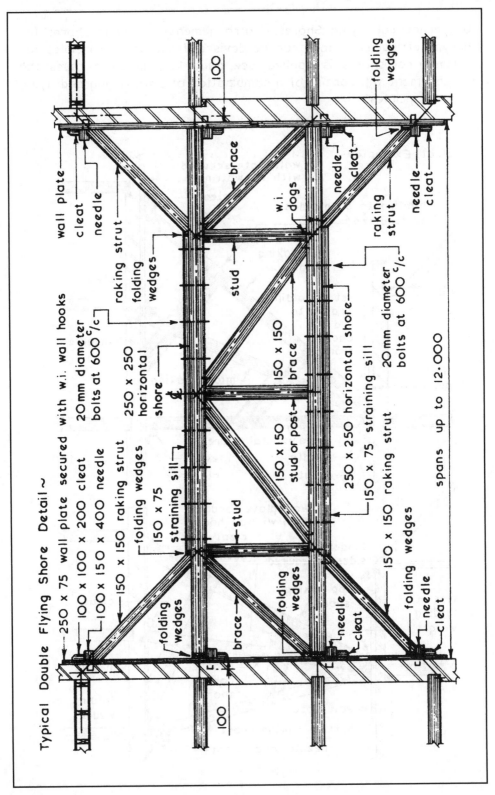

Typical Double Flying Shore Detail ~

250 x 75 wall plate secured with w.i. wall hooks

100 x 100 x 200 cleat

100 x 150 x 400 needle

150 x 150 raking strut

folding wedges

150 x 75 straining sill

folding wedges

brace

needle

cleat

wall plate
cleat
needle

raking strut

folding wedges

250 x 250 horizontal shore

20mm diameter bolts at 600°/c

stud

w.i. dogs

needle
cleat

needle
cleat

folding wedges

raking strut

150 x 150 brace

150 x 150 stud or post

250 x 250 horizontal shore

150 x 75 straining sill

20mm diameter bolts at 600°/c

150 x 150 raking strut

folding wedges

needle
cleat

spans up to 12.000

stud

Unsymmetrical Flying Shores ~ arrangements of flying shores for unsymmetrical situations can be devised if the basic principles for symmetrical shores is applied (see page 156). In some cases the arrangement will consist of a combination of both raking and flying shore principles.

Typical Examples ~

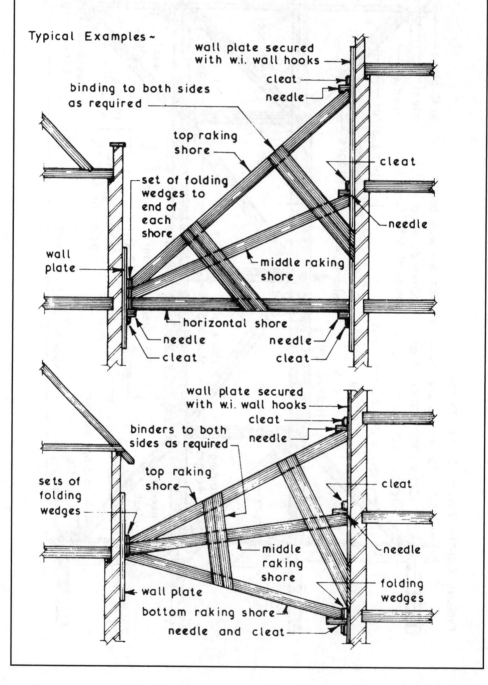

Temporary Support Determination ~ the basic sizing of most temporary supports follows the principles of elementary structural design. Readers with this basic knowledge should be able to calculate such support members which are required, particularly those used in the context of the maintenance and adaptation of buildings such as a dead shoring system.

Typical Example ~

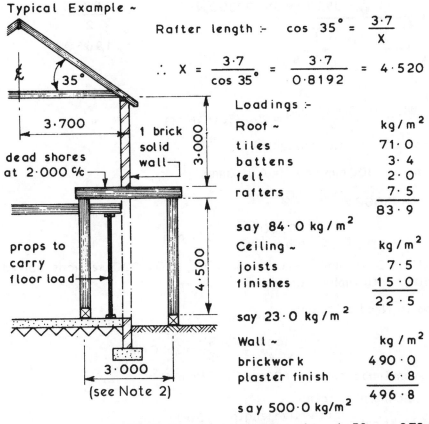

Rafter length :- $\cos 35° = \dfrac{3 \cdot 7}{X}$

$\therefore X = \dfrac{3 \cdot 7}{\cos 35°} = \dfrac{3 \cdot 7}{0 \cdot 8192} = 4 \cdot 520$

Loadings :-

Roof ~	kg/m²
tiles	71·0
battens	3·4
felt	2·0
rafters	7·5
	83·9

say 84·0 kg/m²

Ceiling ~	kg/m²
joists	7·5
finishes	15·0
	22·5

say 23·0 kg/m²

Wall ~	kg/m²
brickwork	490·0
plaster finish	6·8
	496·8

say 500·0 kg/m²

Weight of roof per metre run of wall = 84 × 4·52 = 379·68
Weight of ceiling per metre run of wall = 23 × 3·70 = 85·10
Weight of wall per metre run of wall = 500 × 3·00 = 1500·00
Total weight of wall per meter run = 1964·78

Total weight supported by needle = 1964·78 × shore centres
= 1964·78 × 2·000
= 3929·56 kg
say 3930 kg

For design calculations see next two pages.
Note 1: Typical weights of building materials, pages 35 & 36.
Note 2: Effective span (3.000 c/c) is often preferred to clear span for design of flexural members.

Design calculations reference previous page.

Timber strength class C22. See page 115 for data.

Needle Design :-

$W = 3930\ kg$

hence force

$= 3930 \times 9.81 \doteq 39300\ N$

$L = 3.000$

R_A R_B

$R_A = R_B = \dfrac{W}{2}$

$= \dfrac{39300}{2}$

$= 19650\ N$

$$BM = \frac{WL}{4} = \frac{39300 \times 3000}{4} = 29475000\ Nmm$$

$$MR = stress \times section\ modulus = fZ = f\frac{bd^2}{6}$$

assume $b = 300\ mm$ and $f = 6.8\ N/mm^2$

$$then\ 29475000 = \frac{6.8 \times 300 \times d^2}{6}$$

$$d = \sqrt{\frac{29475000 \times 6}{6.8 \times 300}} = 294\ mm$$

use 300×300 timber section or 2 No. 150×300 sections bolted together with timber connectors.

Props to Needle Design:-

$$area = \frac{load}{stress} = \frac{19650}{7.5} = 2620\ mm^2$$

$\therefore$ minimum timber size $= \sqrt{2620} = 52 \times 52\ mm$

check slenderness ratio:

$$slenderness\ ratio = \frac{effective\ length}{breadth} = \frac{4500}{52} = 86.5$$

the $52 \times 52\ mm$ section is impractically small with a very high slenderness ratio, therefore a more stable section of say, $300 \times 225\ mm$ would be selected giving a slenderness ratio of $4500/225 = 20$ (stability check, next page)

Check crushing at point of loading on needle:-

wall loading on needle $= 3930\ kg = 39300\ N = 39.3\ kN$

area of contact $= width\ of\ wall \times width\ of\ needle$

$\qquad\qquad = 215 \times 300 = 64500\ mm^2$

safe compressive stress perpendicular to grain $= 2.3\ N/mm^2$

$\therefore$ safe load $= \dfrac{64500 \times 2.3}{1000} = 148.3\ kN$ which is $> 39.3\ kN$

Stability check using the example from previous page ~

Timber of strength classification C22 (see page 115):

Modulus of elasticity, 6500 N/mm^2 minimum.

Grade stress in compression parallel to the grain, 7.5 N/mm^2.

Grade stress ratio = 6500 ÷ 7.5 = 867

The grade stress and slenderness ratios are used to provide a modification factor (K_{12}) for the compression parallel to the grain. The following table shows some factors adapted from BS 5268-2:

| | Effective length/breadth of section (slenderness ratio) | | | | | | | | |
	3	6	12	18	24	30	36	42	48
Grade stress ratio									
400	0.95	0.90	0.74	0.51	0.34	0.23	0.17	0.11	0.10
600	0.95	0.90	0.77	0.58	0.41	0.29	0.21	0.16	0.13
800	0.95	0.90	0.78	0.63	0.48	0.36	0.26	0.21	0.16
1000	0.95	0.90	0.79	0.66	0.52	0.41	0.30	0.24	0.19
1200	0.95	0.90	0.80	0.68	0.56	0.44	0.34	0.27	0.22
1400	0.95	0.90	0.80	0.69	0.58	0.47	0.37	0.30	0.24
1600	0.95	0.90	0.81	0.70	0.60	0.49	0.40	0.32	0.27
1800	0.95	0.90	0.81	0.71	0.61	0.51	0.42	0.34	0.29
2000	0.95	0.90	0.81	0.71	0.62	0.52	0.44	0.36	0.31

By interpolation, a grade stress of 867 and a slenderness ratio of 20 indicates that 7.5 N/mm^2 is multiplied by 0.57.

Applied stress should be ≤7.5 × 0.57 = 4.275 N/mm^2.

Applied stress = axial load ÷ prop section area
    = 19650 N ÷ (300 × 225 mm) = 0.291 N/mm^2

0.291 N/mm^2 is well within the allowable stress of 4.275 N/mm^2, therefore 300 × 225 mm props are satisfactory.

Ref. BS 5268-2: Structural use of timber. Code of practice for permissible stress design, materials and workmanship.

Town and Country Planning Act ~ demolition is generally not regarded as development, but planning permission will be required if the site is to have a change of use. Attitudes to demolition can vary between local planning authorities and consultation should be sought.

Planning (Listed Buildings and Conservation Areas) Act ~ listed buildings and those in conservation areas will require local authority approval for any alterations. Consent for change may be limited to partial demolition, particularly where it is necessary to preserve a building frontage for historic reasons. See the arrangements for temporary shoring on the preceding pages.

Building Act ~ intention to demolish a building requires six weeks written notice of intent. The next page shows the typical outline of a standard form for submission to the building control department of the local authority, along with location plans. Notice must also be given to utilities providers and adjoining/ adjacent building owners, particularly where party walls are involved. Small buildings of volume less than $50\,m^3$ are generally exempt. Within six weeks of the notice being submitted, the local authority will specify their requirements for shoring, protection of adjacent buildings, debris disposal and general safety requirements under the HSE.

Public Health Act ~ the local authority can issue a demolition enforcement order to a building owner, where a building is considered to be insecure, a danger to the general public and detrimental to amenities.

Highways Act ~ concerns the protection of the general public using a thoroughfare in or near to an area affected by demolition work. The building owner and demolition contractor are required to ensure that debris and other materials are not deposited in the street unless in a suitable receptacle (skip) and the local authority highways department and police are in agreement with its location. Temporary road works require protective fencing and site hoardings must be robust and secure. All supplementary provisions such as hoardings and skips may also require adequate illumination. Provision must be made for immediate removal of poisonous and hazardous waste.

Anytown Borough Council
Building Control Section
Anytown
UK

Tel: . . . .
Fax: . . . .
Email: . . . .

NOTICE TO LOCAL AUTHORITY TO CARRY OUT DEMOLITION WORKS

THE BUILDING ACT 1984 – SECTION 80

It is my intention to commence demolition of:

. . . .

. . . .

. . . .

(date)
As shown on the attached site plan, on the . . . .

This date is at least six weeks from the date of this notice. Under section 81 of the Building Act, I anticipate your notification within six weeks.

Copies of this notice have been sent to:

• The occupants/owners of any/all buildings adjacent to the proposed demolition.
• The public services/utilities supply companies.

Signed . . . .   Date . . . .

Company name and address . . . .

. . . .

. . . .

Demolition ~ skilled and potentially dangerous work that should only be undertaken by experienced contractors.

Types of demolition ~ partial or complete removal. Partial is less dynamic than complete removal, requiring temporary support to the remaining structure. This may involve window strutting, floor props and shoring. The execution of work is likely to be limited to manual handling with minimal use of powered equipment.

Preliminaries ~ a detailed survey should include:

* an assessment of condition of the structure and the impact of removing parts on the remainder.
* the effect demolition will have on adjacent properties.
* photographic records, particularly of any noticeable defects on adjacent buildings.
* neighbourhood impact, i.e. disruption, disturbance, protection.
* the need for hoardings, see pages 89 to 93.
* potential for salvaging/recycling/re-use of materials.
* extent of basements and tunnels.
* services – need to terminate and protect for future reconnections.
* means for selective removal of hazardous materials.

Insurance ~ general builders are unlikely to find demolition cover in their standard policies. All risks indemnity should be considered to cover claims from site personnel and others accessing the site. Additional third party cover will be required for claims for loss or damage to other property, occupied areas, business, utilities, private and public roads.

Salvage ~ salvaged materials and components can be valuable, bricks, tiles, slates, steel sections and timber are all marketable. Architectural features such as fireplaces and stairs will command a good price. Reclamation costs will be balanced against the financial gain.

Asbestos ~ this banned material has been used in a variety of applications including pipe insulation, fire protection, sheet claddings, linings and roofing. Samples should be taken for laboratory analysis and if necessary, specialist contractors engaged to remove material before demolition commences.

Generally ~ the reverse order of construction to gradually reduce the height. Where space in not confined, overturning or explosives may be considered.

Piecemeal ~ use of hand held equipment such as pneumatic breakers, oxy-acetylene cutters, picks and hammers. Care should be taken when salvaging materials and other reusable components. Chutes should be used to direct debris to a suitable place of collection (see page 185).

Pusher Arm ~ usually attached to a long reach articulated boom fitted to a tracked chassis. Hydraulic movement is controlled from a robust cab structure mounted above the tracks.

Wrecking Ball ~ largely confined to history, as even with safety features such as anti-spin devices, limited control over a heavy weight swinging and slewing from a crane jib will be considered unsafe in many situations.

Impact Hammer ~ otherwise known as a "pecker". Basically a large chisel operated by pneumatic power and fitted to the end of an articulated boom on a tracked chassis.

Nibbler ~ a hydraulically operated grip fitted as above that can be rotated to break brittle materials such as concrete.

Overturning ~ steel wire ropes of at least 38 mm diameter attached at high level and to an anchored winch or heavy vehicle. May be considered where controlled collapse is encouraged by initial removal of key elements of structure, typical of steel framed buildings. Alternative methods should be given preference.

Explosives ~ demolition is specialised work and the use of explosives in demolition is a further specialised practice limited to very few licensed operators. Charges are set to fire in a sequence that weakens the building to a controlled internal collapse.

Some additional references ~

BS 6187: Code of practice for demolition.

The Construction (Health, Safety and Welfare) Regulations.

The Management of Health and Safety at Work Regulations.

Concept ~ to reduce waste by designing for deconstruction.

Linear (wasteful, non-sustainable) process ~

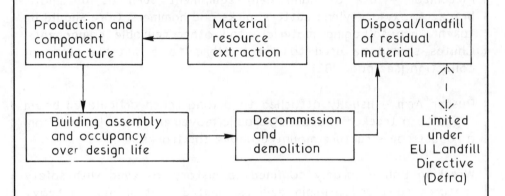

Closed-loop (near zero waste, sustainable) process ~

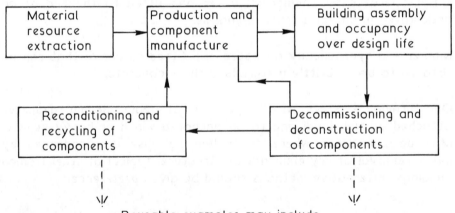

Reusable examples may include ~

- Recycled concrete aggregate (RCA) and broken
  masonry for hardcore, backfill, landscaping, etc.
- Recovered: structural timber joists and joinery.
  architectural components/features.
  slates and tiles.
  structural steel standard sections.

3 BUILDERS PLANT

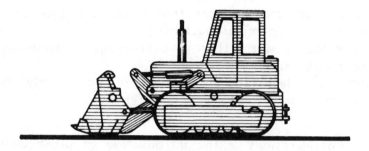

GENERAL CONSIDERATIONS

BULLDOZERS

SCRAPERS

GRADERS

TRACTOR SHOVELS

EXCAVATORS

TRANSPORT VEHICLES

HOISTS

RUBBLE CHUTES AND SKIPS

CRANES

CONCRETING PLANT

General Considerations ~ items of builders plant ranging from small hand held power tools to larger pieces of plant such as mechanical excavators and tower cranes can be considered for use for one or more of the following reasons:-

1. Increased production.
2. Reduction in overall construction costs.
3. Carry out activities which cannot be carried out by the traditional manual methods in the context of economics.
4. Eliminate heavy manual work thus reducing fatigue and as a consequence increasing productivity.
5. Replacing labour where there is a shortage of personnel with the necessary skills.
6. Maintain the high standards required particularly in the context of structural engineering works.

Economic Considerations ~ the introduction of plant does not always result in economic savings since extra temporary site works such as roadworks, hardstandings, foundations and anchorages may have to be provided at a cost which is in excess of the savings made by using the plant. The site layout and circulation may have to be planned around plant positions and movements rather than around personnel and material movements and accommodation. To be economic plant must be fully utilised and not left standing idle since plant, whether hired or owned, will have to be paid for even if it is non-productive. Full utilisation of plant is usually considered to be in the region of 85% of on site time, thus making an allowance for routine, daily and planned maintenance which needs to be carried out to avoid as far as practicable plant breakdowns which could disrupt the construction programme. Many pieces of plant work in conjunction with other items of plant such as excavators and their attendant haulage vehicles therefore a correct balance of such plant items must be obtained to achieve an economic result.

Maintenance Considerations ~ on large contracts where a number of plant items are to be used it may be advantageous to employ a skilled mechanic to be on site to carry out all the necessary daily, preventive and planned maintenance tasks together with any running repairs which could be carried out on site.

Plant Costing ~ with the exception of small pieces of plant, which are usually purchased, items of plant can be bought or hired or where there are a number of similar items a combination of buying and hiring could be considered. The choice will be governed by economic factors and the possibility of using the plant on future sites thus enabling the costs to be apportioned over several contracts.

Advantages of Hiring Plant:-

1. Plant can be hired for short periods.
2. Repairs and replacements are usually the responsibility of the hire company.
3. Plant is returned to the hire company after use thus relieving the building contractor of the problem of disposal or finding more work for the plant to justify its purchase or retention.
4. Plant can be hired with the operator, fuel and oil included in the hire rate.

Advantages of Buying Plant:-

1. Plant availability is totally within the control of the contractor.
2. Hourly cost of plant is generally less than hired plant.
3. Owner has choice of costing method used.

Typical Costing Methods ~

1. Straight Line — simple method

Capital Cost = £ 100 000
Anticipated life = 5 years
Year's working = 1500 hrs
Resale or scrap value = £ 9000
Annual depreciation ~

$$= \frac{100\,000 - 9000}{5} = £\,18\,200$$

Hourly depreciation ~

$$= \frac{18200}{1500} = 12 \cdot 13$$

Add 2% insurance = 0·27
10% maintenance = 1·33
Hourly rate = £13·73

2. Interest on Capital Outlay- widely used more accurate method

Capital Cost	= £ 100 000
C.I. on capital (8% for 5 yrs)	= 46 930
	146 930
Deduct resale value	9 000
	137 930
+ Insurance at 2% =	2 000
+ Maintenance at 10% =	10 000
	149 930

Hourly rate ~

$$= \frac{149\,930}{5 \times 1500} = £\,20 \cdot 00$$

N.B. add to hourly rate running costs

Output and Cycle Times ~ all items of plant have optimum output and cycle times which can be used as a basis for estimating anticipated productivity taking into account the task involved, task efficiency of the machine, operator's efficiency and in the case of excavators the type of soil. Data for the factors to be taken into consideration can be obtained from timed observations, feedback information or published tables contained in manufacturer's literature or reliable textbooks.

Typical Example ~

Backacter with $1m^3$ capacity bucket engaged in normal trench excavation in a clayey soil and discharging directly into an attendant haulage vehicle.

Optimum output	= 60 bucket loads per hour
Task efficiency factor	= 0·8 (from tables)
Operator efficiency factor	= 75% (typical figure)
∴ Anticipated output	= 60 × 0·8 × 0·75
	= 36 bucket loads per hour
	= 36 × 1 = 36 m^3 per hour

An allowance should be made for the bulking or swell of the solid material due to the introduction of air or voids during the excavation process

∴ Net output allowing for a 30% swell = 36 – (36 × 0·3)

$$= say \ 25 \ m^3 \ per \ hr.$$

If the Bill of Quantities gives a total net excavation of $950 \ m^3$

time required = $\dfrac{950}{25}$ = $\underline{38 \ hours}$

or assuming an 8 hour day–1/2 hour maintenance time in

days = $\dfrac{38}{7·5}$ = say $\underline{5 \ days}$

Haulage vehicles required = $1 + \dfrac{round \ trip \ time \ of \ vehicle}{loading \ time \ of \ vehicle}$

If round trip time = 30 minutes and loading time = 10 mins.

number of haulage vehicles required = $1 + \dfrac{30}{10} = 4$

This gives a vehicle waiting overlap ensuring excavator is fully utilised which is economically desirable.

Bulldozers ~ these machines consist of a track or wheel mounted power unit with a mould blade at the front which is controlled by hydraulic rams. Many bulldozers have the capacity to adjust the mould blade to form an angledozer and the capacity to tilt the mould blade about a central swivel point. Some bulldozers can also be fitted with rear attachments such as rollers and scarifiers.

The main functions of a bulldozer are:-

1. Shallow excavations up to 300 m deep either on level ground or sidehill cutting.
2. Clearance of shrubs and small trees.
3. Clearance of trees by using raised mould blade as a pusher arm.
4. Acting as a towing tractor.
5. Acting as a pusher to scraper machines (see next page).

NB. Bulldozers push earth in front of the mould blade with some side spillage whereas angledozers push and cast the spoil to one side of the mould blade.

Typical Bulldozer Details ~

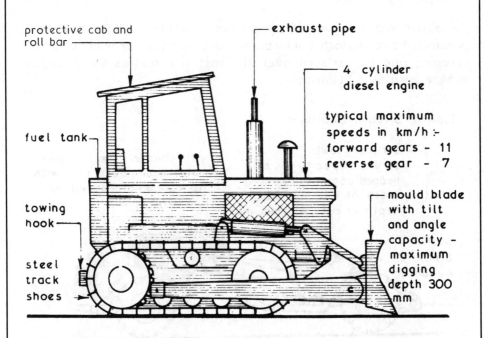

protective cab and roll bar

exhaust pipe

4 cylinder diesel engine

typical maximum speeds in km/h :-
forward gears - 11
reverse gear - 7

fuel tank

towing hook

mould blade with tilt and angle capacity - maximum digging depth 300 mm

steel track shoes

Note: Protective cab/roll bar to be fitted before use.

Scrapers ~ these machines consist of a scraper bowl which is lowered to cut and collect soil where site stripping and levelling operations are required involving large volume of earth. When the scraper bowl is full the apron at the cutting edge is closed to retain the earth and the bowl is raised for travelling to the disposal area. On arrival the bowl is lowered, the apron opened and the spoil pushed out by the tailgate as the machine moves forwards. Scrapers are available in three basic formats:-

1. Towed Scrapers – these consist of a four wheeled scraper bowl which is towed behind a power unit such as a crawler tractor. They tend to be slower than other forms of scraper but are useful for small capacities with haul distances up to 300·00.

2. Two Axle Scrapers – these have a two wheeled scraper bowl with an attached two wheeled power unit. They are very manoeuvrable with a low rolling resistance and very good traction.

3. Three Axle Scrapers – these consist of a two wheeled scraper bowl which may have a rear engine to assist the four wheeled traction engine which makes up the complement. Generally these machines have a greater capacity potential than their counterparts, are easier to control and have a faster cycle time.

To obtain maximum efficiency scrapers should operate downhill if possible, have smooth haul roads, hard surfaces broken up before scraping and be assisted over the last few metres by a pushing vehicle such as a bulldozer.

Typical Scraper Details ~

scraper bowl
struck capacity 14 m^3
heaped capacity 20m^3
width of cut 3·000
depth of cut 450mm max.

8 cylinder diesel engine attached power unit with a top forward speed of 45 km/h

pusher block for bulldozer

tailgate

apron

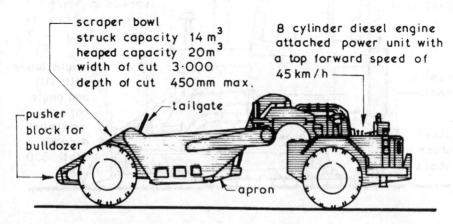

Note: Protective cab/roll bar to be fitted before use.

Graders ~ these machines are similar in concept to bulldozers in that they have a long slender adjustable mould blade, which is usually slung under the centre of the machine. A grader's main function is to finish or grade the upper surface of a large area usually as a follow up operation to scraping or bulldozing. They can produce a fine and accurate finish but do not have the power of a bulldozer therefore they are not suitable for oversite excavation work. The mould blade can be adjusted in both the horizontal and vertical planes through an angle of 300° the latter enabling it to be used for grading sloping banks.

Two basic formats of grader are available:-

1. Four Wheeled – all wheels are driven and steered which gives the machine the ability to offset and crab along its direction of travel.
2. Six Wheeled – this machine has 4 wheels in tandem drive at the rear and 2 front tilting idler wheels giving it the ability to counteract side thrust.

Typical Grader Details ~

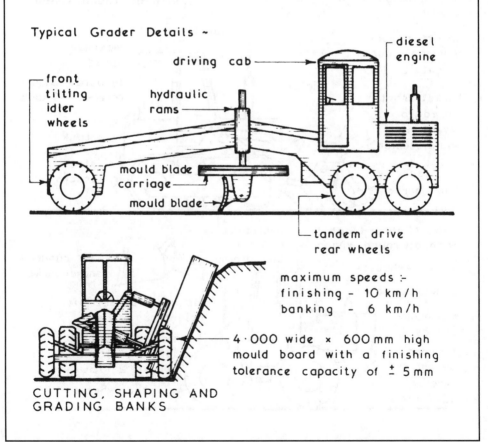

driving cab

diesel engine

front tilting idler wheels

hydraulic rams

mould blade carriage

mould blade

tandem drive rear wheels

maximum speeds :-
finishing - 10 km/h
banking - 6 km/h

4·000 wide × 600 mm high mould board with a finishing tolerance capacity of ± 5 mm

CUTTING, SHAPING AND GRADING BANKS

Tractor Shovels ~ these machines are sometimes called loaders or loader shovels and primary function is to scoop up loose materials in the front mounted bucket, elevate the bucket and manoeuvre into a position to deposit the loose material into an attendant transport vehicle. Tractor shovels are driven towards the pile of loose material with the bucket lowered, the speed and power of the machine will enable the bucket to be filled. Both tracked and wheeled versions are available, the tracked format being more suitable for wet and uneven ground conditions than the wheeled tractor shovel which has greater speed and manoeuvring capabilities. To increase their versatility tractor shovels can be fitted with a 4 in 1 bucket enabling them to carry out bulldozing, excavating, clam lifting and loading activities.

Typical Tractor Shovel Details ~

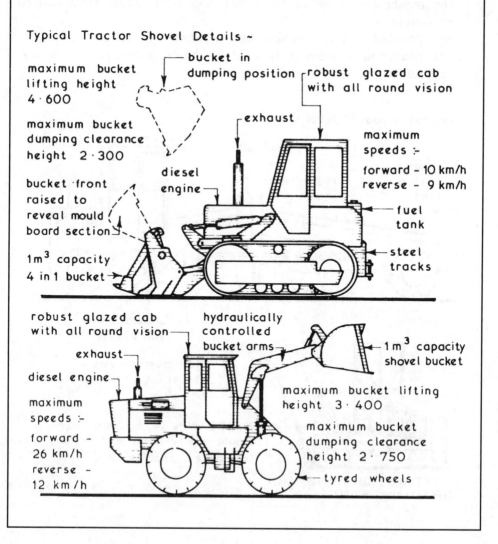

maximum bucket lifting height 4·600

maximum bucket dumping clearance height 2·300

bucket front raised to reveal mould board section

1m³ capacity 4 in 1 bucket

bucket in dumping position

exhaust

diesel engine

robust glazed cab with all round vision

maximum speeds :-
forward – 10 km/h
reverse – 9 km/h

fuel tank

steel tracks

robust glazed cab with all round vision

exhaust

diesel engine

maximum speeds :-
forward – 26 km/h
reverse – 12 km/h

hydraulically controlled bucket arms

1m³ capacity shovel bucket

maximum bucket lifting height 3·400

maximum bucket dumping clearance height 2·750

tyred wheels

Excavating Machines ~ these are one of the major items of builders plant and are used primarily to excavate and load most types of soil. Excavating machines come in a wide variety of designs and sizes but all of them can be placed within one of three categories:-

1. Universal Excavators - this category covers most forms of excavators all of which have a common factor the power unit. The universal power unit is a tracked based machine with a slewing capacity of 360° and by altering the boom arrangement and bucket type different excavating functions can be obtained. These machines are selected for high output requirements and are rope controlled.
2. Purpose Designed Excavators - these are machines which have been designed specifically to carry out one mode of excavation and they usually have smaller bucket capacities than universal excavators; they are hydraulically controlled with a shorter cycle time.
3. Multi-purpose Excavators - these machines can perform several excavating functions having both front and rear attachments. They are designed to carry out small excavation operations of low output quickly and efficiently. Multi-purpose excavators can be obtained with a wheeled or tracked base and are ideally suited for a small building firm with low excavation plant utilisation requirements.

Skimmers ~ these excavators are rigged using a universal power unit for surface stripping and shallow excavation work up to 300 mm deep where a high degree of accuracy is required. They usually require attendant haulage vehicles to remove the spoil and need to be transported between sites on a low-loader. Because of their limitations and the alternative machines available they are seldom used today.

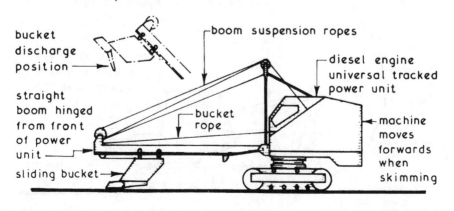

bucket discharge position

boom suspension ropes

diesel engine universal tracked power unit

straight boom hinged from front of power unit

bucket rope

machine moves forwards when skimming

sliding bucket

Face Shovels ~ the primary function of this piece of plant is to excavate above its own track or wheel level. They are available as a universal power unit based machine or as a hydraulic purpose designed unit. These machines can usually excavate any type of soil except rock which needs to be loosened, usually by blasting, prior to excavation. Face shovels generally require attendant haulage vehicles for the removal of spoil and a low loader transport lorry for travel between sites. Most of these machines have a limited capacity of between 300 and 400 mm for excavation below their own track or wheel level.

Typical Face Shovel Details ~

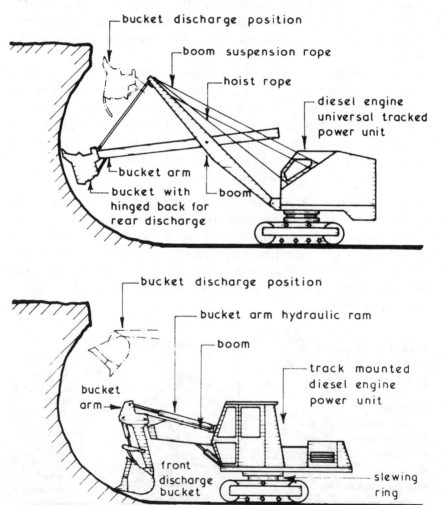

Backacters ~ these machines are suitable for trench, foundation and basement excavations and are available as a universal power unit base machine or as a purpose designed hydraulic unit. They can be used with or without attendant haulage vehicles since the spoil can be placed alongside the excavation for use in backfilling. These machines will require a low loader transport vehicle for travel between sites. Backacters used in trenching operations with a bucket width equal to the trench width can be very accurate with a high output rating.

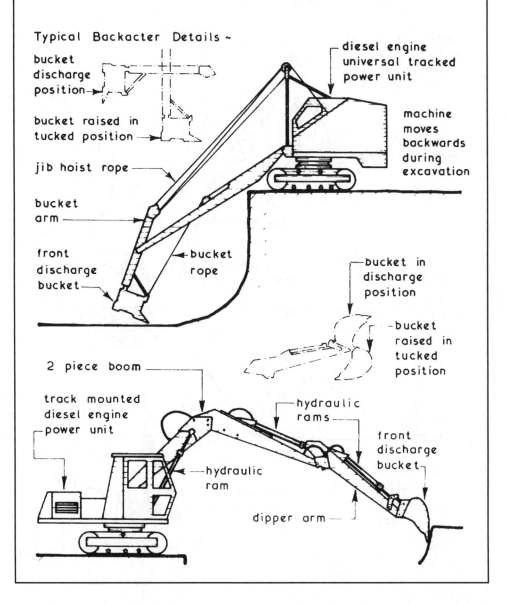

Typical Backacter Details ~

bucket discharge position

bucket raised in tucked position

jib hoist rope

bucket arm

front discharge bucket

bucket rope

diesel engine universal tracked power unit

machine moves backwards during excavation

bucket in discharge position

bucket raised in tucked position

2 piece boom

track mounted diesel engine power unit

hydraulic ram

hydraulic rams

front discharge bucket

dipper arm

Draglines ~ these machines are based on the universal power unit with basic crane rigging to which is attached a drag bucket. The machine is primarily designed for bulk excavation in loose soils up to 3·000 below its own track level by swinging the bucket out to the excavation position and hauling or dragging it back towards the power unit. Dragline machines can also be fitted with a grab or clamshell bucket for excavating in very loose soils.

Typical Dragline Details ~

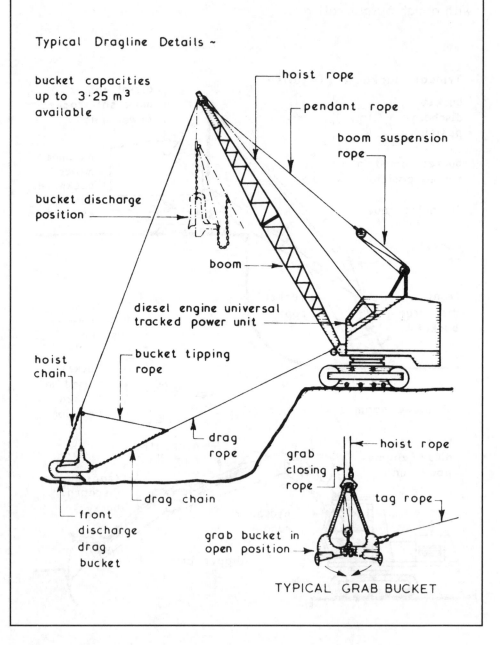

bucket capacities up to 3·25 m³ available

hoist rope

pendant rope

boom suspension rope

bucket discharge position

boom

diesel engine universal tracked power unit

hoist chain

bucket tipping rope

drag rope

grab closing rope

hoist rope

tag rope

drag chain

front discharge drag bucket

grab bucket in open position

TYPICAL GRAB BUCKET

Multi-purpose Excavators ~ these machines are usually based on the agricultural tractor with 2 or 4 wheel drive and are intended mainly for use in conjunction with small excavation works such as those encountered by the small to medium sized building contractor. Most multi-purpose excavators are fitted with a loading/excavating front bucket and a rear backacter bucket both being hydraulically controlled. When in operation using the backacter bucket the machine is raised off its axles by rear mounted hydraulic outriggers or jacks and in some models by placing the front bucket on the ground. Most machines can be fitted with a variety of bucket widths and various attachments such as bulldozer blades, scarifiers, grab buckets and post hole auger borers.

Typical Multi-purpose Excavator Details ~

bucket in raised position

bucket in discharge position

enclosed glazed cab with all round vision

bucket in discharge position

bucket raised in tucked position

pivot connection giving 180° arc of operation

boom

4 cylinder diesel engine tractor

hydraulic outriggers

dipper arm

ram

loading / excavating bucket - capacities up to 1m³ - widths up to 2·000

backacter or backhoe bucket - capacities up to 0·28 m³ - widths up to 900 mm

typical maximum road speed 30 km/h

Transport Vehicles ~ these can be defined as vehicles whose primary function is to convey passengers and/or materials between and around building sites. The types available range from the conventional saloon car to the large low loader lorries designed to transport other items of builders plant between construction sites and the plant yard or depot.

Vans – these transport vehicles range from the small two person plus a limited amount of materials to the large vans with purpose designed bodies such as those built to carry large sheets of glass. Most small vans are usually fitted with a petrol engine and are based on the manufacturer's standard car range whereas the larger vans are purpose designed with either petrol or diesel engines. These basic designs can usually be supplied with an uncovered tipping or non-tipping container mounted behind the passenger cab for use as a `pick-up´ truck.

Passenger Vehicles – these can range from a simple framed cabin which can be placed in the container of a small lorry or `pick-up´ truck to a conventional bus or coach. Vans can also be designed to carry a limited number of seated passengers by having fixed or removable seating together with windows fitted in the van sides thus giving the vehicle a dual function. The number of passengers carried can be limited so that the driver does not have to hold a PSV (public service vehicle) licence.

Lorries – these are sometimes referred to as haul vehicles and are available as road or site only vehicles. Road haulage vehicles have to comply with all the requirements of the Road Traffic Acts which among other requirements limits size and axle loads. The off-highway or site only lorries are not so restricted and can be designed to carry two to three times the axle load allowed on the public highway. Site only lorries are usually specially designed to traverse and withstand the rough terrain encountered on many construction sites. Lorries are available as non-tipping, tipping and special purpose carriers such as those with removable skips and those equipped with self loading and unloading devices. Lorries specifically designed for the transportation of large items of plant are called low loaders and are usually fitted with integral or removable ramps to facilitate loading and some have a winching system to haul the plant onto the carrier platform.

Dumpers ~ these are used for the horizontal transportation of materials on and off construction sites generally by means of an integral tipping skip. Highway dumpers are of a similar but larger design and can be used to carry materials such as excavated spoil along the roads. A wide range of dumpers are available of various carrying capacities and options for gravity or hydraulic discharge control with front tipping, side tipping or elevated tipping facilities. Special format dumpers fitted with flat platforms, rigs to carry materials skips and rigs for concrete skips for crane hoisting are also obtainable. These machines are designed to traverse rough terrain but they are not designed to carry passengers and this misuse is the cause of many accidents involving dumpers.

Typical Dumper Details ~

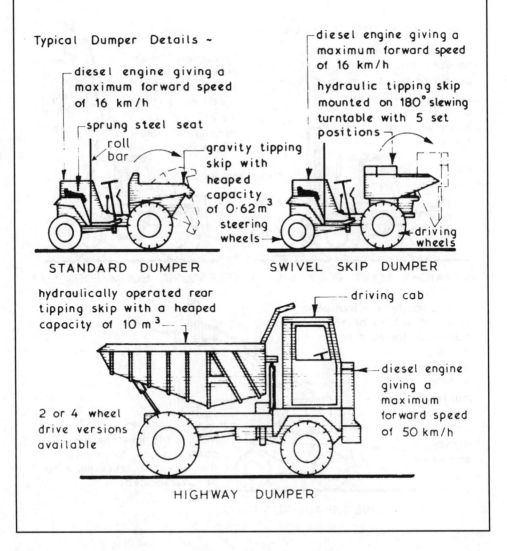

diesel engine giving a maximum forward speed of 16 km/h

sprung steel seat

roll bar

gravity tipping skip with heaped capacity of 0·62 m³

steering wheels

STANDARD DUMPER

diesel engine giving a maximum forward speed of 16 km/h

hydraulic tipping skip mounted on 180° slewing turntable with 5 set positions

driving wheels

SWIVEL SKIP DUMPER

hydraulically operated rear tipping skip with a heaped capacity of 10 m³

driving cab

diesel engine giving a maximum forward speed of 50 km/h

2 or 4 wheel drive versions available

HIGHWAY DUMPER

Fork Lift Trucks ~ these are used for the horizontal and limited vertical transportation of materials positioned on pallets or banded together such as brick packs. They are generally suitable for construction sites where the building height does not exceed three storeys. Although designed to negotiate rough terrain site fork lift trucks have a higher productivity on firm and level soils. Three basic fork lift truck formats are available, namely straight mast, overhead and telescopic boom with various height, reach and lifting capacities. Scaffolds onto which the load(s) are to be placed should be strengthened locally or a specially constructed loading tower could be built as an attachment to or as an integral part of the main scaffold.

Typical Fork Lift Truck Details ~

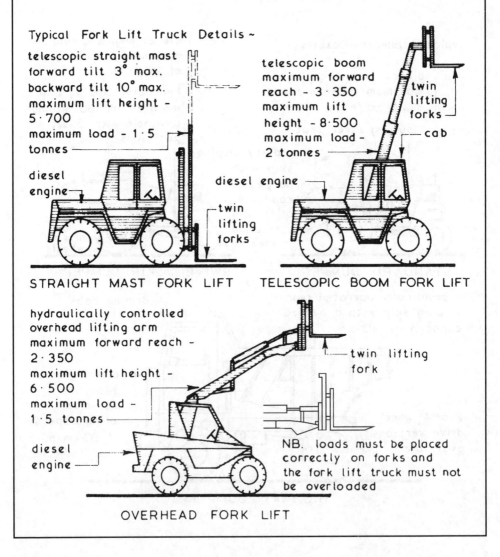

telescopic straight mast
forward tilt 3° max.
backward tilt 10° max.
maximum lift height –
5·700
maximum load – 1·5
tonnes

diesel
engine

twin
lifting
forks

STRAIGHT MAST FORK LIFT

telescopic boom
maximum forward
reach – 3·350
maximum lift
height – 8·500
maximum load –
2 tonnes

diesel engine

twin
lifting
forks

cab

TELESCOPIC BOOM FORK LIFT

hydraulically controlled
overhead lifting arm
maximum forward reach –
2·350
maximum lift height –
6·500
maximum load –
1·5 tonnes

diesel
engine

twin lifting
fork

NB. loads must be placed
correctly on forks and
the fork lift truck must not
be overloaded

OVERHEAD FORK LIFT

Hoists ~ these are designed for the vertical transportation of materials, passengers or materials and passengers (see page 184). Materials hoists are designed for one specific use (i.e. the vertical transportation of materials) and under no circumstances should they be used to transport passengers. Most material hoists are of a mobile format which can be dismantled, folded onto the chassis and moved to another position or site under their own power or towed by a haulage vehicle. When in use material hoists need to be stabilised and/or tied to the structure and enclosed with a protective screen.

Typical Materials Hoist Details ~

top bracket with automatic overrun control

protective screen out of scaffolding placed around mast to form a hoistway fitted gates at least 2·000 high at all landing levels to be supplied and erected by main contractor

lattice hoist mast 7·320 high which can be extended by adding further hoist mast sections to 32·000 high providing tie support is given every 2·750 above the initial 7·320 mast height

hoist rope

control rope operated from outside protective screen

tubular mast support struts

1·500 wide x 1·200 deep two barrow hardwood timber hoist platform with a maximum load capacity of 500kg

diesel or electric power unit

2·000

anti-walk through screen around power unit

timber buffer plate

stabilising jacks or outriggers

Passenger Hoists ~ these are designed to carry passengers although most are capable of transporting a combined load of materials and passengers within the lifting capacity of the hoist. A wide selection of hoists are available ranging from a single cage with rope suspension to twin cages with rack and pinion operation mounted on two sides of a static tower.

Typical Passenger Hoist Details ~

face of structure

standards

ties to structure at 12·000 centres

2·700 high cage to carry 12 persons or a total payload of 1000 kg. at speeds of 40 to 100 metres per minute

landings as required

NB. operation of hoist is from within the cage and the hoist must be fitted to prevent any overrun

passenger hoist tower assembled from 1·500 long sections to a maximum tied height of 240·000

climbing rack

working platform on top of cage for scaffold type crane used to extend hoist tower

electric motor and pinion housed behind cage

1·680 long x 1·370 wide enclosed passenger cage

access gate hoist

2·600 high wire mesh screen enclosure to lowest hoist position

reinforced concrete base

Rubble Chutes ~ these apply to contracts involving demolition, repair, maintenance and refurbishment. The simple concept of connecting several perforated dustbins is reputed to have been conceived by an ingenious site operative for the expedient and safe conveyance of materials.

In purpose designed format, the tapered cylinders are produced from reinforced rubber with chain linkage for continuity. Overall unit lengths are generally 1100 mm, providing an effective length of 1 m. Hoppers and side entry units are made for special applications.

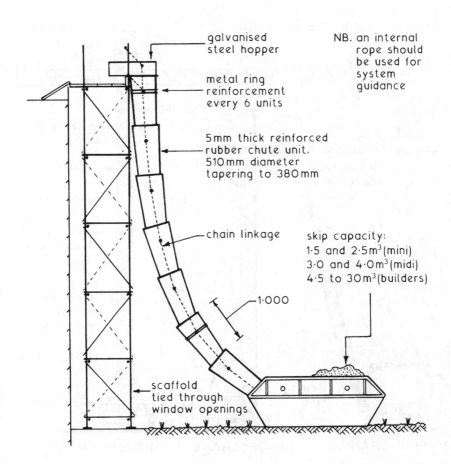

galvanised
steel hopper

metal ring
reinforcement
every 6 units

5mm thick reinforced
rubber chute unit,
510mm diameter
tapering to 380mm

chain linkage

1·000

scaffold
tied through
window openings

NB. an internal
rope should
be used for
system
guidance

skip capacity:
1·5 and 2·5m³(mini)
3·0 and 4·0m³(midi)
4·5 to 30m³(builders)

Ref. Highways Act – written permit (licence) must be obtained from the local authority highways department for use of a skip on a public thoroughfare. It will have to be illuminated at night and may require a temporary traffic light system to regulate vehicles.

Cranes ~ these are lifting devices designed to raise materials by means of rope operation and move the load horizontally within the limitations of any particular machine. The range of cranes available is very wide and therefore choice must be based on the loads to be lifted, height and horizontal distance to be covered, time period(s) of lifting operations, utilisation factors and degree of mobility required. Crane types can range from a simple rope and pulley or gin wheel to a complex tower crane but most can be placed within 1 of 3 groups, namely mobile, static and tower cranes.

Typical Crane Classifications ~

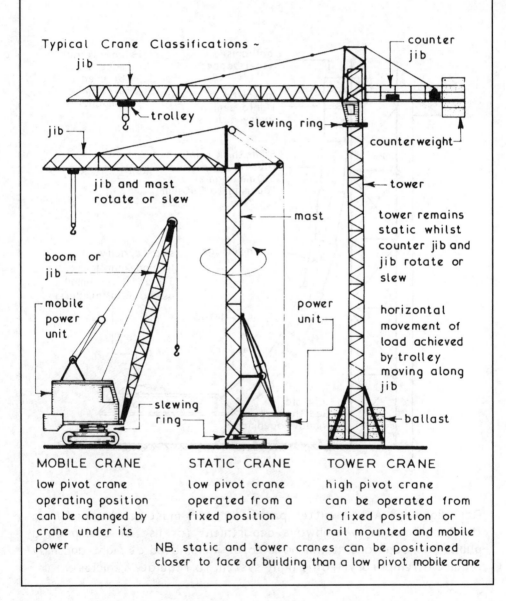

jib

trolley

counter jib

jib

jib and mast rotate or slew

slewing ring

counterweight

boom or jib

mobile power unit

mast

tower

tower remains static whilst counter jib and jib rotate or slew

power unit

power unit

horizontal movement of load achieved by trolley moving along jib

slewing ring

ballast

MOBILE CRANE

low pivot crane operating position can be changed by crane under its power

STATIC CRANE

low pivot crane operated from a fixed position

TOWER CRANE

high pivot crane can be operated from a fixed position or rail mounted and mobile

NB. static and tower cranes can be positioned closer to face of building than a low pivot mobile crane

Self Propelled Cranes ~ these are mobile cranes mounted on a wheeled chassis and have only one operator position from which the crane is controlled and the vehicle driven. The road speed of this type of crane is generally low, usually not exceeding 30 km p.h. A variety of self propelled crane formats are available ranging from short height lifting strut booms of fixed length to variable length lattice booms with a fly jib attachment.

Typical Self Propelled Crane Details ~

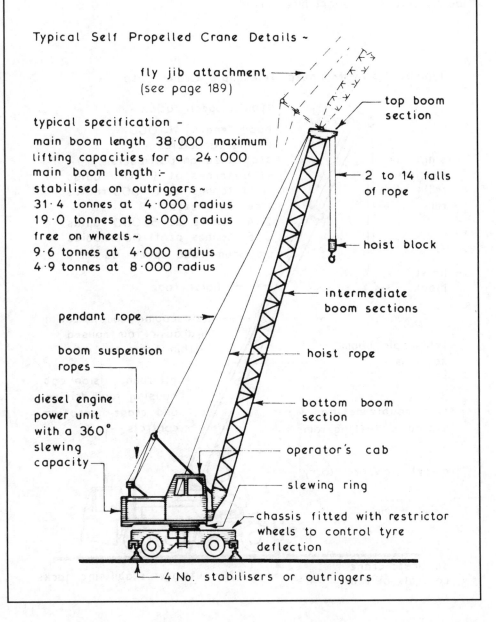

fly jib attachment
(see page 189)

typical specification –

main boom length 38·000 maximum
lifting capacities for a 24·000
main boom length :-
stabilised on outriggers ~
31·4 tonnes at 4·000 radius
19·0 tonnes at 8·000 radius
free on wheels ~
9·6 tonnes at 4·000 radius
4·9 tonnes at 8·000 radius

top boom section

2 to 14 falls of rope

hoist block

intermediate boom sections

pendant rope

boom suspension ropes

diesel engine power unit with a 360° slewing capacity

hoist rope

bottom boom section

operator's cab

slewing ring

chassis fitted with restrictor wheels to control tyre deflection

4 No. stabilisers or outriggers

Lorry Mounted Cranes ~ these mobile cranes consist of a lattice or telescopic boom mounted on a specially adapted truck or lorry. They have two operating positions: the lorry being driven from a conventional front cab and the crane being controlled from a different location. The lifting capacity of these cranes can be increased by using outrigger stabilising jacks and the approach distance to the face of building decreased by using a fly jib. Lorry mounted telescopic cranes require a firm surface from which to operate and because of their short site preparation time they are ideally suited for short hire periods.

Typical Lorry Mounted Telescopic Crane Details ~

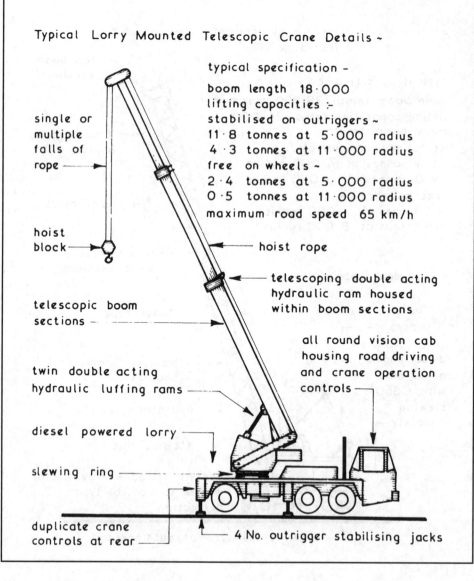

typical specification –

boom length 18·000
lifting capacities :-
stabilised on outriggers ~
11·8 tonnes at 5·000 radius
4·3 tonnes at 11·000 radius
free on wheels ~
2·4 tonnes at 5·000 radius
0·5 tonnes at 11·000 radius
maximum road speed 65 km/h

single or multiple falls of rope

hoist block

hoist rope

telescoping double acting hydraulic ram housed within boom sections

telescopic boom sections –

all round vision cab housing road driving and crane operation controls –

twin double acting hydraulic luffing rams –

diesel powered lorry –

slewing ring –

duplicate crane controls at rear –

4 No. outrigger stabilising jacks

Lorry Mounted Lattice Jib Cranes ~ these cranes follow the same basic principles as the lorry mounted telescopic cranes but they have a lattice boom and are designed as heavy duty cranes with lifting capacities in excess of 100 tonnes. These cranes will require a firm level surface from which to operate and can have a folding or sectional jib which will require the crane to be rigged on site before use.

Typical Lorry Mounted Lattice Jib Crane Details ~

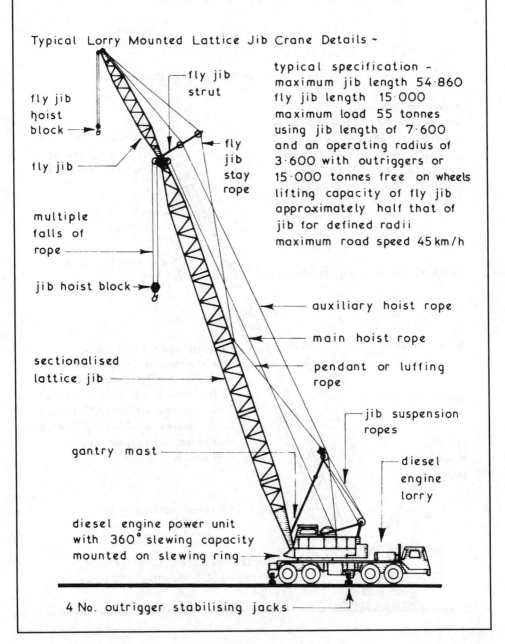

fly jib strut

fly jib hoist block →

fly jib

fly jib stay rope

multiple falls of rope

jib hoist block →

sectionalised lattice jib →

gantry mast

diesel engine power unit with 360° slewing capacity mounted on slewing ring

auxiliary hoist rope

main hoist rope

pendant or luffing rope

jib suspension ropes

diesel engine lorry

4 No. outrigger stabilising jacks

typical specification ~ maximum jib length 54·860 fly jib length 15·000 maximum load 55 tonnes using jib length of 7·600 and an operating radius of 3·600 with outriggers or 15·000 tonnes free on wheels lifting capacity of fly jib approximately half that of jib for defined radii maximum road speed 45 km/h

189

Track Mounted Cranes ~ these machines can be a universal power unit rigged as a crane (see page 178) or a purpose designed track mounted crane with or without a fly jib attachment. The latter type are usually more powerful with lifting capacities up to 45 tonnes. Track mounted cranes can travel and carry out lifting operations on most sites without the need for special road and hardstand provisions but they have to be rigged on arrival after being transported to site on a low loader lorry.

Typical Track Mounted or Crawler Crane Details ~

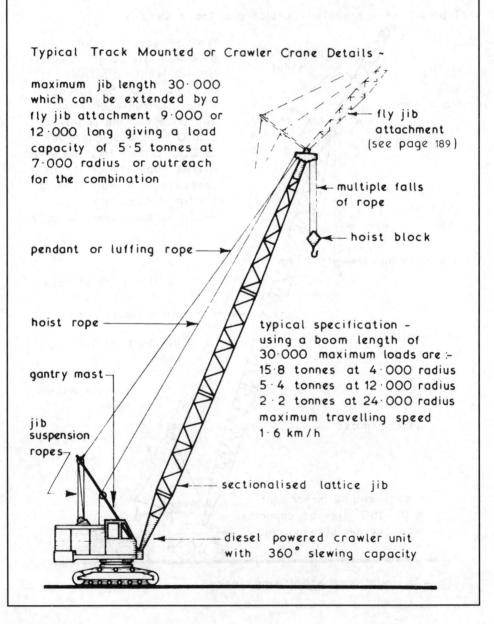

maximum jib length 30·000 which can be extended by a fly jib attachment 9·000 or 12·000 long giving a load capacity of 5·5 tonnes at 7·000 radius or outreach for the combination

fly jib attachment (see page 189)

multiple falls of rope

hoist block

pendant or luffing rope →

hoist rope →

typical specification – using a boom length of 30·000 maximum loads are :– 15·8 tonnes at 4·000 radius 5·4 tonnes at 12·000 radius 2·2 tonnes at 24·000 radius maximum travelling speed 1·6 km/h

gantry mast

jib suspension ropes

sectionalised lattice jib

diesel powered crawler unit with 360° slewing capacity

Gantry Cranes ~ these are sometimes called portal cranes and consist basically of two 'A' frames joined together with a cross member on which transverses the lifting appliance. In small gantry cranes (up to 10 tonnes lifting capacity) the 'A' frames are usually wheel mounted and manually propelled whereas in the large gantry cranes (up to 100 tonnes lifting capacity) the 'A' frames are mounted on powered bogies running on rail tracks with the driving cab and lifting gear mounted on the cross beam or gantry. Small gantry cranes are used primarily for loading and unloading activities in stock yards whereas the medium and large gantry cranes are used to straddle the work area such as in power station construction or in repetitive low to medium rise developments. All gantry cranes have the advantage of three direction movement –

1. Transverse by moving along the cross beam.
2. Vertical by raising and lowering the hoist block.
3. Horizontal by forward and reverse movements of the whole gantry crane.

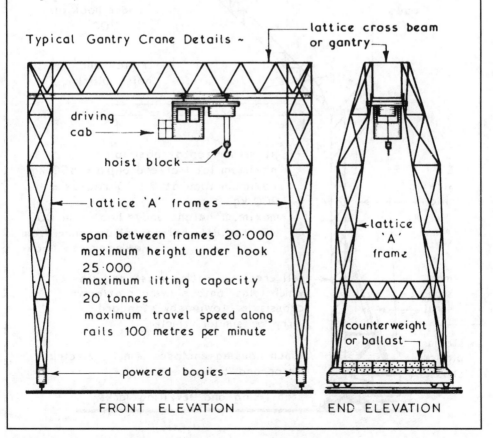

Typical Gantry Crane Details ~

lattice cross beam or gantry

driving cab

hoist block

lattice 'A' frames

span between frames 20·000
maximum height under hook 25·000
maximum lifting capacity 20 tonnes
maximum travel speed along rails 100 metres per minute

powered bogies

lattice 'A' frame

counterweight or ballast

FRONT ELEVATION   END ELEVATION

Mast Cranes ~ these are similar in appearance to the familiar tower cranes but they have one major difference in that the mast or tower is mounted on the slewing ring and thus rotates whereas a tower crane has the slewing ring at the top of the tower and therefore only the jib portion rotates. Mast cranes are often mobile, self erecting, of relatively low lifting capacity and are usually fitted with a luffing jib. A wide variety of models are available and have the advantage over most mobile low pivot cranes of a closer approach to the face of the building.

Typical Mast Crane Details ~

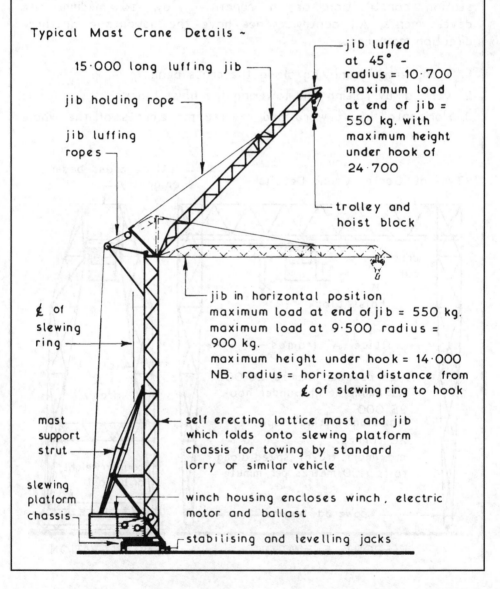

15·000 long luffing jib

jib holding rope

jib luffing ropes

jib luffed at 45° - radius = 10·700 maximum load at end of jib = 550 kg. with maximum height under hook of 24·700

trolley and hoist block

jib in horizontal position maximum load at end of jib = 550 kg. maximum load at 9·500 radius = 900 kg. maximum height under hook = 14·000 NB. radius = horizontal distance from ₵ of slewing ring to hook

₵ of slewing ring

mast support strut

self erecting lattice mast and jib which folds onto slewing platform chassis for towing by standard lorry or similar vehicle

slewing platform chassis

winch housing encloses winch, electric motor and ballast

stabilising and levelling jacks

Tower Cranes ~ most tower cranes have to be assembled and erected on site prior to use and can be equipped with a horizontal or luffing jib. The wide range of models available often make it difficult to choose a crane suitable for any particular site but most tower cranes can be classified into one of four basic groups thus:-

1. Self Supporting Static Tower Cranes - high lifting capacity with the mast or tower fixed to a foundation base - they are suitable for confined and open sites. (see page 194)

2. Supported Static Tower Cranes - similar in concept to self supporting cranes and are used where high lifts are required, the mast or tower being tied at suitable intervals to the structure to give extra stability. (see page 195)

3. Travelling Tower Cranes - these are tower cranes mounted on power bogies running on a wide gauge railway track to give greater site coverage - only slight gradients can be accommodated therefore a reasonably level site or specially constructed railway support trestle is required. (see page 196)

4. Climbing Cranes - these are used in conjunction with tall buildings and structures. The climbing mast or tower is housed within the structure and raised as the height of the structure is increased. Upon completion the crane is dismantled into small sections and lowered down the face of the building. (see page 197)

All tower cranes should be left in an 'out of service' condition when unattended and in high wind conditions, the latter varying with different models but generally wind speeds in excess of 60 km p.h. would require the crane to be placed in an out of service condition thus:-

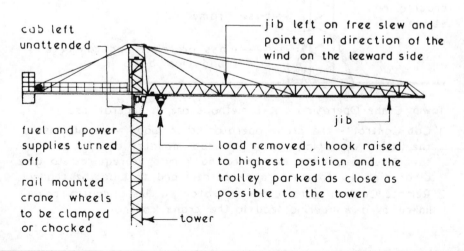

cab left unattended

jib left on free slew and pointed in direction of the wind on the leeward side

jib

fuel and power supplies turned off

rail mounted crane wheels to be clamped or chocked

load removed, hook raised to highest position and the trolley parked as close as possible to the tower

tower

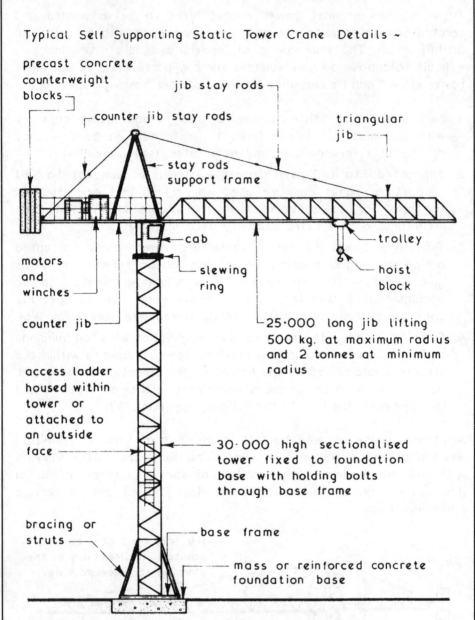

Typical Self Supporting Static Tower Crane Details ~

precast concrete counterweight blocks

counter jib stay rods

jib stay rods

triangular jib

stay rods support frame

motors and winches

cab

trolley

hoist block

slewing ring

counter jib

25·000 long jib lifting 500 kg. at maximum radius and 2 tonnes at minimum radius

access ladder housed within tower or attached to an outside face

30·000 high sectionalised tower fixed to foundation base with holding bolts through base frame

bracing or struts

base frame

mass or reinforced concrete foundation base

Tower Crane Operation ~ two methods are in general use :-

1 Cab Control - the crane operator has a good view of most of the lifting operations from the cab mounted at top of the tower but a second person or banksman is required to give clear signals to the crane operator and to load the crane

2. Remote Control - the crane operator carries a control box linked by a wandering lead to the crane controls.

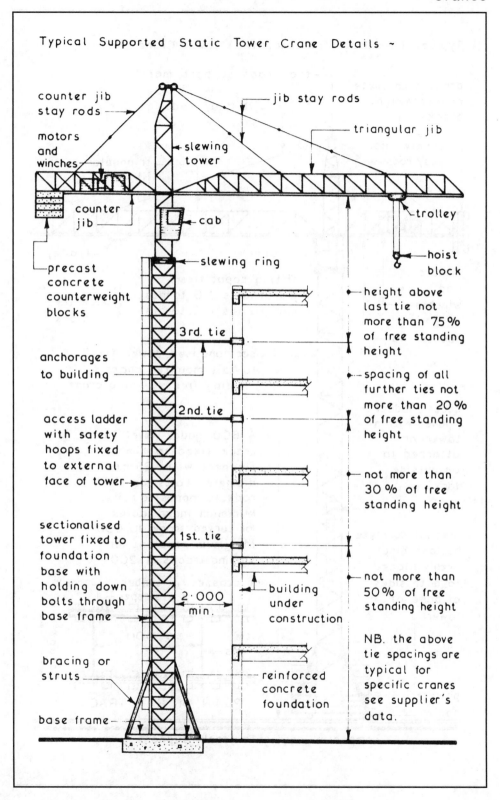

Typical Supported Static Tower Crane Details ~

counter jib stay rods

jib stay rods

motors and winches

slewing tower

triangular jib

counter jib

cab

trolley

hoist block

precast concrete counterweight blocks

slewing ring

height above last tie not more than 75% of free standing height

3rd. tie

anchorages to building

spacing of all further ties not more than 20% of free standing height

2nd. tie

access ladder with safety hoops fixed to external face of tower

not more than 30% of free standing height

1st. tie

sectionalised tower fixed to foundation base with holding down bolts through base frame

2·000 min.

building under construction

not more than 50% of free standing height

bracing or struts

reinforced concrete foundation

NB. the above tie spacings are typical for specific cranes see supplier's data.

base frame

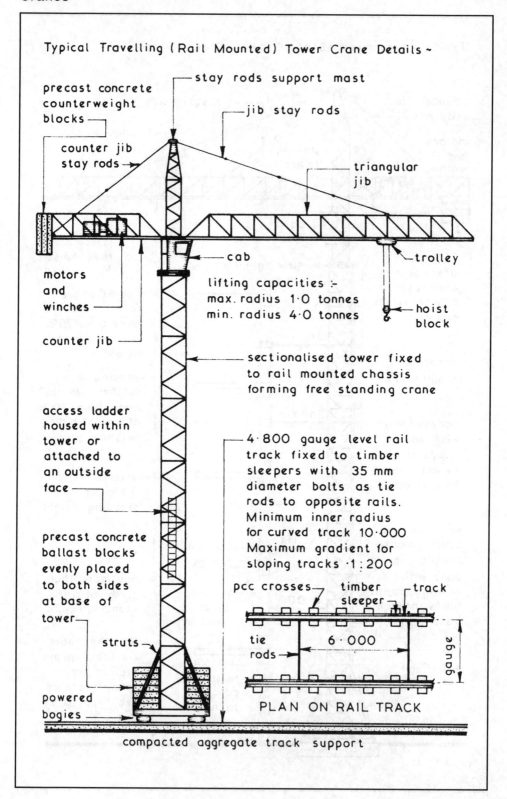

Typical Travelling (Rail Mounted) Tower Crane Details ~

stay rods support mast

precast concrete counterweight blocks

jib stay rods

counter jib stay rods →

triangular jib

cab

trolley

motors and winches

lifting capacities :-
max. radius 1·0 tonnes
min. radius 4·0 tonnes

hoist block

counter jib

sectionalised tower fixed to rail mounted chassis forming free standing crane

access ladder housed within tower or attached to an outside face

4·800 gauge level rail track fixed to timber sleepers with 35 mm diameter bolts as tie rods to opposite rails. Minimum inner radius for curved track 10·000 Maximum gradient for sloping tracks ·1 : 200

precast concrete ballast blocks evenly placed to both sides at base of tower

pcc crosses

timber sleeper

track

tie rods →

6 · 000

gauge

struts

powered bogies

PLAN ON RAIL TRACK

compacted aggregate track support

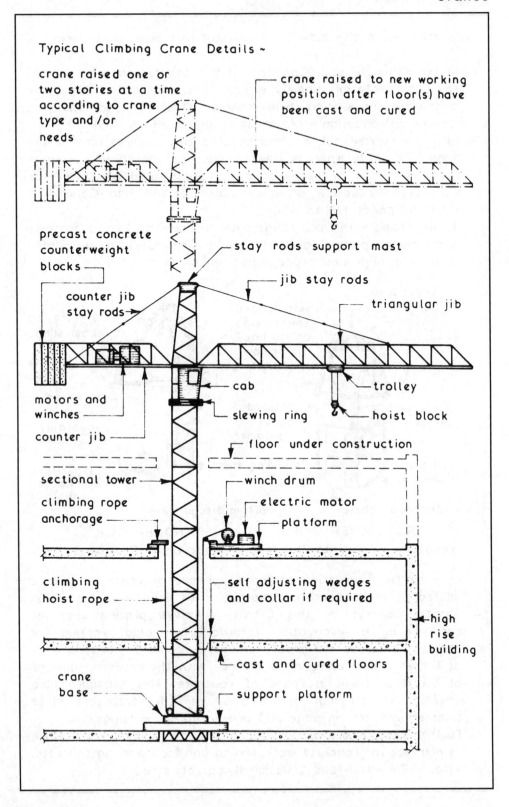

Typical Climbing Crane Details ~

crane raised one or two stories at a time according to crane type and /or needs

crane raised to new working position after floor(s) have been cast and cured

precast concrete counterweight blocks

stay rods support mast

jib stay rods

counter jib stay rods

triangular jib

motors and winches

cab

counter jib

slewing ring

trolley

hoist block

floor under construction

sectional tower

winch drum

climbing rope anchorage

electric motor

platform

climbing hoist rope

self adjusting wedges and collar if required

high rise building

crane base

cast and cured floors

support platform

Concreting ~ this site activity consists of four basic procedures –

1. Material Supply and Storage – this is the receiving on site of the basic materials namely cement, fine aggregate and coarse aggregate and storing them under satisfactory conditions. (see Concrete Production – Materials on pages 284 & 285)
2. Mixing – carried out in small batches this requires only simple hand held tools whereas when demand for increased output is required mixers or ready mixed supplies could be used. (see Concrete Production on pages 286 to 289 and Concreting Plant on pages 199 to 204)
3. Transporting – this can range from a simple bucket to barrows and dumpers for small amounts. For larger loads, especially those required at high level, crane skips could be used:-

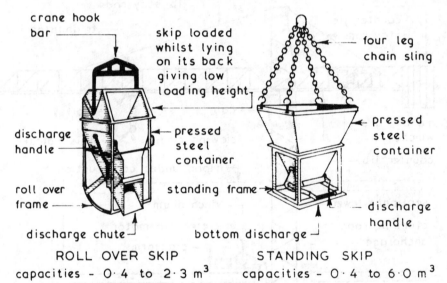

ROLL OVER SKIP
capacities – 0·4 to 2·3 m³

STANDING SKIP
capacities – 0·4 to 6·0 m³

For the transportation of large volumes of concrete over a limited distance concrete pumps could be used. (see page 202)
4. Placing Concrete – this activity involves placing the wet concrete in the excavation, formwork or mould; working the concrete between and around any reinforcement; vibrating and/or tamping and curing in accordance with the recommendations of BS 8110: Structural use of concrete. This standard also covers the striking or removal of the formwork. (see Concreting Plant on page 203 and Formwork on page 514)
Further ref. BS 8000-2.1: Workmanship on building sites. Code of practice for concrete work. Mixing and transporting concrete. Also, BS EN 1992-1-1 and -2: Design of concrete structures.

Concrete Mixers ~ apart from the very large output mixers most concrete mixers in general use have a rotating drum designed to produce a concrete without segregation of the mix.

Concreting Plant ~ the selection of concreting plant can be considered under three activity headings –
1. Mixing. 2. Transporting. 3. Placing.

Choice of Mixer ~ the factors to be taken into consideration when selecting the type of concrete mixer required are –

1. Maximum output required (m³/hour).
2. Total output required (m³).
3. Type or method of transporting the mixed concrete.
4. Discharge height of mixer (compatibility with transporting method).

Concrete mixer types are generally related to their designed output performance, therefore when the answer to the question 'How much concrete can be placed in a given time period?' or alternatively 'What mixing and placing methods are to be employed to mix and place a certain amount of concrete in a given time period?' has been found the actual mixer can be selected. Generally a batch mixing time of 5 minutes per cycle or 12 batches per hour can be assumed as a reasonable basis for assessing mixer output.

Small Batch Mixers ~ these mixers have outputs of up to 200 litres per batch with wheelbarrow transportation an hourly placing rate of 2 to 3 m³ can be achieved. Most small batch mixers are of the tilting drum type. Generally these mixers are hand loaded which makes the quality control of successive mixes difficult to regulate.

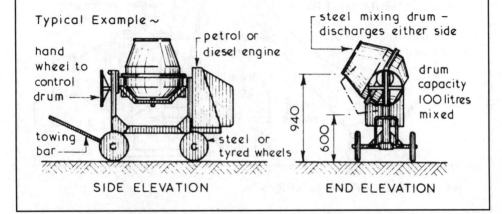

Typical Example ~

hand wheel to control drum

towing bar

petrol or diesel engine

steel or tyred wheels

SIDE ELEVATION

steel mixing drum – discharges either side

drum capacity 100 litres mixed

940

600

END ELEVATION

Medium Batch Mixers ~ outputs of these mixers range from 200 to 750 litres and can be obtained at the lower end of the range as a tilting drum mixer or over the complete range as a non-tilting drum mixer with either reversing drum or chute discharge. The latter usually having a lower discharge height. These mixers usually have integral weight batching loading hoppers, scraper shovels and water tanks thus giving better quality control than the small batch mixers. Generally they are unsuitable for wheelbarrow transportation because of their high output.

Typical Examples ~

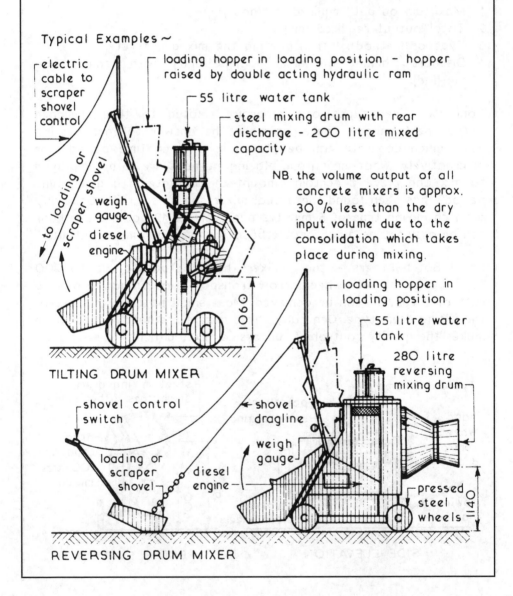

electric cable to scraper shovel control

loading hopper in loading position - hopper raised by double acting hydraulic ram

55 litre water tank

steel mixing drum with rear discharge - 200 litre mixed capacity

NB. the volume output of all concrete mixers is approx. 30 % less than the dry input volume due to the consolidation which takes place during mixing.

to loading or scraper shovel

weigh gauge

diesel engine

1060

TILTING DRUM MIXER

loading hopper in loading position

55 litre water tank

280 litre reversing mixing drum

shovel control switch

shovel dragline

loading or scraper shovel

diesel engine

weigh gauge

pressed steel wheels

1140

REVERSING DRUM MIXER

Transporting Concrete ~ the usual means of transporting mixed concrete produced in a small capacity mixer is by wheelbarrow. The run between the mixing and placing positions should be kept to a minimum and as smooth as possible by using planks or similar materials to prevent segregation of the mix within the wheelbarrow.

Dumpers ~ these can be used for transporting mixed concrete from mixers up to 600 litre capacity when fitted with an integral skip and for lower capacities when designed to take a crane skip.

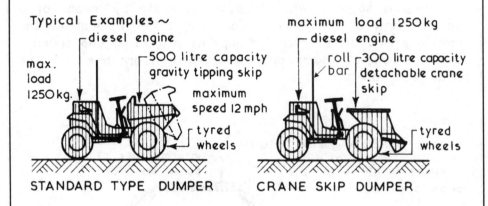

Typical Examples ~
diesel engine
max. load 1250 kg.
500 litre capacity gravity tipping skip
maximum speed 12 mph
tyred wheels

STANDARD TYPE DUMPER

maximum load 1250 kg
diesel engine
roll bar
300 litre capacity detachable crane skip
tyred wheels

CRANE SKIP DUMPER

Ready Mixed Concrete Trucks ~ these are used to transport mixed concrete from a mixing plant or depot to the site. Usual capacity range of ready mixed concrete trucks is 4 to 6 m³. Discharge can be direct into placing position via a chute or into some form of site transport such as a dumper, crane skip or concrete pump.

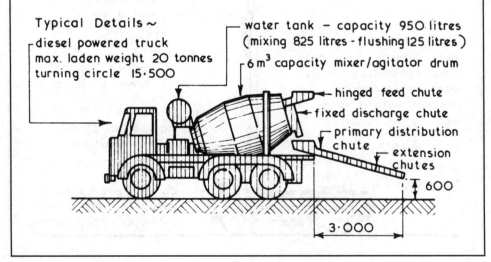

Typical Details ~
diesel powered truck
max. laden weight 20 tonnes
turning circle 15·500

water tank – capacity 950 litres (mixing 825 litres - flushing 125 litres)
6 m³ capacity mixer/agitator drum
hinged feed chute
fixed discharge chute
primary distribution chute
extension chutes
600
3·000

Concrete Pumps ~ these are used to transport large volumes of concrete in a short time period (up to 100 m³ per hour) in both the vertical and horizontal directions from the pump position to the point of placing. Concrete pumps can be trailer or lorry mounted and are usually of a twin cylinder hydraulically driven format with a small bore pipeline (100 mm diameter) with pumping ranges of up to 85·000 vertically and 200·000 horizontally depending on the pump model and the combination of vertical and horizontal distances. It generally requires about 45 minutes to set up a concrete pump on site including coating the bore of the pipeline with a cement grout prior to pumping the special concrete mix. The pump is supplied with pumpable concrete by means of a constant flow of ready mixed concrete lorries throughout the pumping period after which the pipeline is cleared and cleaned. Usually a concrete pump and its operator(s) are hired for the period required.

Typical Concrete Pump Details ~

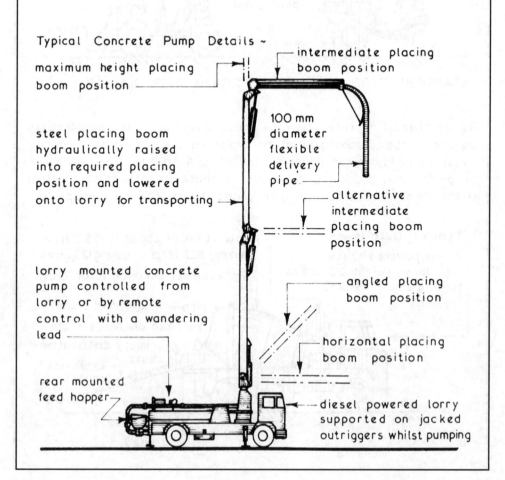

maximum height placing boom position

intermediate placing boom position

steel placing boom hydraulically raised into required placing position and lowered onto lorry for transporting

100 mm diameter flexible delivery pipe

alternative intermediate placing boom position

lorry mounted concrete pump controlled from lorry or by remote control with a wandering lead

angled placing boom position

horizontal placing boom position

rear mounted feed hopper

diesel powered lorry supported on jacked outriggers whilst pumping

Placing Concrete ~ this activity is usually carried out by hand with the objectives of filling the mould, formwork or excavated area to the correct depth, working the concrete around any inserts or reinforcement and finally compacting the concrete to the required consolidation. The compaction of concrete can be carried out using simple tamping rods or boards or alternatively it can be carried out with the aid of plant such as vibrators.

Poker Vibrators ~ these consist of a hollow steel tube casing in which is a rotating impeller which generates vibrations as its head comes into contact with the casing ~

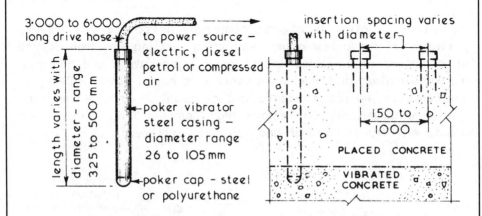

Poker vibrators should be inserted vertically and allowed to penetrate 75 mm into any previously vibrated concrete.

Clamp or Tamping Board Vibrators ~ clamp vibrators are powered either by compressed air or electricity whereas tamping board vibrators are usually petrol driven ~

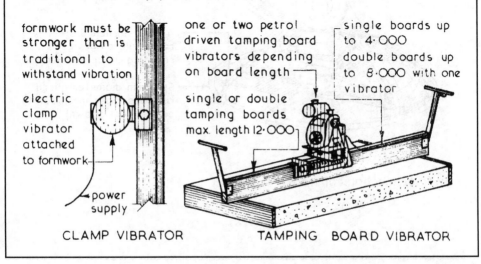

CLAMP VIBRATOR   TAMPING BOARD VIBRATOR

203

Power Float – a hand-operated electric motor or petrol engine, surmounted over a mechanical surface skimmer. Machines are provided with an interchangeable revolving disc and a set of blades. These are used in combination to produce a smooth, dense and level surface finish to in-situ concrete beds.

The advantages offset against the cost of plant hire are:

* Eliminates the time and materials needed to apply a finishing screed.
* A quicker process and less labour-intensive than hand troweling.

Application – after transverse tamping, the concrete is left to partially set for a few hours. Amount of setting time will depend on a number of variables, including air temperature and humidity, mix specification and machine weight. As a rough guide, walking on the concrete will leave indentations of about 3–4 mm. A surfacing disc is used initially to remove high tamping lines, before two passes with blades to finish and polish the surface.

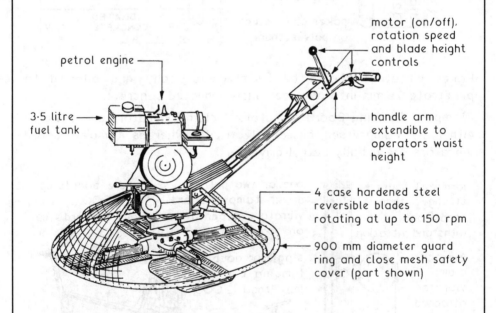

petrol engine

3·5 litre fuel tank

motor (on/off), rotation speed and blade height controls

handle arm extendible to operators waist height

4 case hardened steel reversible blades rotating at up to 150 rpm

900 mm diameter guard ring and close mesh safety cover (part shown)

Power or mechanical float

4 SUBSTRUCTURE

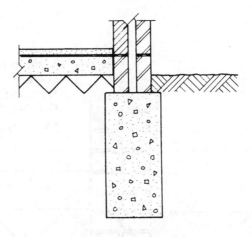

FOUNDATIONS – FUNCTION, MATERIALS AND SIZING
FOUNDATION BEDS
SHORT BORED PILE FOUNDATIONS
FOUNDATION TYPES AND SELECTION
PILED FOUNDATIONS
RETAINING WALLS
GABIONS AND MATTRESSES
BASEMENT CONSTRUCTION
WATERPROOFING BASEMENTS
EXCAVATIONS
CONCRETE PRODUCTION
COFFERDAMS
CAISSONS
UNDERPINNING
GROUND WATER CONTROL
SOIL STABILISATION AND IMPROVEMENT
RECLAMATION OF WASTE LAND
CONTAMINATED SUBSOIL TREATMENT

Foundations ~ the function of any foundation is to safely sustain and transmit to the ground on which it rests the combined dead, imposed and wind loads in such a manner as not to cause any settlement or other movement which would impair the stability or cause damage to any part of the building.

Example ~

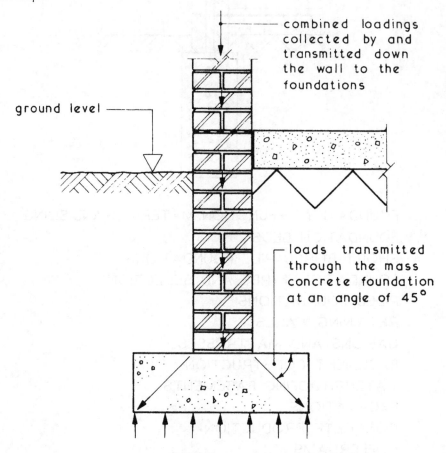

combined loadings collected by and transmitted down the wall to the foundations

ground level

loads transmitted through the mass concrete foundation at an angle of 45°

Subsoil beneath foundation is compressed and reacts by exerting an upward pressure to resist foundation loading. If foundation load exceeds maximum passive pressure of ground (i.e. bearing capacity) a downward movement of the foundation could occur. Remedy is to increase plan size of foundation to reduce the load per unit area or alternatively reduce the loadings being carried by the foundations.

Subsoil Movements ~ these are due primarily to changes in volume when the subsoil becomes wet or dry and occurs near the upper surface of the soil. Compact granular soils such as gravel suffer very little movement whereas cohesive soils such as clay do suffer volume changes near the upper surface. Similar volume changes can occur due to water held in the subsoil freezing and expanding – this is called Frost Heave.

Typical Examples ~

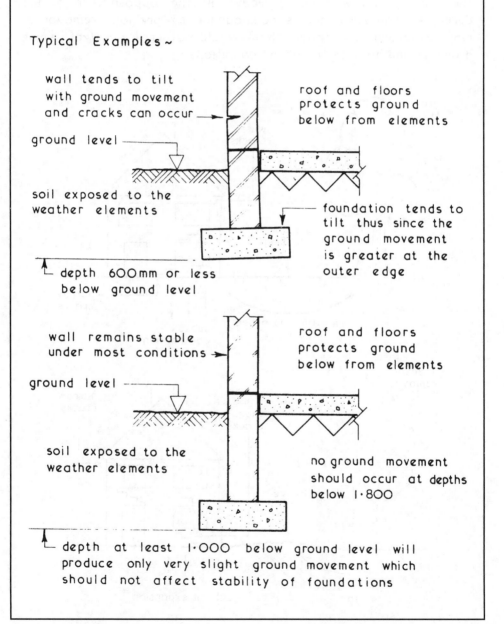

wall tends to tilt with ground movement and cracks can occur →

ground level

soil exposed to the weather elements

depth 600mm or less below ground level

roof and floors protects ground below from elements

foundation tends to tilt thus since the ground movement is greater at the outer edge

wall remains stable under most conditions →

ground level

soil exposed to the weather elements

depth at least 1·000 below ground level will produce only very slight ground movement which should not affect stability of foundations

roof and floors protects ground below from elements

no ground movement should occur at depths below 1·800

Trees ~ damage to foundations. Substructural damage to buildings can occur with direct physical contact by tree roots. More common is the indirect effect of moisture shrinkage or heave, particularly apparent in clay subsoils.

Shrinkage is most evident in long periods of dry weather, compounded by moisture abstraction from vegetation. Notably broad leaved trees such as oak, elm and poplar in addition to the thirsty willow species. Heave is the opposite. It occurs during wet weather and is compounded by previous removal of moisture-dependent trees that would otherwise effect some drainage and balance to subsoil conditions.

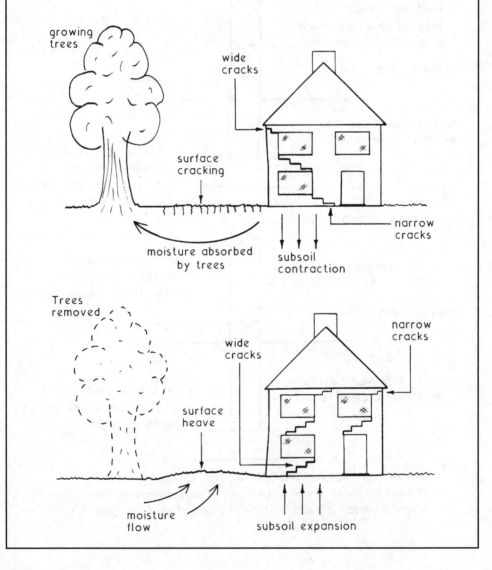

Trees ~ effect on foundations. Trees up to 30 m distance may have an effect on foundations, therefore reference to local authority building control policy should be undertaken before specifying construction techniques.

Traditional strip foundations are practically unsuited, but at excavation depths up to 2·5 or 3·0 m, deep strip or trench fill (preferably reinforced) may be appropriate. Short bored pile foundations are likely to be more economical and particularly suited to depths exceeding 3·0 m.

For guidance only, the illustration and table provide an indication of foundation depths in shrinkable subsoils.

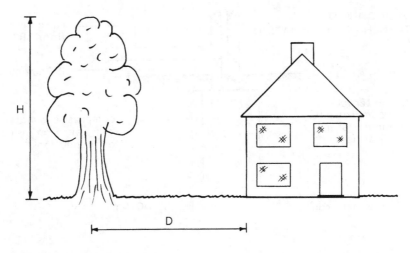

H = Mature height of tree
D = Distance to centre of tree

D/H – Distance from tree/Height of tree

Tree species	0·10	0·25	0·33	0·50	0·66	0·75	1·00
Oak, elm, poplar and willow	3·00	2·80	2·60	2·30	2·10	1·90	1·50
All others	2·80	2·40	2·10	1·80	1·50	1·20	1·00

Minimum foundation depth (m)

Trees ~ preservation orders (see page 123) may be waived by the local planning authority. Permission for tree felling is by formal application and will be considered if the proposed development is in the economic and business interests of the community. However, tree removal is only likely to be acceptable if there is an agreement for replacement stock being provided elsewhere on the site.

In these circumstances, there is potential for ground heave within the 'footprint' of felled trees. To resist this movement, foundations must incorporate an absorbing layer or compressible filler with ground floor suspended above the soil.

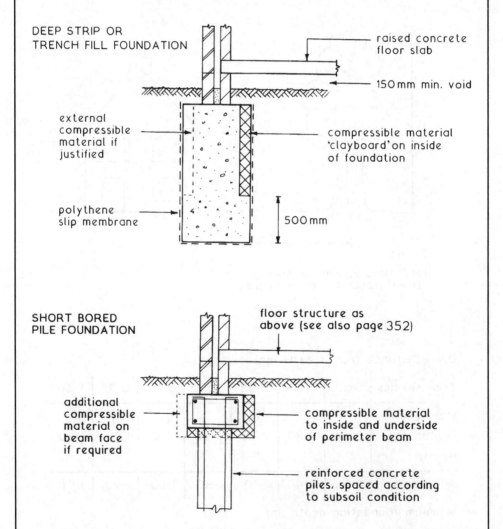

DEEP STRIP OR
TRENCH FILL FOUNDATION

raised concrete floor slab

150 mm min. void

external compressible material if justified

compressible material 'clayboard' on inside of foundation

polythene slip membrane

500 mm

SHORT BORED PILE FOUNDATION

floor structure as above (see also page 352)

additional compressible material on beam face if required

compressible material to inside and underside of perimeter beam

reinforced concrete piles, spaced according to subsoil condition

Cracking in Walls ~ cracks are caused by applied forces which exceed those that the building can withstand. Most cracking is superficial, occurring as materials dry out and subsequently shrink to reveal minor surface fractures of < 2 mm. These insignificant cracks can be made good with proprietary fillers.

Severe cracking in walls may result from foundation failure, due to inadequate design or physical damage. Further problems could include:

* Structural instability
* Air infiltration
* Sound insulation reduction
* Rain penetration
* Heat loss
* Visual depreciation

A survey should be undertaken to determine:
1. The cause of cracking, i.e.
    * Loads applied externally (tree roots, subsoil movement).
    * Climate/temperature changes (thermal movement).
    * Moisture content change (faulty dpc, building leakage).
    * Vibration (adjacent work, traffic).
    * Changes in physical composition (salt or ice formation).
    * Chemical change (corrosion, sulphate attack).
    * Biological change (timber decay).
2. The effect on a building's performance (structural and environmental).
3. The nature of movement - completed, ongoing or intermittent (seasonal).

Observations over a period of several months, preferably over a full year, will determine whether the cracking is new or established and whether it is progressing.

Simple method for monitoring cracks -

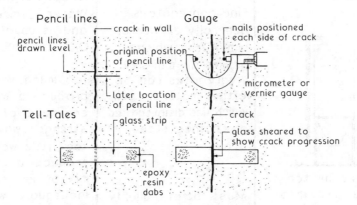

Further reading - BRE Digest 251: Assessment of damage in low rise buildings.

Foundation Materials ~ from page 190 one of the functions of a foundation can be seen to be the ability to spread its load evenly over the ground on which it rests. It must of course be constructed of a durable material of adequate strength. Experience has shown that the most suitable material is concrete.

Concrete is a mixture of cement + aggregates + water in controlled proportions.

CEMENT

Manufactured from clay and chalk and is the matrix or binder of the concrete mix. Cement powder can be supplied in bags or bulk —

Bags ~

25 kg.

air-tight sealed bags requiring a dry damp free store.

Bulk ~

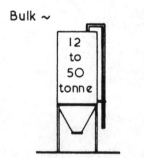

12 to 50 tonne

delivered by tanker and pumped into storage silo.

AGGREGATES

Coarse aggregate is generally defined as a material which is retained on a 4mm sieve.

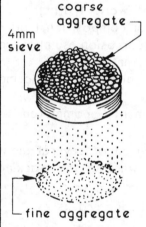

coarse aggregate

4mm sieve

fine aggregate

Fine aggregate is generally defined as a material which passes a 4mm sieve. Aggregates can be either natural rock which has disintegrated or crushed stone or gravel.

WATER

Must be of a quality fit for drinking.

MIXES

These are expressed as a ratio thus:~
1:3:6/20mm
which means —
1 part cement.
3 parts of fine aggregate.
6 parts of coarse aggregate
20mm — maximum size of coarse aggregate for the mix.

Water is added to start the chemical reaction and to give the mix workability ~ the amount used is called the — Water/Cement Ratio and is usually about 0·4 to 0·5.

Too much water will produce a weak concrete of low strength whereas too little water will produce a concrete mix of low and inadequate work-ability.

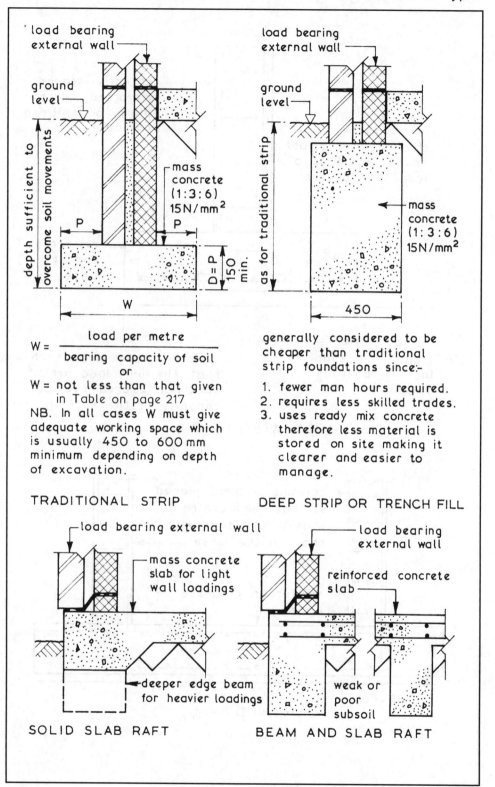

W = $\dfrac{\text{load per metre}}{\text{bearing capacity of soil}}$

or

W = not less than that given in Table on page 217

NB. In all cases W must give adequate working space which is usually 450 to 600 mm minimum depending on depth of excavation.

TRADITIONAL STRIP

generally considered to be cheaper than traditional strip foundations since:-

1. fewer man hours required.
2. requires less skilled trades.
3. uses ready mix concrete therefore less material is stored on site making it clearer and easier to manage.

DEEP STRIP OR TRENCH FILL

SOLID SLAB RAFT

BEAM AND SLAB RAFT

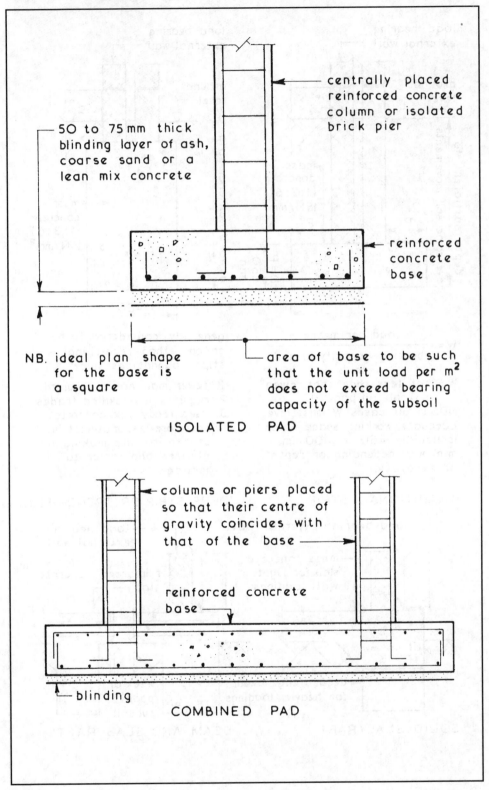

50 to 75mm thick blinding layer of ash, coarse sand or a lean mix concrete

centrally placed reinforced concrete column or isolated brick pier

reinforced concrete base

NB. ideal plan shape for the base is a square

area of base to be such that the unit load per m² does not exceed bearing capacity of the subsoil

ISOLATED PAD

columns or piers placed so that their centre of gravity coincides with that of the base

reinforced concrete base

blinding

COMBINED PAD

Bed ~ a concrete slab resting on and supported by the subsoil, usually forming the ground floor surface. Beds (sometimes called oversite concrete) are usually cast on a layer of hardcore which is used to make up the reduced level excavation and thus raise the level of the concrete bed to a position above ground level.

Typical Example ~

mass concrete bed (1:3:6/20mm mix 15N/mm^2). Thickness for domestic work is usually 100 to 150mm and the bed is constructed so as to prevent the passage of moisture from the ground to the upper surface of the floor – this is usually achieved by incorporating into the design a damp-proof membrane ~ for details see page 648

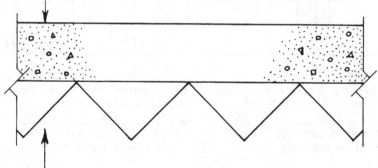

100 to 150 mm thick layer of hardcore ~ material used should be inert and not affected by water. Suitable materials are gravel; crushed rock; quarry waste; concrete rubble; brick or tile rubble; blast furnace slag and pulverised fuel ash (fly ash). The hardcore material should be laid evenly and well compacted with the upper surface blinded with fine grade material such as sand. Sand blinding fills the gaps in the hardcore to prevent concrete wastage and to provide a relatively smooth and level surface for a 0.3 mm LDPE (1200 gauge polythene) dpm where required.

Basic Sizing ~ the size of a foundation is basically dependent on two factors –

1. Load being transmitted, max 70 kN/m (dwellings up to 3 storeys).
2. Bearing capacity of subsoil under proposed foundation.

Bearing capacities for different types of subsoils may be obtained from tables such as those in BS 8004: Code of practice for foundations and BS 8103-1: Structural design of low rise buildings. Also, directly from soil investigation results.

Typical Examples ~

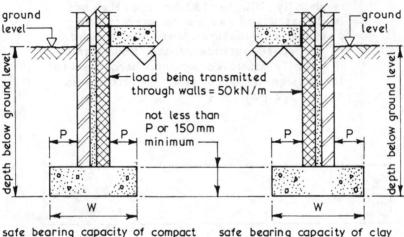

safe bearing capacity of compact gravel subsoil = 100 kN/m²

$$W = \frac{load}{bearing\ capacity} = \frac{50}{100}$$

= 500 mm minimum

safe bearing capacity of clay subsoil = 80 kN/m²

$$W = \frac{load}{bearing\ capacity} = \frac{50}{80}$$

= 625 mm minimum

The above widths may not provide adequate working space within the excavation and can be increased to give required space. Guidance on the minimum width for a limited range of applications can be taken from the table on the next page.

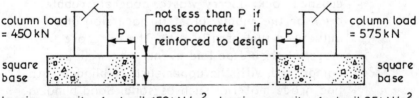

bearing capacity of subsoil 150 kN/m²

$$area\ of\ base = \frac{load}{bc} = \frac{450}{150}$$

= 3 m² ∴ side = $\sqrt{3}$

= 1·732 min.

bearing capacity of subsoil 85 kN/m²

$$area\ of\ base = \frac{load}{bc} = \frac{575}{85}$$

= 6·765 m² ∴ side = $\sqrt{6·765}$

= 2·6 min.

Ground type	Ground condition	Field test	Max. total load on load bearing wall (kN/m)					
			20	30	40	50	60	70
			Minimum width (mm)					
Rock	Not inferior to sandstone, limestone or firm chalk.	Requires a mechanical device to excavate.	At least equal to the width of the wall					
Gravel Sand	Medium density Compact	Pick required to excavate. 50 mm square peg hard to drive beyond 150 mm.	250	300	400	500	600	650
Clay Sandy clay	Stiff Stiff	Requires pick or mechanical device to aid removal. Can be indented slightly with thumb.	250	300	400	500	600	650
Clay Sandy clay	Firm Firm	Can be moulded under substantial pressure by fingers.	300	350	450	600	750	850
Sand Silty sand Clayey sand	Loose Loose Loose	Can be excavated by spade. 50 mm square peg easily driven.	400	600	Conventional strip foundations unsuitable for a total load exceeding 30 kN/m.			
Silt Clay Sandy clay Silty clay	Soft Soft Soft Soft	Finger pushed in up to 10 mm. Easily moulded with fingers.	450	650				
Silt Clay Sandy clay Silty clay	Very soft Very soft Very soft Very soft	Finger easily pushed in up to 25 mm. Wet sample exudes between fingers when squeezed.	Conventional strip inappropriate. Steel reinforced wide strip, deep strip or piled foundation selected subject to specialist advice.					

Adapted from Table 10 in the Bldg. Regs., A.D: A – Structure.

Typical procedure (for guidance only) –

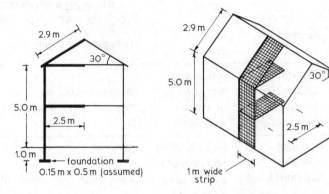

Dead load per m run (see pages 35 and 36)

Substructure brickwork, 1 m × 1 m × 476 kg/m²	=	476 kg
.... cavity conc. (50 mm), 1 m × 1 m × 2300 kg/m³	=	115 kg
Foundation concrete, 0·15 m × 1 m × 0·5 m × 2300 kg/m³	=	173 kg
Superstructure brickwork, 5 m × 1 m × 221 kg/m²	=	1105 kg
.... blockwork & ins., 5 m × 1 m × 79 kg/m²	=	395 kg
.... 2 coat plasterwork, 5 m × 1 m × 22 kg/m²	=	110 kg
Floor joists/boards/plstrbrd., 2·5 m × 1 m × 42·75 kg/m²	=	107 kg
Ceiling joists/plstrbrd/ins., 2·5 m × 1 m × 19·87 kg/m²	=	50 kg
Rafters, battens & felt, 2·9 m × 1 m × 12·10 kg/m²	=	35 kg
Single lap tiling, 2·9 m × 1 m × 49 kg/m²	=	142 kg
		2708 kg

Note: kg × 9·81 = Newtons

Therefore: 2708 kg × 9·81 = 26565 N or 26·56 kN

Imposed load per m run (see BS 6399-1: Code of practice for dead and imposed loads) –

Floor, 2·5 m × 1 m × 1·5 kN/m²   = 3·75 kN
Roof, 2·9 m × 1 m × 1·5 kN/m² (snow) = <u>4·05 kN</u>
                                     7·80 kN

Note: For roof pitch >30°, snow load = 0·75 kN/m²

Dead + imposed load is, 26·56 kN + 7·80 kN = 34·36 kN

Given that the subsoil has a safe bearing capacity of 75 kN/m²,

W = load ÷ bearing capacity = 34·36 ÷ 75 = 0·458 m or 458 mm

Therefore a foundation width of 500 mm is adequate.

Note: This example assumes the site is sheltered. If it is necessary to make allowance for wind loading, reference should be made to BS 6399-2: Code of practice for wind loads.

Stepped Foundations ~ these are usually considered in the context of strip foundations and are used mainly on sloping sites to reduce the amount of excavation and materials required to produce an adequate foundation.

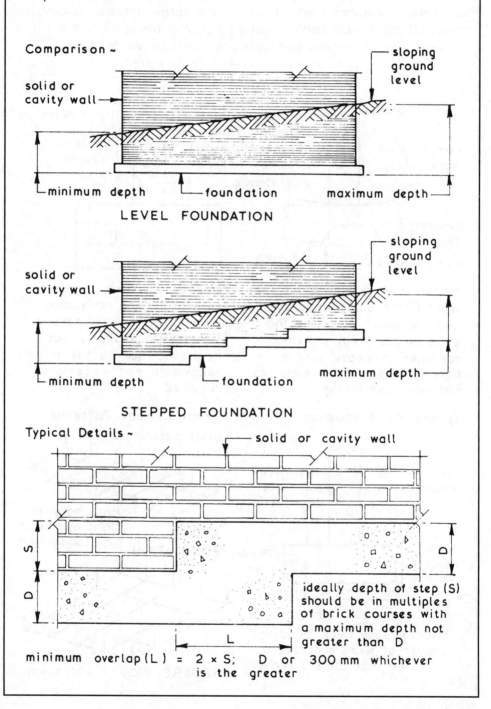

Comparison ~

solid or cavity wall →

sloping ground level

minimum depth — foundation — maximum depth

LEVEL FOUNDATION

solid or cavity wall →

sloping ground level

minimum depth — foundation — maximum depth

STEPPED FOUNDATION

Typical Details ~

solid or cavity wall

ideally depth of step (S) should be in multiples of brick courses with a maximum depth not greater than D

minimum overlap (L) = 2 × S; D or 300 mm whichever is the greater

Concrete Foundations ~ concrete is a material which is strong in compression but weak in tension. If its tensile strength is exceeded cracks will occur resulting in a weak and unsuitable foundation. One method of providing tensile resistance is to include in the concrete foundation bars of steel as a form of reinforcement to resist all the tensile forces induced into the foundation. Steel is a material which is readily available and has high tensile strength.

Comparisons ~

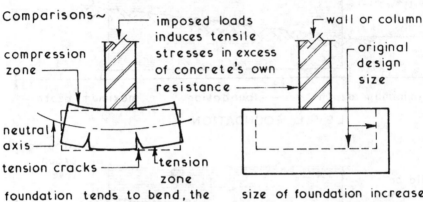

foundation tends to bend, the upper fibres being compressed and the lower fibres being stretched and put in tension- remedies increase size of base or design as a reinforced concrete foundation

size of foundation increased to provide the resistance against the induced tensile stresses - generally not economic due to the extra excavation and materials required

Typical RC Foundation

Reinforcement Patterns

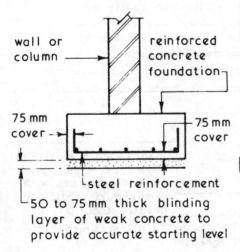

Short Bored Piles ~ these are a form of foundation which are suitable for domestic loadings and clay subsoils where ground movements can occur below the 1·000 depth associated with traditional strip and trench fill foundations. They can be used where trees are planted close to a new building since the trees may eventually cause damaging ground movements due to extracting water from the subsoil and root growth. Conversely where trees have been removed this may lead to ground swelling.

Typical Details ~

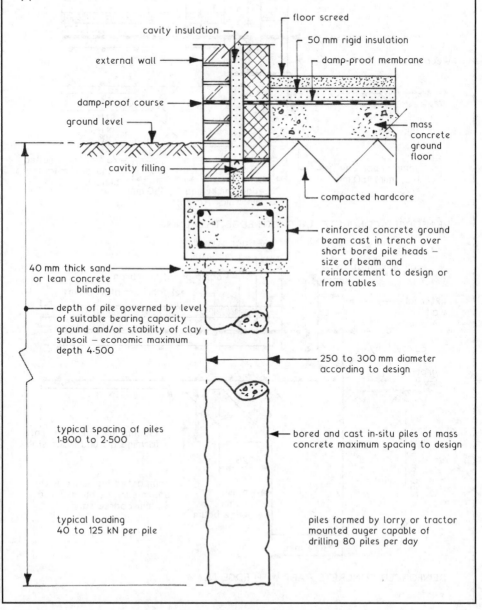

cavity insulation

floor screed

50 mm rigid insulation

external wall

damp-proof membrane

damp-proof course

ground level

mass concrete ground floor

cavity filling

compacted hardcore

reinforced concrete ground beam cast in trench over short bored pile heads — size of beam and reinforcement to design or from tables

40 mm thick sand or lean concrete blinding

depth of pile governed by level of suitable bearing capacity ground and/or stability of clay subsoil — economic maximum depth 4·500

250 to 300 mm diameter according to design

typical spacing of piles 1·800 to 2·500

bored and cast in-situ piles of mass concrete maximum spacing to design

typical loading 40 to 125 kN per pile

piles formed by lorry or tractor mounted auger capable of drilling 80 piles per day

Simple RC Raft Foundations

Simple Raft Foundations ~ these can be used for lightly loaded buildings on poor soils or where the top 450 to 600 mm of soil is overlaying a poor quality substrata.

Typical Details ~

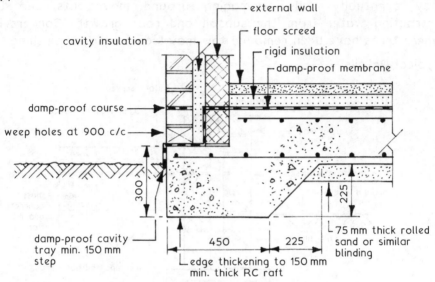

REINFORCED CONCRETE RAFT WITH EDGE THICKENING

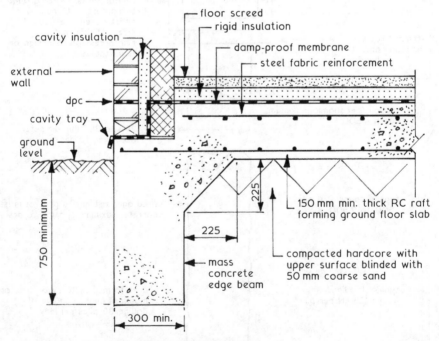

REINFORCED CONCRETE RAFT WITH EDGE BEAM

Foundation Design Principles ~ the main objectives of foundation design are to ensure that the structural loads are transmitted to the subsoil(s) safely, economically and without any unacceptable movement during the construction period and throughout the anticipated life of the building or structure.

Basic Design Procedure ~ this can be considered as a series of steps or stages −

1. Assessment of site conditions in the context of the site and soil investigation report.

2. Calculation of anticipated structural loading(s).

3. Choosing the foundation type taking into consideration −
    a. Soil conditions;
    b. Type of structure;
    c. Structural loading(s);
    d. Economic factors;
    e. Time factors relative to the proposed contract period;
    f. Construction problems.

4. Sizing the chosen foundation in the context of loading(s), ground bearing capacity and any likely future movements of the building or structure.

Foundation Types ~ apart from simple domestic foundations most foundation types are constructed in reinforced concrete and may be considered as being shallow or deep. Most shallow types of foundation are constructed within 2·000 of the ground level but in some circumstances it may be necessary to take the whole or part of the foundations down to a depth of 2·000 to 5·000 as in the case of a deep basement where the structural elements of the basement are to carry the superstructure loads. Generally foundations which need to be taken below 5·000 deep are cheaper when designed and constructed as piled foundations and such foundations are classified as deep foundations. (For piled foundation details see pages 228 to 247.)

Foundations are usually classified by their type such as strips, pads, rafts and piles. It is also possible to combine foundation types such as strip foundations connected by beams to and working in conjunction with pad foundations.

Strip Foundations ~ these are suitable for most subsoils and light structural loadings such as those encountered in low to medium rise domestic dwellings where mass concrete can be used. Reinforced concrete is usually required for all other situations.

Typical Strip Foundation Types ~

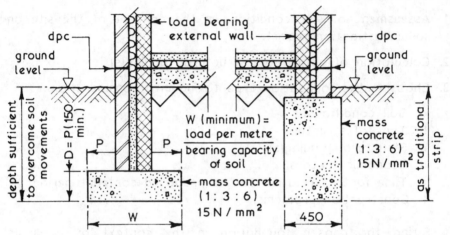

$$W \text{ (minimum)} = \frac{\text{load per metre}}{\text{bearing capacity of soil}}$$

TRADITIONAL STRIP
low rise domestic dwellings or similar buildings

DEEP STRIP or TRENCH FILL
alternative to traditional strip

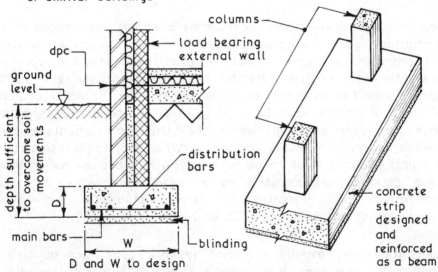

REINFORCED CONCRETE STRIP
used where induced tension exceeds concrete's own tensile resistance

CONTINUOUS COLUMN
used for closely spaced or close to boundary columns

Pad Foundations ~ suitable for most subsoils except loose sands, loose gravels and filled areas. Pad foundations are usually constructed of reinforced concrete and where possible are square in plan.

Typical Pad Foundation Types ~

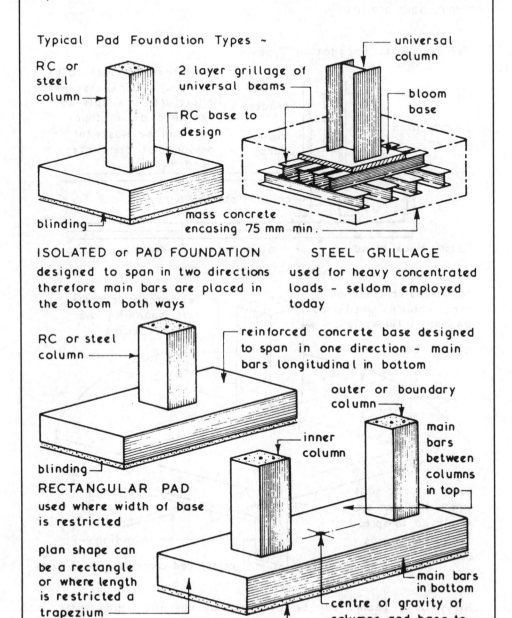

RC or steel column

2 layer grillage of universal beams

RC base to design

universal column

bloom base

blinding

mass concrete encasing 75 mm min.

ISOLATED or PAD FOUNDATION
designed to span in two directions therefore main bars are placed in the bottom both ways

STEEL GRILLAGE
used for heavy concentrated loads - seldom employed today

RC or steel column

reinforced concrete base designed to span in one direction - main bars longitudinal in bottom

outer or boundary column

main bars between columns in top

inner column

blinding

RECTANGULAR PAD
used where width of base is restricted

plan shape can be a rectangle or where length is restricted a trapezium

main bars in bottom

centre of gravity of columns and base to coincide

blinding

COMBINED COLUMN FOUNDATIONS - outer column close to boundary or existing wall

Raft Foundations ~ these are used to spread the load of the superstructure over a large base to reduce the load per unit area being imposed on the ground and this is particularly useful where low bearing capacity soils are encountered and where individual column loads are heavy.

Typical Raft Foundation Types ~

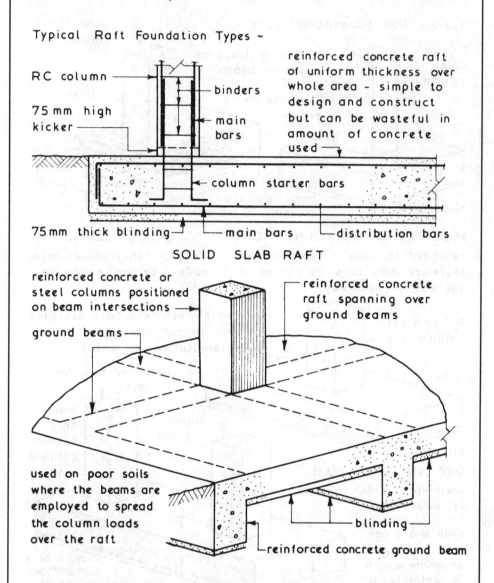

reinforced concrete raft of uniform thickness over whole area - simple to design and construct but can be wasteful in amount of concrete used

RC column

binders

75 mm high kicker

main bars

column starter bars

75mm thick blinding

main bars

distribution bars

SOLID SLAB RAFT

reinforced concrete or steel columns positioned on beam intersections

reinforced concrete raft spanning over ground beams

ground beams

used on poor soils where the beams are employed to spread the column loads over the raft

blinding

reinforced concrete ground beam

NB. Ground beams can be designed as upstand beams with a precast concrete suspended floor at ground level thus creating a void space between raft and ground floor.

BEAM AND SLAB RAFT

Cantilever Foundations ~ these can be used where it is necessary to avoid imposing any pressure on an adjacent foundation or underground service.

Typical Cantilever Foundation Types ~

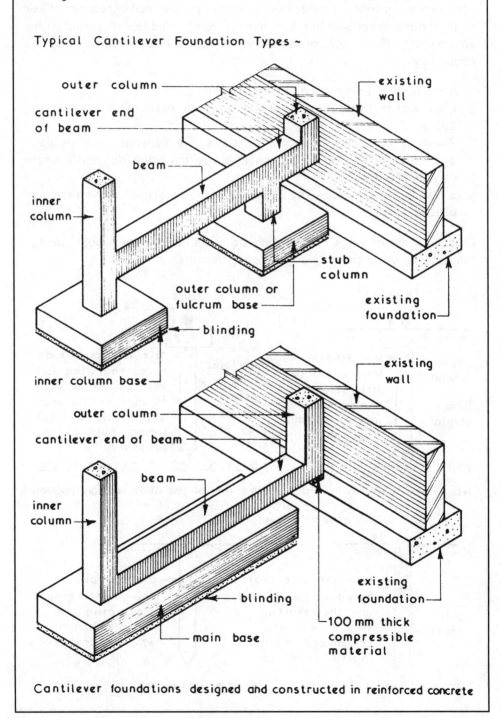

outer column

cantilever end of beam

beam

inner column

existing wall

stub column

outer column or fulcrum base

existing foundation

blinding

inner column base

outer column

cantilever end of beam

beam

inner column

existing wall

existing foundation

100 mm thick compressible material

blinding

main base

Cantilever foundations designed and constructed in reinforced concrete

Piled Foundations ~ these can be defined as a series of columns constructed or inserted into the ground to transmit the load(s) of a structure to a lower level of subsoil. Piled foundations can be used when suitable foundation conditions are not present at or near ground level making the use of deep traditional foundations uneconomic. The lack of suitable foundation conditions may be caused by:-

1. Natural low bearing capacity of subsoil.
2. High water table – giving rise to high permanent dewatering costs.
3. Presence of layers of highly compressible subsoils such as peat and recently placed filling materials which have not sufficiently consolidated.
4. Subsoils which may be subject to moisture movement or plastic failure.

Classification of Piles ~ piles may be classified by their basic design function or by their method of construction:-

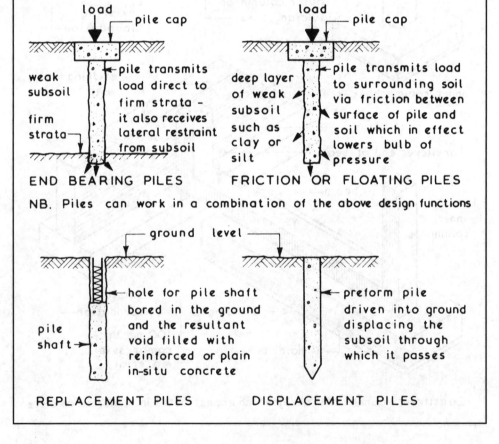

load — pile cap

weak subsoil

firm strata

pile transmits load direct to firm strata – it also receives lateral restraint from subsoil

END BEARING PILES

load — pile cap

deep layer of weak subsoil such as clay or silt

pile transmits load to surrounding soil via friction between surface of pile and soil which in effect lowers bulb of pressure

FRICTION OR FLOATING PILES

NB. Piles can work in a combination of the above design functions

ground level

pile shaft

hole for pile shaft bored in the ground and the resultant void filled with reinforced or plain in-situ concrete

REPLACEMENT PILES

preform pile driven into ground displacing the subsoil through which it passes

DISPLACEMENT PILES

Replacement Piles ~ these are often called bored piles since the removal of the spoil to form the hole for the pile is always carried out by a boring technique. They are used primarily in cohesive subsoils for the formation of friction piles and when forming pile foundations close to existing buildings where the allowable amount of noise and/or vibration is limited.

Replacement Pile Types ~

PERCUSSION BORED

small or medium size contracts with up to 300 piles

load range - 300 to 1300 kN

length range - up to 24·000

diameter range - 300 to 900

may have to be formed as a pressure pile in waterlogged subsoils - see page 230

FLUSH BORED

large projects - these are basically a rotary bored pile using bentonite as a drilling fluid

load range - 1000 to 5000 kN

length range - up to 30·000

diameter range - 600 to 1500

see page 231

ROTARY BORED

Small Diameter - <600 mm

light loadings - can also be used in groups or clusters with a common pile cap to receive heavy loads

load range - 50 to 400 kN

length range - up to 15·000

diameter range - 240 to 600

see pages 232 and 234

Large Diameter - >600 mm

heavy concentrated loadings - may have an underreamed or belled toe

load range - 800 to 15 000 kN

length range - up to 60·000

diameter range - 600 to 2400

see page 233

NB. The above given data depicts typical economic ranges. More than one pile type can be used on a single contract.

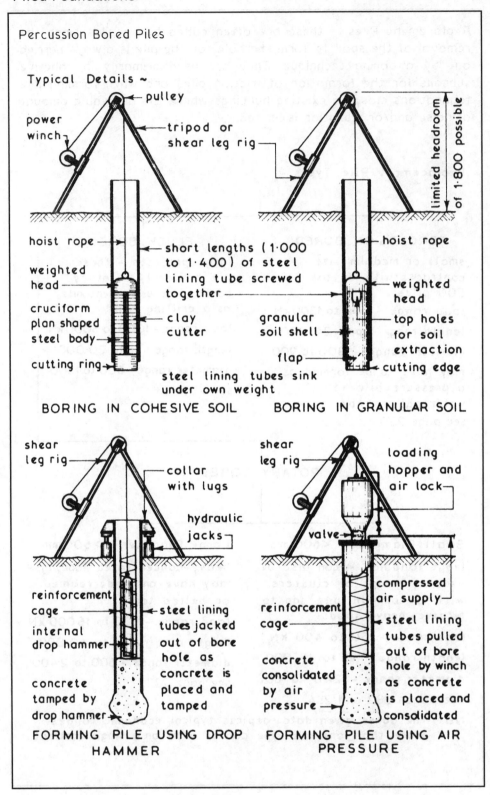

Percussion Bored Piles

Typical Details ~

pulley

power winch

tripod or shear leg rig

limited headroom of 1·800 possible

hoist rope

weighted head

cruciform plan shaped steel body

cutting ring

short lengths (1·000 to 1·400) of steel lining tube screwed together

clay cutter

steel lining tubes sink under own weight

BORING IN COHESIVE SOIL

hoist rope

weighted head

granular soil shell

flap

top holes for soil extraction

cutting edge

BORING IN GRANULAR SOIL

shear leg rig

collar with lugs

hydraulic jacks

reinforcement cage

internal drop hammer

concrete tamped by drop hammer

steel lining tubes jacked out of bore hole as concrete is placed and tamped

FORMING PILE USING DROP HAMMER

shear leg rig

loading hopper and air lock

valve

reinforcement cage

concrete consolidated by air pressure

compressed air supply

steel lining tubes pulled out of bore hole by winch as concrete is placed and consolidated

FORMING PILE USING AIR PRESSURE

230

Flush Bored Piles

Typical Details ~

Stage 1

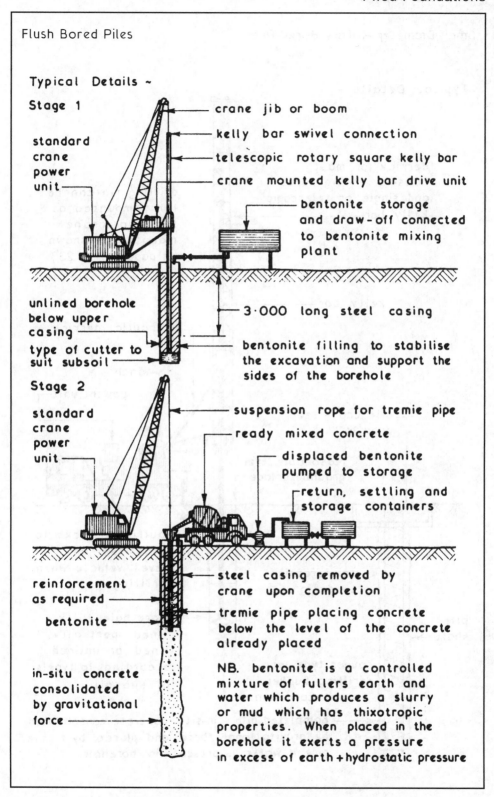

standard
crane
power
unit

crane jib or boom

kelly bar swivel connection

telescopic rotary square kelly bar

crane mounted kelly bar drive unit

bentonite storage
and draw-off connected
to bentonite mixing
plant

unlined borehole
below upper
casing

type of cutter to
suit subsoil

3·000 long steel casing

bentonite filling to stabilise
the excavation and support the
sides of the borehole

Stage 2

standard
crane
power
unit

suspension rope for tremie pipe

ready mixed concrete

displaced bentonite
pumped to storage

return, settling and
storage containers

reinforcement
as required

bentonite

in-situ concrete
consolidated
by gravitational
force

steel casing removed by
crane upon completion

tremie pipe placing concrete
below the level of the concrete
already placed

NB. bentonite is a controlled
mixture of fullers earth and
water which produces a slurry
or mud which has thixotropic
properties. When placed in the
borehole it exerts a pressure
in excess of earth + hydrostatic pressure

231

Small Diameter Rotary Bored Piles

Typical Details ~

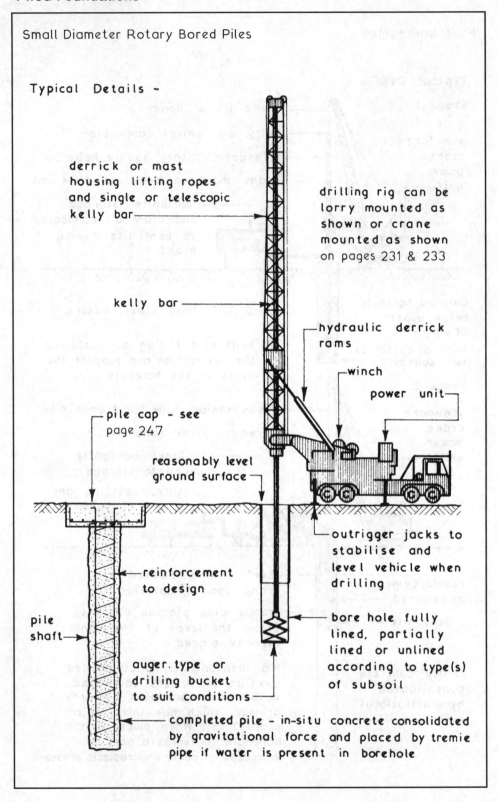

derrick or mast housing lifting ropes and single or telescopic kelly bar

drilling rig can be lorry mounted as shown or crane mounted as shown on pages 231 & 233

kelly bar

hydraulic derrick rams

winch

power unit

pile cap - see page 247

reasonably level ground surface

outrigger jacks to stabilise and level vehicle when drilling

reinforcement to design

pile shaft

bore hole fully lined, partially lined or unlined according to type(s) of subsoil

auger type or drilling bucket to suit conditions

completed pile - in-situ concrete consolidated by gravitational force and placed by tremie pipe if water is present in borehole

Large Diameter Rotary Bored Piles

Typical Details ~

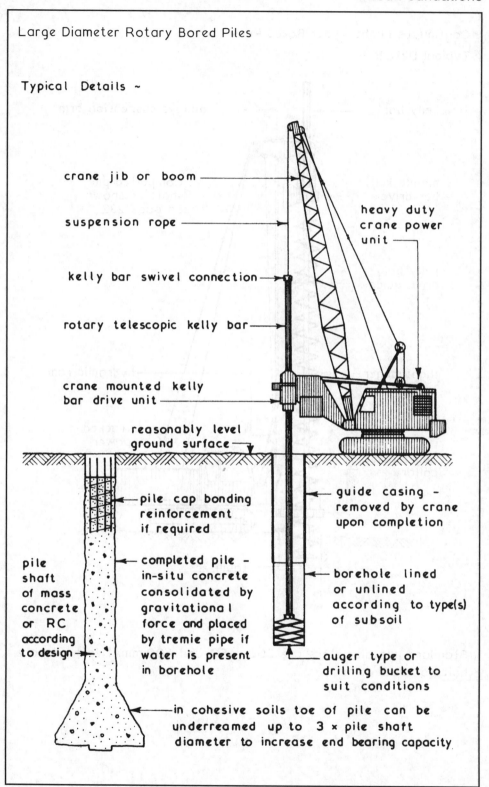

crane jib or boom

heavy duty
crane power
unit

suspension rope

kelly bar swivel connection

rotary telescopic kelly bar

crane mounted kelly
bar drive unit

reasonably level
ground surface

pile cap bonding
reinforcement
if required

guide casing –
removed by crane
upon completion

pile
shaft
of mass
concrete
or RC
according
to design

completed pile –
in-situ concrete
consolidated by
gravitational
force and placed
by tremie pipe if
water is present
in borehole

borehole lined
or unlined
according to type(s)
of subsoil

auger type or
drilling bucket to
suit conditions

in cohesive soils toe of pile can be
underreamed up to 3 × pile shaft
diameter to increase end bearing capacity.

Continuous Flight Auger Bored Piles
Typical Details ~

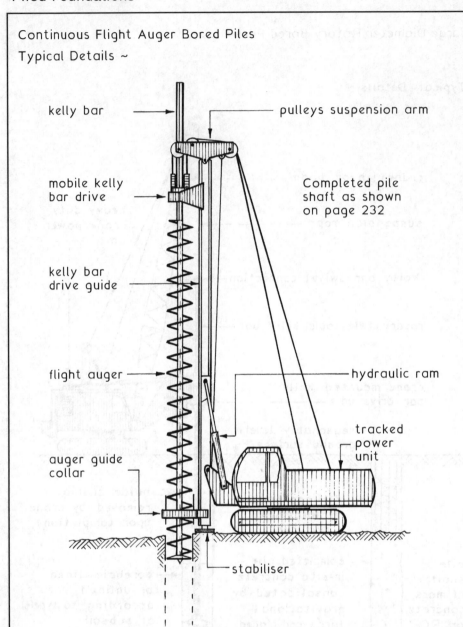

kelly bar

pulleys suspension arm

mobile kelly
bar drive

Completed pile
shaft as shown
on page 232

kelly bar
drive guide

flight auger

hydraulic ram

tracked
power
unit

auger guide
collar

stabiliser

Standard pile diameters are 300, 450 and 600 mm. Depth of bore
hole to about 15 m.

Grout Injection Piling ~

A variation of continuous flight auger bored piling that uses an open ended hollow core to the flight. After boring to the required depth, high slump concrete is pumped through the hollow stem as the auger is retracted. Spoil is displaced at the surface and removed manually. In most applications there is no need to line the boreholes, as the subsoil has little time to be disturbed. A preformed reinforcement cage is pushed into the wet concrete.

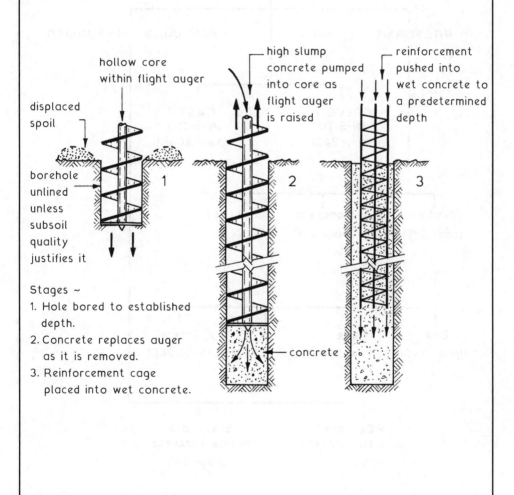

hollow core within flight auger

displaced spoil

borehole unlined unless subsoil quality justifies it

1

high slump concrete pumped into core as flight auger is raised

2

concrete

reinforcement pushed into wet concrete to a predetermined depth

3

Stages ~
1. Hole bored to established depth.
2. Concrete replaces auger as it is removed.
3. Reinforcement cage placed into wet concrete.

Displacement Piles ~ these are often called driven piles since they are usually driven into the ground displacing the earth around the pile shaft. These piles can be either preformed or partially preformed if they are not cast in-situ and are available in a wide variety of types and materials. The pile or forming tube is driven into the required position to a predetermined depth or to the required 'set' which is a measure of the subsoils resistance to the penetration of the pile and hence its bearing capacity by noting the amount of penetration obtained by a fixed number of hammer blows.

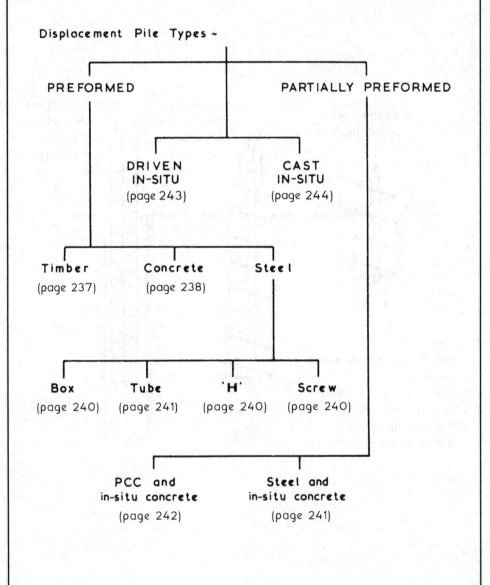

Displacement Pile Types ~

PREFORMED   PARTIALLY PREFORMED

DRIVEN   CAST
IN-SITU   IN-SITU
(page 243)   (page 244)

Timber   Concrete   Steel
(page 237)   (page 238)

Box   Tube   'H'   Screw
(page 240)   (page 241)   (page 240)   (page 240)

PCC and   Steel and
in-situ concrete   in-situ concrete
(page 242)   (page 241)

Timber Piles ~ these are usually square sawn and can be used for small contracts on sites with shallow alluvial deposits overlying a suitable bearing strata (e.g. river banks and estuaries.) Timber piles are percussion driven.

Typical Example ~

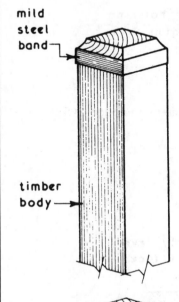

mild
steel
band

timber
body

Typical Data :-

load range - 50 to 350 kN

length range - up to 12·000
                without splicing

size range -   225 × 225
               300 × 300 *
               350 × 350 *
               400 × 400 *
               450 × 450
               600 × 600

* common sizes

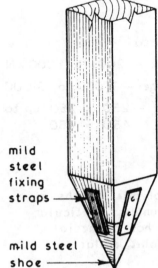

mild
steel
fixing
straps

mild steel
shoe

NB. timber piles are not easy
    to splice and are liable to
    attack by marine borers
    when set in water therefore
    such piles should always
    be treated with a suitable
    preservative before being
    driven.

Preformed Concrete Piles ~ variety of types available which are generally used on medium to large contracts of not less than one hundred piles where soft soil deposits overlie a firmer strata. These piles are percussion driven using a drop or single acting hammer.

Typical Example [West's Hardrive Precast Modular Pile] ~

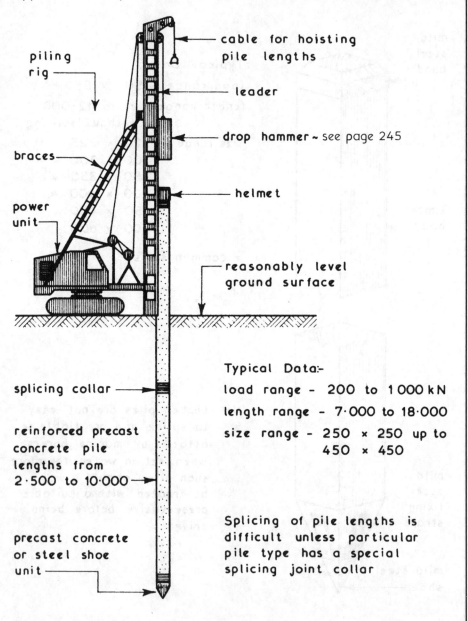

cable for hoisting pile lengths

piling rig

leader

drop hammer ~ see page 245

braces

helmet

power unit

reasonably level ground surface

splicing collar

Typical Data:-
load range - 200 to 1000 kN
length range - 7·000 to 18·000
size range - 250 × 250 up to 450 × 450

reinforced precast concrete pile lengths from 2·500 to 10·000

precast concrete or steel shoe unit

Splicing of pile lengths is difficult unless particular pile type has a special splicing joint collar

Preformed Concrete Piles – jointing with a peripheral steel splicing collar as shown on the preceding page is adequate for most concentrically or directly loaded situations. Where very long piles are to be used and/or high stresses due to compression, tension and bending from the superstructure or the ground conditions are anticipated, the 4 or 8 lock pile joint [AARSLEFF PILING] may be considered.

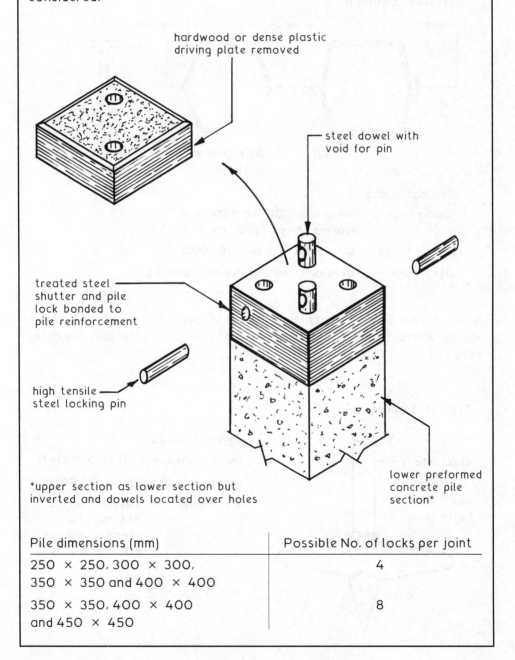

hardwood or dense plastic driving plate removed

steel dowel with void for pin

treated steel shutter and pile lock bonded to pile reinforcement

high tensile steel locking pin

*upper section as lower section but inverted and dowels located over holes

lower preformed concrete pile section*

Pile dimensions (mm)	Possible No. of locks per joint
250 × 250, 300 × 300, 350 × 350 and 400 × 400	4
350 × 350, 400 × 400 and 450 × 450	8

Steel Box and 'H' Sections ~ standard steel sheet pile sections can be used to form box section piles whereas the 'H' section piles are cut from standard rolled sections. These piles are percussion driven and are used mainly in connection with marine structures.

Typical Examples ~

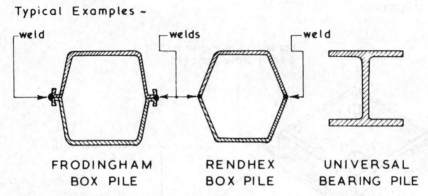

FRODINGHAM   RENDHEX   UNIVERSAL
BOX PILE   BOX PILE   BEARING PILE

Typical Data :-

load range - box piles 300 to 1 500 kN
               bearing piles 300 to 1 700 kN

length range - all types up to 36·000

size range - various sizes and profiles available

Steel Screw Piles ~ rotary driven and used for dock and jetty works where support at shallow depths in soft silts and sands is required.

Typical Example ~

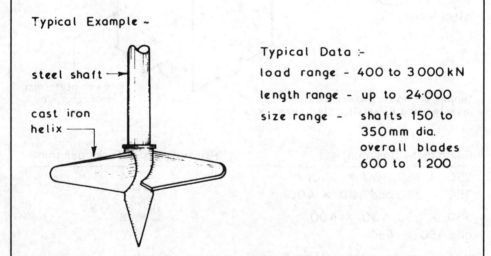

steel shaft →

cast iron
helix

Typical Data :-

load range - 400 to 3 000 kN

length range - up to 24·000

size range - shafts 150 to
               350 mm dia.
               overall blades
               600 to 1 200

Steel Tube Piles ~ used on small to medium size contracts for marine structures and foundations in soft subsoils over a suitable bearing strata. Tube piles are usually bottom driven with an internal drop hammer. The loading can be carried by the tube alone but it is usual to fill the tube with mass concrete to form a composite pile. Reinforcement, except for pile cap bonding bars, is not normally required.

Typical Example [BSP Cased Pile] ~

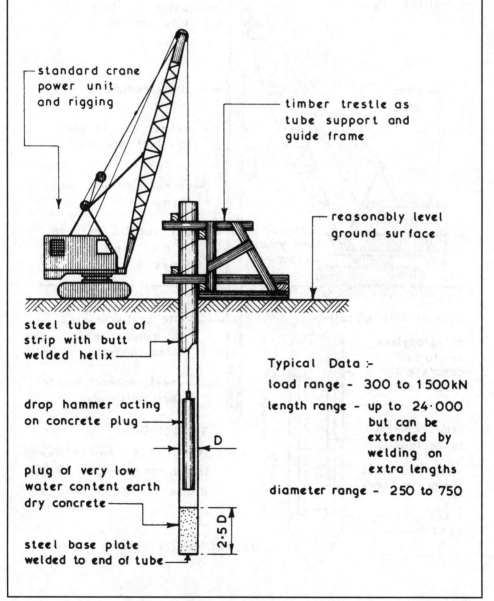

standard crane power unit and rigging

timber trestle as tube support and guide frame

reasonably level ground surface

steel tube out of strip with butt welded helix

drop hammer acting on concrete plug

D

plug of very low water content earth dry concrete

steel base plate welded to end of tube

2·5 D

Typical Data :-

load range - 300 to 1500kN

length range - up to 24·000 but can be extended by welding on extra lengths

diameter range - 250 to 750

Partially Preformed Piles ~ these are composite piles of precast concrete and in-situ concrete or steel and in-situ concrete (see page 241). These percussion driven piles are used on medium to large contracts where bored piles would not be suitable owing to running water or very loose soils.

Typical Example [West's Shell Pile] ~

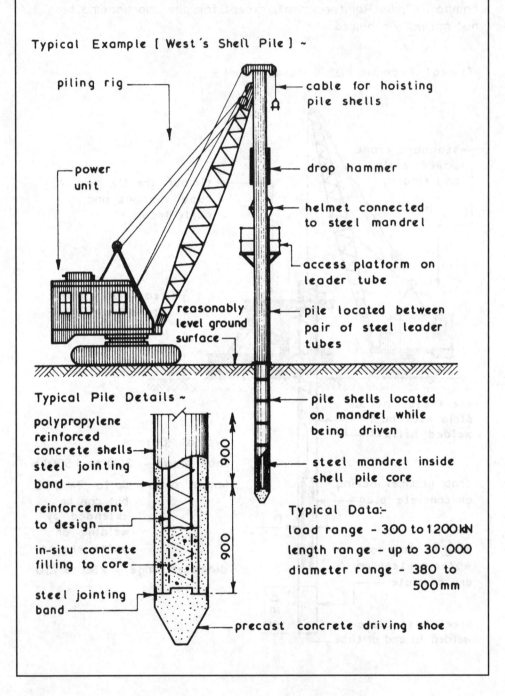

piling rig

power unit

reasonably level ground surface

cable for hoisting pile shells

drop hammer

helmet connected to steel mandrel

access platform on leader tube

pile located between pair of steel leader tubes

Typical Pile Details ~

polypropylene reinforced concrete shells

steel jointing band

reinforcement to design

in-situ concrete filling to core

steel jointing band

pile shells located on mandrel while being driven

steel mandrel inside shell pile core

900

900

Typical Data:-
load range - 300 to 1200 kN
length range - up to 30·000
diameter range - 380 to 500 mm

precast concrete driving shoe

Driven In-situ Piles ~ used on medium to large contracts as an alternative to preformed piles particularly where final length of pile is a variable to be determined on site.

Typical Example [Franki Driven Insitu Pile] ~

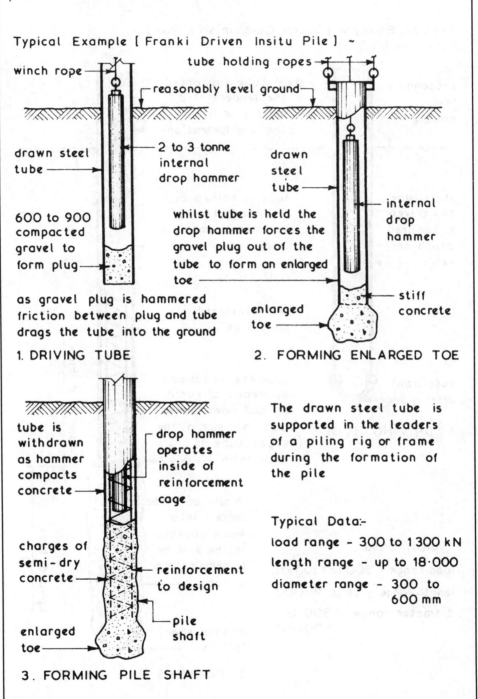

winch rope

tube holding ropes

reasonably level ground

drawn steel tube

2 to 3 tonne internal drop hammer

drawn steel tube

internal drop hammer

600 to 900 compacted gravel to form plug

whilst tube is held the drop hammer forces the gravel plug out of the tube to form an enlarged toe

stiff concrete

enlarged toe

as gravel plug is hammered friction between plug and tube drags the tube into the ground

1. DRIVING TUBE

2. FORMING ENLARGED TOE

tube is withdrawn as hammer compacts concrete

drop hammer operates inside of reinforcement cage

The drawn steel tube is supported in the leaders of a piling rig or frame during the formation of the pile

charges of semi-dry concrete

reinforcement to design

Typical Data:-
load range - 300 to 1 300 kN
length range - up to 18·000
diameter range - 300 to 600 mm

enlarged toe

pile shaft

3. FORMING PILE SHAFT

Cast In-situ Piles ~ an alternative to the driven in-situ piles (see page 243)

Typical Example [Vibro Cast In-situ Pile] ~

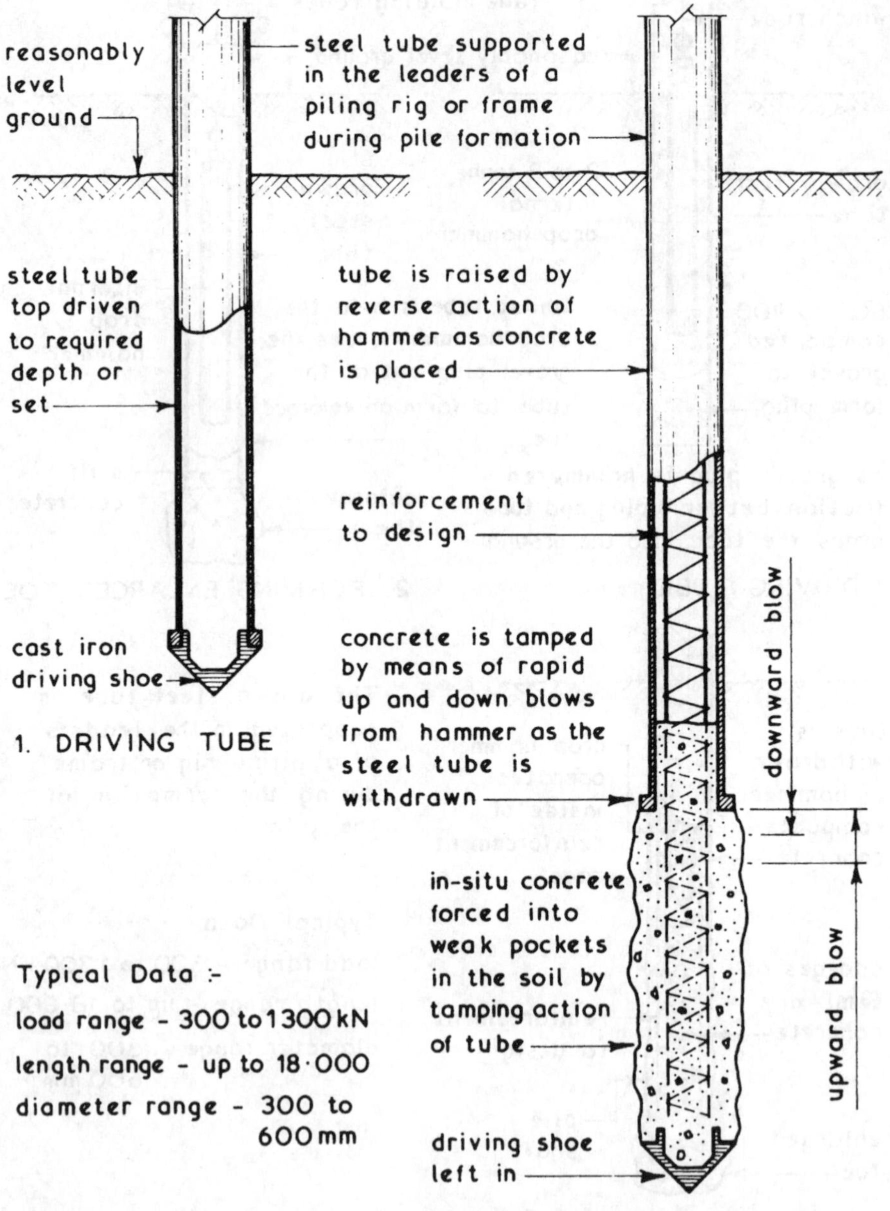

reasonably
level
ground

steel tube supported
in the leaders of a
piling rig or frame
during pile formation

steel tube
top driven
to required
depth or
set

tube is raised by
reverse action of
hammer as concrete
is placed

reinforcement
to design

cast iron
driving shoe

concrete is tamped
by means of rapid
up and down blows
from hammer as the
steel tube is
withdrawn

downward blow

1. DRIVING TUBE

in-situ concrete
forced into
weak pockets
in the soil by
tamping action
of tube

upward blow

Typical Data :-

load range - 300 to 1300 kN

length range - up to 18·000

diameter range - 300 to
                 600 mm

driving shoe
left in

2. FORMING PILE SHAFT

Piling Hammers ~ these are designed to deliver an impact blow to the top of the pile to be driven. The hammer weight and drop height is chosen to suit the pile type and nature of subsoil(s) through which it will be driven. The head of the pile being driven is protected against damage with a steel helmet which is padded with a sand bed or similar material and is cushioned with a plastic or hardwood block called a dolly.

Drop Hammers ~ these are blocks of iron with a rear lug(s) which locate in the piling rig guides or leaders and have a top eye for attachment of the winch rope. The number of blows which can be delivered with a free fall of 1·200 to 1·500 ranges from 10 to 20 per minute. The weight of the hammer should be not less than 50% of the concrete or steel pile weight and 1 to 1·5 times the weight of a timber pile.

Single Acting Hammers ~ these consist of a heavy falling cylinder raised by steam or compressed air sliding up and down a fixed piston. Guide lugs or rollers are located in the piling frame leaders to maintain the hammer position relative to the pile head. The number of blows delivered ranges from 36 to 75 per minute with a total hammer weight range of 2 to 15 tonnes.

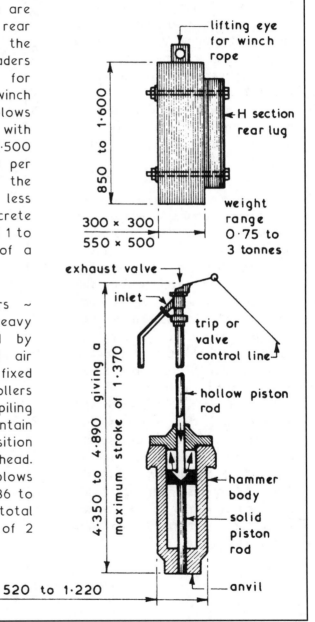

Double Acting Hammers ~ these consist of a cast iron cylinder which remains stationary on the pile head whilst a ram powered by steam or compressed air for both up and down strokes delivers a series of rapid blows which tends to keep the pile on the move during driving. The blow delivered is a smaller force than that from a drop or single acting hammer. The number of blows delivered ranges from 95 to 300 per minute with a total hammer weight range of 0·7 to 6·5 tonnes. Diesel powered double acting hammers are also available.

Diesel Hammers ~ these are self contained hammers which are located in the leaders of a piling rig and rest on the head of the pile. The driving action is started by raising the ram within the cylinder which activates the injection of a measured amount of fuel. The free falling ram compresses the fuel above the anvil causing the fuel to explode and expand resulting in a downward force on the anvil and upward force which raises the ram to recommence the cycle which is repeated until the fuel is cut off. The number of blows delivered ranges from 40 to 60 per minute with a total hammer weight range of 1·0 to 4·5 tonnes.

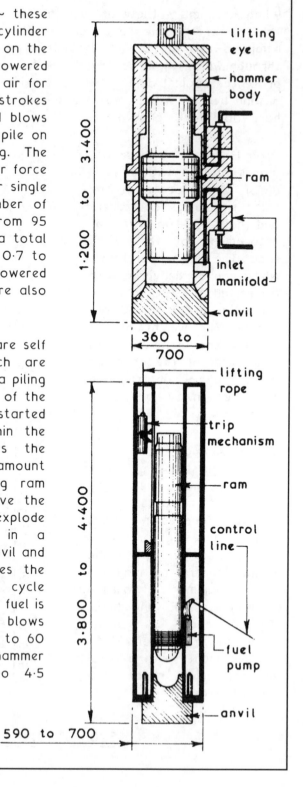

Pile Caps ~ piles can be used singly to support the load but often it is more economical to use piles in groups or clusters linked together with a reinforced concrete cap. The pile caps can also be linked together with reinforced concrete ground beams.

The usual minimum spacing for piles is:-

1. Friction Piles – 1·100 or not less than 3 × pile diameter, whichever is the greater.
2. Bearing Piles – 750 mm or not less than 2 × pile diameter, whichever is the greater.

Typical Examples ~

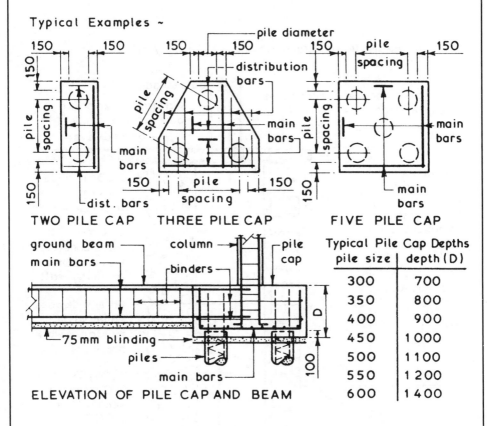

TWO PILE CAP   THREE PILE CAP   FIVE PILE CAP

ELEVATION OF PILE CAP AND BEAM

Typical Pile Cap Depths

pile size	depth (D)
300	700
350	800
400	900
450	1000
500	1100
550	1200
600	1400

Pile Testing ~ it is advisable to test load at least one pile per scheme. The test pile should be overloaded by at least 50% of its working load and this load should be held for 24 hours. The test pile should not form part of the actual foundations. Suitable testing methods are:-

1. Jacking against kentledge placed over test pile.
2. Jacking against a beam fixed to anchor piles driven in on two sides of the test pile.

Retaining Walls ~ the major function of any retaining wall is to act as on earth retaining structure for the whole or part of its height on one face, the other being exposed to the elements. Most small height retaining walls are built entirely of brickwork or a combination of brick facing and blockwork or mass concrete backing. To reduce hydrostatic pressure on the wall from ground water an adequate drainage system in the form of weep holes should be used, alternatively subsoil drainage behind the wall could be employed.

Typical Example of Combination Retaining Wall ~

precast concrete weathered coping stone

balustrade

pervious membrane over granular backfill

ground level

facings of dense clay engineering bricks tied to concrete wall with wall ties at 900 c/c horizontally and 450 c/c vertically

50

1·000

ground level

75

250

half round channel laid to fall to outlet

300

900

200mm wide 'no-fines' granular backfill

300 mm wide mass concrete 1:2:4 /20mm ag. retaining wall

12mm wide gap filled with mortar as work proceeds

75 mm diameter PVC sleeved weepholes at 2·000 c/c

20mm diameter x 600mm long dowel bars at 450 c/c

mass concrete 1:2:4 /20mm ag. foundation

expansion joints required every 30·000

Small Height Retaining Walls ~ retaining walls must be stable and the usual rule of thumb for small height brick retaining walls is for the height to lie between 2 and 4 times the wall thickness. Stability can be checked by applying the middle third rule –

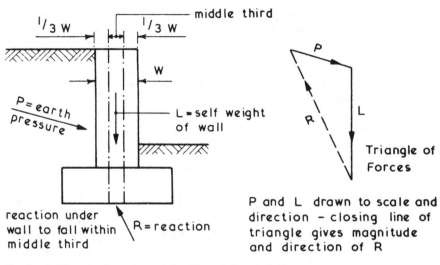

P and L drawn to scale and direction – closing line of triangle gives magnitude and direction of R

Typical Example of Brick Retaining Wall ~

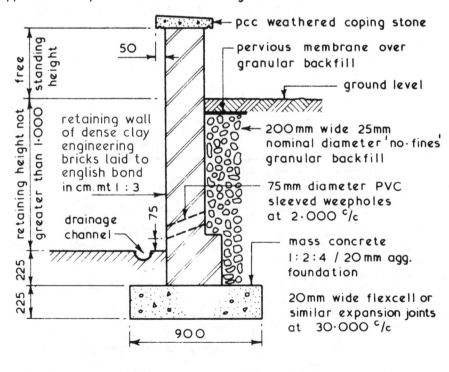

Retaining Walls up to 6·000 high ~ these can be classified as medium height retaining walls and have the primary function of retaining soils at an angle in excess of the soil's natural angle of repose. Walls within this height range are designed to provide the necessary resistance by either their own mass or by the principles of leverage.

Design ~ the actual design calculations are usually carried out by a structural engineer who endeavours to ensure that:-

1. Overturning of the wall does not occur.
2. Forward sliding of the wall does not occur.
3. Materials used are suitable and not overstressed.
4. The subsoil is not overloaded.
5. In clay subsoils slip circle failure does not occur.

The factors which the designer will have to take into account:-

1. Nature and characteristics of the subsoil(s).
2. Height of water table – the presence of water can create hydrostatic pressure on the rear face of the wall, it can also affect the bearing capacity of the subsoil together with its shear strength, reduce the frictional resistance between the underside of the foundation and the subsoil and reduce the passive pressure in front of the toe of the wall.
3. Type of wall.
4. Material(s) to be used in the construction of the wall.

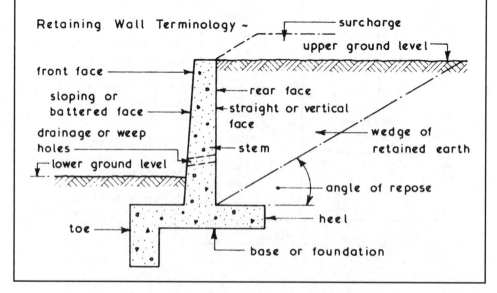

Retaining Wall Terminology ~

Earth Pressures ~ these can take one of two forms namely:-

1. Active Earth Pressures – these are those pressures which tend to move the wall at all times and consist of the wedge of earth retained plus any hydrostatic pressure. The latter can be reduced by including a subsoil drainage system behind and/or through the wall.
2. Passive Earth Pressures ~ these are a reaction of an equal and opposite force to any imposed pressure thus giving stability by resisting movement.

Typical Examples ~

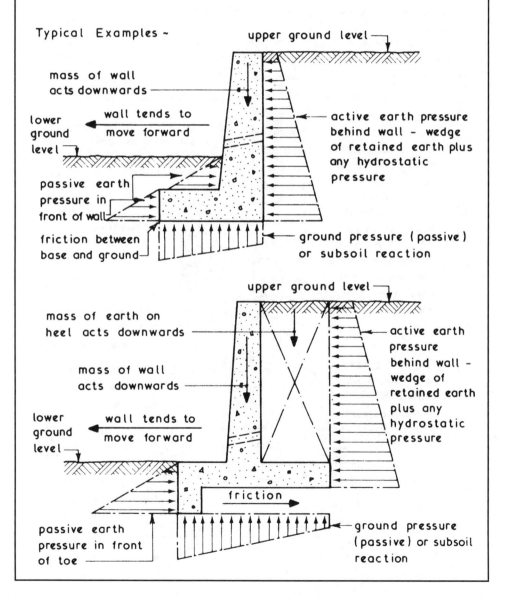

upper ground level

mass of wall acts downwards

wall tends to move forward

lower ground level

active earth pressure behind wall – wedge of retained earth plus any hydrostatic pressure

passive earth pressure in front of wall

friction between base and ground

ground pressure (passive) or subsoil reaction

upper ground level

mass of earth on heel acts downwards

active earth pressure behind wall – wedge of retained earth plus any hydrostatic pressure

mass of wall acts downwards

lower ground level

wall tends to move forward

friction

passive earth pressure in front of toe

ground pressure (passive) or subsoil reaction

Mass Retaining Walls ~ these walls rely mainly on their own mass to overcome the tendency to slide forwards. Mass retaining walls are not generally considered to be economic over a height of 1·800 when constructed of brick or concrete and 1·000 high in the case of natural stonework. Any mass retaining wall can be faced with another material but generally any applied facing will not increase the strength of the wall and is therefore only used for aesthetic reasons.

Typical Brick Mass Retaining Wall Details ~

precast concrete
weathered coping
stone

dpc

ground level

bricks to have a
crushing strength
of not less than
20·5 MN/m² and
to be laid with a
mortar mix of
1:¼:3 (cement :
lime : sand) –
vertical movement
joints should be
provided at not
more than 15·000
centres

back of wall to be coated
with bituminous paint or
lined with heavy duty
polythene sheet

COHESIVE
SUBSOIL

1·800 maximum

40

75 mm diameter
weep holes at
1·800 c/c

ground level

rubble filling
behind wall
and weep holes

PVC or similar pipe
lining to weep holes

225

225

mass concrete
foundation

450

225

890

900

525

300

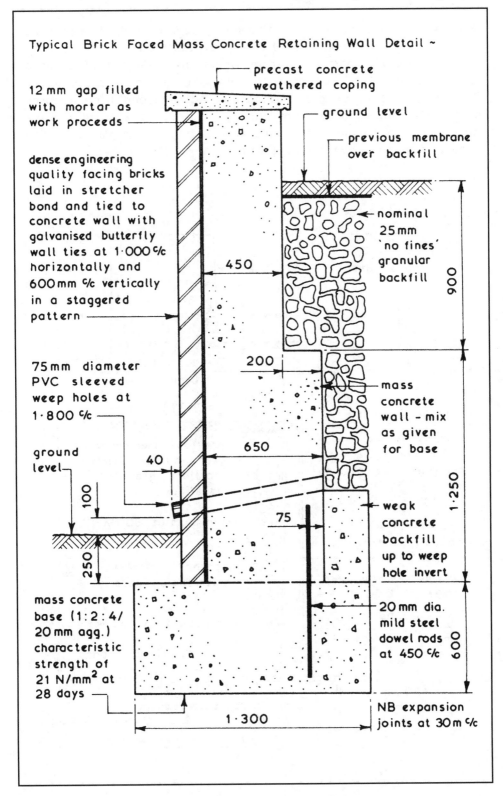

Typical Brick Faced Mass Concrete Retaining Wall Detail ~

precast concrete weathered coping

12 mm gap filled with mortar as work proceeds

ground level

previous membrane over backfill

dense engineering quality facing bricks laid in stretcher bond and tied to concrete wall with galvanised butterfly wall ties at 1·000 ℅ horizontally and 600 mm ℅ vertically in a staggered pattern

nominal 25 mm 'no fines' granular backfill

450

900

75 mm diameter PVC sleeved weep holes at 1·800 ℅

200

mass concrete wall – mix as given for base

ground level

40

650

100

250

75

weak concrete backfill up to weep hole invert

1·250

mass concrete base (1:2:4/ 20 mm agg.) characteristic strength of 21 N/mm² at 28 days

20 mm dia. mild steel dowel rods at 450 ℅

600

1·300

NB expansion joints at 30 m ℅

253

Cantilever Retaining Walls ~ these are constructed of reinforced concrete with an economic height range of 1·200 to 6·000. They work on the principles of leverage where the stem is designed as a cantilever fixed at the base and base is designed as a cantilever fixed at the stem. Several formats are possible and in most cases a beam is placed below the base to increase the total passive resistance to sliding. Facing materials can be used in a similar manner to that shown on page 253.

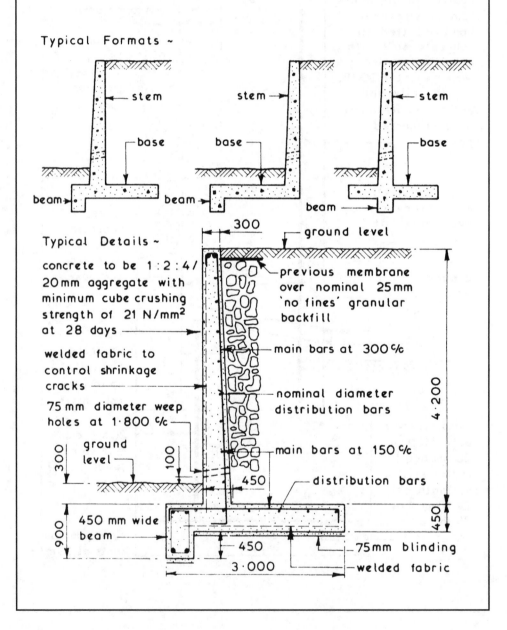

Typical Formats ~

stem — base — beam

stem — base — beam

stem — base — beam

Typical Details ~

concrete to be 1:2:4/ 20mm aggregate with minimum cube crushing strength of 21 N/mm² at 28 days

welded fabric to control shrinkage cracks

75 mm diameter weep holes at 1·800 c/c

ground level

450 mm wide beam

900

300

ground level

300

100

450

previous membrane over nominal 25mm 'no fines' granular backfill

main bars at 300 c/c

nominal diameter distribution bars

main bars at 150 c/c

distribution bars

4·200

450

75mm blinding

welded fabric

450

3·000

Formwork ~ concrete retaining walls can be cast in one of three ways – full height; climbing (page 256) or against earth face (page 257).

Full Height Casting ~ this can be carried out if the wall is to be cast as a freestanding wall and allowed to cure and gain strength before the earth to be retained is backfilled behind the wall. Considerations are the height of the wall, anticipated pressure of wet concrete, any strutting requirements and the availability of suitable materials to fabricate the formwork. As with all types of formwork a traditional timber format or a patent system using steel forms could be used.

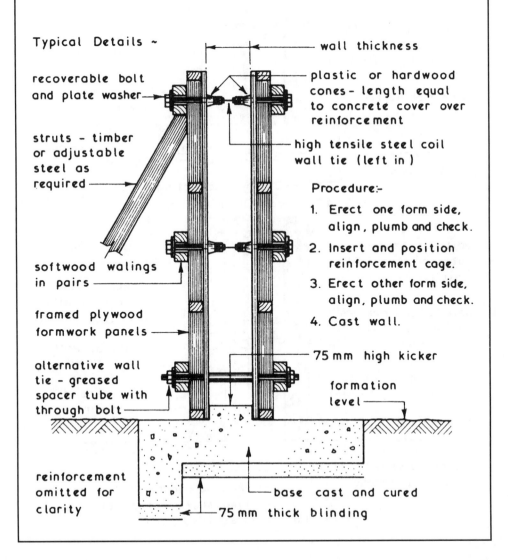

Typical Details ~

recoverable bolt and plate washer

struts - timber or adjustable steel as required

softwood walings in pairs

framed plywood formwork panels

alternative wall tie - greased spacer tube with through bolt

reinforcement omitted for clarity

wall thickness

plastic or hardwood cones- length equal to concrete cover over reinforcement

high tensile steel coil wall tie (left in)

Procedure:-

1. Erect one form side, align, plumb and check.

2. Insert and position reinforcement cage.

3. Erect other form side, align, plumb and check.

4. Cast wall.

75 mm high kicker

formation level

base cast and cured

75 mm thick blinding

Climbing Formwork or Lift Casting ~ this method can be employed on long walls, high walls or where the amount of concrete which can be placed in a shift is limited.

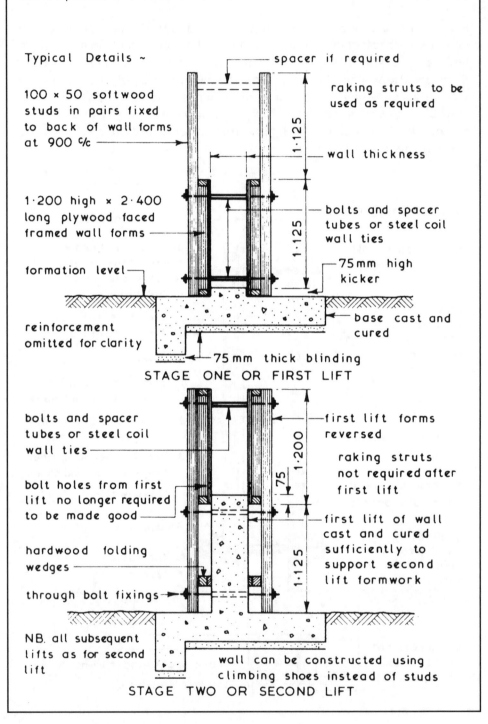

Typical Details ~

100 × 50 softwood studs in pairs fixed to back of wall forms at 900 ℃

spacer if required

raking struts to be used as required

1·125

wall thickness

1·200 high × 2·400 long plywood faced framed wall forms

1·125

bolts and spacer tubes or steel coil wall ties

formation level

75mm high kicker

reinforcement omitted for clarity

base cast and cured

75mm thick blinding

STAGE ONE OR FIRST LIFT

bolts and spacer tubes or steel coil wall ties

first lift forms reversed

1·200

raking struts not required after first lift

75

bolt holes from first lift no longer required to be made good

first lift of wall cast and cured sufficiently to support second lift formwork

hardwood folding wedges

1·125

through bolt fixings

NB. all subsequent lifts as for second lift

wall can be constructed using climbing shoes instead of studs

STAGE TWO OR SECOND LIFT

Casting Against Earth Face ~ this method can be an adaptation of the full height or climbing formwork systems. The latter uses a steel wire loop tie fixing to provide the support for the second and subsequent lifts.

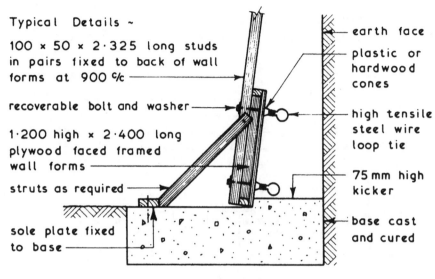

Typical Details ~

100 × 50 × 2·325 long studs in pairs fixed to back of wall forms at 900 ℅

recoverable bolt and washer

1·200 high × 2·400 long plywood faced framed wall forms

struts as required

sole plate fixed to base

earth face

plastic or hardwood cones

high tensile steel wire loop tie

75 mm high kicker

base cast and cured

STAGE ONE OR FIRST LIFT

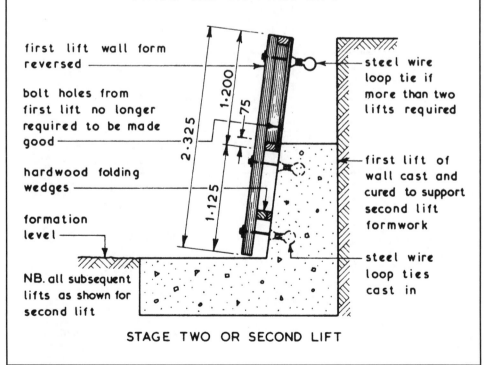

first lift wall form reversed

bolt holes from first lift no longer required to be made good

hardwood folding wedges

formation level

NB. all subsequent lifts as shown for second lift

steel wire loop tie if more than two lifts required

first lift of wall cast and cured to support second lift formwork

steel wire loop ties cast in

STAGE TWO OR SECOND LIFT

257

Masonry units – these are an option where it is impractical or cost-ineffective to use temporary formwork to in-situ concrete. Exposed brick or blockwork may also be a preferred finish. In addition to being a structural component, masonry units provide permanent formwork to reinforced concrete poured into the voids created by:

* Quetta bonded standard brick units, OR
* Stretcher bonded standard hollow dense concrete blocks.

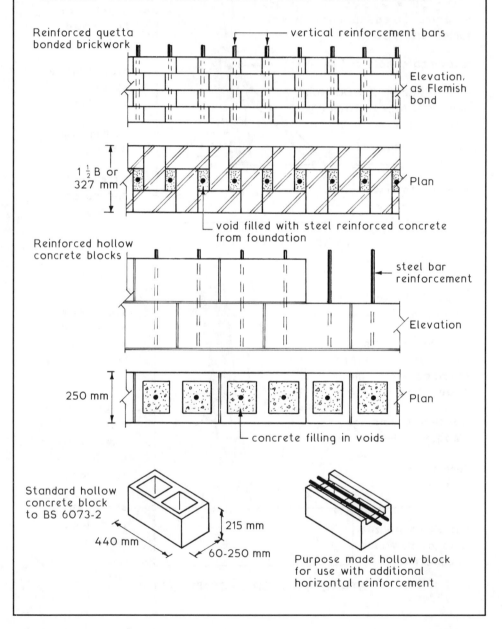

Reinforced quetta bonded brickwork

vertical reinforcement bars

Elevation, as Flemish bond

$1\frac{1}{2}$ B or 327 mm

Plan

void filled with steel reinforced concrete from foundation

Reinforced hollow concrete blocks

steel bar reinforcement

Elevation

250 mm

Plan

concrete filling in voids

Standard hollow concrete block to BS 6073-2

215 mm

440 mm

60-250 mm

Purpose made hollow block for use with additional horizontal reinforcement

Construction – a reinforced concrete base is cast with projecting steel bars accurately located for vertical continuity. The wall may be built solid, e.g. Quetta bond, with voids left around the bars for subsequent grouting. Alternatively, the wall may be of wide cavity construction, where the exposed reinforcement is wrapped in 'denso' grease tape for protection against corrosion. Steel bars are threaded at the top to take a tensioning nut over a bearing plate.

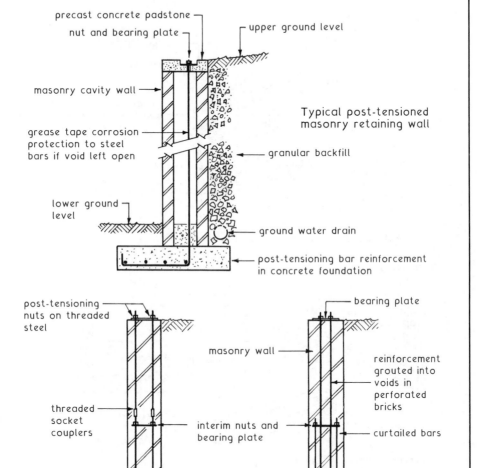

precast concrete padstone
nut and bearing plate
upper ground level
masonry cavity wall
grease tape corrosion protection to steel bars if void left open
Typical post-tensioned masonry retaining wall
granular backfill
lower ground level
ground water drain
post-tensioning bar reinforcement in concrete foundation

post-tensioning nuts on threaded steel
bearing plate
masonry wall
threaded socket couplers
reinforcement grouted into voids in perforated bricks
interim nuts and bearing plate
curtailed bars
continuity reinforcement from base
base retention plate

Staged post-tensioning to high masonry retaining walls

Ref. BS 5628-2: Code of practice for use of masonry. Structural use of reinforced and prestressed masonry.

Crib Retaining Walls – a system of pre-cast concrete or treated timber components comprising headers and stretchers which interlock to form a three-dimensional framework. During assembly the framework is filled with graded stone to create sufficient mass to withstand ground pressures.

Principle –

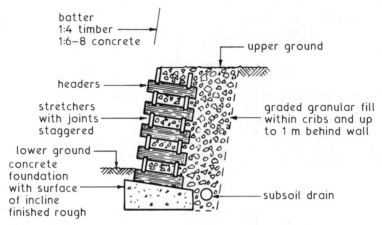

batter
1:4 timber
1:6–8 concrete

upper ground

headers

stretchers with joints staggered

graded granular fill within cribs and up to 1 m behind wall

lower ground

concrete foundation with surface of incline finished rough

subsoil drain

Note: height limited to 10 m with timber

Components –

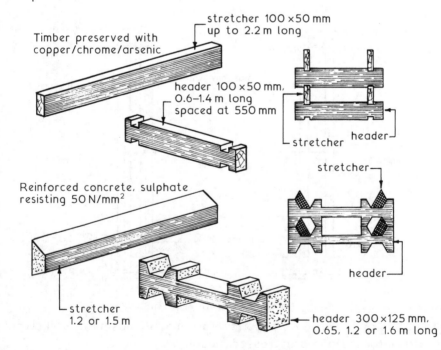

stretcher 100 × 50 mm up to 2.2 m long

Timber preserved with copper/chrome/arsenic

header 100 × 50 mm, 0.6–1.4 m long spaced at 550 mm

stretcher

header

Reinforced concrete, sulphate resisting 50 N/mm²

stretcher

header

stretcher 1.2 or 1.5 m

header 300 × 125 mm, 0.65, 1.2 or 1.6 m long

Soil Nailing ~ a cost effective geotechnic process used for retaining large soil slopes, notably highway and railway embankments.

Function ~ after excavating and removing the natural slope support, the remaining wedge of exposed unstable soil is pinned or nailed back with tendons into stable soil behind the potential slip plane.

Types of Soil Nails or Tendons ~

• Solid deformed steel rods up to 50 mm in diameter, located in bore holes up to 100 mm in diameter. Cement grout is pressurised into the void around the rods.
• Hollow steel, typically 100 mm diameter tubes with an expendable auger attached. Cement grout is injected into the tube during boring to be ejected through purpose-made holes in the auger.
• Solid glass reinforced plastic (GRP) with resin grouts.

Embankment Treatment ~ the exposed surface is faced with a plastic coated wire mesh to fit over the ends of the tendons. A steel head plate is fitted over and centrally bolted to each projecting tendon, followed by spray concreting to the whole face.

Typical Application ~

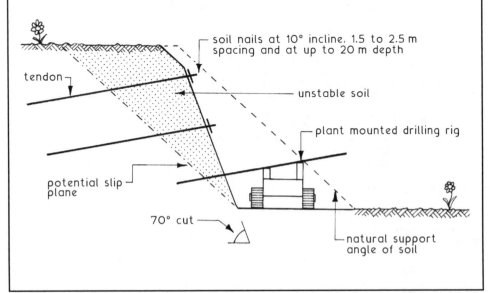

261

Gabion ~ a type of retaining wall produced from individual rectangular boxes made from panels of wire mesh, divided internally and filled with stones. These units are stacked and overlapped (like stretcher bonded masonry) and applied in several layers or courses to retained earth situations. Typical sizes, 1·0 m long x 0·5 m wide x 0·5 m high, up to 4·0 m long x 1·0 m wide x 1·0 m high.

Mattress ~ unit fabrication is similar to a gabion but of less thickness, smaller mesh and stone size to provide some flexibility and shaping potential. Application is at a much lower incline. Generally used next to waterways for protection against land erosion where tidal movement and/or water level differentials could scour embankments. Typical sizes, 3·0 m long x 2·0 m wide x 0·15 m thick, up to 6·0 m long x 2·0 m wide x 0·3 m thick.

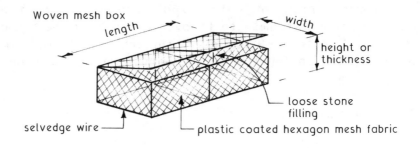

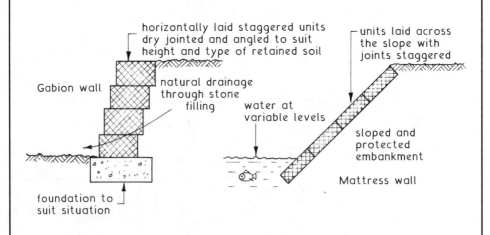

Design of Retaining Walls ~ this should allow for the effect of hydrostatics or water pressure behind the wall and the pressure created by the retained earth (see page 251). Calculations are based on a 1m unit length of wall, from which it is possible to ascertain:

1. The resultant thrust

2. The overturning or bending moment

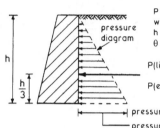

P = resultant thrust
w = density of material
h = height of wall
θ = soil angle of repose (see page 281)

$$P(\text{liquid}) = \frac{wh^2}{2}$$

$$P(\text{earth}) = \frac{wh^2}{2} \times \frac{1-\sin\theta}{1+\sin\theta} \quad \left[\begin{array}{l}\text{Rankine's}\\\text{formula}\end{array}\right]$$

pressure at base (water) = wh kg/m^2

pressure at base (earth) = $wh\left(\frac{1-\sin\theta}{1+\sin\theta}\right)$ kg/m^2

P, the resultant thrust, will act through the centre of gravity of the pressure diagram, i.e. at h/3.

The overturning moment due to water is therefore:

$$\frac{wh^2}{2} \times \frac{h}{3} \text{ or } \frac{wh^3}{6}$$

and for earth:

$$\frac{wh^2}{2} \times \frac{1-\sin\theta}{1+\sin\theta} \times \frac{h}{3} \quad \text{or} \quad \frac{wh^3}{6} \times \frac{1-\sin\theta}{1+\sin\theta}$$

Typical example ~

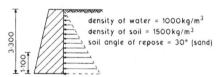

density of water = 1000kg/m^3
density of soil = 1500kg/m^3
soil angle of repose = 30° (sand)

For water:

$$p = \frac{wh^2}{2} = \frac{1000 \times (3\cdot3)^2}{2} = 5445 \text{ kg}$$

NB. kg × gravity = Newtons

Therefore, 5445 kg × 9·81 = 53·42 kN

The overturning or bending moment will be: P × h/3 = 53·42 kN x 1.1 m = 58·8 kNm

For earth:

$$p = \frac{wh^2}{2} \times \frac{1-\sin\theta}{1+\sin\theta}$$

$$p = \frac{1500 \times (3\cdot3)^2}{2} \times \frac{1-\sin 30°}{1+\sin 30°} = 2723 \text{ kg} \quad \text{or} \quad 26\cdot7 \text{ kN}$$

The overturning or bending moment will be:
P × h/3 = 26·7 kN × 1·1m = 29·4 kNm

A graphical design solution, to determine the earth thrust (P) behind a retaining wall. Data from previous page:

h = 3·300 m

θ = 30°

w = 1500 kg/m^3

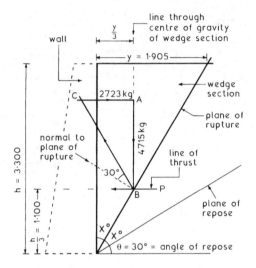

Wall height is drawn to scale and plane of repose plotted. The wedge section is obtained by drawing the plane of rupture through an angle bisecting the plane of repose and vertical back of the wall. Dimension 'y' can be scaled or calculated:

Tangent x = $\dfrac{y}{3\cdot3}$ x = 30°, and tan 30° = 0·5774

therefore, y = 3·3 × 0·5774 = 1·905 m

Area of wedge section = $\dfrac{3\cdot3}{2}$ × 1·905 m = 3·143 m^2

Volume of wedge per metre run of wall = 3·143 × 1 = 3·143 m^3

Weight .. .. = 3·143 × 1500 = 4715 kg

Vector line A – B is drawn to a scale through centre of gravity of wedge section, line of thrust and plane of rupture to represent 4715 kg.

Vector line B – C is drawn at the angle of earth friction (usually same as angle of repose, i.e. 30° in this case), to the normal to the plane of rupture until it meets the horizontal line C – A.

Triangle ABC represents the triangle of forces for the wedge section of earth, so C – A can be scaled at 2723 kg to represent (P), the earth thrust behind the retaining wall.

Open Excavations ~ one of the main problems which can be encountered with basement excavations is the need to provide temporary support or timbering to the sides of the excavation. This can be intrusive when the actual construction of the basement floor and walls is being carried out. One method is to use battered excavation sides cut back to a safe angle of repose thus eliminating the need for temporary support works to the sides of the excavation.

Typical Example of Open Basement Excavations ~

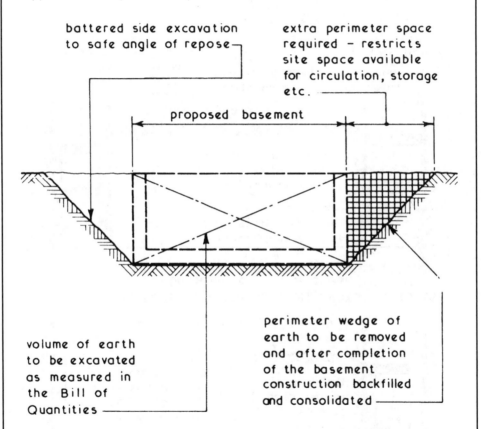

battered side excavation to safe angle of repose

extra perimeter space required – restricts site space available for circulation, storage etc.

proposed basement

volume of earth to be excavated as measured in the Bill of Quantities

perimeter wedge of earth to be removed and after completion of the basement construction backfilled and consolidated

In economic terms the costs of plant and manpower to cover the extra excavation, backfilling and consolidating must be offset by the savings made by omitting the temporary support works to the sides of the excavation. The main disadvantage of this method is the large amount of free site space required.

Perimeter Trench Excavations ~ in this method a trench wide enough for the basement walls to be constructed is excavated and supported with timbering as required. It may be necessary for runners or steel sheet piling to be driven ahead of the excavation work. This method can be used where weak subsoils are encountered so that the basement walls act as permanent timbering whilst the mound or dumpling is excavated and the base slab cast. Perimeter trench excavations can also be employed in firm subsoils when the mechanical plant required for excavating the dumpling is not available at the right time.

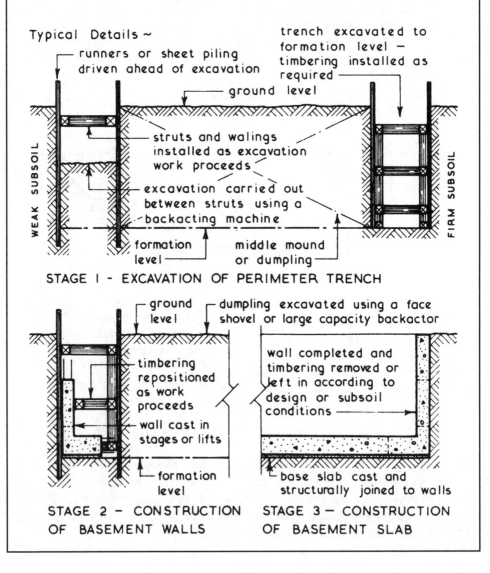

Typical Details ~

runners or sheet piling driven ahead of excavation

trench excavated to formation level — timbering installed as required

ground level

struts and walings installed as excavation work proceeds

excavation carried out between struts using a backacting machine

WEAK SUBSOIL

FIRM SUBSOIL

formation level

middle mound or dumpling

STAGE 1 - EXCAVATION OF PERIMETER TRENCH

ground level

dumpling excavated using a face shovel or large capacity backactor

timbering repositioned as work proceeds

wall cast in stages or lifts

wall completed and timbering removed or left in according to design or subsoil conditions

formation level

base slab cast and structurally joined to walls

STAGE 2 - CONSTRUCTION OF BASEMENT WALLS

STAGE 3 - CONSTRUCTION OF BASEMENT SLAB

Complete Excavation ~ this method can be used in firm subsoils where the centre of the proposed basement can be excavated first to enable the basement slab to be cast thus giving protection to the subsoil at formation level. The sides of excavation to the perimeter of the basement can be supported from the formation level using raking struts or by using raking struts pitched from the edge of the basement slab.

Typical Details ~

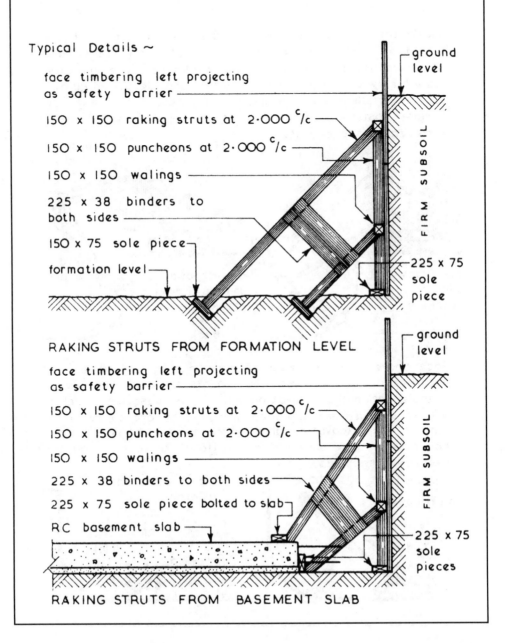

face timbering left projecting
as safety barrier

150 x 150 raking struts at 2·000 $^c/_c$

150 x 150 puncheons at 2·000 $^c/_c$

150 x 150 walings

225 x 38 binders to both sides

150 x 75 sole piece

formation level

ground level

FIRM SUBSOIL

225 x 75 sole piece

RAKING STRUTS FROM FORMATION LEVEL

face timbering left projecting
as safety barrier

150 x 150 raking struts at 2·000 $^c/_c$

150 x 150 puncheons at 2·000 $^c/_c$

150 x 150 walings

225 x 38 binders to both sides

225 x 75 sole piece bolted to slab

RC basement slab

ground level

FIRM SUBSOIL

225 x 75 sole pieces

RAKING STRUTS FROM BASEMENT SLAB

Excavating Plant ~ the choice of actual pieces of plant to be used in any construction activity is a complex matter taking into account many factors. Specific details of various types of excavators are given on pages 175 to 179. At this stage it is only necessary to consider basic types for particular operations. In the context of basement excavation two forms of excavator could be considered.

1. Backactors – these machines are available as cable rigged or hydraulic excavators suitable for trench and bulk excavating. Cable rigged backactors are usually available with larger bucket sizes and deeper digging capacities than the hydraulic machines but these have a more positive control and digging operation and are also easier to operate.

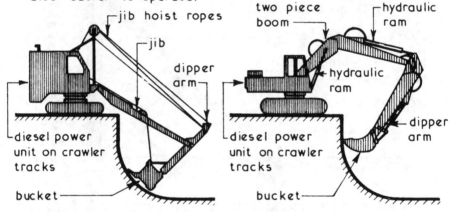

2. Face Shovels – these are robust machines designed to excavate above their own wheel or track level and are suitable for bulk excavation work. In basement work they will require a ramp approach unless they are to be lifted out of the excavation area by means of a crane. Like backactors face shovels are available as cable rigged or hydraulic machines.

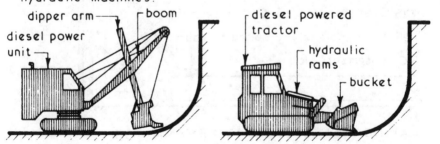

Basement Construction ~ in the general context of buildings a basement can be defined as a storey which is below the ground storey and is therefore constructed below ground level. Most basements can be classified into one of three groups:-

1. Retaining Wall and Raft Basements - this is the general format for basement construction and consists of a slab raft foundation which forms the basement floor and helps to distribute the structural loads transmitted down the retaining walls.

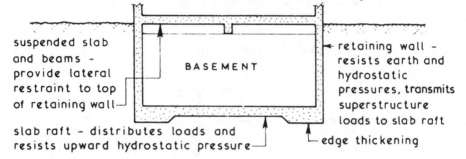

suspended slab and beams - provide lateral restraint to top of retaining wall

BASEMENT

retaining wall - resists earth and hydrostatic pressures, transmits superstructure loads to slab raft

slab raft - distributes loads and resists upward hydrostatic pressure

edge thickening

2. Box and Cellular Raft Basements - similar method to above except that internal walls are used to transmit and spread loads over raft as well as dividing basement into cells.

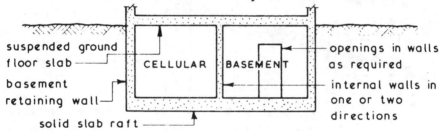

suspended ground floor slab

basement retaining wall

CELLULAR

BASEMENT

openings in walls as required

internal walls in one or two directions

solid slab raft

3. Piled Basements - the main superstructure loads are carried to the basement floor level by columns where they are finally transmitted to the ground via pile caps and bearing piles. This method can be used where low bearing capacity soils are found at basement floor level.

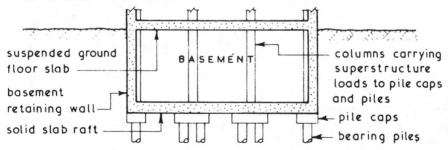

suspended ground floor slab

basement retaining wall

solid slab raft

BASEMENT

columns carrying superstructure loads to pile caps and piles

pile caps

bearing piles

Deep Basement Construction ~ basements can be constructed within a cofferdam or other temporary supported excavation (see Basement Excavations on pages 265 to 267) up to the point when these methods become uneconomic, unacceptable or both due to the amount of necessary temporary support work. Deep basements can be constructed by installing diaphragm walls within a trench and providing permanent support with ground anchors or by using the permanent lateral support given by the internal floor during the excavation period (see next page). Temporary lateral support during the excavation period can be provided by lattice beams spanning between the diaphragm walls (see next page).

Typical Ground Anchor Support Details ~

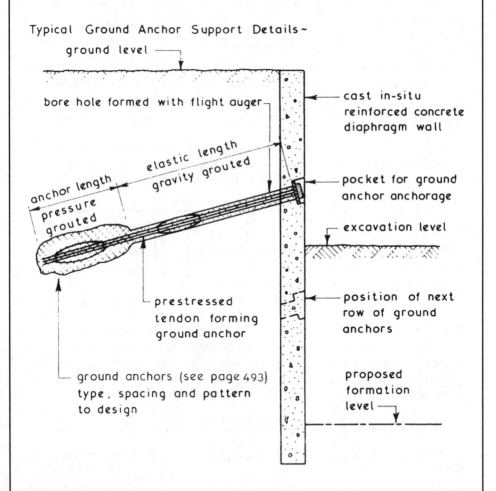

ground level

bore hole formed with flight auger

cast in-situ reinforced concrete diaphragm wall

elastic length gravity grouted

anchor length

pressure grouted

pocket for ground anchor anchorage

excavation level

prestressed tendon forming ground anchor

position of next row of ground anchors

ground anchors (see page 493) type, spacing and pattern to design

proposed formation level

NB. vertical ground anchors installed through the lowest floor can be used to overcome any tendency to flotation during the construction period

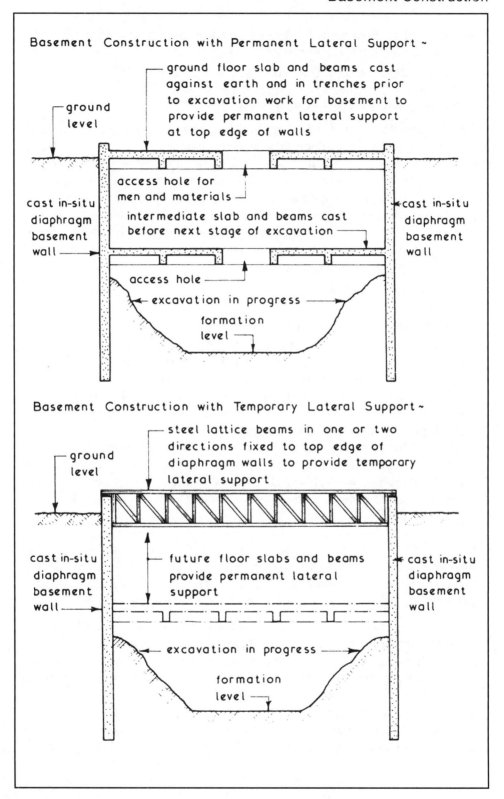

Basement Construction with Permanent Lateral Support ~

ground floor slab and beams cast against earth and in trenches prior to excavation work for basement to provide permanent lateral support at top edge of walls

ground level

cast in-situ diaphragm basement wall

cast in-situ diaphragm basement wall

access hole for men and materials

intermediate slab and beams cast before next stage of excavation

access hole

excavation in progress

formation level

Basement Construction with Temporary Lateral Support ~

steel lattice beams in one or two directions fixed to top edge of diaphragm walls to provide temporary lateral support

ground level

cast in-situ diaphragm basement wall

cast in-situ diaphragm basement wall

future floor slabs and beams provide permanent lateral support

excavation in progress

formation level

Waterproofing Basements ~ basements can be waterproofed by one of three basic methods namely:-

1. Use of dense monolithic concrete walls and floor
2. Tanking techniques (see pages 274 & 275)
3. Drained cavity system (see page 276)

Dense Monolithic Concrete – the main objective is to form a watertight basement using dense high quality reinforced or prestressed concrete by a combination of good materials, good workmanship, attention to design detail and on site construction methods. If strict control of all aspects is employed a sound watertight structure can be produced but it should be noted that such structures are not always water vapourproof. If the latter is desirable some waterproof coating, lining or tanking should be used. The watertightness of dense concrete mixes depends primarily upon two factors:-

1. Water/cement ratio.
2. Degree of compaction.

The hydration of cement during the hardening process produces heat therefore to prevent early stage cracking the temperature changes within the hardening concrete should be kept to a minimum. The greater the cement content the more is the evolution of heat therefore the mix should contain no more cement than is necessary to fulfil design requirements. Concrete with a free water/cement ratio of 0.5 is watertight and although the permeability is three time more at a ratio of 0.6 it is for practical purposes still watertight but above this ratio the concrete becomes progressively less watertight. For lower water/cement ratios the workability of the mix would have to be increased, usually by adding more cement, to enable the concrete to be fully compacted.

Admixtures – if the ingredients of good design, materials and workmanship are present watertight concrete can be produced without the use of admixtures. If admixtures are used they should be carefully chosen and used to obtain a specific objective:-

1. Water-reducing admixtures – used to improve workability
2. Retarding admixtures – slow down rate of hardening
3. Accelerating admixtures – increase rate of hardening – useful for low temperatures – calcium chloride not suitable for reinforced concrete.
4. Water-repelling admixtures – effective only with low water head, will not improve poor quality or porous mixes.
5. Air-entraining admixtures – increases workability – lowers water content.

Joints ~ in general these are formed in basement constructions to provide for movement accommodation (expansion joints) or to create a convenient stopping point in the construction process (construction joints). Joints are lines of weakness which will leak unless carefully designed and constructed therefore they should be simple in concept and easy to construct.

Basement slabs ~ these are usually designed to span in two directions and as a consequence have relatively heavy top and bottom reinforcement. To enable them to fulfil their basic functions they usually have a depth in excess of 250 mm. The joints, preferably of the construction type, should be kept to a minimum and if waterbars are specified they must be placed to ensure that complete compaction of the concrete is achieved.

Typical Basement Slab Joint Details ~

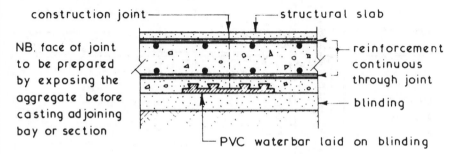

Basement Walls ~ joints can be horizontal and/or vertical according to design requirements. A suitable waterbar should be incorporated in the joint to prevent the ingress of water. The top surface of a kicker used in conjunction with single lift pouring if adequately prepared by exposing the aggregate should not require a waterbar but if one is specified it should be either placed on the rear face or consist of a centrally placed mild steel strip inserted into the kicker whilst the concrete is still in a plastic state.

Typical Basement Wall Joint Details ~

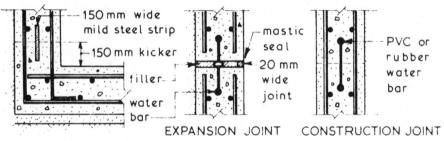

273

Mastic Asphalt Tanking ~ the objective of tanking is to provide a continuous waterproof membrane which is applied to the base slab and walls with complete continuity between the two applications. The tanking can be applied externally or internally according to the circumstances prevailing on site. Alternatives to mastic asphalt are polythene sheeting: bituminous compounds: epoxy resin compounds and bitumen laminates.

External Mastic Asphalt Tanking ~ this is the preferred method since it not only prevents the ingress of water it also protects the main structure of the basement from aggressive sulphates which may be present in the surrounding soil or ground water.

Typical External Tanking Details ~

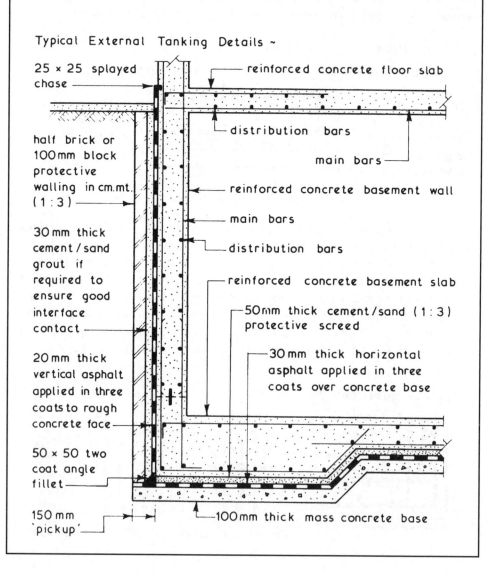

25 × 25 splayed chase

reinforced concrete floor slab

half brick or 100mm block protective walling in cm.mt. (1:3)

distribution bars

main bars

reinforced concrete basement wall

main bars

30 mm thick cement/sand grout if required to ensure good interface contact

distribution bars

reinforced concrete basement slab

50mm thick cement/sand (1:3) protective screed

20 mm thick vertical asphalt applied in three coats to rough concrete face

30 mm thick horizontal asphalt applied in three coats over concrete base

50 × 50 two coat angle fillet

150 mm 'pickup'

100mm thick mass concrete base

Internal Mastic Asphalt Tanking ~ this method should only be adopted if external tanking is not possible since it will not give protection to the main structure and unless adequately loaded may be forced away from the walls and/or floor by hydrostatic pressure. To be effective the horizontal and vertical coats of mastic asphalt must be continuous.

Typical Internal Tanking Details ~

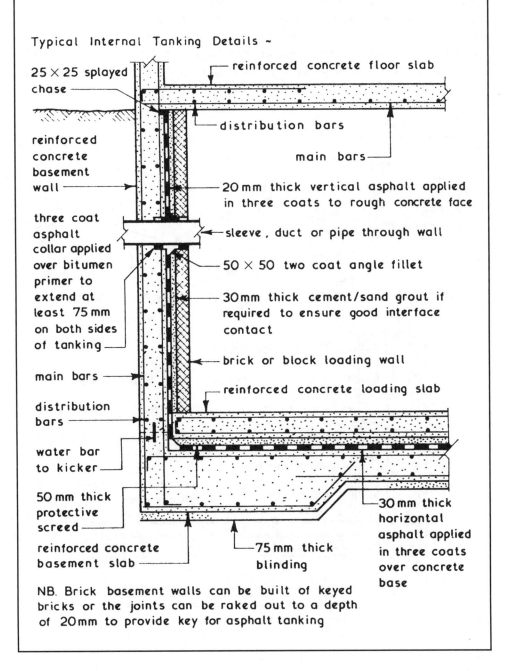

25 × 25 splayed chase

reinforced concrete floor slab

reinforced concrete basement wall

distribution bars

main bars

20 mm thick vertical asphalt applied in three coats to rough concrete face

three coat asphalt collar applied over bitumen primer to extend at least 75 mm on both sides of tanking

sleeve, duct or pipe through wall

50 × 50 two coat angle fillet

30 mm thick cement/sand grout if required to ensure good interface contact

brick or block loading wall

reinforced concrete loading slab

main bars

distribution bars

water bar to kicker

50 mm thick protective screed

30 mm thick horizontal asphalt applied in three coats over concrete base

reinforced concrete basement slab

75 mm thick blinding

NB. Brick basement walls can be built of keyed bricks or the joints can be raked out to a depth of 20mm to provide key for asphalt tanking

Drained Cavity System ~ this method of waterproofing basements can be used for both new and refurbishment work. The basic concept is very simple in that it accepts that a small amount of water seepage is possible through a monolithic concrete wall and the best method of dealing with such moisture is to collect it and drain it away. This is achieved by building an inner non-load bearing wall to form a cavity which is joined to a floor composed of special triangular tiles laid to falls which enables the moisture to drain away to a sump from which it is either discharged direct or pumped into the surface water drainage system. The inner wall should be relatively vapour tight or alternatively the cavity should be ventilated.

Typical Details ~

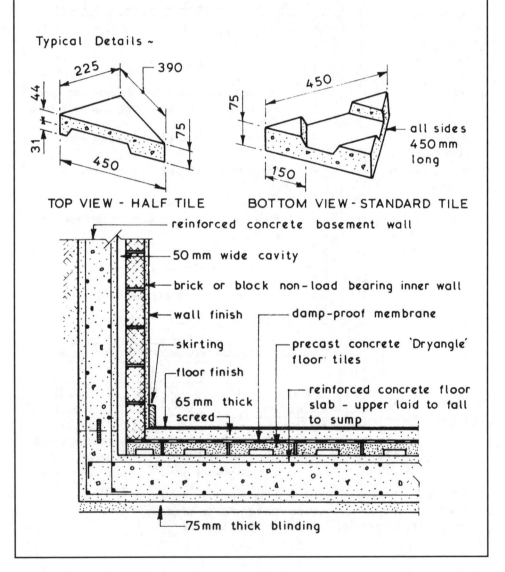

TOP VIEW – HALF TILE   BOTTOM VIEW – STANDARD TILE

reinforced concrete basement wall

50 mm wide cavity

brick or block non-load bearing inner wall

wall finish

damp-proof membrane

skirting

precast concrete 'Dryangle' floor tiles

floor finish

65 mm thick screed

reinforced concrete floor slab – upper laid to fall to sump

75mm thick blinding

Basements benefit considerably from the insulating properties of the surrounding soil. However, that alone is insufficient to satisfy the typical requirements for wall and floor U-values of 0·35 and 0·30 W/m²K, respectively.

Refurbishment of existing basements may include insulation within dry lined walls and under the floor screed or particle board overlay. This should incorporate an integral vapour control layer to minimise risk of condensation.

External insulation of closed cell rigid polystyrene slabs is generally applied to new construction. These slabs combine low thermal conductivity with low water absorption and high compressive strength. The external face of insulation is grooved to encourage moisture run off. It is also filter faced to prevent clogging of the grooves. Backfill is granular.

Typical application~

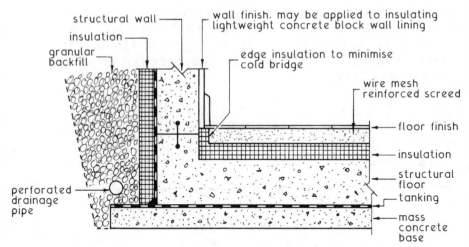

Note: reinforcement in concrete omitted, see details on previous pages.

Tables and calculations to determine U-values for basements are provided in the Building Regulations, Approved Document L and in BS EN ISO 13370: Thermal performance of buildings. Heat transfer via the ground. Calculation methods.

Excavation ~ to hollow out – in building terms to remove earth to form a cavity in the ground.

Types of Excavation ~

Oversite – the removal of top soil (Building Regulations requirement.)

depth varies from site to site but is usually in a 150 to 300 mm range. Top soil contains plant life animal life and decaying matter which makes the soil compressible and therefore unsuitable for supporting buildings.

s u b s o i l

Reduce Level – carried out below oversite level to form a level surface on which to build and can consist of both cutting and filling operations. The level to which the ground is reduced is called the formation level.

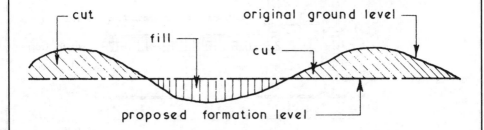

cut

fill

original ground level

cut

proposed formation level

NB. Water in Excavations – this should be removed since it can:~

1. Undermine sides of excavation.
2. Make it impossible to adequately compact bottom of excavation to receive foundations.
3. Cause puddling which can reduce the bearing capacity of the subsoil.

Trench Excavations ~ narrow excavations primarily for strip foundations and buried services – excavation can be carried out by hand or machine.

Typical Examples ~

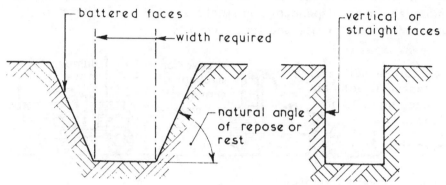

Disadvantage ~ extra cost of over excavating and extra backfilling.

Advantage ~ no temporary support required to sides of excavation.

Disadvantage ~ sides of excavation may require some degree of temporary support.

Advantage ~ minimum amount of soil removed and therefore minimum amount of backfilling.

Pier Holes ~ isolated pits primarily used for foundation pads for columns and piers or for the construction of soakaways.

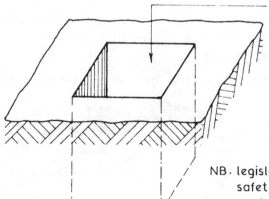

sides of excavation can be battered or straight as described above – deep pier holes may have to be over excavated in plan to provide good access to and good egress from the working area for both men and materials.

NB. legislation affecting safety in excavation is contained in the Construction (Health, Safety and Welfare) Regulations 1996.

Site Clearance and Removal of Top Soil ~

On small sites this could be carried out by manual means using hand held tools such as picks, shovels and wheelbarrows.

On all sites mechanical methods could be used the actual plant employed being dependent on factors such as volume of soil involved, nature of site and time elements.

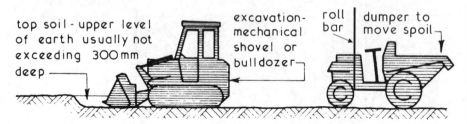

top soil - upper level of earth usually not exceeding 300mm deep

excavation-mechanical shovel or bulldozer

roll bar

dumper to move spoil

Reduced Level Excavations ~
On small sites — hand processes as given above
On all sites mechanical methods could be used dependent on factors given above.

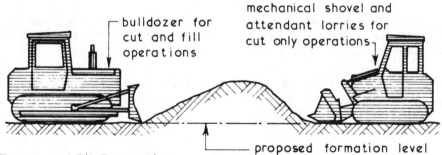

bulldozer for cut and fill operations

mechanical shovel and attendant lorries for cut only operations

proposed formation level

Trench and Pit Excavations ~
On small sites — hand processes as given above but if depth of excavation exceeds 1·200 some method of removing spoil from the excavation will have to be employed.

On all sites mechanical methods could be used dependent on factors given above.

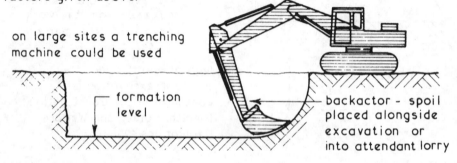

on large sites a trenching machine could be used

formation level

backactor - spoil placed alongside excavation or into attendant lorry

All subsoils have different abilities in remaining stable during excavation works. Most will assume a natural angle of repose or rest unless given temporary support. The presence of ground water apart from creating difficult working conditions can have an adverse effect on the subsoil's natural angle of repose.

Typical Angles of Repose ~
Excavations cut to a natural angle of repose are called battered.

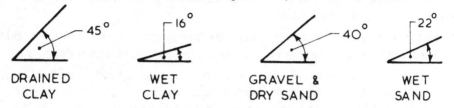

DRAINED CLAY — 45°
WET CLAY — 16°
GRAVEL & DRY SAND — 40°
WET SAND — 22°

Factors for Temporary Support of Excavations ~

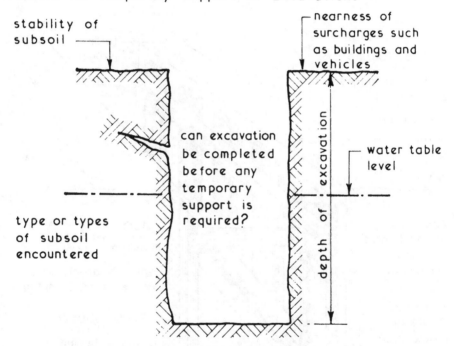

stability of subsoil

nearness of surcharges such as buildings and vehicles

can excavation be completed before any temporary support is required?

water table level

type or types of subsoil encountered

depth of excavation

Time factors such as period during which excavation will remain open and the time of year when work is carried out.

The need for an assessment of risk with regard to the support of excavations and protection of people within, is contained in the Construction (Health, Safety and Welfare) Regulations 1996.

Temporary Support ~ in the context of excavations this is called timbering irrespective of the actual materials used. If the sides of the excavation are completely covered with timbering it is known as close timbering whereas any form of partial covering is called open timbering.

An adequate supply of timber or other suitable material must be available and used to prevent danger to any person employed in an excavation from a fall or dislodgement of materials forming the sides of an excavation.

A suitable barrier or fence must be provided to the sides of all excavations or alternatively they must be securely covered.

Materials must not be placed near to the edge of any excavation, nor must plant be placed or moved near to any excavation so that persons employed in the excavation are endangered.

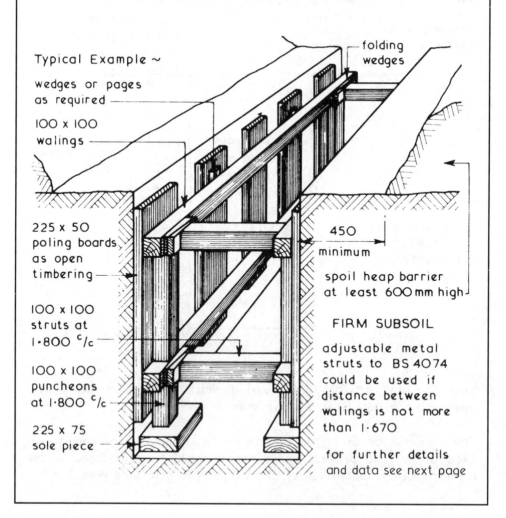

Typical Example ~

wedges or pages as required

100 x 100 walings

folding wedges

225 x 50 poling boards as open timbering

100 x 100 struts at 1·800 c/c

100 x 100 puncheons at 1·800 c/c

225 x 75 sole piece

450 minimum

spoil heap barrier at least 600 mm high

FIRM SUBSOIL

adjustable metal struts to BS 4074 could be used if distance between walings is not more than 1·670

for further details and data see next page

Poling Boards ~ a form of temporary support which is placed in position against the sides of excavation after the excavation work has been carried out. Poling boards are placed at centres according to the stability of the subsoils encountered.

Runners ~ a form of temporary support which is driven into position ahead of the excavation work either to the full depth or by a drive and dig technique where the depth of the runner is always lower than that of the excavation.

Trench Sheeting ~ form of runner made from sheet steel with a trough profile – can be obtained with a lapped joint or an interlocking joint.

Water ~ if present or enters an excavation, a pit or sump should be excavated below the formation level to act as collection point from which the water can be pumped away.

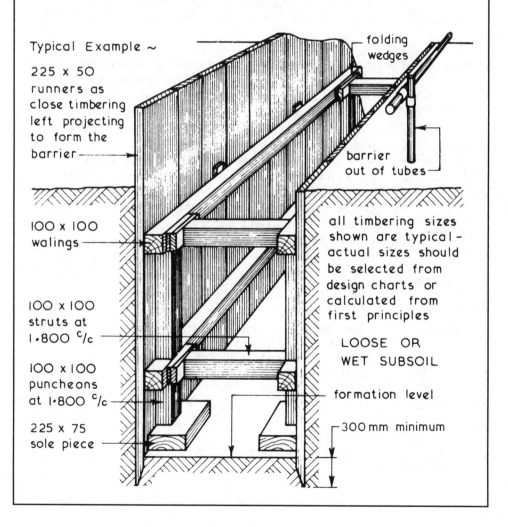

Typical Example ~

225 x 50 runners as close timbering left projecting to form the barrier

folding wedges

barrier out of tubes

100 x 100 walings

all timbering sizes shown are typical - actual sizes should be selected from design charts or calculated from first principles

100 x 100 struts at 1·800 c/c

LOOSE OR WET SUBSOIL

100 x 100 puncheons at 1·800 c/c

formation level

225 x 75 sole piece

300 mm minimum

Concrete ~ a mixture of cement + fine aggregate + coarse aggregate + water in controlled proportions and of a suitable quality.

Cement ~ powder produced from clay and chalk or limestone. In general most concrete is made with ordinary or rapid hardening Portland cement, both types being manufactured to the recommendations of BS EN 197-1. Ordinary Portland cement is adequate for most purposes but has a low resistance to attack by acids and sulphates. Rapid hardening Portland cement does not set faster than ordinary Portland cement but it does develop its working strength at a faster rate. For a concrete which must have an acceptable degree of resistance to sulphate attack sulphate resisting Portland cement made to the recommendations of BS 4027 could be specified.

BAGS

SILOS

Aggregates ~ shape, surface texture and grading (distribution of particle sizes) are factors which influence the workability and strength of a concrete mix. Fine aggregates are generally regarded as those materials which pass through a 4mm sieve whereas coarse aggregates are retained on a 4mm sieve. Dense aggregates have a density of more than 1200 kg/m^3 for coarse aggregates and more than 1250 kg/m^3 for fine aggregates. These are detailed in BS EN 12620: Aggregates for concrete. Lightweight aggregates include clinker; foamed or expanded blastfurance slag and exfoliated and expanded materials such as vermiculite, perlite, clay and sintered pulverized-fuel ash to BS EN 13055-1.

coarse aggregate

fine aggregate

Water ~ must be clean and free from impurities which are likely to affect the quality or strength of the resultant concrete. Pond, river, canal and sea water should not be used and only water which is fit for drinking should be specified.

drinking water quality

Cement ~ whichever type of cement is being used it must be properly stored on site to keep it in good condition. The cement must be kept dry since contact with any moisture whether direct or airborne could cause it to set. A rotational use system should be introduced to ensure that the first batch of cement delivered is the first to be used.

Typical Storage Methods ~

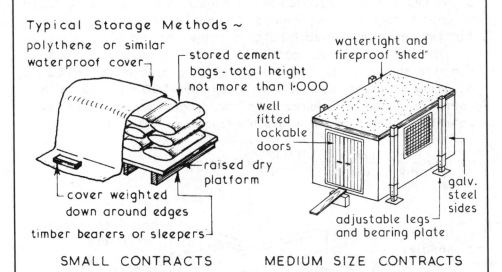

polythene or similar waterproof cover

stored cement bags - total height not more than 1·000

watertight and fireproof "shed"

well fitted lockable doors

raised dry platform

cover weighted down around edges

timber bearers or sleepers

adjustable legs and bearing plate

galv. steel sides

SMALL CONTRACTS   MEDIUM SIZE CONTRACTS

LARGE CONTRACTS — for bagged cement watertight container as above. For bulk delivery loose cement, a cement storage silo.

Aggregates ~ essentials of storage are to keep different aggregate types and/or sizes separate, store on a clean, hard, free draining surface and to keep the stored aggregates clean and free of leaves and rubbish.

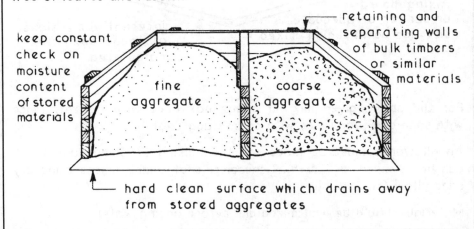

keep constant check on moisture content of stored materials

retaining and separating walls of bulk timbers or similar materials

fine aggregate   coarse aggregate

hard clean surface which drains away from stored aggregates

285

Concrete Batching ~ a batch is one mixing of concrete and can be carried out by measuring the quantities of materials required by volume or weight. The main aim of both methods is to ensure that all consecutive batches are of the same standard and quality.

Volume Batching ~ concrete mixes are often quoted by ratio such as 1:2:4 (cement : fine aggregate or sand : coarse aggregate). Cement weighing 50 kg has a volume of 0·033 m³ therefore for the above mix 2 × 0·033 (0·066 m³) of sand and 4 × 0·033 (0·132 m³) of coarse aggregate is required. To ensure accurate amounts of materials are used for each batch a gauge box should be employed its size being based on convenient handling. Ideally a batch of concrete should be equated to using 50 kg of cement per batch. Assuming a gauge box 300 mm deep and 300 mm wide with a volume of half the required sand the gauge box size would be –
volume = length × width × depth = length × 300 × 300

$$\text{length} = \frac{\text{volume}}{\text{width} \times \text{depth}} = \frac{0 \cdot 033}{0 \cdot 3 \times 0 \cdot 3} = 0 \cdot 366 \text{ m}$$

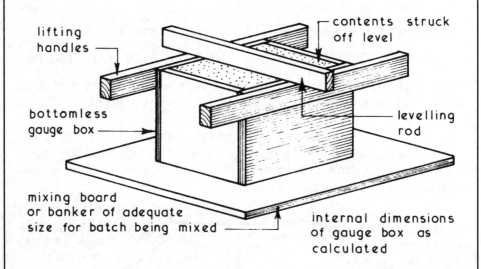

lifting handles

contents struck off level

bottomless gauge box

levelling rod

mixing board or banker of adequate size for batch being mixed

internal dimensions of gauge box as calculated

For the above given mix fill gauge box once with cement, twice with sand and four times with coarse aggregate.

An allowance must be made for the bulking of damp sand which can be as much as 33⅓%. General rule of thumb unless using dry sand allow for 25% bulking.

Materials should be well mixed dry before adding water.

Weight or Weigh Batching ~ this is a more accurate method of measuring materials for concrete than volume batching since it reduces considerably the risk of variation between different batches. The weight of sand is affected very little by its dampness which in turn leads to greater accuracy in proportioning materials. When loading a weighing hopper the materials should be loaded in a specific order –

1. Coarse aggregates – tends to push other materials out and leaves the hopper clean.
2. Cement – this is sandwiched between the other materials since some of the fine cement particles could be blown away if cement is put in last.
3. Sand or fine Aggregates – put in last to stabilise the fine lightweight particles of cement powder.

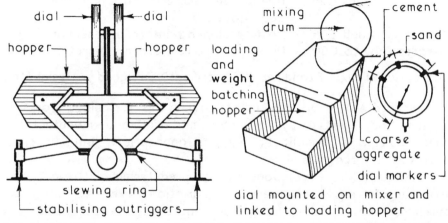

INDEPENDENT WEIGHT BATCHER   INTEGRAL WEIGHT BATCHER

Typical Densities ~ cement – 1440 kg/m^3 sand – 1600 kg/m^3 coarse aggregate – 1440 kg/m^3

Water/Cement Ratio ~ water in concrete has two functions –

1. Start the chemical reaction which causes the mixture to set into a solid mass.
2. Give the mix workability so that it can be placed, tamped or vibrated into the required position.

Very little water is required to set concrete (approximately 0·2 w/c ratio) the surplus evaporates leaving minute voids therefore the more water added to the mix to increase its workability the weaker is the resultant concrete. Generally w/c ratios of 0·4 to 0·5 are adequate for most purposes.

Concrete ~ a composite with many variables, represented by numerous gradings which indicate components, quality and manufacturing control.

Grade mixes: C7.5, C10, C15, C20, C25, C30, C35, C40, C45, C50, C55, and C60; F3, F4 and F5; IT2, IT2.5, and IT3.

C = Characteristic compressive

F = Flexural $\Big\}$ strengths at 28 days (N/mm^2)

IT = Indirect tensile

NB. If the grade is followed by a 'P', e.g. C30P, this indicates a prescribed mix (see below).

Grades C7.5 and C10 – Unreinforced plain concrete.
Grades C15 and C20 – Plain concrete or if reinforced containing lightweight aggregate.
Grades C25 – Reinforced concrete containing dense aggregate.
Grades C30 and C35 – Post-tensioned reinforced concrete.
Grades C40 to C60 – Pre-tensioned reinforced concrete.

Categories of mix: 1. Standard; 2. Prescribed; 3. Designed; 4. Designated.

1. Standard Mix – BS guidelines provide this for minor works or in situations limited by available material and manufacturing data. Volume or weight batching is appropriate, but no grade over C30 is recognised.

2. Prescribed Mix – components are predetermined (to a recipe) to ensure strength requirements. Variations exist to allow the purchaser to specify particular aggregates, admixtures and colours. All grades permitted.

3. Designed Mix – concrete is specified to an expected performance. Criteria can include characteristic strength, durability and workability, to which a concrete manufacturer will design and supply an appropriate mix. All grades permitted.

4. Designated Mix – selected for specific applications. General (GEN) graded 0–4, 7.5–25 N/mm^2 for foundations, floors and external works. Foundations (FND) graded 2, 3, 4A and 4B, 35 N/mm^2 mainly for sulphate resisting foundations.

Paving (PAV) graded 1 or 2, 35 or 45 N/mm^2 for roads and drives.

Reinforced (RC) graded 30, 35, 40, 45 and 50 N/mm^2 mainly for prestressing.

See also BS EN 206-1: Concrete. Specification, performance, production and conformity, and BS's 8500-1 and -2: Concrete.

Concrete Supply ~ this is usually geared to the demand or the rate at which the mixed concrete can be placed. Fresh concrete should always be used or placed within 30 minutes of mixing to prevent any undue drying out. Under no circumstances should more water be added after the initial mixing.

Small Batches ~ small easily transported mixers with 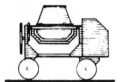 output capacities of up to 100 litres can be used for small and intermittent batche⸱ These mixers are versatile and robust machines which can be used for mixing mortars and plasters as well as concrete.

Medium to Large Batches ~ mixers with output capacities 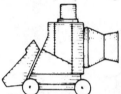 from 100 litres to 10 m³ with either diesel or electric motors. Many models are available with tilting or reversing drum discharge, integral weigh batching and loading hopper and a controlled water supply.

Ready Mixed Concrete ~ used mainly for large concrete batches of up to 6 m³. This method of concrete supply has the advantages of eliminating the need for site space to accommodate storage of materials, mixing plant and the need to employ adequately trained site staff who can constantly produce reliable and consistent concrete mixes.  Ready mixed concrete supply depots also have better facilities and arrangements for producing and supplying mixed concrete in winter or inclement weather conditions. In many situations it is possible to place the ready mixed concrete into the required position direct from the delivery lorry via the delivery chute or by feeding it into a concrete pump. The site must be capable of accepting the 20 tonnes laden weight of a typical ready mixed concrete lorry with a turning circle of about 15·000. The supplier will want full details of mix required and the proposed delivery schedule.

Ref. BS EN 206-1: Concrete. Specification, performance, production and conformity.

Cofferdams ~ these are temporary enclosures installed in soil or water to prevent the ingress of soil and/or water into the working area with the cofferdam. They are usually constructed from interlocking steel sheet piles which are suitably braced or tied back with ground anchors. Alternatively a cofferdam can be installed using any structural material which will fulfil the required function.

Typical Cofferdam Details ~

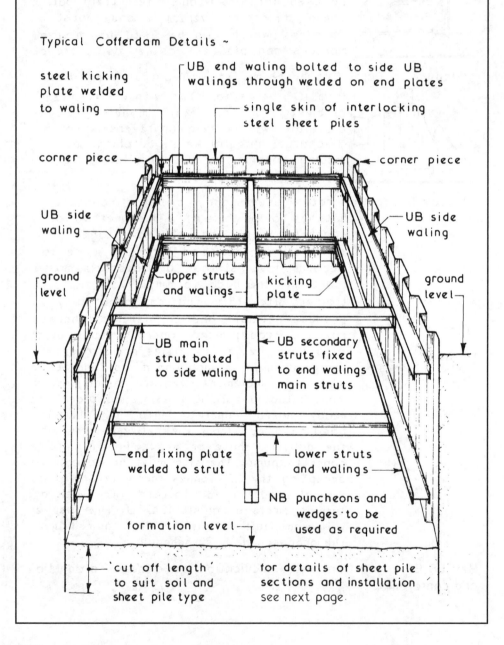

steel kicking plate welded to waling

UB end waling bolted to side UB walings through welded on end plates

single skin of interlocking steel sheet piles

corner piece

corner piece

UB side waling

UB side waling

ground level

upper struts and walings

kicking plate

ground level

UB main strut bolted to side waling

UB secondary struts fixed to end walings main struts

end fixing plate welded to strut

lower struts and walings

formation level

NB puncheons and wedges to be used as required

'cut off length' to suit soil and sheet pile type

for details of sheet pile sections and installation see next page.

Steel Sheet Piling ~ apart from cofferdam work steel sheet can be used as a conventional timbering material in excavations and to form permanent retaining walls. Three common formats of steel sheet piles with interlocking joints are available with a range of section sizes and strengths up to a usual maximum length of 18·000:-

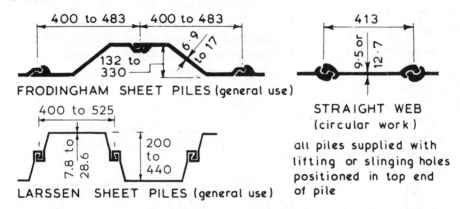

FRODINGHAM SHEET PILES (general use)

STRAIGHT WEB (circular work)

LARSSEN SHEET PILES (general use)

all piles supplied with lifting or slinging holes positioned in top end of pile

Installing Steel Sheet Piles ~ to ensure that the sheet piles are pitched and installed vertically a driving trestle or guide frame is used. These are usually purpose built to accommodate a panel of 10 to 12 pairs of piles. The piles are lifted into position by a crane and driven by means of percussion piling hammer or alternatively they can be pushed into the ground by hydraulic rams acting against the weight of the power pack which is positioned over the heads of the pitched piles.

Typical Installation Details ~

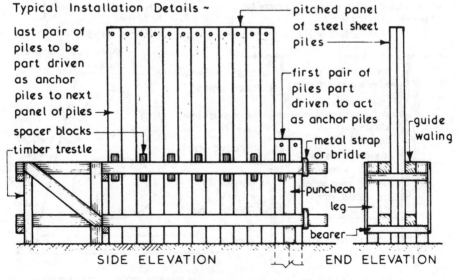

Note: Rot-proof PVC sheet piling is also available.

Caissons ~ these are box-like structures which are similar in concept to cofferdams but they usually form an integral part of the finished structure. They can be economically constructed and installed in water or soil where the depth exceeds 18·000. There are 4 basic types of caisson namely:-

1. Box Caissons
2. Open Caissons
3. Monolithic Caissons
} usually of precast concrete and used in water being towed or floated into position and sunk – land caissons are of the open type and constructed in-situ.

4. Pneumatic Caissons – used in water – see next page.

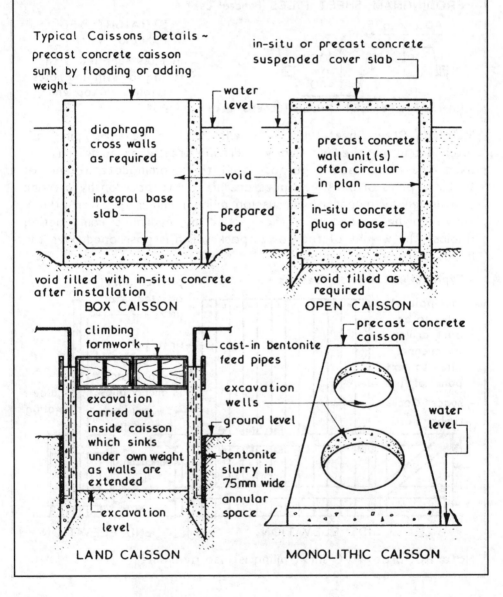

Typical Caissons Details ~
precast concrete caisson sunk by flooding or adding weight

in-situ or precast concrete suspended cover slab

water level

diaphragm cross walls as required

void

integral base slab

prepared bed

precast concrete wall unit(s) – often circular in plan

in-situ concrete plug or base

void filled with in-situ concrete after installation
BOX CAISSON

void filled as required
OPEN CAISSON

climbing formwork

cast-in bentonite feed pipes

excavation carried out inside caisson which sinks under own weight as walls are extended

ground level

bentonite slurry in 75mm wide annular space

excavation level

LAND CAISSON

precast concrete caisson

excavation wells

water level

MONOLITHIC CAISSON

Pneumatic Caissons ~ these are sometimes called compressed air caissons and are similar in concept to open caissons. They can be used in difficult subsoil conditions below water level and have a pressurised lower working chamber to provide a safe dry working area. Pneumatic caissons can be made of concrete whereby they sink under their own weight or they can be constructed from steel with hollow walls which can be filled with water to act as ballast. These caissons are usually designed to form part of the finished structure.

Typical Pneumatic Caisson Details ~

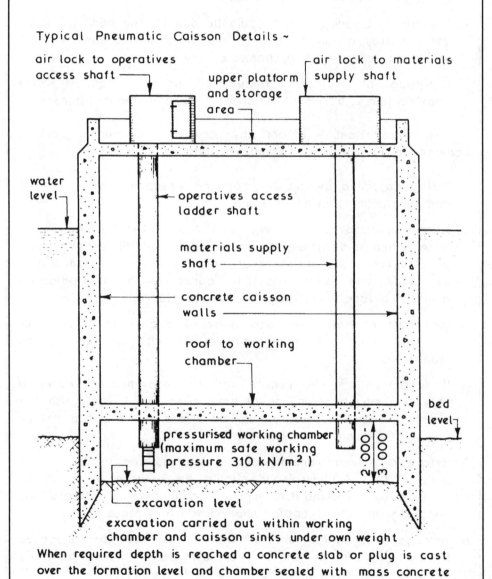

air lock to operatives access shaft

air lock to materials supply shaft

upper platform and storage area

water level

operatives access ladder shaft

materials supply shaft

concrete caisson walls

roof to working chamber

bed level

pressurised working chamber (maximum safe working pressure 310 kN/m²)

2·000 - 3·000

excavation level

excavation carried out within working chamber and caisson sinks under own weight

When required depth is reached a concrete slab or plug is cast over the formation level and chamber sealed with mass concrete

Underpinning ~ the main objective of most underpinning work is to transfer the load carried by a foundation from its existing bearing level to a new level at a lower depth. Underpinning techniques can also be used to replace an existing weak foundation. An underpinning operation may be necessary for one or more of the following reasons:-

1. Uneven Settlement – this could be caused by uneven loading of the building, unequal resistance of the soil action of tree roots or cohesive soil settlement.

2. Increase in Loading – this could be due to the addition of an extra storey or an increase in imposed loadings such as that which may occur with a change of use.

3. Lowering of Adjacent Ground – usually required when constructing a basement adjacent to existing foundations.

General Precautions ~ before any form of underpinning work is commenced the following precautions should be taken:-

1. Notify adjoining owners of proposed works giving full details and temporary shoring or tying.

2. Carry out a detailed survey of the site, the building to be underpinned and of any other adjoining or adjacent building or structures. A careful record of any defects found should be made and where possible agreed with the adjoining owner(s) before being lodged in a safe place.

3. Indicators or 'tell tales' should be fixed over existing cracks so that any subsequent movements can be noted and monitored.

4. If settlement is the reason for the underpinning works a thorough investigation should be carried out to establish the cause and any necessary remedial work put in hand before any underpinning works are started.

5. Before any underpinning work is started the loads on the building to be underpinned should be reduced as much as possible by removing the imposed loads from the floors and installing any props and/or shoring which is required.

6. Any services which are in the vicinity of the proposed underpinning works should be identified, traced, carefully exposed, supported and protected as necessary.

Underpinning to Walls ~ to prevent fracture, damage or settlement of the wall(s) being underpinned the work should always be carried out in short lengths called legs or bays. The length of these bays will depend upon the following factors:-

1. Total length of wall to be underpinned.

2. Wall loading.

3. General state of repair and stability of wall and foundation to be underpinned.

4. Nature of subsoil beneath existing foundation.

5. Estimated spanning ability of existing foundation.

Generally suitable bay lengths are:-

1·000 to 1·500 for mass concrete strip foundations supporting walls of traditional construction.

1·500 to 3·000 for reinforced concrete strip foundations supporting walls of moderate loading.

In all the cases the total sum of the unsupported lengths of wall should not exceed 25% of the total wall length.

The sequence of bays should be arranged so that working in adjoining bays is avoided until one leg of underpinning has been completed, pinned and cured sufficiently to support the wall above.

Typical   Underpinning Schedule ~

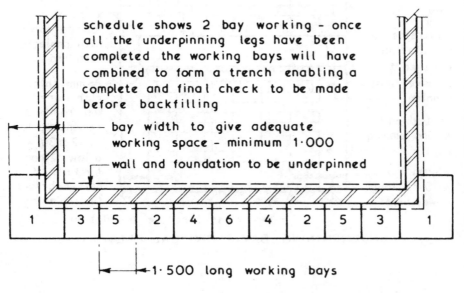

schedule shows 2 bay working - once all the underpinning legs have been completed the working bays will have combined to form a trench enabling a complete and final check to be made before backfilling

bay width to give adequate working space - minimum 1·000

wall and foundation to be underpinned

| 1 | 3 | 5 | 2 | 4 | 6 | 4 | 2 | 5 | 3 | 1 |

1·500 long working bays

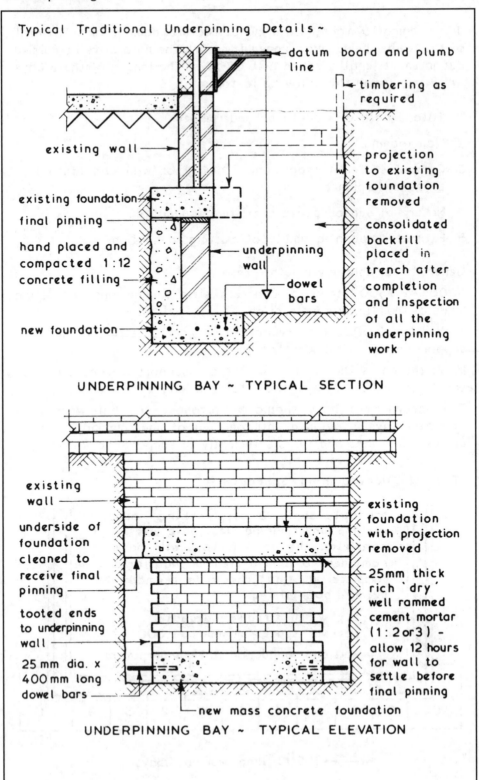

Typical Traditional Underpinning Details ~

datum board and plumb line

timbering as required

existing wall

projection to existing foundation removed

existing foundation

final pinning

hand placed and compacted 1:12 concrete filling

underpinning wall

dowel bars

consolidated backfill placed in trench after completion and inspection of all the underpinning work

new foundation

UNDERPINNING BAY ~ TYPICAL SECTION

existing wall

underside of foundation cleaned to receive final pinning

tooted ends to underpinning wall

25 mm dia. x 400 mm long dowel bars

existing foundation with projection removed

25mm thick rich 'dry' well rammed cement mortar (1:2 or 3) - allow 12 hours for wall to settle before final pinning

new mass concrete foundation

UNDERPINNING BAY ~ TYPICAL ELEVATION

Jack Pile Underpinning ~ this method can be used when the depth of a suitable bearing capacity subsoil is too deep to make traditional underpinning uneconomic. Jack pile underpinning is quiet, vibration free and flexible since the pile depth can be adjusted to suit subsoil conditions encountered. The existing foundations must be in a good condition since they will have to span over the heads of the pile caps which are cast onto the jack pile heads after the hydraulic jacks have been removed.

Typical Details ~

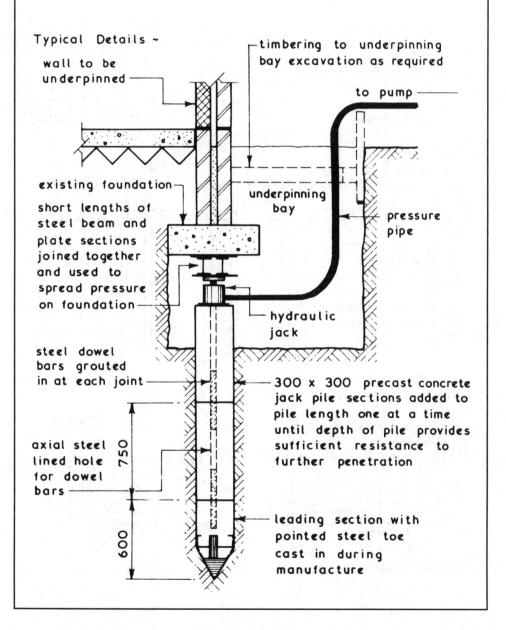

wall to be underpinned

timbering to underpinning bay excavation as required

to pump

existing foundation

short lengths of steel beam and plate sections joined together and used to spread pressure on foundation

underpinning bay

pressure pipe

hydraulic jack

steel dowel bars grouted in at each joint

300 x 300 precast concrete jack pile sections added to pile length one at a time until depth of pile provides sufficient resistance to further penetration

axial steel lined hole for dowel bars

750

600

leading section with pointed steel toe cast in during manufacture

Needle and Pile Underpinning ~ this method of underpinning can be used where the condition of the existing foundation is unsuitable for traditional or jack pile underpinning techniques. The brickwork above the existing foundation must be in a sound condition since this method relies on the 'arching effect' of the brick bonding to transmit the wall loads onto the needles and ultimately to the piles. The piles used with this method are usually small diameter bored piles – see page 230.

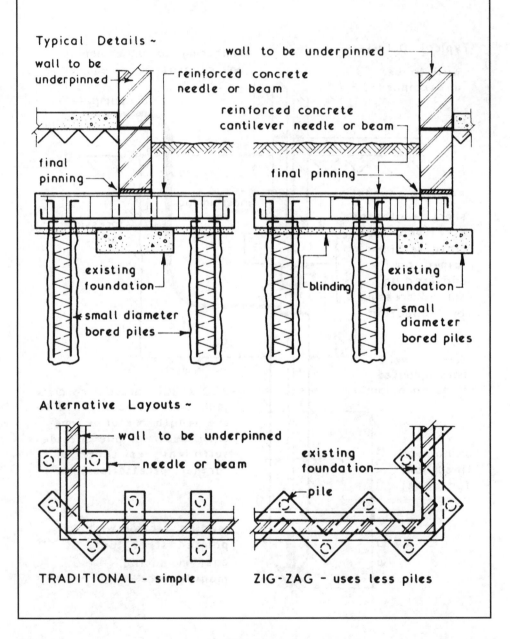

Typical Details ~

wall to be underpinned

wall to be underpinned

reinforced concrete needle or beam

reinforced concrete cantilever needle or beam

final pinning

final pinning

existing foundation

existing foundation

blinding

small diameter bored piles

small diameter bored piles

Alternative Layouts ~

wall to be underpinned

needle or beam

existing foundation

pile

TRADITIONAL - simple

ZIG-ZAG - uses less piles

'Pynford' Stool Method of Underpinning ~ this method can be used where the existing foundations are in a poor condition and it enables the wall to be underpinned in a continuous run without the need for needles or shoring. The reinforced concrete beam formed by this method may well be adequate to spread the load of the existing wall or it may be used in conjunction with other forms of underpinning such as traditional and jack pile.

Typical Details ~

Stage 1 - holes formed in wall to receive steel or precast concrete stools

Stage 2 - stools inserted and pinned to soffit of brickwork over opening

Stage 3 - brickwork between pinned stools removed to leave wall supported on pinned stools

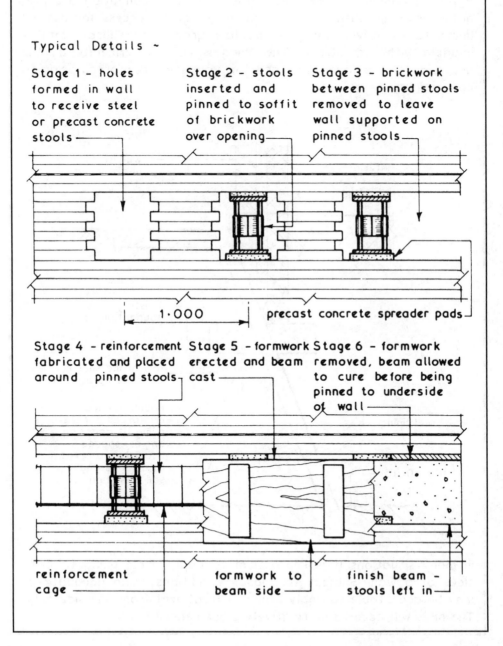

1·000   precast concrete spreader pads

Stage 4 - reinforcement fabricated and placed around pinned stools

Stage 5 - formwork erected and beam cast

Stage 6 - formwork removed, beam allowed to cure before being pinned to underside of wall

reinforcement cage

formwork to beam side

finish beam - stools left in

Root Pile or Angle Piling ~ this is a much simpler alternative to traditional underpinning techniques, applying modern concrete drilling equipment to achieve cost benefits through time saving. The process is also considerably less disruptive, as large volumes of excavation are avoided. Where sound bearing strata can be located within a few metres of the surface, wall stability is achieved through lined reinforced concrete piles installed in pairs, at opposing angles. The existing floor, wall and foundation are pre-drilled with air flushed percussion auger, giving access for a steel lining to be driven through the low grade/clay subsoil until it impacts with firm strata. The lining is cut to terminate at the underside of the foundation and the void steel reinforced prior to concreting.

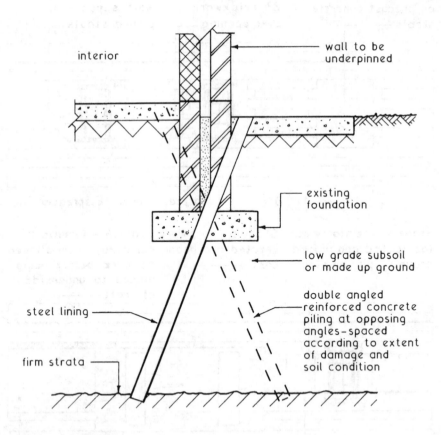

interior

wall to be underpinned

existing foundation

low grade subsoil or made up ground

double angled reinforced concrete piling at opposing angles-spaced according to extent of damage and soil condition

steel lining

firm strata

In many situations it is impractical to apply angle piling to both sides of a wall. Subject to subsoil conditions being adequate, it may be acceptable to apply remedial treatment from one side only. The piles will need to be relatively close spaced.

Underpinning Columns ~ columns can be underpinned in the some manner as walls using traditional or jack pile methods after the columns have been relieved of their loadings. The beam loads can usually be transferred from the columns by means of dead shores and the actual load of the column can be transferred by means of a pair of beams acting against a collar attached to the base of the column shaft.

Typical Details ~

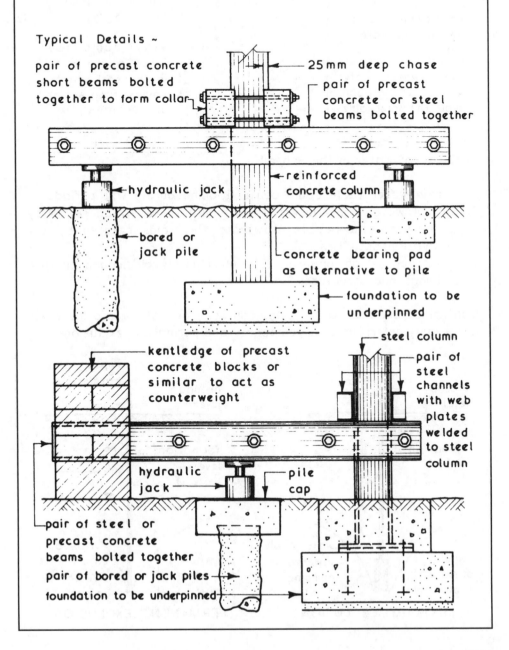

pair of precast concrete short beams bolted together to form collar

25mm deep chase

pair of precast concrete or steel beams bolted together

hydraulic jack

reinforced concrete column

bored or jack pile

concrete bearing pad as alternative to pile

foundation to be underpinned

kentledge of precast concrete blocks or similar to act as counterweight

steel column

pair of steel channels with web plates welded to steel column

hydraulic jack

pile cap

pair of steel or precast concrete beams bolted together

pair of bored or jack piles

foundation to be underpinned

Classification of Water ~ water can be classified by its relative position to or within the ground thus –

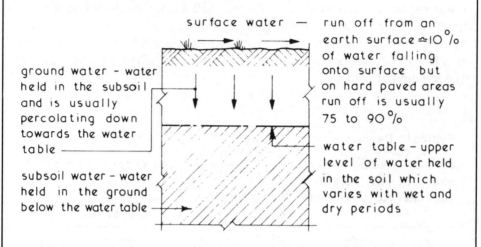

surface water — run off from an earth surface $\simeq 10\%$ of water falling onto surface but on hard paved areas run off is usually 75 to 90%

ground water - water held in the subsoil and is usually percolating down towards the water table

subsoil water - water held in the ground below the water table

water table – upper level of water held in the soil which varies with wet and dry periods

Problems of Water in the Subsoil ~

1. A high water table could cause flooding during wet periods.
2. Subsoil water can cause problems during excavation works by its natural tendency to flow into the voids created by the excavation activities.
3. It can cause an unacceptable humidity level around finished buildings and structures.

Control of Ground Water ~ this can take one of two forms which are usually referred to as temporary and permanent exclusion –

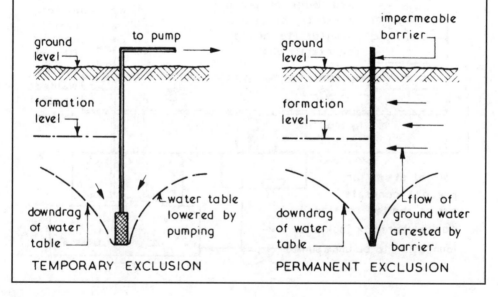

ground level

to pump

formation level

downdrag of water table

water table lowered by pumping

TEMPORARY EXCLUSION

impermeable barrier

ground level

formation level

downdrag of water table

flow of ground water arrested by barrier

PERMANENT EXCLUSION

Permanent Exclusion ~ this can be defined as the insertion of an impermeable barrier to stop the flow of water within the ground.

Temporary Exclusion ~ this can be defined as the lowering of the water table and within the economic depth range of 1·500 can be achieved by subsoil drainage methods, for deeper treatment a pump or pumps are usually involved.

Simple Sump Pumping ~ suitable for trench work and/or where small volumes of water are involved.

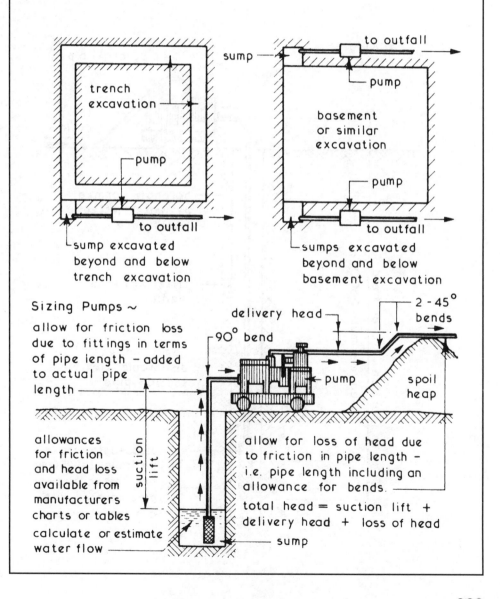

Jetted Sumps ~ this method achieves the same objectives as the simple sump methods of dewatering (previous page) but it will prevent the soil movement associated with this and other open sump methods. A borehole is formed in the subsoil by jetting a metal tube into the ground by means of pressurised water, to a depth within the maximum suction lift of the extract pump. The metal tube is withdrawn to leave a void for placing a disposable wellpoint and plastic suction pipe. The area surrounding the pipe is filled with coarse sand to function as a filtering media.

Typical Example ~

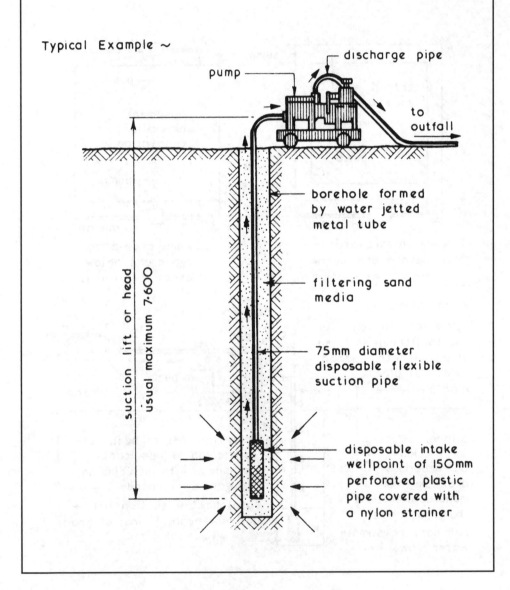

discharge pipe

pump

to outfall

suction lift or head
usual maximum 7·600

borehole formed by water jetted metal tube

filtering sand media

75mm diameter disposable flexible suction pipe

disposable intake wellpoint of 150mm perforated plastic pipe covered with a nylon strainer

Wellpoint Systems ~ method of lowering the water table to a position below the formation level to give a dry working area. The basic principle is to jet into the subsoil a series of wellpoints which are connected to a common header pipe which is connected to a vacuum pump. Wellpoint systems are suitable for most subsoils and can encircle an excavation or be laid progressively alongside as in the case of a trench excavation. If the proposed formation level is below the suction lift capacity of the pump a multi-stage system can be employed – see next page.

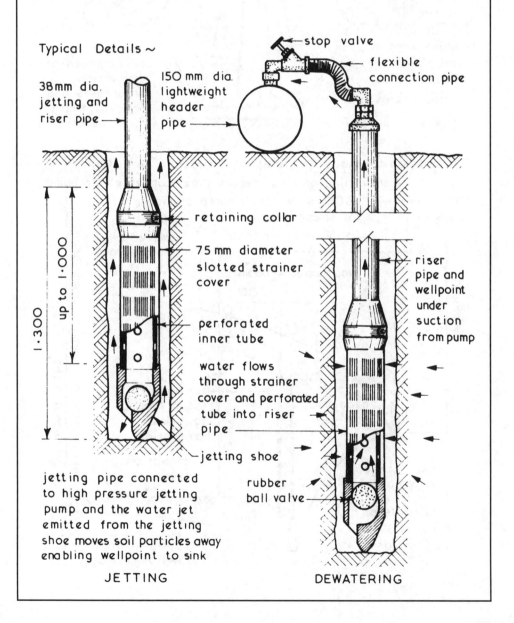

Typical Details ~

38mm dia. jetting and riser pipe

150 mm dia. lightweight header pipe

stop valve

flexible connection pipe

retaining collar

75 mm diameter slotted strainer cover

perforated inner tube

water flows through strainer cover and perforated tube into riser pipe

jetting shoe

riser pipe and wellpoint under suction from pump

rubber ball valve

up to 1·000

1·300

jetting pipe connected to high pressure jetting pump and the water jet emitted from the jetting shoe moves soil particles away enabling wellpoint to sink

JETTING

DEWATERING

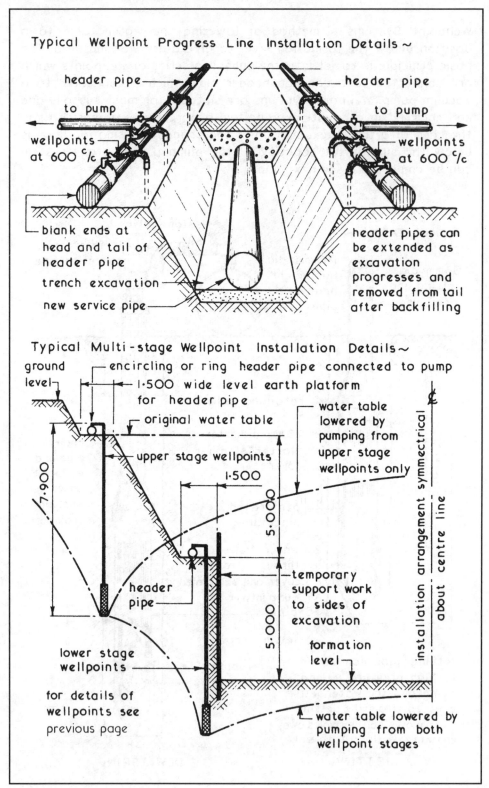

Typical Wellpoint Progress Line Installation Details ~

header pipe

header pipe

to pump

to pump

wellpoints at 600 ^{c}/c

wellpoints at 600 ^{c}/c

blank ends at head and tail of header pipe

header pipes can be extended as excavation progresses and removed from tail after backfilling

trench excavation

new service pipe

Typical Multi-stage Wellpoint Installation Details ~

ground level

encircling or ring header pipe connected to pump

1·500 wide level earth platform for header pipe

original water table

water table lowered by pumping from upper stage wellpoints only

upper stage wellpoints

1·500

7·900

5·000

header pipe

temporary support work to sides of excavation

5·000

lower stage wellpoints

formation level

for details of wellpoints see previous page

water table lowered by pumping from both wellpoint stages

installation arrangement symmetrical about centre line

Thin Grouted Membranes ~ these are permanent curtain or cut-off non-structural walls or barriers inserted in the ground to enclose the proposed excavation area. They are suitable for silts and sands and can be installed rapidly but they must be adequately supported by earth on both sides. The only limitation is the depth to which the formers can be driven and extracted.

Typical Details ~

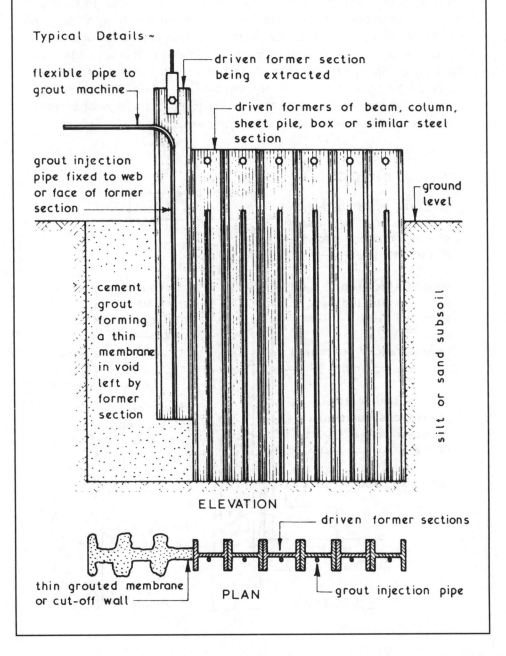

flexible pipe to grout machine

driven former section being extracted

driven formers of beam, column, sheet pile, box or similar steel section

grout injection pipe fixed to web or face of former section

ground level

cement grout forming a thin membrane in void left by former section

silt or sand subsoil

ELEVATION

driven former sections

thin grouted membrane or cut-off wall

PLAN

grout injection pipe

Contiguous or Secant Piling ~ this forms a permanent structural wall of interlocking bored piles. Alternate piles are bored and cast by traditional methods and before the concrete has fully hardened the interlocking piles are bored using a toothed flight auger. This system is suitable for most types of subsoil and has the main advantages of being economical on small and confined sites; capable of being formed close to existing foundations and can be installed with the minimum of vibration and noise. Ensuring a complete interlock of all piles over the entire length may be difficult to achieve in practice therefore the exposed face of the piles is usually covered with a mesh or similar fabric and face with rendering or sprayed concrete. Alternatively a reinforced concrete wall could be cast in front of the contiguous piling. This method of ground water control is suitable for structures such as basements, road underpasses and underground car parks.

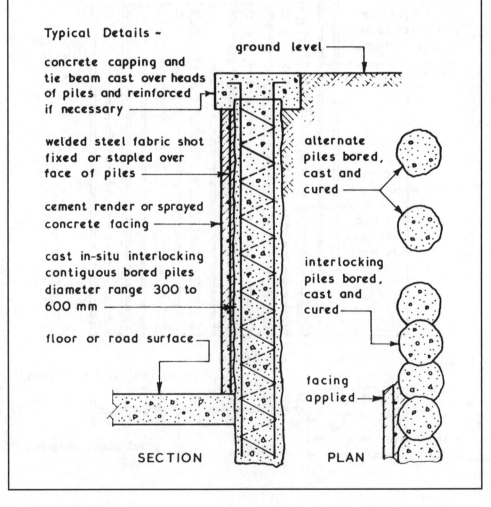

Typical Details ~

concrete capping and tie beam cast over heads of piles and reinforced if necessary

welded steel fabric shot fixed or stapled over face of piles

cement render or sprayed concrete facing

cast in-situ interlocking contiguous bored piles diameter range 300 to 600 mm

floor or road surface

ground level

alternate piles bored, cast and cured

interlocking piles bored, cast and cured

facing applied

SECTION   PLAN

Diaphragm Walls ~ these are structural concrete walls which can be cast in-situ (usually by the bentonite slurry method) or constructed using precast concrete components (see next page). They are suitable for most subsoils and their installation generates only a small amount of vibration and noise making them suitable for works close to existing buildings. The high cost of these walls makes them uneconomic unless they can be incorporated into the finished structure. Diaphragm walls are suitable for basements, underground car parks and similar structures.

Typical Cast In-situ Concrete Diaphragm Wall Details ~

ready mixed concrete supply
tremie pipe placing
bentonite return, storage and draw off tanks
displaced bentonite pumped to storage
kelly bar
crane boom
standard crane power unit
pcc lined guide trench
reinforcement
stop end pipes as formers for interlocking joint between panels
hydraulic grab
placed concrete
bentonite slurry filling excavation void

panel 1   panel 7   panel 2   panel 9
length of panel = 3 × hydraulic grab width

NB. Bentonite is a controlled mixture of fullers earth and water which produces a mud or slurry which has thixotropic properties and exerts a pressure in excess of earth +hydrostatic pressure present on sides of excavation.

309

Precast Concrete Diaphragm Walls ~ these walls have the some applications as their in-situ counterparts and have the advantages of factory produced components but lack the design flexibility of cast in-situ walls. The panel or post and panel units are installed in a trench filled with a special mixture of bentonite and cement with a retarder to control the setting time. This mixture ensures that the joints between the wall components are effectively sealed. To provide stability the panels or posts are tied to the retained earth with ground anchors.

Typical Precast Concrete Diaphragm Wall Details ~

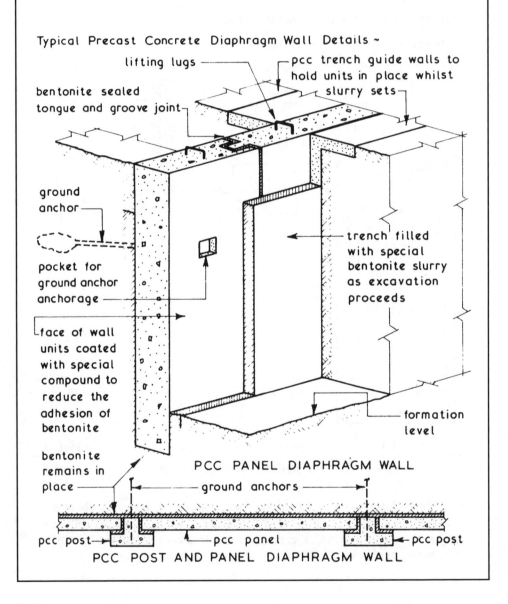

lifting lugs

pcc trench guide walls to hold units in place whilst slurry sets

bentonite sealed tongue and groove joint

ground anchor

pocket for ground anchor anchorage

trench filled with special bentonite slurry as excavation proceeds

face of wall units coated with special compound to reduce the adhesion of bentonite

formation level

bentonite remains in place

PCC PANEL DIAPHRAGM WALL

ground anchors

pcc post

pcc panel

pcc post

PCC POST AND PANEL DIAPHRAGM WALL

Grouting Methods ~ these techniques are used to form a curtain or cut off wall in high permeability soils where pumping methods could be uneconomic. The curtain walls formed by grouting methods are non-structural therefore adequate earth support will be required and in some cases this will be a distance of at least 4·000 from the face of the proposed excavation. Grout mixtures are injected into the soil by pumping the grout at high pressure through special injection pipes inserted in the ground. The pattern and spacing of the injection pipes will depend on the grout type and soil conditions.

Grout Types ~

1. Cement Grouts – mixture of neat cement and water cement sand up to 1:4 or PFA (pulverized fuel ash) cement to a 1:1 ratio. Suitable for coarse grained soils and fissured and jointed rock strata.

2. Chemical Grouts – one shot (premixed) of two shot (first chemical is injected followed immediately by second chemical resulting in an immediate reaction) methods can be employed to form a permanent gel in the soil to reduce its permeability and at the same time increase the soil's strength. Suitable for medium to coarse sands and gravels.

3. Resin Grouts – these are similar in application to chemical grouts but have a low viscosity and can therefore penetrate into silty fine sands.

Typical Cement Grouting Details ~

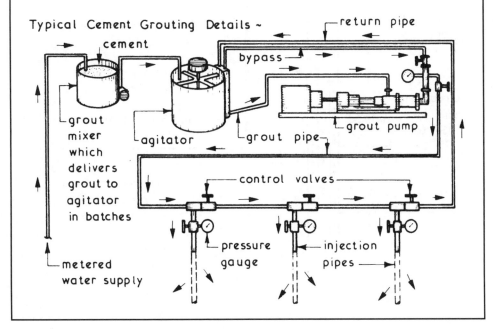

311

Ground Freezing Techniques ~ this method is suitable for all types of saturated soils and rock and for soils with a moisture content in excess of 8% of the voids. The basic principle is to insert into the ground a series of freezing tubes to form an ice wall thus creating an impermeable barrier. The treatment takes time to develop and the initial costs are high, therefore it is only suitable for large contracts of reasonable duration. The freezing tubes can be installed vertically for conventional excavations and horizontally for tunnelling works. The usual circulating brines employed are magnesium chloride and calcium chloride with a temperature of –15° to –25°C which would take 10 to 17 days to form an ice wall 1·000 thick. Liquid nitrogen could be used as the freezing medium to reduce the initial freezing period if the extra cost can be justified.

Typical Ground Freezing Details ~

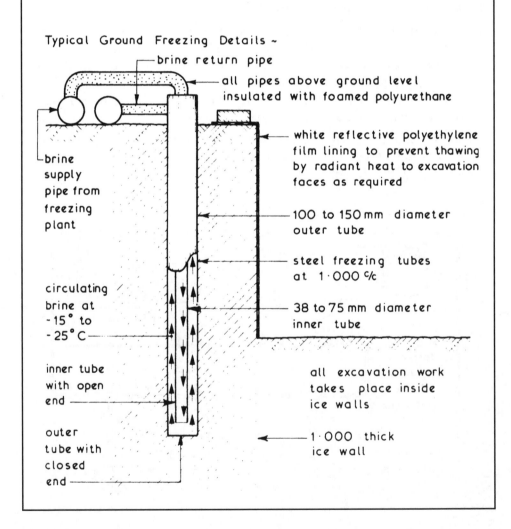

brine return pipe

all pipes above ground level insulated with foamed polyurethane

brine supply pipe from freezing plant

white reflective polyethylene film lining to prevent thawing by radiant heat to excavation faces as required

100 to 150 mm diameter outer tube

steel freezing tubes at 1·000 % c

circulating brine at –15° to –25°C

38 to 75 mm diameter inner tube

inner tube with open end

all excavation work takes place inside ice walls

outer tube with closed end

1·000 thick ice wall

Soil Investigation ~ before a decision is made as to the type of foundation which should be used on any particular site a soil investigation should be carried out to establish existing ground conditions and soil properties. The methods which can be employed together with other sources of information such as local knowledge, ordnance survey and geological maps, mining records and aerial photography should be familiar to students at this level. If such an investigation reveals a naturally poor subsoil or extensive filling the designer has several options:-

1. Not to Build – unless a new and suitable site can be found building is only possible if the poor ground is localised and the proposed foundations can be designed around these areas with the remainder of the structure bridging over these positions.

2. Remove and Replace – the poor ground can be excavated, removed and replaced by compacted fills. Using this method there is a risk of differential settlement and generally for depths over 4·000 it is uneconomic.

3. Surcharging – this involves preloading the poor ground with a surcharge of aggregate or similar material to speed up settlement and thereby improve the soil's bearing capacity. Generally this method is uneconomic due to the time delay before actual building operations can commence which can vary from a few weeks to two or more years.

4. Vibration – this is a method of strengthening ground by vibrating a granular soil into compacted stone columns either by using the natural coarse granular soil or by replacement – see pages 314 and 315.

5. Dynamic Compaction – this is a method of soil improvement which consists of dropping a heavy weight through a considerable vertical distance to compact the soil and thus improve its bearing capacity and is especially suitable for granular soils – see page 316.

6. Jet Grouting – this method of consolidating ground can be used in all types of subsoil and consists of lowering a monitor probe into a 150 mm diameter prebored guide hole. The probe has two jets the upper of which blasts water, concentrated by compressed air to force any loose material up the guide to ground level. The lower jet fills the void with a cement slurry which sets into a solid mass – see page 317.

Ground Vibration ~ the objective of this method is to strengthen the existing soil by rearranging and compacting coarse granular particles to form stone columns with the ground. This is carried out by means of a large poker vibrator which has an effective compacting radius of 1·500 to 2·700. On large sites the vibrator is inserted on a regular triangulated grid pattern with centres ranging from 1·500 to 3·000. In coarse grained soils extra coarse aggregate is tipped into the insertion positions to make up levels as required whereas in clay and other fine particle soils the vibrator is surged up and down enabling the water jetting action to remove the surrounding soft material thus forming a borehole which is backfilled with a coarse granular material compacted in-situ by the vibrator. The backfill material is usually of 20 to 70 mm size of uniform grading within the chosen range. Ground vibration is not a piling system but a means of strengthening ground to increase the bearing capacity within a range of 200 to 500 kN/m².

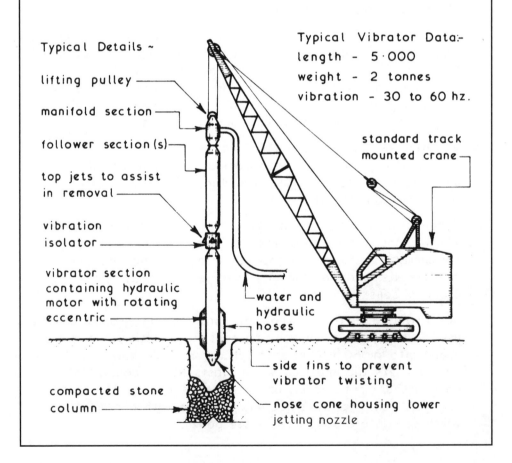

Typical Details ~

lifting pulley

manifold section

follower section(s)

top jets to assist in removal

vibration isolator

vibrator section containing hydraulic motor with rotating eccentric

compacted stone column

Typical Vibrator Data:-
length - 5·000
weight - 2 tonnes
vibration - 30 to 60 hz.

standard track mounted crane

water and hydraulic hoses

side fins to prevent vibrator twisting

nose cone housing lower jetting nozzle

Sand Compaction – applied to non-cohesive subsoils where the granular particles are rearranged into a denser condition by poker vibration.

The crane-suspended vibrating poker is water-jetted into the ground using a combination of self weight and water displacement of the finer soil particles to penetrate the ground. Under this pressure, the soil granules compact to increase in density as the poker descends. At the appropriate depth, which may be determined by building load calculations or the practical limit of plant (generally 30 m max.), jetting ceases and fine aggregates or sand are infilled around the poker. The poker is then gradually withdrawn compacting the granular fill in the process. Compaction continues until sand fill reaches ground level. Spacing of compaction boreholes is relatively close to ensure continuity and an integral ground condition.

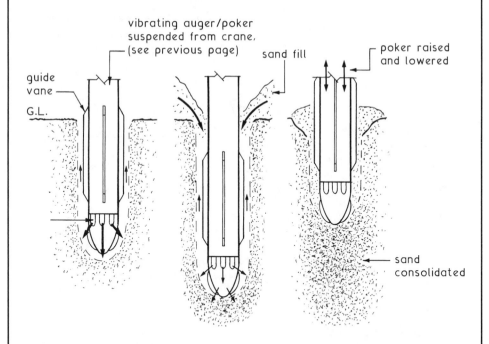

1. Vibrating poker penetrates ground under full water jet pressure.

2. At design depth, water pressure is reduced and sand fill introduced and compacted.

3. With resistance to compaction, poker is raised and lowered to consolidate further sand.

Sand compaction procedure

Dynamic Compaction ~ this method of ground improvement consists of dropping a heavy weight from a considerable height and is particularly effective in granular soils. Where water is present in the subsoil, trenches should be excavated to allow the water to escape and not collect in the craters formed by the dropped weight. The drop pattern, size of weight and height of drop are selected to suit each individual site but generally 3 or 4 drops are made in each position forming a crater up to 2·500 deep and 5·000 in diameter. Vibration through the subsoil can be a problem with dynamic compaction operations therefore the proximity and condition of nearby buildings must be considered together with the depth position and condition of existing services on site.

Typical Details ~

NB. Final ground level after compaction treatment and final levelling could be up to 1·500 lower than original ground level

heavy duty track mounted crane

weight range 10 to 20 tonnes

depth up to 2·500 after 4 blows

crater up to 5·000 in diameter

free fall distance range 15·000 to 25·000

compacted soil

20° to 40° spread

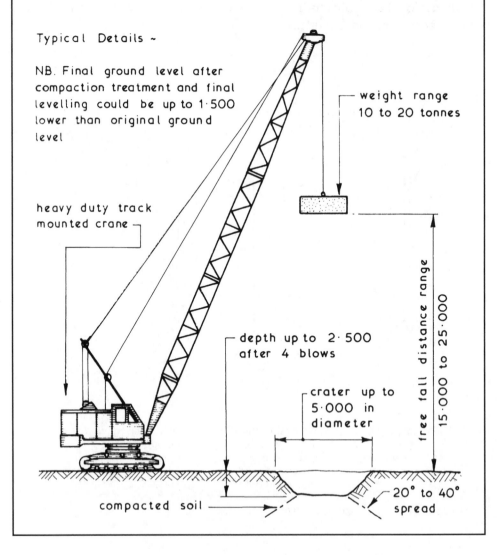

Jet Grouting ~ this is a means of consolidating ground by lowering into preformed bore holes a monitor probe. The probe is rotated and the sides of the bore hole are subjected to a jet of pressurised water and air from a single outlet which enlarges and compacts the bore hole sides. At the same time a cement grout is being introduced under pressure to fill the void being created. The water used by the probe and any combined earth is forced up to the surface in the form of a sludge. If the monitor probe is not rotated grouted panels can be formed. The spacing, depth and layout of the bore holes is subject to specialist design.

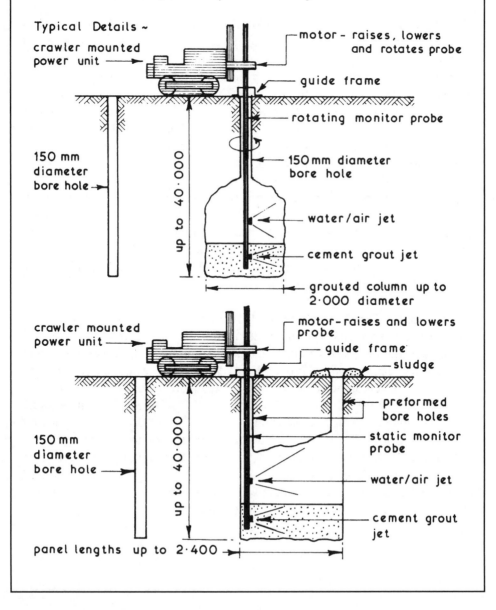

Green-Field – land not previously built upon. Usually part of the 'green-belt' surrounding urban areas, designated inappropriate for development in order to preserve the countryside. Limited development for agricultural purposes only may be permitted on 'green-belt' land.

Brown-Field – derelict land formerly a developed site and usually associated with previous construction of industrial buildings. UK government has set an objective to build 60% of the 4 million new homes required by 2016 on these sites.

Site Survey – essential that a geo-technical survey is undertaken to determine whether contaminants are in the soil and ground water. Of particular concern are: acids, salts, heavy metals, cyanides and coal tars, in addition to organic materials which decompose to form the highly explosive gas, methane. Analysis of the soil will determine a 'trigger threshold value', above which it will be declared sensitive to the end user. For example, a domestic garden or children's play area will have a low value relative to land designated for a commercial car park.

Site Preparation – when building on sites previously infilled with uncontaminated material, a reinforced raft type foundation may be adequate for light structures. Larger buildings will justify soil consolidation and compaction processes to improve the bearing capacity. Remedial measures for subsoils containing chemicals or other contaminants are varied.

Legislation – the Environment Protection Act of 1990 attempted to enforce responsibility on local authorities to compile a register of all potentially contaminated land. This proved unrealistic and too costly due to inherent complexities. Since then, requirements under the Environment Act 1995, the Pollution Prevention and Control Act 1999, the PPC Regulations 2000 and the subsequent DCLG Planning Policy Statement (PPS 23, 2004): Planning and Pollution Control (Annex 2: Development on land affected by contamination), have made this more of a planning issue. It has become the responsibility of developers to conduct site investigations and to present details of proposed remedial measures as part of their planning application.

The traditional low-technology method for dealing with contaminated sites has been to excavate the soil and remove it to places licensed for depositing. However, with the increase in building work on brown-field sites, suitable dumps are becoming scarce. Added to this is the reluctance of ground operators to handle large volumes of this type of waste. Also, where excavations exceed depths of about 5 m, it becomes less practical and too expensive. Alternative physical, biological or chemical methods of soil treatment may be considered.

Encapsulation – in-situ enclosure of the contaminated soil. A perimeter trench is taken down to rock or other sound strata and filled with an impervious agent such as Bentonite clay. An impermeable horizontal capping is also required to link with the trenches. A high-specification barrier is necessary where liquid or gas contaminants are present as these can migrate quite easily. A system of monitoring soil condition is essential as the barrier may decay in time. Suitable for all types of contaminant.

Soil washing – involves extraction of the soil, sifting to remove large objects and placing it in a scrubbing unit resembling a huge concrete mixer. Within this unit water and detergents are added for a basic wash process, before pressure spraying to dissolve pollutants and to separate clay from silt. Eliminates fuels, metals and chemicals.

Vapour extraction – used to remove fuels or industrial solvents and other organic deposits. At variable depths, small diameter boreholes are located at frequent intervals. Attached to these are vacuum pipes to draw air through the contaminated soil. The contaminants are collected at a vapour treatment processing plant on the surface, treated and evaporated into the atmosphere. This is a slow process and it may take several months to cleanse a site.

Electrolysis – use of low voltage d.c. in the presence of metals. Electricity flows between an anode and cathode, where metal ions in water accumulate in a sump before pumping to the surface for treatment.

BIOLOGICAL

Phytoremediation – the removal of contaminants by plants which will absorb harmful chemicals from the ground. The plants are subsequently harvested and destroyed. A variant uses fungal degradation of the contaminants.

Bioremediation – stimulating the growth of naturally occurring microbes. Microbes consume petrochemicals and oils, converting them to water and carbon dioxide. Conditions must be right, i.e. a temperature of at least 10°C with an adequate supply of nutrients and oxygen. Untreated soil can be excavated and placed over perforated piping, through which air is pumped to enhance the process prior to the soil being replaced.

CHEMICAL

Oxidation – sub-soil boreholes are used for the pumped distribution of liquid hydrogen peroxide or potassium permanganate. Chemicals and fuel deposits convert to water and carbon dioxide.

Solvent extraction – the sub-soil is excavated and mixed with a solvent to break down oils, grease and chemicals that do not dissolve in water.

THERMAL

Thermal treatment (off site) – an incineration process involving the use of a large heating container/oven. Soil is excavated, dried and crushed prior to heating to 2500°C, where harmful chemicals are removed by evaporation or fusion.

Thermal treatment (in-situ) – steam, hot water or hot air is pressure-injected through the soil. Variations include electric currents and radio waves to heat water in the ground to become steam. Evaporates chemicals.

Ref. Building Regulations, Approved Document, C1: Site preparation and resistance to contaminants. Section 1: Clearance or treatment of unsuitable material. Section 2: Resistance to contaminants.

5 SUPERSTRUCTURE — 1

CHOICE OF MATERIALS

BRICK AND BLOCK WALLS

BRICK BONDING

SPECIAL BRICKS AND APPLICATIONS

CAVITY WALLS

DAMP-PROOF COURSES

GAS RESISTANT MEMBRANES

ARCHES AND OPENINGS

WINDOWS, GLASS AND GLAZING

DOMESTIC AND INDUSTRIAL DOORS

TIMBER FRAME CONSTRUCTION

RENDERING AND CLADDING EXTERNAL WALLS

TIMBER PITCHED AND FLAT ROOFS

GREEN ROOFS

THERMAL INSULATION

U-VALUE CALCULATION

THERMAL BRIDGING

ACCESS FOR THE DISABLED

STAGE 1

Consideration to be given to the following:~

1. Building type and usage.
2. Building owner's requirements and preferences.
3. Local planning restrictions.
4. Legal restrictions and requirements.
5. Site restrictions.
6. Capital resources.
7. Future policy in terms of maintenance and adaptation.

STAGE 2

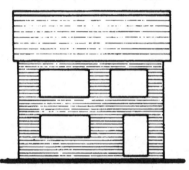

Decide on positions, sizes and shapes of openings.

STAGE 3

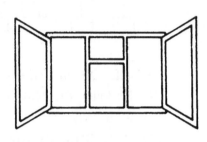

Decide on style, character and materials for openings

STAGE 4

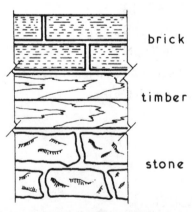

brick

timber

stone

Decide on basic materials for fabric of roof and walls

STAGE 5

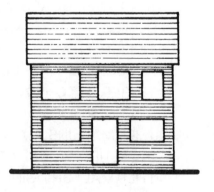

Review all decisions and make changes if required

Bricks ~ these are walling units within a length of 337·5mm, a width of 225mm and a height of 112·5mm. The usual size of bricks in common use is length 215mm, width 102·5mm and height 65mm and like blocks they must be laid in a definite pattern or bond if they are to form a structural wall. Bricks are usually made from clay (BS EN 772-1, BS EN 772-3 and BS EN 772-7) or from sand and lime (BS EN 771-2) and are available in a wide variety of strengths, types, textures, colours and special shaped bricks to BS 4729.

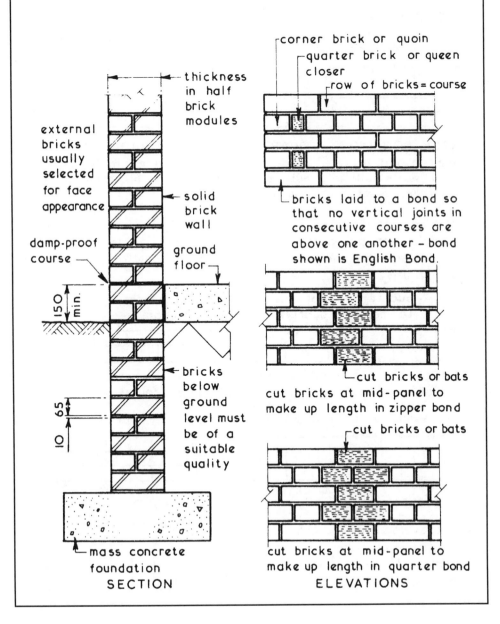

thickness in half brick modules

external bricks usually selected for face appearance

solid brick wall

damp-proof course

ground floor

150 min.

ground floor

65

10

bricks below ground level must be of a suitable quality

mass concrete foundation

SECTION

corner brick or quoin

quarter brick or queen closer

row of bricks = course

bricks laid to a bond so that no vertical joints in consecutive courses are above one another - bond shown is English Bond.

cut bricks or bats

cut bricks at mid-panel to make up length in zipper bond

cut bricks or bats

cut bricks at mid-panel to make up length in quarter bond

ELEVATIONS

Typical Details ~

Bonding ~ an arrangement of bricks in a wall, column or pier laid to a set pattern to maintain an adequate lap.

Purposes of Brick Bonding ~

1. Obtain maximum strength whilst distributing the loads to be carried throughout the wall, column or pier.
2. Ensure lateral stability and resistance to side thrusts.
3. Create an acceptable appearance.

Lap Forms ~

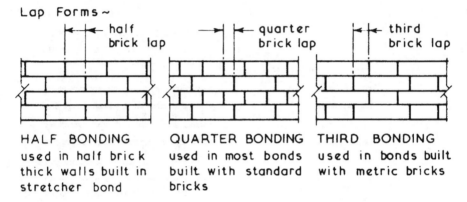

HALF BONDING	QUARTER BONDING	THIRD BONDING
used in half brick thick walls built in stretcher bond	used in most bonds built with standard bricks	used in bonds built with metric bricks

Simple Bonding Rules ~

1. Bond is set out along length of wall working from each end to ensure that no vertical joints are above one another in consecutive courses.

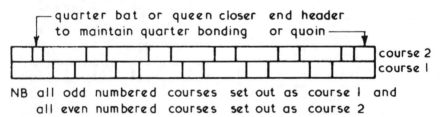

NB all odd numbered courses set out as course I and all even numbered courses set out as course 2

2. Walls which are not in exact bond length can be set out thus –

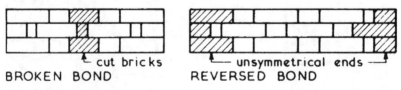

BROKEN BOND   REVERSED BOND

3. Transverse or cross joints continue unbroken across the width of wall unless stopped by a face stretcher.

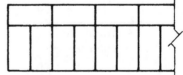

English Bond ~ formed by laying alternate courses of stretchers and headers it is one of the strongest bonds but it will require more facing bricks than other bonds (89 facing bricks per m²)

Typical Example ~

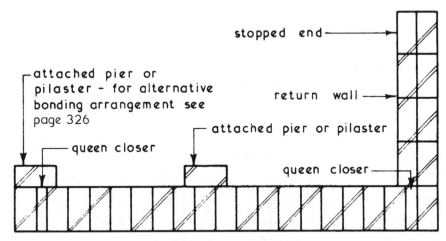

stopped end

attached pier or pilaster - for alternative bonding arrangement see page 326

return wall

queen closer

attached pier or pilaster

queen closer

PLAN ON ODD NUMBERED COURSES

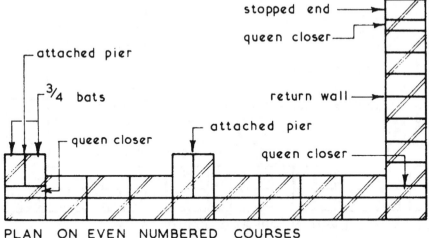

stopped end

queen closer

attached pier

return wall

³⁄₄ bats

attached pier

queen closer

queen closer

PLAN ON EVEN NUMBERED COURSES

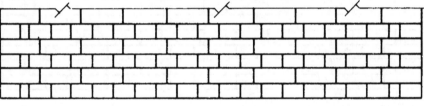

ELEVATION

325

Flemish Bond ~ formed by laying headers and stretchers alternately in each course. Not as strong as English bond but is considered to be aesthetically superior uses less facing bricks. (79 facing bricks per m²)

Typical Example

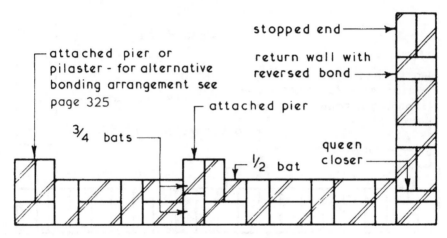

PLAN ON ODD NUMBERED COURSES

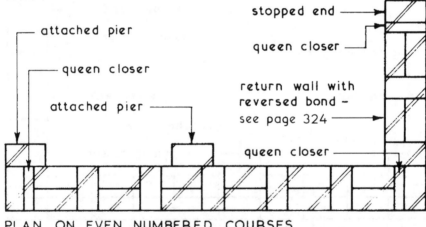

PLAN ON EVEN NUMBERED COURSES

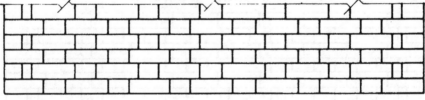

ELEVATION

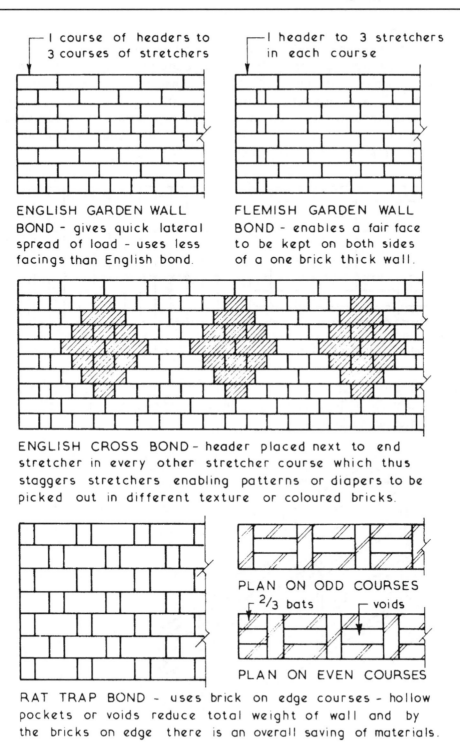

I course of headers to 3 courses of stretchers

I header to 3 stretchers in each course

ENGLISH GARDEN WALL BOND - gives quick lateral spread of load - uses less facings than English bond.

FLEMISH GARDEN WALL BOND - enables a fair face to be kept on both sides of a one brick thick wall.

ENGLISH CROSS BOND - header placed next to end stretcher in every other stretcher course which thus staggers stretchers enabling patterns or diapers to be picked out in different texture or coloured bricks.

PLAN ON ODD COURSES

2/3 bats   voids

PLAN ON EVEN COURSES

RAT TRAP BOND - uses brick on edge courses - hollow pockets or voids reduce total weight of wall and by the bricks on edge there is an overall saving of materials.

Stack Bonding – the quickest, easiest and most economical bond to lay, as there is no need to cut bricks or to provide special sizes. Visually the wall appears unbonded as continuity of vertical joints is structurally unsound, unless wire bed-joint reinforcement is placed in every horizontal course, or alternate courses where loading is moderate. In cavity walls, wall ties should be closer than normal at 600mm max. spacing horizontally and 225mm max. spacing vertically and staggered.

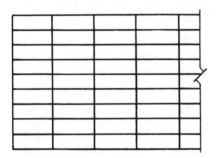

Horizontal stack bond

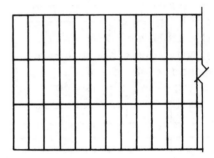

Vertical stack bond

Application – this distinctive uniform pattern is popular as non-structural infill panelling to framed buildings and for non-load bearing exposed brickwork partitions.

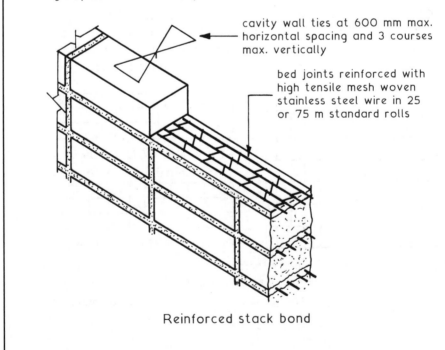

cavity wall ties at 600 mm max. horizontal spacing and 3 courses max. vertically

bed joints reinforced with high tensile mesh woven stainless steel wire in 25 or 75 m standard rolls

Reinforced stack bond

Attached Piers ~ the main function of an attached pier is to give lateral support to the wall of which it forms part from the base to the top of the wall. It also has the subsidiary function of dividing a wall into distinct lengths whereby each length can be considered as a wall. Generally walls must be tied at end to an attached pier, buttressing or return wall.

Typical Examples ~

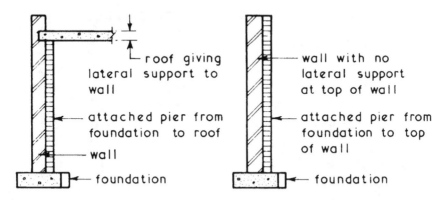

roof giving lateral support to wall

attached pier from foundation to roof

wall

foundation

wall with no lateral support at top of wall

attached pier from foundation to top of wall

foundation

Requirements for the external wall of a small single storey non-residential building or annex exceeding 2.5 m in length or height and of floor area not exceeding 36 m² ~

- Minimum thickness, 90 mm, i.e. 102.5 mm brick or 100 mm block.
- Built solid of bonded brick or block masonry and bedded in cement mortar.
- Surface mass of masonry, minimum 130 kg/m² where floor area exceeds 10 m².
- No lateral loading permitted excepting wind loads.
- Maximum length or width not greater than 9 m.
- Maximum height as shown on page 331.
- Lateral restraint provided by direct bearing of roof and as shown on page 462.
- Maximum of two major openings in one wall of the building. Height maximum 2.1 m, width maximum 5 m (if 2 openings, total width maximum 5 m).
- Other small openings permitted, as shown on next page.
- Bonded or connected to piers of minimum size 390 × 190 mm at maximum 3 m centres for the full wall height as shown above. Pier connections are with pairs of wall ties of 20 × 3 mm flat stainless steel type at 300 mm vertical spacing.

Attached piers as applied to 1/2 brick (90 mm min.) thick walls ~

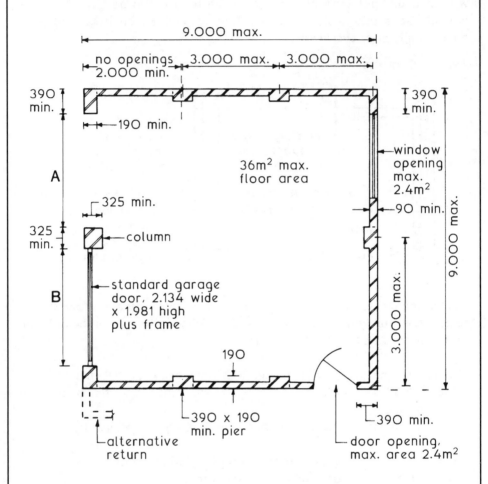

- Major openings A and B are permitted in one wall only. Aggregate width is 5 m maximum. Height not greater than 2.1 m. No other openings within 2 m.
- Other walls not containing a major opening can have smaller openings of maximum aggregate area 2.4 m².
- Maximum of only one opening between piers.
- Distance from external corner of a wall to an opening at least 390 mm unless the corner contains a pier.
- The minimum pier dimension of 390 × 190 mm can be varied to 327 × 215 mm to suit brick sizes.

Construction of half-brick and 100mm thick solid concrete block walls (90mm min.) with attached piers, has height limitations to maintain stability. The height of these buildings will vary depending on the roof profile; it should not exceed the lesser value in the following examples ~

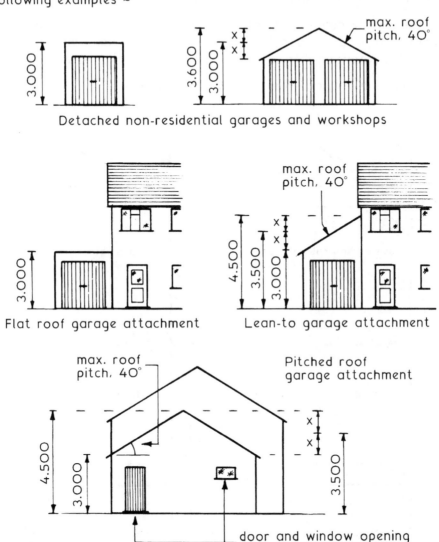

Detached non-residential garages and workshops

Flat roof garage attachment

Lean-to garage attachment

Pitched roof garage attachment

door and window opening limitations as previous page

Note: All dimensions are maximum.

Height is measured from top of foundation to top of wall except where shown at an intermediate position. Where the underside of the floor slab provides an effective lateral restraint, measurements may be taken from here.

The appearance of a building can be significantly influenced by the mortar finishing treatment to masonry. Finishing may be achieved by jointing or pointing.

Jointing – the finish applied to mortar joints as the work proceeds.

Pointing – the process of removing semi-set mortar to a depth of about 20 mm and replacing it with fresh mortar. Pointing may contain a colouring pigment to further enhance the masonry.

Finish profiles, typical examples shown pointed –

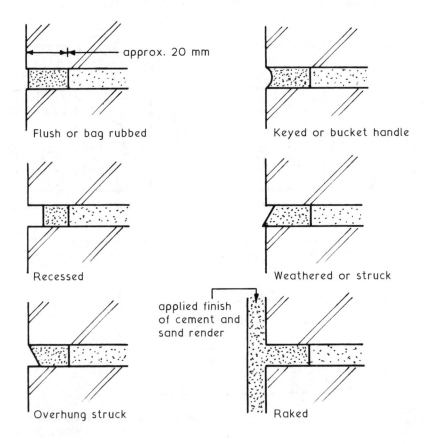

approx. 20 mm

Flush or bag rubbed

Keyed or bucket handle

Recessed

Weathered or struck

Overhung struck

applied finish of cement and sand render

Raked

Examples of pointing to masonry

Note: Recessed and overhung finishes should not be used in exposed situations, as rainwater can be detained. This could encourage damage by frost action and growth of lichens.

Specials – these are required for feature work and application to various bonds, as shown on the preceding pages. Bonding is not solely for aesthetic enhancement. In many applications, e.g. English bonded manhole walls, the disposition of bricks is to maximise wall strength and integrity. In a masonry wall the amount of overlap should not be less than one quarter of a brick length. Specials may be machine or hand cut from standard bricks, or they may be purchased as purpose-made. These purpose-made bricks are relatively expensive as they are individually manufactured in hardwood moulds.

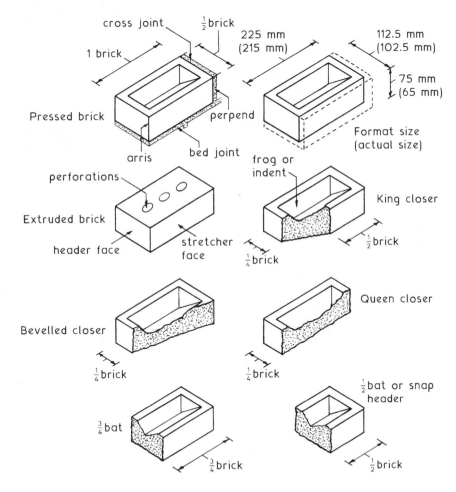

Standard bricks and cut specials

Ref. BS 4729: Clay and calcium silicate bricks of special shapes and sizes. Recommendations.

Brickwork can be repetitive and monotonous, but with a little imagination and skilled application it can be a highly decorative art form. Artistic potential is made possible by the variety of naturally occurring brick colours, textures and finishes, the latter often applied as a sanding to soft clay prior to baking. Furthermore, the range of pointing techniques, mortar colourings, brick shapes and profiles can combine to create countless possibilities for architectural expression.

Bricks are manufactured from baked clay, autoclaved sand/lime or concrete. Clay is ideally suited to hand making special shapes in hardwood moulds. Some popular formats are shown below, but there is no limit to creative possibilities.

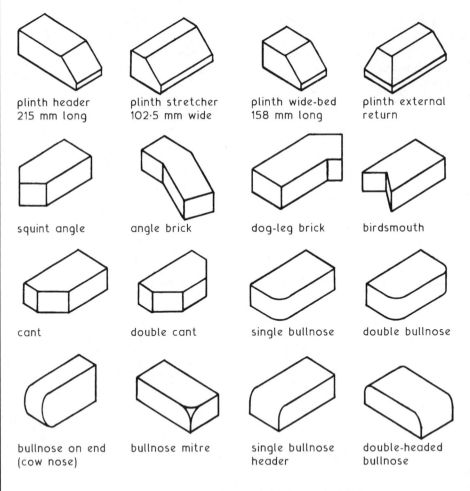

plinth header
215 mm long

plinth stretcher
102·5 mm wide

plinth wide-bed
158 mm long

plinth external
return

squint angle

angle brick

dog-leg brick

birdsmouth

cant

double cant

single bullnose

double bullnose

bullnose on end
(cow nose)

bullnose mitre

single bullnose
header

double-headed
bullnose

Purpose-made and special shape bricks

Plinths – used as a projecting feature to enhance external wall appearance at its base. The exposed projection determines that only frost-proof quality bricks are suitable and that recessed or raked out joints which could retain water must be avoided.

Typical external wall base –

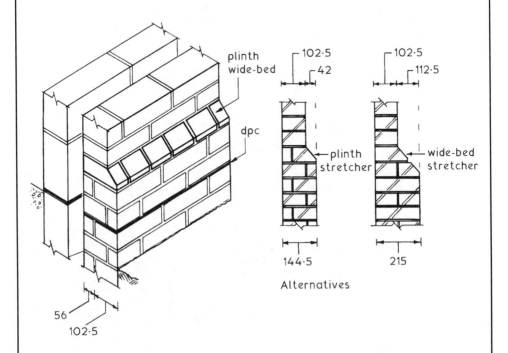

Alternatives

Corbel – a projecting feature at higher levels of a building. This may be created by using plinth bricks laid upside down with header and stretcher formats maintaining bond. For structural integrity, the amount of projection (P) must not exceed one third of the overall wall thickness (T). Some other types of corbel are shown on the next page.

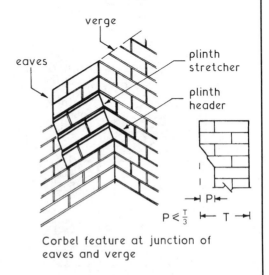

$P \leqslant \frac{T}{3}$

Corbel feature at junction of eaves and verge

Corbel – a type of inverted plinth, generally located at the higher levels of a building to create a feature. A typical example is quarter bonded headers as a detail below window openings.

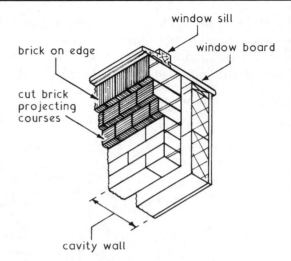

Corbelled sill

Dentil Coursing – a variation on continuous corbelling where alternative headers project. This is sometimes referred to as table corbelling.

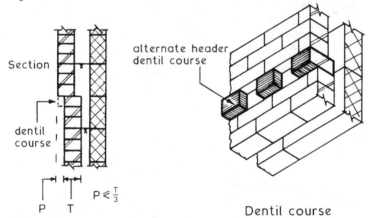

$P \leqslant \frac{T}{3}$

Dentil course

Dog Toothing – a variation on a dentil course created by setting the feature bricks at 45°.

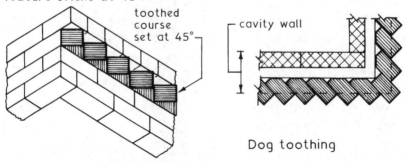

Dog toothing

Note: Cavity insulated as required.

Blocks ~ these are walling units exceeding in length, width or height the dimensions specified for bricks in BS EN 772-16. Precast concrete blocks should comply with the recommendations set out in BS 6073-2 and BS EN 771-3. Blocks suitable for external solid walls are classified as loadbearing and are required to have a minimum average crushing strength of 2·8 N/mm^2.

Typical Details ~

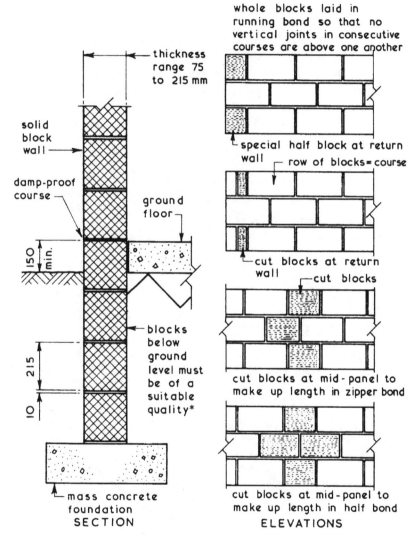

SECTION

ELEVATIONS

*See pages 339 and 340

Refs. BS 6073-2: Precast concrete masonry units.

BS EN 772-16: Methods of test for masonry units.

BS EN 771-3: Specification for masonry units.

Cavity Walls ~ these consist of an outer brick or block leaf or skin separated from an inner brick or block leaf or skin by an air space called a cavity. These walls have better thermal insulation and weather resistance properties than a comparable solid brick or block wall and therefore are in general use for the enclosing walls of domestic buildings. The two leaves of a cavity wall are tied together with wall ties located at 2.5/m², or at equivalent spacings shown below and as given in Section 2C of Approved Document A – Building Regulations.

With butterfly type ties the width of the cavity should be between 50 and 75mm. Where vertical twist type ties are used the cavity width can be between 75 and 300mm. Cavities are not normally ventilated and are closed by roof insulation at eaves level.

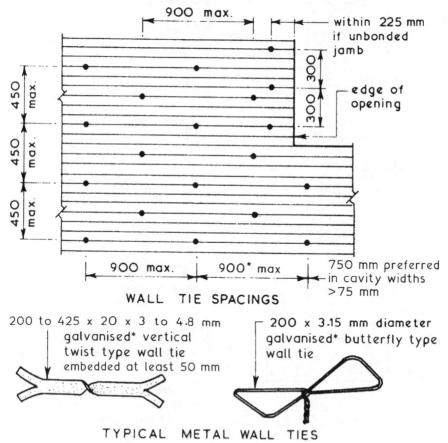

WALL TIE SPACINGS

200 to 425 x 20 x 3 to 4.8 mm galvanised* vertical twist type wall tie embedded at least 50 mm

200 x 3.15 mm diameter galvanised* butterfly type wall tie

TYPICAL METAL WALL TIES

Ref: BS EN 845-1: Specification for ancillary components for masonry. Ties, tension straps, hangers and brackets.

* Note: Stainless steel or non-ferrous ties are now preferred.

338

Minimum requirements ~

Thickness of each leaf, 90 mm.

Width of cavity, 50 mm.

Wall ties at 2.5/m^2 (see previous page).

Compressive strength of bricks, 5 N/mm^2 up to two storeys.*

Compressive strength of blocks, 2.8 N/mm^2 up to two storeys.*

* For work between the foundation and the surface a 7 N/mm^2 minimum brick and block strength is normally specified. This is also a requirement where the foundation to underside of the ground floor structure exceeds 1.0 m.

Combined thickness of each leaf + 10 mm whether used as an external wall, a separating wall or a compartment wall, should be not less than 1/16 of the storey height** which contains the wall.

** Generally measured between the undersides of lateral supports, eg. undersides of floor or ceiling joists, or from the underside of upper floor joists to half way up a laterally restrained gable wall. See Approved Document A, Section 2C for variations.

Wall dimensions for minimum combined leaf thicknesses of 90 mm + 90 mm ~

Height	Length
3.5 m max.	12.0 m max.
3.5 m – 9.0 m	9.0 m max.

Wall dimensions for minimum combined leaf thickness of 280 mm, eg. 190 mm + 90 mm for one storey height and a minimum 180 mm combined leaf thickness, ie. 90 mm + 90 mm for the remainder of its height ~

Height	Length
3.5 – 9.0 m	9.0 - 12.0 m
9.0 m – 12.0 m	9.0 m max.

Wall dimensions for minimum combined leaf thickness of 280 mm for two storey heights and a minimum 180 mm combined leaf thickness for the remainder of its height ~

Height	Length
9.0 m – 12.0 m	9.0 m – 12.0 m

Wall length is measured from centre to centre of restraints by buttress walls, piers or chimneys.

For other wall applications, see the reference to calculated brickwork on page 355.

Cavity Walls

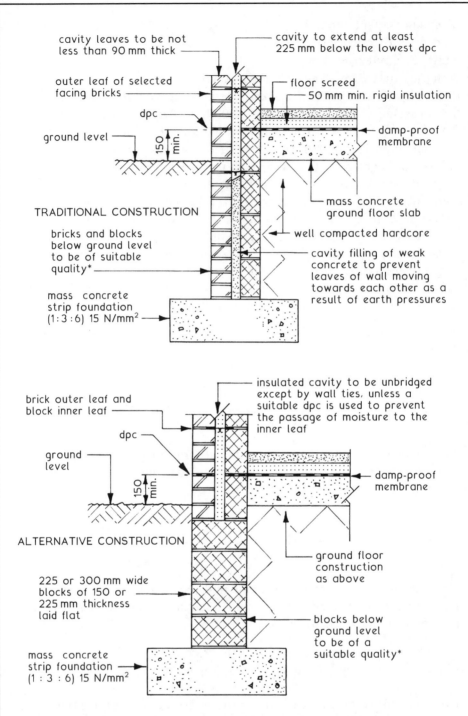

cavity leaves to be not less than 90 mm thick

outer leaf of selected facing bricks

dpc

ground level

150 min.

TRADITIONAL CONSTRUCTION

bricks and blocks below ground level to be of suitable quality*

mass concrete strip foundation (1:3:6) 15 N/mm²

cavity to extend at least 225 mm below the lowest dpc

floor screed

50 mm min. rigid insulation

damp-proof membrane

mass concrete ground floor slab

well compacted hardcore

cavity filling of weak concrete to prevent leaves of wall moving towards each other as a result of earth pressures

brick outer leaf and block inner leaf

dpc

ground level

150 min.

ALTERNATIVE CONSTRUCTION

225 or 300 mm wide blocks of 150 or 225 mm thickness laid flat

mass concrete strip foundation (1 : 3 : 6) 15 N/mm²

insulated cavity to be unbridged except by wall ties, unless a suitable dpc is used to prevent the passage of moisture to the inner leaf

damp-proof membrane

ground floor construction as above

blocks below ground level to be of a suitable quality*

*Min. compressive strength depends on building height and loading. See Building Regulations AD A: Section 2C (Diagram 9).

Parapet ~ a low wall projecting above the level of a roof, bridge or balcony forming a guard or barrier at the edge. Parapets are exposed to the elements justifying careful design and construction for durability.

Typical Details ~

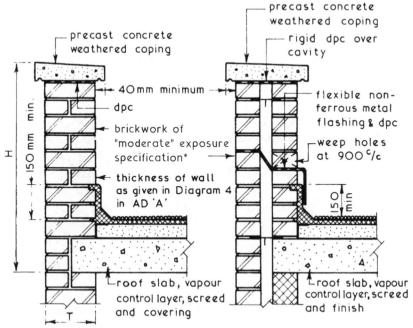

precast concrete weathered coping

precast concrete weathered coping

rigid dpc over cavity

40mm minimum

dpc

brickwork of "moderate" exposure specification*

thickness of wall as given in Diagram 4 in AD 'A'

flexible non-ferrous metal flashing & dpc

weep holes at 900 °/c

H

150 mm min

150 min

roof slab, vapour control layer, screed and covering

roof slab, vapour control layer, screed and finish

T (1 bk. solid and 1/2 bk. cavity wall), H ⩾860mm.

SOLID WALL - HIGH LEVEL   CAVITY WALL - HIGH LEVEL

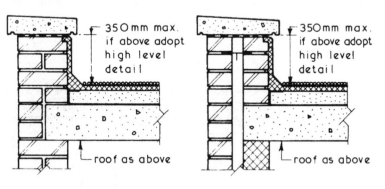

350mm max. if above adopt high level detail

350mm max. if above adopt high level detail

roof as above

roof as above

SOLID WALL - LOW LEVEL   CAVITY WALL - LOW LEVEL

Ref. BS EN 771-1: Specification for (clay) masonry units.

*"severe" exposure specification in the absence of a protective coping.

Historically, finned or buttressed walls have been used to provide lateral support to tall single storey masonry structures such as churches and cathedrals. Modern applications are similar in principle and include theatres, gymnasiums, warehouses, etc. Where space permits, they are an economic alternative to masonry cladding of steel or reinforced concrete framed buildings. The fin or pier is preferably brick bonded to the main wall. It may also be connected with horizontally bedded wall ties, sufficient to resist vertical shear stresses between fin and wall.

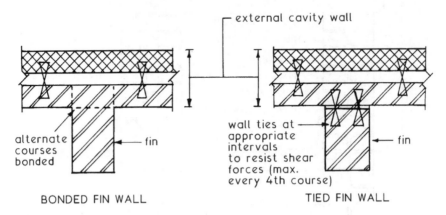

external cavity wall

alternate courses bonded — fin

wall ties at appropriate intervals to resist shear forces (max. every 4th course) — fin

BONDED FIN WALL   TIED FIN WALL

Structurally, the fins are deep piers which reinforce solid or cavity masonry walls. For design purposes the wall may be considered as a series of 'T' sections composed of a flange and a pier. If the wall is of cavity construction, the inner leaf is not considered for bending moment calculations, although it does provide stiffening to the outer leaf or flange.

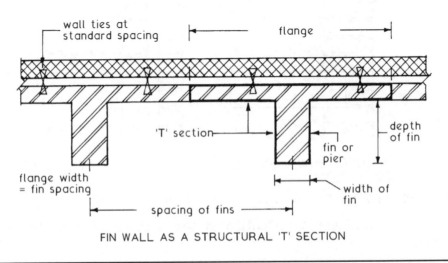

wall ties at standard spacing

flange

'T' section

fin or pier

depth of fin

flange width = fin spacing

width of fin

spacing of fins

FIN WALL AS A STRUCTURAL 'T' SECTION

Masonry diaphragm walls are an alternative means of constructing tall, single storey buildings such as warehouses, sports centres, churches, assembly halls, etc. They can also be used as retaining and boundary walls with planting potential within the voids. These voids may also be steel reinforced and concrete filled to resist the lateral stresses in high retaining walls.

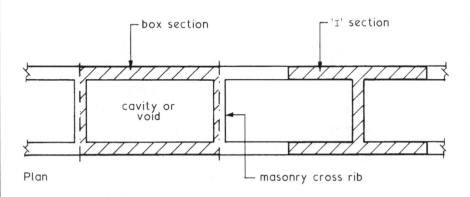

A diaphragm wall is effectively a cavity wall where the two leaves of masonry are bonded together with cross ribs and not wall ties. It is stronger than a conventionally tied cavity wall and for structural purposes may be considered as a series of bonded 'I' sections or box sections. The voids may be useful for housing services, but any access holes in the construction must not disturb the integrity of the wall. The voids may also be filled with insulation to reduce heat energy losses from the building, and to prevent air circulatory heat losses within the voids. Where thermal insulation standards apply, this type of wall will have limitations as the cross ribs will provide a route for cold bridging. U values will increase by about 10% compared with conventional cavity wall construction of the same materials.

Ref. BS 5628-1: Code of practice for use of masonry. Structural use of unreinforced masonry.
BS 5628-3: Code of practice for use of masonry. Materials and components, design and workmanship.

Damp-proof Courses and Membranes

Function – the primary function of any damp-proof course (dpc) or damp-proof membrane (dpm) is to provide an impermeable barrier to the passage of moisture. The three basic ways in which damp-proof courses are used is to:-

1. Resist moisture penetration from below (rising damp).
2. Resist moisture penetration from above.
3. Resist moisture penetration from horizontal entry.

Typical examples ~

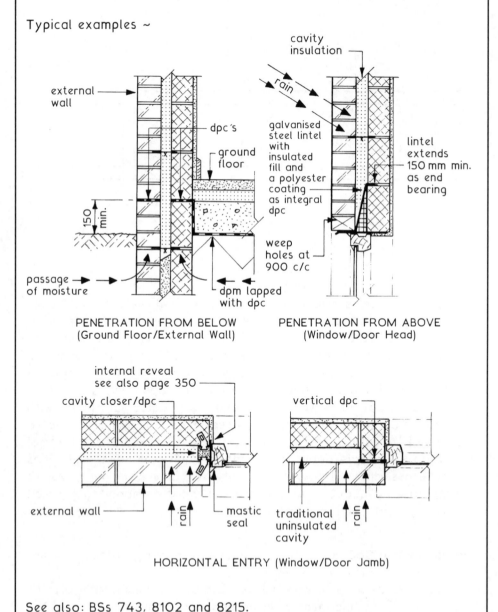

PENETRATION FROM BELOW
(Ground Floor/External Wall)

PENETRATION FROM ABOVE
(Window/Door Head)

HORIZONTAL ENTRY (Window/Door Jamb)

See also: BSs 743, 8102 and 8215.

Building Regulations, Approved Document C2, Section 5:
A wall may be built with a 'damp-proof course of bituminous material, polyethylene, engineering bricks or slates in cement mortar, or any other material that will prevent the passage of moisture.'

Material			Remarks
Lead	BS EN 12588	Code 4 (1·8 mm)	May corrode in the presence of mortar. Both surfaces to be coated with bituminous paint. Workable for application to cavity trays, etc.
Copper	BS EN 1172	0·25 mm	Can cause staining to adjacent masonry. Resistant to corrosion.
Bitumen	BS 6398		Hessian or fibre may decay with age, but this will not affect efficiency. Tearable if not protected. Lead bases are suited where there may be a high degree of movement in the wall. Asbestos is now prohibited.
in various bases:			
Hessian		3·8 kg/m^2	
Fibre		3·3	
Asbestos		3·8	
Hessian & lead		4·4	
Fibre & lead		4·4	
LDPE (polyethylene)	BS 6515	0·46 mm	No deterioration likely, but may be difficult to bond, hence the profiled surface finish. Not suited under light loads.
Bitumen polymer and pitch polymer		1·10 mm	Absorbs movement well. Joints and angles made with product manufacturer's adhesive tape.
Polypropylene BS 5139 1.5 to 2.0 mm			Preformed dpc for cavity trays, cloaks, direction changes and over lintels.

Note: All the above dpcs to be lapped at least 100 mm at joints and adhesive sealed. Dpcs should be continuous with any dpm in the floor.

Material			Remarks
Mastic asphalt	BS 6925	12 kg/m²	Does not deteriorate. Requires surface treatment with sand or scoring to effect a mortar key.
Engineering bricks	BS EN 771-1 BS EN 772-7	<4·5% absorption	Min. 2 courses laid breaking joint in cement mortar 1:3. No deterioration, but may not blend with adjacent facings.
Slate	BS EN 12326-1	4 mm	Min. 2 courses laid as above. Will not deteriorate, but brittle so may fracture if building settles.

Refs:

BS 743: Specification for materials for damp-proof courses.

BS 5628-3: Code of practice for the use of masonry. Materials and components, design and workmanship.

BS 8102: Code of practice for protection of structures against water from the ground.

BS 8215: Code of practice for design and installation of damp-proof courses in masonry construction.

BRE Digest 380: Damp-proof courses.

Note: It was not until the Public Health Act of 1875, that it became mandatory to instal damp-proof courses in new buildings. Structures constructed before that time, and those since, which have suffered dpc failure due to deterioration or incorrect installation, will require remedial treatment. This could involve cutting out the mortar bed joint two brick courses above ground level in stages of about 1m in length. A new dpc can then be inserted with mortar packing, before proceeding to the next length. No two adjacent sections should be worked consecutively. This process is very time consuming and may lead to some structural settlement. Therefore, the measures explained on the following two pages are usually preferred.

Materials – Silicone solutions in organic solvent.
          Aluminium stearate solutions.
          Water soluble silicone formulations (siliconates).

Methods – High pressure injection (0·70 – 0·90 MPa) solvent based.
          Low pressure injection (0·15 – 0·30 MPa) water based.
          Gravity feed, water based.
          Insertion/injection, mortar based.

Pressure injection – 12 mm diameter holes are bored to about two-thirds the depth of masonry, at approximately 150 mm horizontal intervals at the appropriate depth above ground (normally 2–3 brick courses). These holes can incline slightly downwards. With high (low) pressure injection, walls in excess of 120 mm (460 mm) thickness should be drilled from both sides. The chemical solution is injected by pressure pump until it exudes from the masonry. Cavity walls are treated as each leaf being a solid wall.

Gravity feed – 25 mm diameter holes are bored as above. Dilute chemical is transfused from containers which feed tubes inserted in the holes. This process can take from a few hours to several days to effect. An alternative application is insertion of frozen pellets placed in the bore holes. On melting, the solution disperses into the masonry to be replaced with further pellets until the wall is saturated.

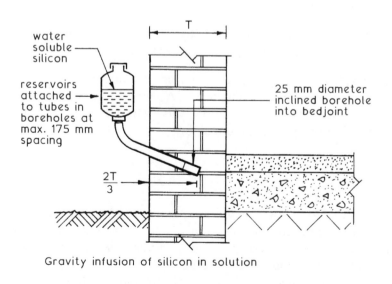

water soluble silicon

reservoirs attached to tubes in boreholes at max. 175 mm spacing

T

25 mm diameter inclined borehole into bedjoint

$\dfrac{2T}{3}$

Gravity infusion of silicon in solution

Injection mortars – 19 mm diameter holes are bored from both sides of a wall, at the appropriate level and no more than 230 mm apart horizontally, to a depth equating to three-fifths of the wall thickness. They should be inclined downwards at an angle of 20 to 30°. The drill holes are flushed out with water, before injecting mortar from the base of the hole and outwards. This can be undertaken with a hand operated caulking gun. Special cement mortars contain styrene butadiene resin (SDR) or epoxy resin and must be mixed in accordance with the manufacturer's guidance.

Notes relating to all applications of chemical dpcs:

* Before commencing work, old plasterwork and rendered undercoats are removed to expose the masonry. This should be to a height of at least 300 mm above the last detectable (moisture meter reading) signs of rising dampness (1 metre min.).

* If the wall is only accessible from one side and both sides need treatment, a second deeper series of holes may be bored from one side, to penetrate the inaccessible side.

* On completion of work, all boreholes are made good with cement mortar. Where dilute chemicals are used for the dpc, the mortar is rammed the full length of the hole with a piece of timber dowelling.

* The chemicals are effective by bonding to, and lining the masonry pores by curing and solvent evaporation.

* The process is intended to provide an acceptable measure of control over rising dampness. A limited amount of water vapour may still rise, but this should be dispersed by evaporation in a heated building.

Refs.

BS 6576: Code of practice for diagnosis of rising damp in walls of buildings and installation of chemical damp-proof courses.
BRE Digest 245: Rising damp in walls: diagnosis and treatment.
BRE Digest 380: Damp-proof courses.
BRE Good Repair Guide 6: Treating rising damp in houses.

In addition to damp-proof courses failing due to deterioration or damage, they may be bridged as a result of:

* Faults occurring during construction.
* Work undertaken after construction, with disregard for the damp-proof course.

Typical examples ~

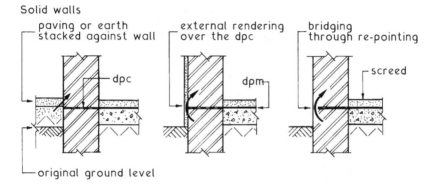

Solid walls

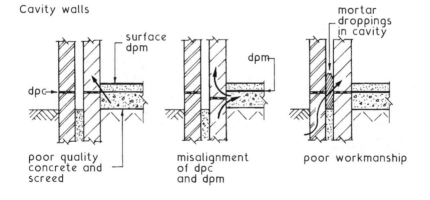

Cavity walls

Thermal insulation regulations may require insulating dpcs to prevent cold bridging around window and door openings in cavity wall construction (see pages 488 and 489). By locating a vertical dpc with a bonded insulant at the cavity closure, the dpc prevents penetration of dampness from the outside, and the insulation retains the structural temperature of the internal reveal. This will reduce heat losses by maintaining the temperature above dewpoint, preventing condensation, wall staining and mould growth.

Application ~

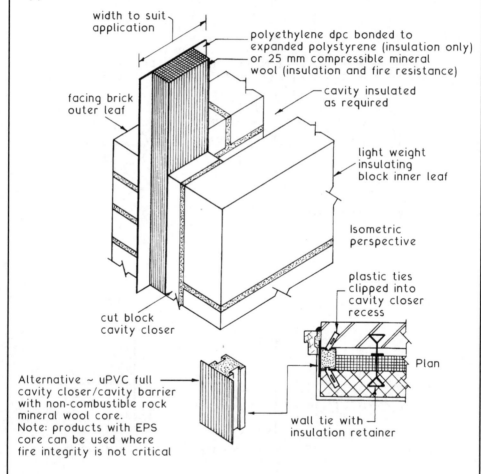

width to suit application

polyethylene dpc bonded to expanded polystyrene (insulation only) or 25 mm compressible mineral wool (insulation and fire resistance)

facing brick outer leaf

cavity insulated as required

light weight insulating block inner leaf

Isometric perspective

plastic ties clipped into cavity closer recess

cut block cavity closer

Plan

Alternative ~ uPVC full cavity closer/cavity barrier with non-combustible rock mineral wool core.
Note: products with EPS core can be used where fire integrity is not critical

wall tie with insulation retainer

Refs. Building Regulations, Approved Document L: Conservation of fuel and power.
BRE Report – Thermal Insulation: avoiding risks (3rd. ed.).
Building Regulations, Approved Document B3, (Vol. 1), Section 6: Concealed spaces (cavities).

Penetrating Gases ~ Methane and Radon

Methane – methane is produced by deposited organic material decaying in the ground. It often occurs with carbon dioxide and traces of other gases to form a cocktail known as landfill gas. It has become an acute problem in recent years, as planning restrictions on 'green-field' sites have forced development of derelict and reclaimed 'brown-field' land.

The gas would normally escape to the atmosphere, but under a building it pressurizes until percolating through cracks, cavities and junctions with services. Being odourless, it is not easily detected until contacting a naked flame, then the result is devastating!

Radon ~ a naturally occurring colour/odourless gas produced by radioactive decay of radium. It originates in uranium deposits of granite subsoils as far apart as the south-west and north of England and the Grampian region of Scotland. Concentrations of radon are considerably increased if the building is constructed of granite masonry. The combination of radon gas and the tiny radioactive particles known as radon daughters are inhaled. In some people with several years' exposure, research indicates a high correlation with cancer related illness and death.

Protection of buildings and the occupants from subterranean gases can be achieved by passive or active measures incorporated within the structure.

1. Passive protection consists of a complete airtight seal integrated within the ground floor and walls. A standard LDPE damp proof membrane of 0.3mm thickness should be adequate if carefully sealed at joints, but thicknesses up to 1mm are preferred, combined with foil and/or wire reinforcement.

2. Active protection requires installation of a permanently running extract fan connected to a gas sump below the ground floor. It is an integral part of the building services system and will incur operating and maintenance costs throughout the building's life.

(See next page for construction details)

Gas Resistant Construction

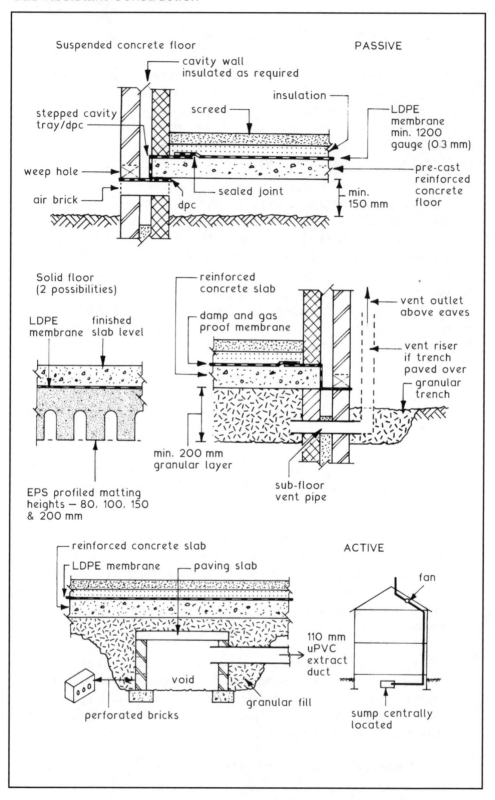

Suspended concrete floor — PASSIVE

cavity wall insulated as required

stepped cavity tray/dpc

screed

insulation

LDPE membrane min. 1200 gauge (0.3 mm)

weep hole

air brick

sealed joint

dpc

min. 150 mm

pre-cast reinforced concrete floor

Solid floor (2 possibilities)

LDPE membrane

finished slab level

EPS profiled matting heights — 80, 100, 150 & 200 mm

reinforced concrete slab

damp and gas proof membrane

min. 200 mm granular layer

sub-floor vent pipe

vent outlet above eaves

vent riser if trench paved over

granular trench

reinforced concrete slab

LDPE membrane

paving slab

ACTIVE

fan

void

perforated bricks

granular fill

110 mm uPVC extract duct

sump centrally located

Calculated Brickwork ~ for small and residential buildings up to three storeys high the sizing of load bearing brick walls can be taken from data given in Section 2C of Approved Document A. The alternative methods for these and other load bearing brick walls are given in:

BS 5628-1: Code of practice for the use of masonry. Structural use of unreinforced masonry, and

BS 8103-2: Structural design of low rise buildings. Code of practice for masonry walls for housing.

The main factors governing the loadbearing capacity of brick walls and columns are:-

1. Thickness of wall.
2. Strength of bricks used.
3. Type of mortar used.
4. Slenderness ratio of wall or column.
5. Eccentricity of applied load.

Thickness of wall ~ this must always be sufficient throughout its entire body to carry the design loads and induced stresses. Other design requirements such as thermal and sound insulation properties must also be taken into account when determining the actual wall thickness to be used.

Effective Thickness ~ this is the assumed thickness of the wall or column used for the purpose of calculating its slenderness ratio – see page 355.

Typical Examples ~

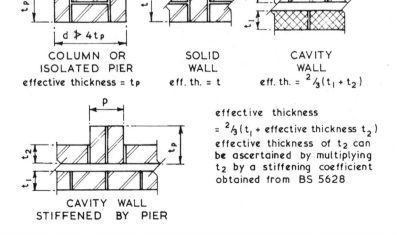

COLUMN OR
ISOLATED PIER
effective thickness = t_p

SOLID
WALL
eff. th. = t

CAVITY
WALL
eff. th. = $\frac{2}{3}(t_1 + t_2)$

CAVITY WALL
STIFFENED BY PIER

effective thickness
= $\frac{2}{3}(t_1 +$ effective thickness $t_2)$
effective thickness of t_2 can be ascertained by multiplying t_2 by a stiffening coefficient obtained from BS 5628

353

Strength of Bricks ~ due to the wide variation of the raw materials and methods of manufacture bricks can vary greatly in their compressive strength. The compressive strength of a particular type of brick or batch of bricks is taken as the arithmetic mean of a sample of ten bricks tested in accordance with the appropriate British Standard. A typical range for clay bricks would be from 20 to 170 MN/m^2 the majority of which would be in the 20 to 90 MN/m^2 band. Generally calcium silicate bricks have a lower compressive strength than clay bricks with a typical strength range of 10 to 65 MN/m^2.

Strength of Mortars ~ mortars consist of an aggregate (sand) and a binder which is usually cement; cement plus additives to improve workability; or cement and lime. The factors controlling the strength of any particular mix are the ratio of binder to aggregate plus the water:cement ratio. The strength of any particular mix can be ascertained by taking the arithmetic mean of a series of test cubes or prisms – see page 357.

Wall Design Strength ~ the basic stress of any brickwork depends on the crushing strength of the bricks and the type of mortar used to form the wall unit. This relationship can be plotted on a graph using data given in BS 5628 as shown below:-

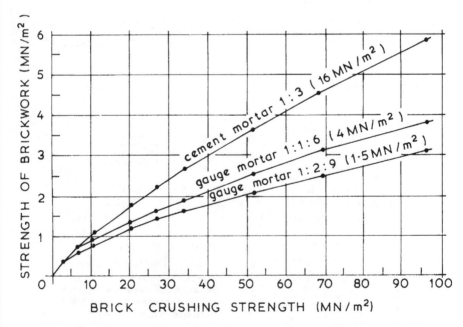

Note: 1 MN/m^2 equates to 1 N/mm^2

Slenderness Ratio ~ this is the relationship of the effective height to the effective thickness thus:-

$$\text{Slenderness ratio} = \frac{\text{effective height}}{\text{effective thickness}} = \frac{h}{t} \ngtr 27 \text{ see BS 5628}$$

Effective Height ~ this is the dimension taken to calculate the slenderness ratio as opposed to the actual height.

Typical Examples – actual height = H   effective height = h

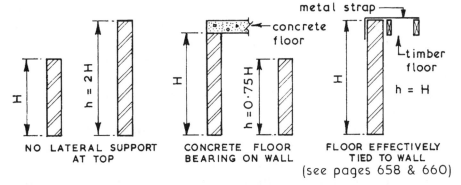

NO LATERAL SUPPORT AT TOP   **CONCRETE FLOOR BEARING ON WALL**   **FLOOR EFFECTIVELY TIED TO WALL**
(see pages 658 & 660)

Effective Thickness ~ this is the dimension taken to calculate the slenderness ratio as opposed to the actual thickness.

Typical Examples – actual thickness = T effective thickness = t

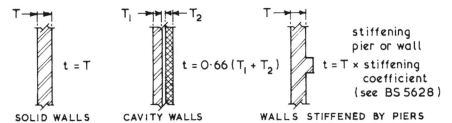

$t = T$ $t = 0.66\,(T_1 + T_2)$ $t = T \times$ stiffening coefficient (see BS 5628)

SOLID WALLS   **CAVITY WALLS**   **WALLS STIFFENED BY PIERS**

Stress Reduction ~ the permissible stress for a wall is based on the basic stress multiplied by a reduction factor related to the slenderness factor and the eccentricity of the load:-

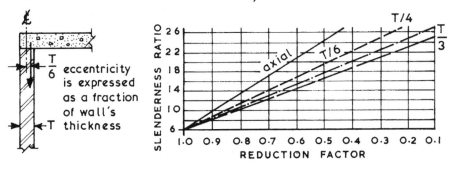

Lime ~ traditional mortars are a combination of lime, sand and water. These mixes are very workable and have sufficient flexibility to accommodate a limited amount of wall movement due to settlement, expansion and contraction. The long term durability of lime mortars is poor as they can break down in the presence of atmospheric contaminants and surface growths. Nevertheless, lime is frequently specified as a supplementary binder with cement, to increase mix workability and to reduce the possibility of joint shrinkage and cracking, a characteristic of stronger cement mortars.

Cement ~ the history of cement type mortar products is extensive. Examples dating back to the Mesopotamians and the Egyptians are not unusual; one of the earliest examples from over 10000 years ago has been found in Galilee, Israel. Modern mortars are made with Portland cement, the name attributed to a bricklayer named Joseph Aspdin. In 1824 he patented his improved hydraulic lime product as Portland cement, as it resembled Portland stone in appearance. It was not until the 1920s that Portland cement, as we now know it, was first produced commercially by mixing a slurry of clay (silica, alumina and iron-oxides) with limestone (calcium carbonate). The mix is burnt in a furnace (calcinated) and the resulting clinker crushed and bagged.

Mortar ~ mixes for masonry should have the following properties:

* Adequate strength
* Workability
* Water retention during laying
* Plasticity during application
* Adhesion or bond
* Durability
* Good appearance ~ texture and colour

Modern mortars are a combination of cement, lime and sand plus water. Liquid plasticisers exist as a substitute for lime, to improve workability and to provide some resistance to frost when used during winter.

Masonry cement ~ these proprietary cements generally contain about 75% Portland cement and about 25% of fine limestone filler with an air entraining plasticiser. Allowance must be made when specifying the mortar constituents to allow for the reduced cement content. These cements are not suitable for concrete.

Refs. BS 6463-101, 102 and 103: Quicklime, hydrated lime and natural calcium carbonate.
BS EN 197-1: Cement. Composition, specifications and conformity criteria for common cements.

Ready mixed mortar ~ this is delivered dry for storage in purpose made silos with integral mixers as an alternative to site blending and mixing. This ensures:

* Guaranteed factory quality controlled product
* Convenience
* Mix consistency between batches
* Convenient facility for satisfying variable demand
* Limited wastage
* Optimum use of site space

Mortar and cement strength ~ see also page 354. Test samples are made in prisms of 40×40 mm cross section, 160 mm long. At 28 days samples are broken in half to test for flexural strength. The broken pieces are subject to a compression test across the 40 mm width. An approximate comparison between mortar strength (MN/m^2 or N/mm^2), mortar designations (i to v) and proportional mix ratios is shown in the classification table below. Included is guidance on application.

Proportional mixing of mortar constituents by volume is otherwise known as a prescribed mix or simply a recipe.

Mortar classification ~

Traditional designation	BS EN 998-2 Strength	Proportions by volume cement/lime/sand	cement/sand	Application
i	12	1:0.25:3	1:3	Exposed external
ii	6	1:0.5:4-4.5	1:3-4	General external
iii	4	1:1:5-6	1:5-6	Sheltered internal
iv	2	1:2:8-9	1:7-8	General internal
v	-	1:3:10-12	1:9-10	Internal, grouting

Relevant standards;

BS 5628-3: Code of practice for use of masonry. Materials and components, design and workmanship.
BS EN 196: Methods of testing cement.
BS EN 998-2: Specification for mortar for masonry. Masonry mortar.
PD 6678: Guide to the specification of masonry mortar.
BS EN 1015: Methods of test for mortar for masonry.

Supports Over Openings ~ the primary function of any support over an opening is to carry the loads above the opening and transmit them safely to the abutments, jambs or piers on both sides. A support over an opening is usually required since the opening infilling such as a door or window frame will not have sufficient strength to carry the load through its own members.

Type of Support ~

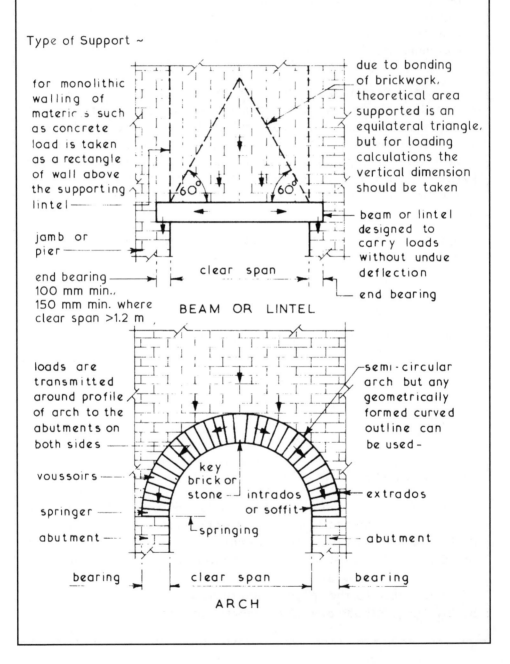

for monolithic walling of material s such as concrete load is taken as a rectangle of wall above the supporting lintel

jamb or pier

end bearing 100 mm min., 150 mm min. where clear span >1.2 m

clear span

due to bonding of brickwork, theoretical area supported is an equilateral triangle, but for loading calculations the vertical dimension should be taken

beam or lintel designed to carry loads without undue deflection

end bearing

BEAM OR LINTEL

loads are transmitted around profile of arch to the abutments on both sides

voussoirs

springer

abutment

key brick or stone

springing

intrados or soffit

semi-circular arch but any geometrically formed curved outline can be used-

extrados

abutment

bearing   clear span   bearing

ARCH

358

Arch Construction ~ by the arrangement of the bricks or stones in an arch over an opening it will be self supporting once the jointing material has set and gained adequate strength. The arch must therefore be constructed over a temporary support until the arch becomes self supporting. The traditional method is to use a framed timber support called a centre. Permanent arch centres are also available for small spans and simple formats.

Typical Arch Formats ~

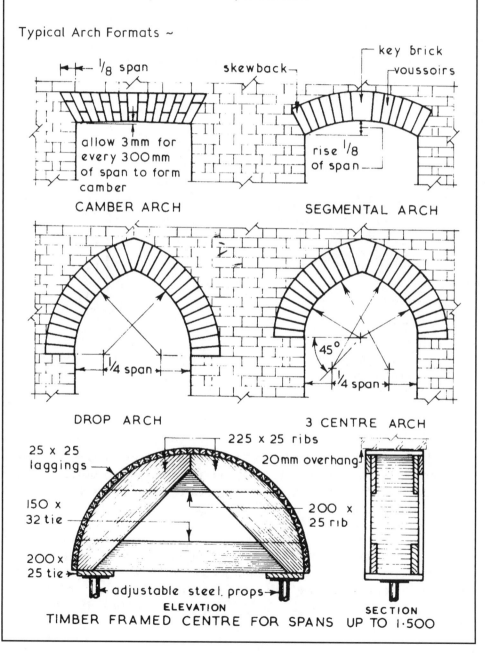

CAMBER ARCH

SEGMENTAL ARCH

DROP ARCH

3 CENTRE ARCH

TIMBER FRAMED CENTRE FOR SPANS UP TO 1·500

The profile of an arch does not lend itself to simple positioning of a damp proof course. At best, it can be located horizontally at upper extrados level. This leaves the depth of the arch and masonry below the dpc vulnerable to dampness. Proprietary galvanised or stainless steel cavity trays resolve this problem by providing:

* Continuity of dpc around the extrados.

* Arch support/centring during construction.

* Arch and wall support after construction.

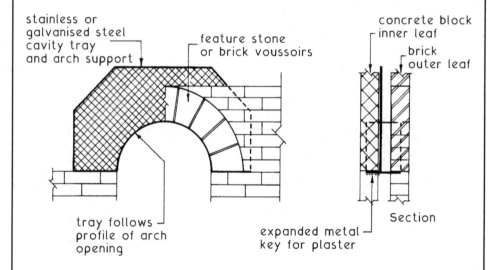

stainless or galvanised steel cavity tray and arch support

feature stone or brick voussoirs

concrete block inner leaf

brick outer leaf

tray follows profile of arch opening

expanded metal key for plaster

Section

Standard profiles are made to the traditional outlines shown on the previous two pages, in spans up to 2 m. Other options may also be available from some manufacturers. Irregular shapes and spans can be made to order.

Note: Arches in semi-circular, segmental or parabolic form up to 2m span can be proportioned empirically. For integrity of structure it is important to ensure sufficient provision of masonry over and around any arch, see BS 5628: Code of practice for use of masonry.

The example in steel shown on the preceding page combines structural support with a damp proof course, without the need for temporary support from a centre. Where traditional centring is retained, a lightweight preformed polypropylene cavity tray/dpc can be used. These factory made plastic trays are produced in various thicknesses of 1.5 to 3mm relative to spans up to about 2 m. Arch centres are made to match the tray profile and with care can be reused several times.

An alternative material is code 4 lead sheet*. Lead is an adaptable material but relatively heavy. Therefore, its suitability is limited to small spans particularly with non-standard profiles.

*BS EN 12588: Lead and lead alloys. Rolled lead sheet for building purposes. Lead sheet is coded numerically from 3 to 8 (1.25 to 3.50mm – see page 452), which closely relates to the traditional specification in lbs./sq. ft.

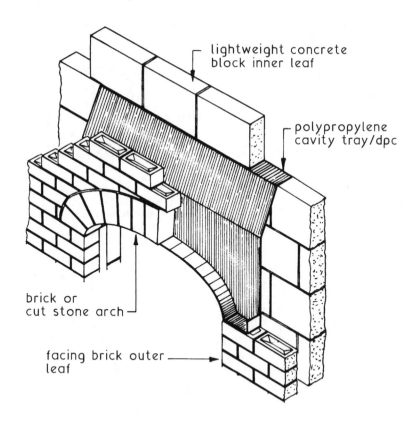

lightweight concrete block inner leaf

polypropylene cavity tray/dpc

brick or cut stone arch

facing brick outer leaf

Ref. BS 5628-3: Code of practice for the use of masonry. Materials and components, design and workmanship.

Openings ~ these consist of a head, jambs and sill. Different methods can be used in their formation, all with the primary objective of adequate support around the void. Details relate to older/existing construction and where thermal insulation is not critical. Application limited – see pages 488 and 489.

Typical Head Details ~

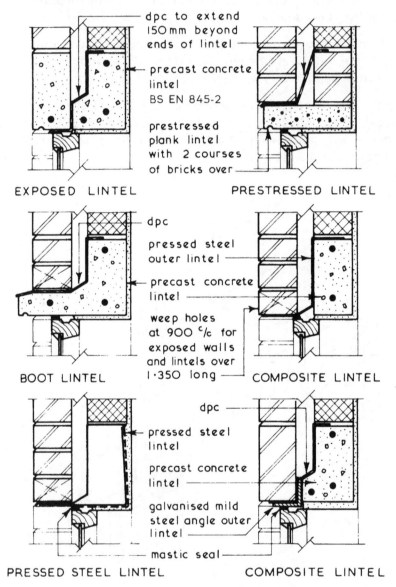

dpc to extend 150mm beyond ends of lintel

precast concrete lintel BS EN 845-2

prestressed plank lintel with 2 courses of bricks over

EXPOSED LINTEL   PRESTRESSED LINTEL

dpc

pressed steel outer lintel

precast concrete lintel

weep holes at 900 c/c for exposed walls and lintels over 1·350 long

BOOT LINTEL   COMPOSITE LINTEL

dpc

pressed steel lintel

precast concrete lintel

galvanised mild steel angle outer lintel

mastic seal

PRESSED STEEL LINTEL   COMPOSITE LINTEL

Ref. BS EN 845-2: Specification for ancillary components for masonry. Lintels.

Jambs ~ these may be bonded as in solid walls or unbonded as in cavity walls. The latter must have some means of preventing the ingress of moisture from the outer leaf to the inner leaf and hence the interior of the building. Details as preceding page.

Application limited – see pages 488 and 489.

Typical Jamb Details ~

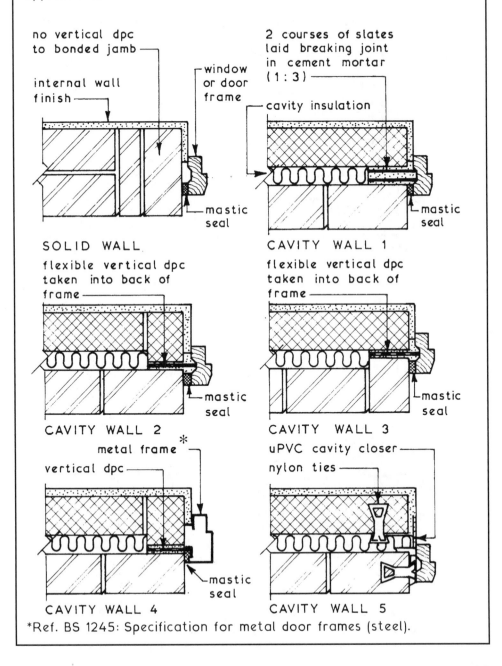

SOLID WALL

CAVITY WALL 1

CAVITY WALL 2

CAVITY WALL 3

CAVITY WALL 4

CAVITY WALL 5

*Ref. BS 1245: Specification for metal door frames (steel).

Sills ~ the primary function of any sill is to collect the rainwater which has run down the face of the window or door and shed it clear of the wall below.

Timber Sill 1, Cast Stone Subsill and Slate Sill have applications limited – see pages 488 and 489.
Typical Sill details ~

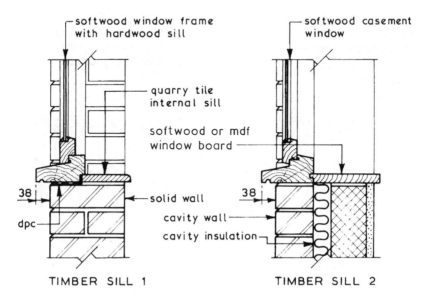

TIMBER SILL 1

TIMBER SILL 2

- softwood window frame with hardwood sill
- quarry tile internal sill
- softwood or mdf window board
- softwood casement window
- solid wall
- cavity wall
- cavity insulation
- dpc
- 38

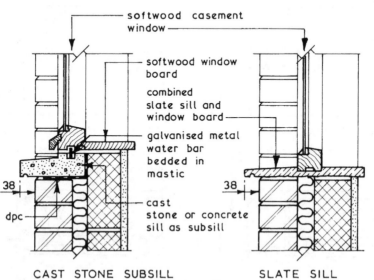

CAST STONE SUBSILL

SLATE SILL

- softwood casement window
- softwood window board
- combined slate sill and window board
- galvanised metal water bar bedded in mastic
- cast stone or concrete sill as subsill
- dpc
- 38

Ref. BS 5642-1: Sills and copings. Specification for window sills of precast concrete, cast stone, clayware, slate and natural stone.

Traditional Construction – checked rebates or recesses in masonry solid walls were often provided at openings to accommodate door and window frames. This detail was used as a means to complement frame retention and prevent weather intrusion.

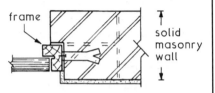

Typical checked masonry (pre 1940s)

Exposure Zones – checked reveal treatment is now required mainly where wind-driven rain will have most impact. This is primarily in the south west and west coast areas of the British Isles, plus some isolated inland parts that will be identified by their respective local authorities.

Typical Checked Opening Details –

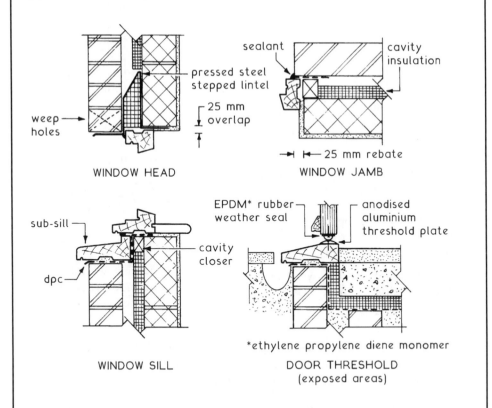

Ref. Building Regulations, Approved Document C. Section 5: Walls. Driving rain exposure zones 3 and 4.

A window must be aesthetically acceptable in the context of building design and surrounding environment

glass and glazing to be suitable for window position and type

suitable and durable materials required for framing

thermal and sound insulation properties to be acceptable to client and within Building Regs.

sizing of openings to meet requirements of Building Regulations for limiting heat losses and fire escape

windows should be weather tight when opening lights are closed

perimeter joint to be adequately sealed

Windows should be selected or designed to resist wind loadings, be easy to clean and provide for safety and security. They should be sited to provide visual contact with the outside.

Habitable upper floor rooms should have a window for emergency escape. Min. opening area, 0.330 m². Min. height and width, 0.450 m. Max height of opening, 1.100 m above floor.

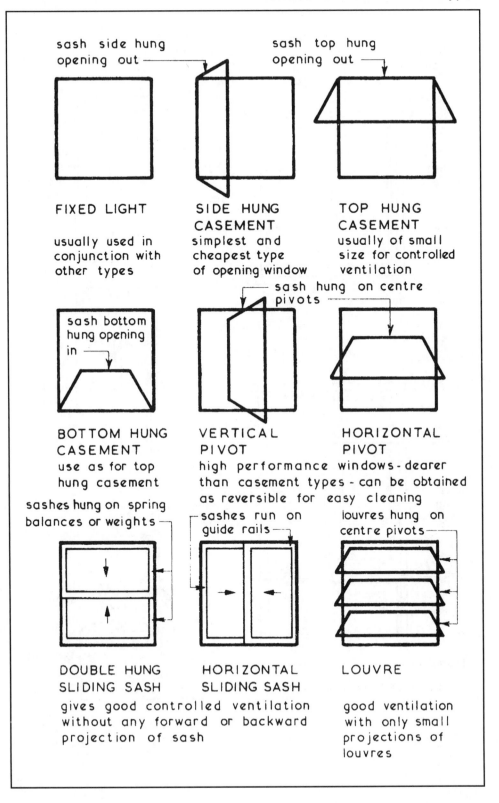

sash side hung
opening out

sash top hung
opening out

FIXED LIGHT

usually used in
conjunction with
other types

**SIDE HUNG
CASEMENT**
simplest and
cheapest type
of opening window

**TOP HUNG
CASEMENT**
usually of small
size for controlled
ventilation

sash bottom
hung opening
in

sash hung on centre
pivots

**BOTTOM HUNG
CASEMENT**
use as for top
hung casement

**VERTICAL
PIVOT**

**HORIZONTAL
PIVOT**

high performance windows - dearer
than casement types - can be obtained
as reversible for easy cleaning

sashes hung on spring
balances or weights

sashes run on
guide rails

louvres hung on
centre pivots

**DOUBLE HUNG
SLIDING SASH**

**HORIZONTAL
SLIDING SASH**

LOUVRE

gives good controlled ventilation
without any forward or backward
projection of sash

good ventilation
with only small
projections of
louvres

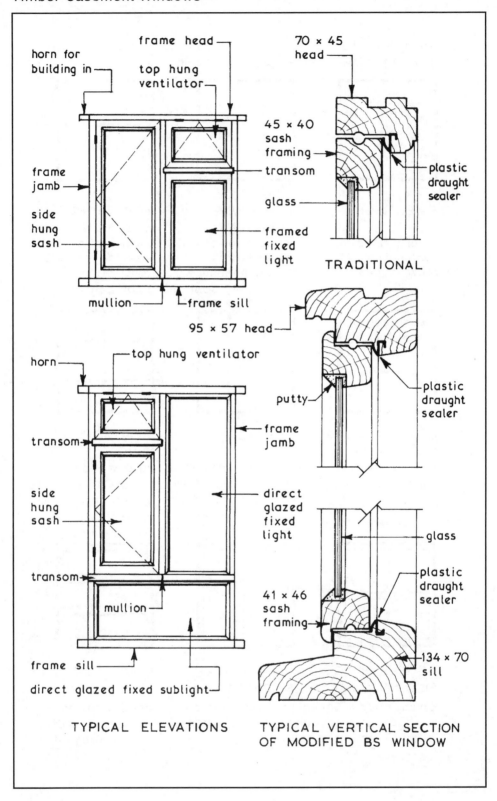

horn for building in
frame head
top hung ventilator
frame jamb
side hung sash
mullion
frame sill

70 × 45 head
45 × 40 sash framing
transom
glass
plastic draught sealer
framed fixed light

TRADITIONAL

horn
top hung ventilator
transom
side hung sash
transom
mullion
frame sill
direct glazed fixed sublight

95 × 57 head
putty
plastic draught sealer
frame jamb
direct glazed fixed light
41 × 46 sash framing
glass
plastic draught sealer
134 × 70 sill

TYPICAL ELEVATIONS

TYPICAL VERTICAL SECTION OF MODIFIED BS WINDOW

The standard range of casement windows used in the UK was derived from the English Joinery Manufacturer's Association (EJMA) designs of some 50 years ago. These became adopted in BS 644: Timber windows. Fully finished factory assembled windows of various types. Specification. A modified type is shown on the preceding page. Contemporary building standards require higher levels of performance in terms of thermal and sound insulation (Bldg. Regs. Pt. L and E), air permeability, water tightness and wind resistance (BS ENs 1026, 1027 and 12211, respectively). This has been achieved by adapting Scandinavian designs with double and triple glazing to attain U values as low as 1·2 W/m²K and a sound reduction of 50 dB.

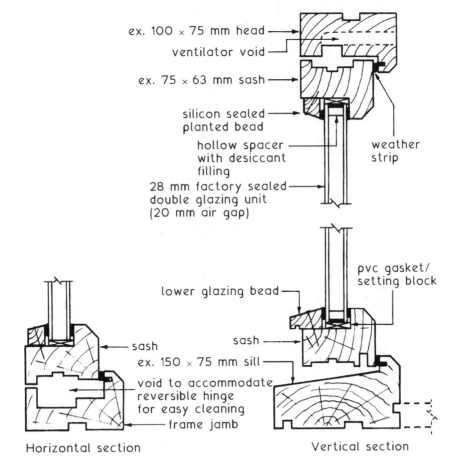

ex. 100 × 75 mm head
ventilator void
ex. 75 × 63 mm sash
silicon sealed planted bead
hollow spacer with desiccant filling
28 mm factory sealed double glazing unit (20 mm air gap)
weather strip
pvc gasket/ setting block
lower glazing bead
sash
sash
ex. 150 × 75 mm sill
void to accommodate reversible hinge for easy cleaning
frame jamb

Horizontal section   Vertical section

Further refs:
BS 6375 series: Performance of windows and doors.
BS 6375-1: Classification for weather tightness.
BS 6375-2: Classification for operation and strength characteristics.
BS 7950: Specification for enhanced security performance.

Metal Windows ~ these can be obtained in steel (BS 6510) or in aluminium alloy (BS 4873). Steel windows are cheaper in initial cost than aluminium alloy but have higher maintenance costs over their anticipated life, both can be obtained fitted into timber subframes. Generally they give a larger glass area for any given opening size than similar timber windows but they can give rise to condensation on the metal components. Page 384 shows an example of an energy efficient improvement.

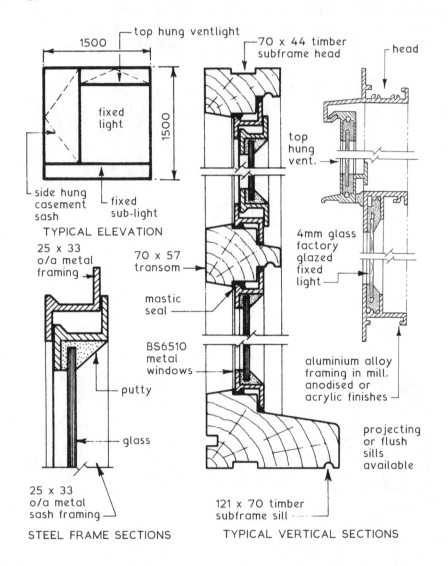

top hung ventlight

1500

1500

fixed light

side hung casement sash

fixed sub-light

TYPICAL ELEVATION

25 x 33 o/a metal framing

70 x 57 transom

70 x 44 timber subframe head

head

top hung vent.

4mm glass factory glazed fixed light

mastic seal

BS6510 metal windows

aluminium alloy framing in mill, anodised or acrylic finishes

putty

glass

projecting or flush sills available

25 x 33 o/a metal sash framing

STEEL FRAME SECTIONS

121 x 70 timber subframe sill

TYPICAL VERTICAL SECTIONS

Refs.:
BS 4873: Aluminium alloy windows and doorsets. Specification.
BS 6510: Steel-framed windows and glazed doors.

Timber Windows ~ wide range of ironmongery available which can be factory fitted or supplied and fixed on site.

Metal Windows ~ ironmongery usually supplied with and factory fitted to the windows.

Typical Examples ~

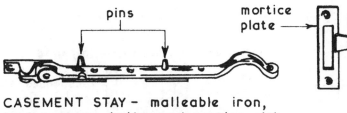

pins

mortice plate

malleable iron, curly tail pattern

CASEMENT STAY - malleable iron, leaf pattern, half round section with two pins
Sizes: 200; 250 and 300 mm

CASEMENT FASTENER

hot pressed aluminium, plain end pattern

wedge plate

CASEMENT STAY - cast aluminium, plain end pattern with one pin
Sizes: 250 and 300 mm

CASEMENT FASTENER

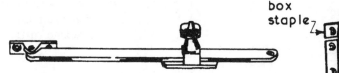

box staple

hot pressed brass

CASEMENT STAY - steel and brass, sliding screw down pattern
Sizes: 250 and 300 mm

VENTLIGHT CATCH
used with bottom hung ventlights

CASEMENT STAY - steel, stayput pattern
Arm Sizes: 100; 140 and 175 mm

malleable iron or brass
Sizes: 150 175 and 200mm

QUADRANT STAY

Sliding Sash Windows ~ these are an alternative format to the conventional side hung casement windows and can be constructed as a vertical or double hung sash window or as a horizontal sliding window in timber, metal, plastic or in any combination of these materials. The performance and design functions of providing daylight, ventilation, vision out, etc., are the same as those given for traditional windows in Windows – Performance Requirements on page 366.

Typical Double Hung Weight Balanced Window Details ~

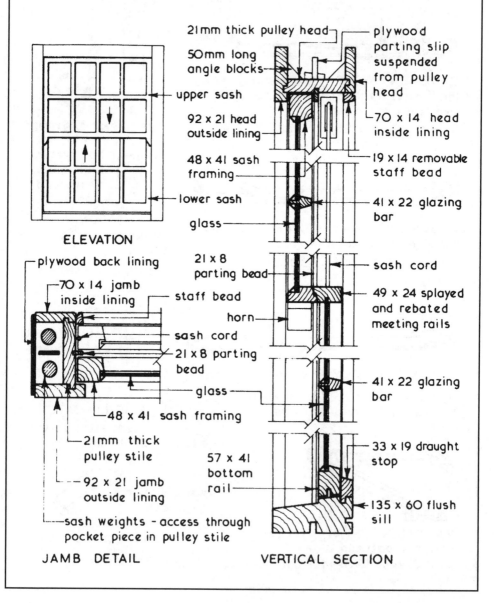

ELEVATION

JAMB DETAIL

VERTICAL SECTION

Double Hung Sash Windows ~ these vertical sliding sash windows come in two formats when constructed in timber. The weight balanced format is shown on the preceding page, the alternative spring balanced type is illustrated below. Both formats are usually designed and constructed to the recommendations set out in BS 644.

Typical Double Hung Spring Balanced Window Details ~

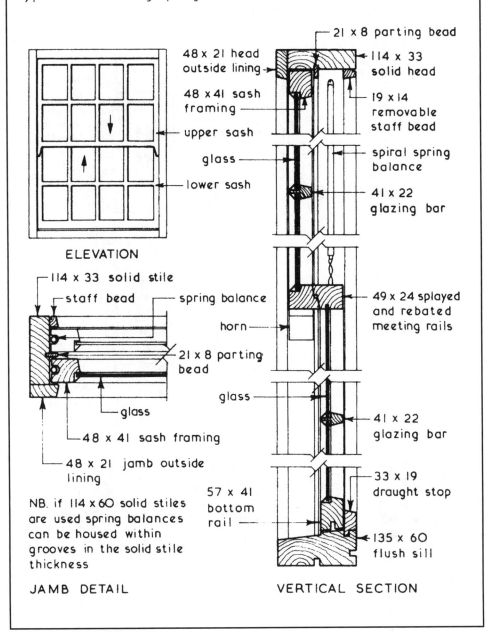

ELEVATION

114 x 33 solid stile
staff bead
spring balance
21 x 8 parting bead
glass
48 x 41 sash framing
48 x 21 jamb outside lining

NB. if 114 x 60 solid stiles are used spring balances can be housed within grooves in the solid stile thickness

JAMB DETAIL

21 x 8 parting bead
48 x 21 head outside lining
48 x 41 sash framing
upper sash
glass
lower sash

114 x 33 solid head
19 x 14 removable staff bead
spiral spring balance
41 x 22 glazing bar

49 x 24 splayed and rebated meeting rails
horn

glass
41 x 22 glazing bar

57 x 41 bottom rail
33 x 19 draught stop
135 x 60 flush sill

VERTICAL SECTION

Horizontally Sliding Sash Windows ~ these are an alternative format to the vertically sliding or double hung sash windows shown on pages 372 & 373 and can be constructed in timber, metal, plastic or combinations of these materials with single or double glazing. A wide range of arrangements are available with two or more sliding sashes which can have a ventlight incorporated in the outer sliding sash.

Typical Horizontally Sliding Sash Window Details ~

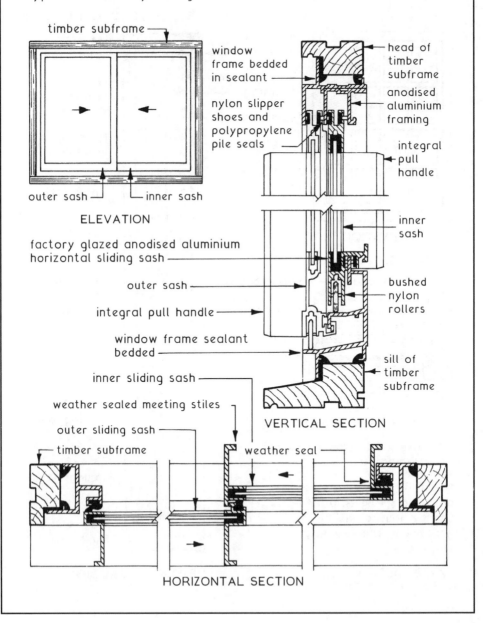

timber subframe

window frame bedded in sealant

nylon slipper shoes and polypropylene pile seals

outer sash — — inner sash

ELEVATION

head of timber subframe

anodised aluminium framing

integral pull handle

inner sash

factory glazed anodised aluminium horizontal sliding sash

outer sash

integral pull handle

bushed nylon rollers

window frame sealant bedded

inner sliding sash

weather sealed meeting stiles

sill of timber subframe

outer sliding sash

timber subframe

weather seal

VERTICAL SECTION

HORIZONTAL SECTION

Pivot Windows ~ like other windows these are available in timber, metal, plastic or in combinations of these materials.

They can be constructed with centre jamb pivots enabling the sash to pivot or rotate in the horizontal plane or alternatively the pivots can be fixed in the head and sill of the frame so that the sash rotates in the vertical plane.

Typical Example ~

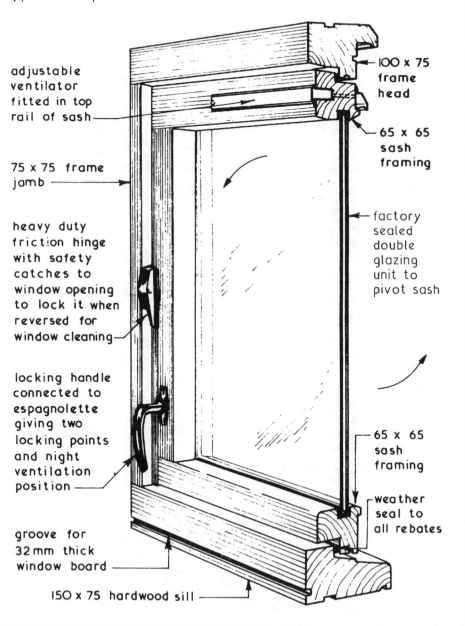

adjustable ventilator fitted in top rail of sash

75 x 75 frame jamb

heavy duty friction hinge with safety catches to window opening to lock it when reversed for window cleaning

locking handle connected to espagnolette giving two locking points and night ventilation position

groove for 32 mm thick window board

150 x 75 hardwood sill

100 x 75 frame head

65 x 65 sash framing

factory sealed double glazing unit to pivot sash

65 x 65 sash framing

weather seal to all rebates

Bay Windows ~ these can be defined as any window with side lights which projects in front of the external wall and is supported by a sill height wall. Bay windows not supported by a sill height wall are called oriel windows. They can be of any window type, constructed from any of the usual window materials and are available in three plan formats namely square, splay and circular or segmental. Timber corner posts can be boxed, solid or jointed the latter being the common method.

Typical Examples ~

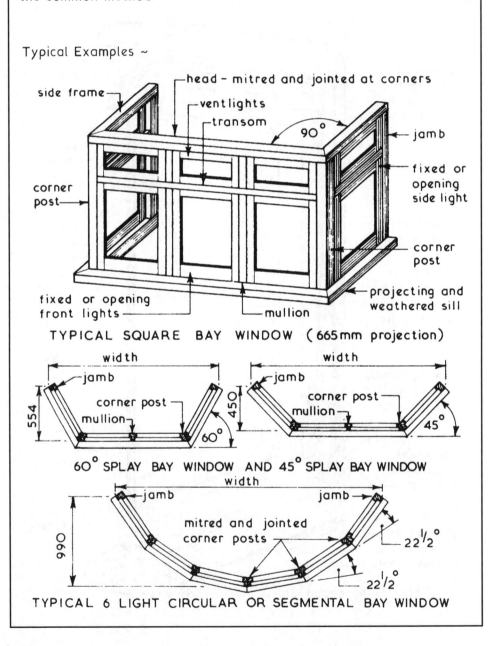

TYPICAL SQUARE BAY WINDOW (665mm projection)

60° SPLAY BAY WINDOW AND 45° SPLAY BAY WINDOW

TYPICAL 6 LIGHT CIRCULAR OR SEGMENTAL BAY WINDOW

Schedules ~ the main function of a schedule is to collect together all the necessary information for a particular group of components such as windows, doors and drainage inspection chambers. There is no standard format for schedules but they should be easy to read, accurate and contain all the necessary information for their purpose. Schedules are usually presented in a tabulated format which can be related to and read in conjunction with the working drawings.

Typical Example ~

WINDOW SCHEDULE – Sheet 1 of 1		Drawn By: RC	Date: 14/4/01	Rev.				
Contract Title & Number: Lane End Farm – H 341/80						Drg. Nos. C(31) 450–7		
Number	Type or catalogue ref.	Material	Overall Size w x h	Glass	Ironmongery	Sill		
						External	Internal	
2	213 CV	hardwood	1200 x 1350	sealed units as supplied with frames	supplied with casements	2 cos. plain tiles subsill	150×150×15 quarry tiles	
4	309 CVC	ditto	1770 x 900	ditto	ditto	ditto	25 mm thick softwood	
4	313 CVC	ditto	1770 x 1350	ditto	ditto	sill of frame	ditto	

Window manufacturers identify their products with a notation that combines figures with numbers. The objective is to simplify catalogue entries, specification clauses and schedules. For example:

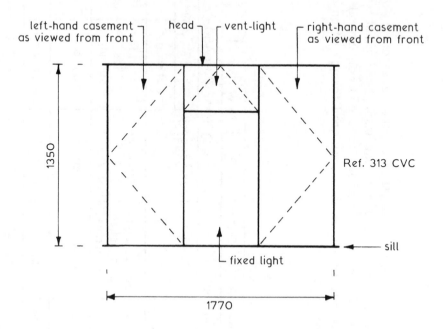

left-hand casement as viewed from front

head

vent-light

right-hand casement as viewed from front

1350

Ref. 313 CVC

sill

fixed light

1770

Notation will vary to some extent between the different joinery producers. The example of 313 CVC translates to:

3 = width divided into three units.
13 = first two dimensions of standard height, ie. 1350 mm.
C = casement.
V = ventlight.

Other common notations include:

N = narrow light.
P = plain (picture type window, ie. no transom or mullion).
T = through transom.
S = sub-light, fixed.
VS = vent-light and sub-light.
F = fixed light.
B = bottom casement opening inwards.
RH/LH = right or left-hand as viewed from the outside.

Glass ~ this material is produced by fusing together soda, lime and silica with other minor ingredients such as magnesia and alumina. A number of glass types are available for domestic work and these include:-

Clear Float ~ used where clear undistorted vision is required. Available thicknesses range from 3mm to 25mm.

Clear Sheet ~ suitable for all clear glass areas but because the two faces of the glass are never perfectly flat or parallel some distortion of vision usually occurs. This type of glass is gradually being superseded by the clear float glass. Available thicknesses range from 3mm to 6mm.

Translucent Glass ~ these are patterned glasses most having one patterned surface and one relatively flat surface. The amount of obscurity and diffusion obtained depend on the type and nature of pattern. Available thicknesses range from 4mm to 6mm for patterned glasses and from 5mm to 10mm for rough cast glasses.

Wired Glass ~ obtainable as a clear polished wired glass or as a rough cast wired glass with a nominal thickness of 7mm. Generally used where a degree of fire resistance is required. Georgian wired glass has a 12mm square mesh whereas the hexagonally wired glass has a 20mm mesh.

Choice of Glass ~ the main factors to be considered are:-
1. Resistance to wind loadings. 2. Clear vision required.
3. Privacy. 4. Security. 5. Fire resistance. 6. Aesthetics.

Glazing Terminology ~

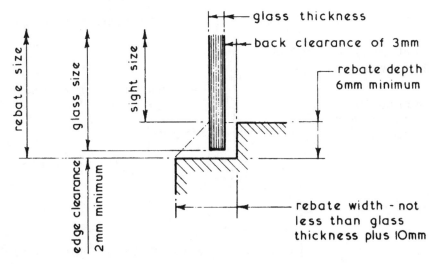

Glazing ~ the act of fixing glass into a frame or surround. In domestic work this is usually achieved by locating the glass in a rebate and securing it with putty or beading and should be carried out in accordance with the recommendations contained in the BS 6262 series: Glazing for buildings.

Timber Surrounds ~ linseed oil putty to BS 544: Specification for linseed oil putty for use in wooden frames. Composed of crushed chalk and linseed oil (whiting). Rebate to be clean, dry and primed before glazing is carried out. Putty should be protected with paint within two weeks of application.

Metal Surrounds ~ metal casement putty if metal surround is to be painted – if surround is not to be painted a non-setting compound should be used.

A general purpose putty is also available. This combines the properties of the two types.

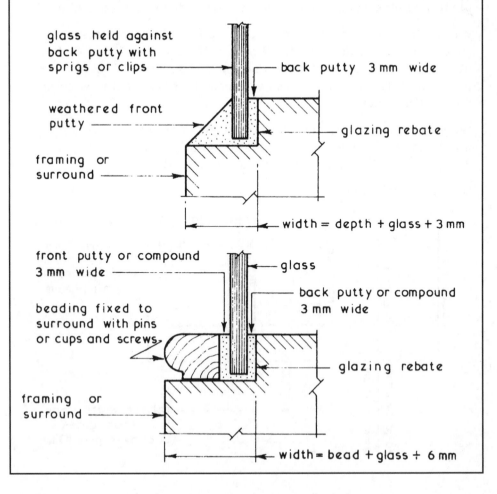

Double Glazing ~ as its name implies this is where two layers of glass are used instead of the traditional single layer. Double glazing can be used to reduce the rate of heat loss through windows and glazed doors or it can be employed to reduce the sound transmission through windows. In the context of thermal insulation this is achieved by having a small air or argon gas filled space within the range of 6 to 20mm between the two layers of glass. The sealed double glazing unit will also prevent internal misting by condensation. If metal frames are used these should have a thermal break incorporated in their design. All opening sashes in a double glazing system should be fitted with adequate weather seals to reduce the rate of heat loss through the opening clearance gap.

In the context of sound insulation three factors affect the performance of double glazing. Firstly good installation to ensure airtightness, secondly the weight of glass used and thirdly the size of air space between the layers of glass. The heavier the glass used the better the sound insulation and the air space needs to be within the range of 50 to 300mm. Absorbent lining to the reveals within the air space will also improve the sound insulation properties of the system.

Typical Examples ~

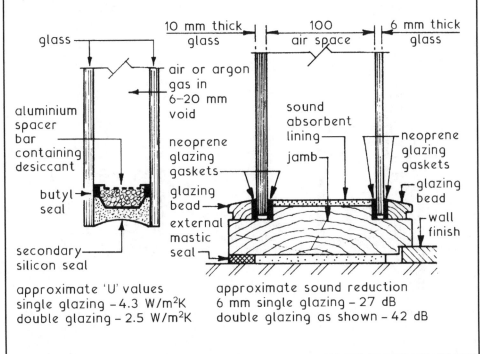

approximate 'U' values
single glazing – 4.3 W/m²K
double glazing – 2.5 W/m²K

approximate sound reduction
6 mm single glazing – 27 dB
double glazing as shown – 42 dB

Secondary glazing of existing windows is an acceptable method for reducing heat energy losses at wall openings. Providing the existing windows are in a good state of repair, this is a cost effective, simple method for upgrading windows to current energy efficiency standards. In addition to avoiding the disruption of removing existing windows, further advantages of secondary glazing include, retention of the original window features, reduction in sound transmission and elimination of draughts. Applications are manufactured for all types of window, with sliding or hinged variations. The following details are typical of horizontal sliding sashes -

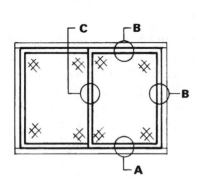

Elevation of frame

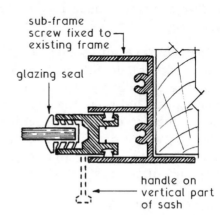

Detail B - Head and jamb

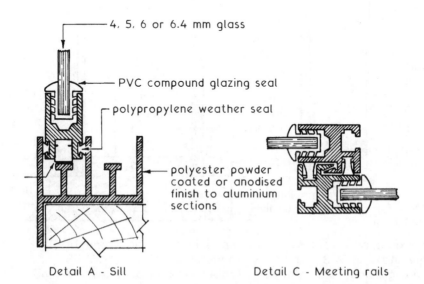

Detail A - Sill

Detail C - Meeting rails

Low emissivity or "Low E" glass is specially manufactured with a surface coating to significantly improve its thermal performance. The surface coating has a dual function:

1. Allows solar short wave light radiation to penetrate a building.
2. Reflects long wave heat radiation losses back into a building.

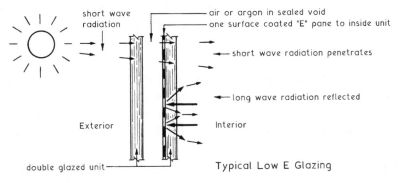

Typical Low E Glazing

Manufacturing processes:

1. Pyrolitic hard coat, applied on-line as the glass is made. Emissivity range, 0.15-0.20, e.g. Pilkington 'K'.
2. A sputtered soft coat applied after glass manufacture. Emissivity range, 0.05-0.10, e.g. Pilkington 'Kappafloat' and 'Suncool High Performance'.

Note: In relative terms, uncoated glass has a normal emissivity of about 0.90.

Indicative U-values for multi-glazed windows of 4 mm glass with a 16 mm void width:

Glazing type	uPVC or wood frame	metal frame
Double, air filled	2.7	3.3
Double, argon filled	2.6	3.2
Double, air filled Low E (0.20)	2.1	2.6
Double, argon filled Low E (0.20)	2.0	2.5
Double, air filled Low E (0.05)	2.0	2.3
Double, argon filled Low E (0.05)	1.7	2.1
Triple, air filled	2.0	2.5
Triple, argon filled	1.9	2.4
Triple, air filled Low E (0.20)	1.6	2.0
Triple, argon filled Low E (0.20)	1.5	1.9
Triple, air filled Low E (0.05)	1.4	1.8
Triple, argon filled Low E (0.05)	1.3	1.7

Notes:

1. A larger void and thicker glass will reduce the U-value, and vice-versa.

2. Data for metal frames assumes a thermal break of 4 mm (see next page).

3. Hollow metal framing units can be filled with a closed cell insulant foam to considerably reduce U-values.

Extruded aluminium profiled sections are designed and manufactured to create lightweight hollow window (and door) framing members.

Finish – untreated aluminium is prone to surface oxidisation. This can be controlled by paint application, but most manufacturers provide a variable colour range of polyester coatings finished gloss, satin or matt.

Thermal insulation – poor insulation and high conductivity are characteristics of solid profile metal windows. This is much less apparent with hollow profile outer members, as they can be considerably enhanced by a thermal infilling of closed cell foam.

Condensation – a high strength 2-part polyurethane resin thermal break between internal and external profiles inhibits cold bridging. This reduces the opportunity for condensation to form on the surface. The indicative U-values given on the preceding page are based on a thermal break of 4 mm. If this is increased to 16 mm, the values can be reduced by up to 0.2 W/m² K.

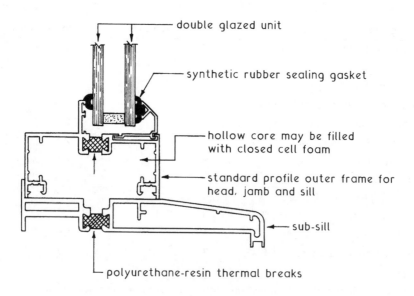

double glazed unit

synthetic rubber sealing gasket

hollow core may be filled with closed cell foam

standard profile outer frame for head, jamb and sill

sub-sill

polyurethane-resin thermal breaks

Hollow Core Aluminium Profiled Window Section

Inert gas fills ~ argon or krypton. Argon is generally used as it is the least expensive and more readily available. Where krypton is used, the air gap need only be half that with argon to achieve a similar effect. Both gases have a higher insulating value than air due to their greater density.

Densities (kg/m^3):
    Air = 1.20
    Argon = 1.66
    Krypton = 3.49

Argon and krypton also have a lower thermal conductivity than air.

Spacers ~ generally hollow aluminium with a desiccant or drying agent fill. The filling absorbs the initial moisture present in between the glass layers. Non-metallic spacers are preferred as aluminium is an effective heat conductor.

Approximate solar gains with ordinary float glass ~

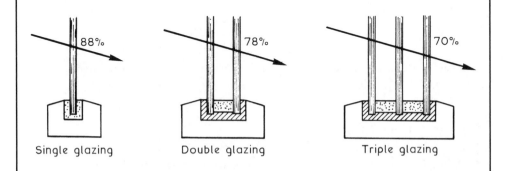

Single glazing   Double glazing   Triple glazing

"Low E" invisible coatings reduce the solar gain by up to one-third. Depending on the glass quality and cleanliness, about 10 to 15% of visible light reduction applies for each pane of glass.

Triple Glazing

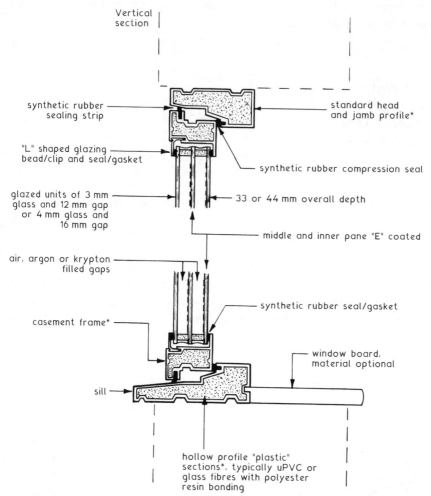

Typical application ~

Vertical section

synthetic rubber sealing strip

standard head and jamb profile*

"L" shaped glazing bead/clip and seal/gasket

synthetic rubber compression seal

glazed units of 3 mm glass and 12 mm gap or 4 mm glass and 16 mm gap

33 or 44 mm overall depth

middle and inner pane "E" coated

air, argon or krypton filled gaps

casement frame*

synthetic rubber seal/gasket

window board, material optional

sill

hollow profile "plastic" sections*, typically uPVC or glass fibres with polyester resin bonding

* Hollow profiles manufactured with a closed cell insulant foam/expanded polystyrene core.

Further considerations ~
* U value potential, less than 1.0 W/m²K.
* "Low E" invisible metallic layer on one pane of double glazing gives a similar insulating value to standard triple glazing (see page 383).
* Performance enhanced with blinds between wide gap panes.
* High quality ironmongery required due to weight of glazed frames.
* Improved sound insulation, particularly with heavier than air gap fill.

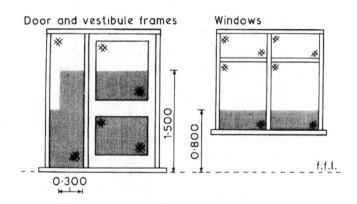

Door and vestibule frames   Windows

In these critical locations, glazing must satisfy one of the following:-

1. Breakage to leave only a small opening with small detachable particles without sharp edges.
2. Disintegrating glass must leave only small detached pieces.
3. Inherent robustness, e.g. polycarbonate composition. Annealed glass acceptable but with the following limitations:-

Thickness of annealed glass (mm)	Max. glazed area. Height (m)	Width(m)
8	1·100	1·100
10	2·250	2·250
12	3·000	4·500
15	no limit	

4. Panes in small areas, <250 mm wide and <0.5 m² area. e.g. leaded lights (4 mm annealed glass) and Georgian pattern (6 mm annealed glass).
5. Protective screening as shown:

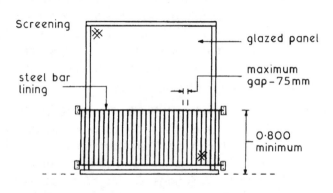

Manifestation or Marking of Glass ~ another aspect of the critical location concept which frequently occurs with contemporary glazed features in a building. Commercial premises such as open plan offices, shops and showrooms often incorporate large walled areas of uninterrupted glass to promote visual depth, whilst dividing space or forming part of the exterior envelope. To prevent collision, glazed doors and walls must have prominent framing or intermediate transoms and mullions. An alternative is to position obvious markings at 1000 and 1500mm above floor level. Glass doors could have large pull/push handles and/or IN and OUT signs in bold lettering. Other areas may be adorned with company logos, stripes, geometric shape, etc.

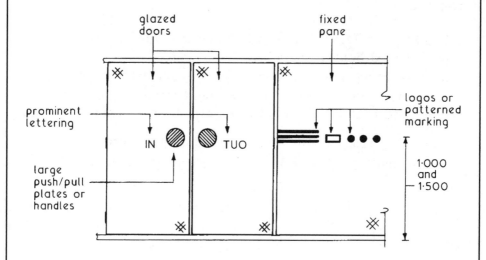

Critical Locations ~ The Building Regulations, Approved Document ~ N, determines positions where potential personal impact and injury with glazed doors and windows are most critical. In these situations the glazing specification must incorporate a degree of safety such that any breakage would be relatively harmless. Additional measures in British Standard 6206 complement the Building Regulations and provide test requirements and specifications for impact performance for different classes of glazing material. See also BS 6262.

Refs. Building Regulations, A.D. N1: Protection against impact.
A.D. N2: Manifestation of glazing.
BS 6206: Specification for impact performance requirements for flat safety glass and safety plastics for use in buildings.
BS 6262 series: Glazing for buildings. Codes of practice.

Glass blocks have been used for some time as internal feature partitioning. They now include a variety of applications in external walls, where they combine the benefits of a walling unit with a natural source of light. They have also been used in paving to allow natural light penetration into basements.

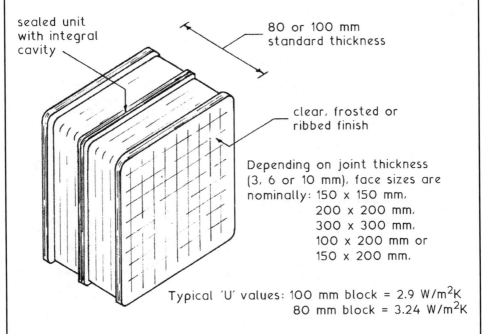

sealed unit with integral cavity

80 or 100 mm standard thickness

clear, frosted or ribbed finish

Depending on joint thickness (3, 6 or 10 mm), face sizes are nominally: 150 x 150 mm,
200 x 200 mm,
300 x 300 mm,
100 x 200 mm or
150 x 200 mm.

Typical 'U' values: 100 mm block = 2.9 W/m^2K
80 mm block = 3.24 W/m^2K

Fire resistance, BS 476-22 - 1 hour integrity (load bearing capacity and fire containment).
Maximum panel size is 9m². Maximum panel dimension is 3 m

Laying – glass blocks can be bonded like conventional brickwork, but for aesthetic reasons are usually laid with continuous vertical and horizontal joints.

Jointing – blocks are bedded in mortar with reinforcement from two, 9 gauge galvanised steel wires in horizontal joints. Every 3rd. course for 150mm units, every 2nd. course for 200mm units and every course for 300mm units. First and last course to be reinforced.

Ref: BS 476-22: Fire tests on building materials and structures. Methods for determination of the fire resistance of non-loadbearing elements of construction.

Mortar – dryer than for bricklaying as the blocks are non-absorbent. The general specification will include: White Portland Cement (BS EN 197-1), High Calcium Lime (BS EN 459-1) and Sand. The sand should be white quartzite or silica type. Fine silver sand is acceptable. An integral waterproofing agent should also be provided. Recommended mix ratios – 1 part cement: 0.5 part lime: 4 parts sand.

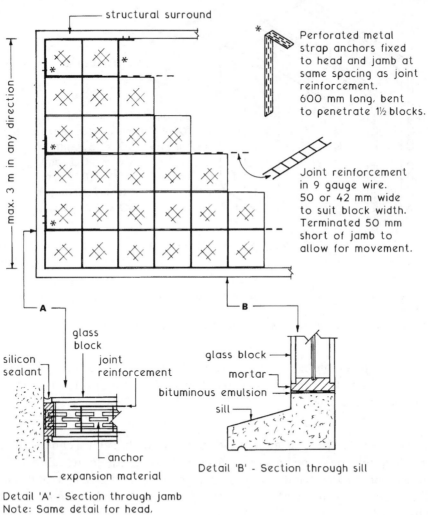

structural surround

max. 3 m in any direction

* Perforated metal strap anchors fixed to head and jamb at same spacing as joint reinforcement.
600 mm long, bent to penetrate 1½ blocks.

Joint reinforcement in 9 gauge wire. 50 or 42 mm wide to suit block width. Terminated 50 mm short of jamb to allow for movement.

A

glass block

silicon sealant
joint reinforcement

anchor
expansion material

Detail 'A' - Section through jamb
Note: Same detail for head, except omit reinforcement

B

glass block
mortar
bituminous emulsion
sill

Detail 'B' - Section through sill

Ref. BS EN 1051-1: Glass in building. Glass blocks and glass pavers. Definitions and description.

Doors ~ can be classed as external or internal. External doors are usually thicker and more robust in design than internal doors since they have more functions to fulfil.

Typical Functions ~

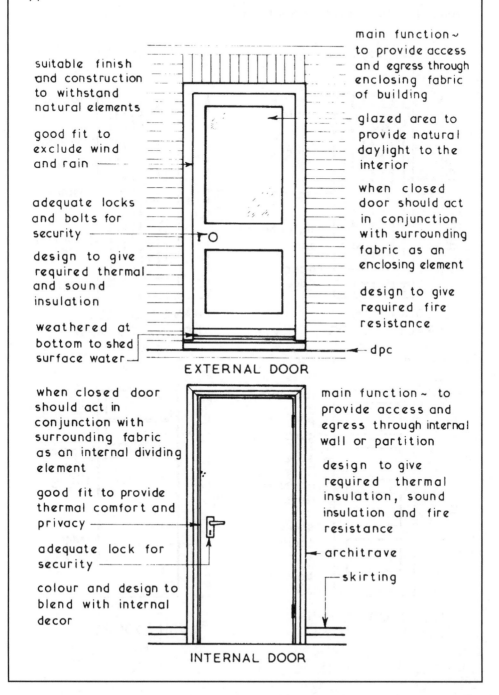

main function ~ to provide access and egress through enclosing fabric of building

suitable finish and construction to withstand natural elements

good fit to exclude wind and rain

glazed area to provide natural daylight to the interior

adequate locks and bolts for security

when closed door should act in conjunction with surrounding fabric as an enclosing element

design to give required thermal and sound insulation

design to give required fire resistance

weathered at bottom to shed surface water

dpc

EXTERNAL DOOR

when closed door should act in conjunction with surrounding fabric as an internal dividing element

main function ~ to provide access and egress through internal wall or partition

good fit to provide thermal comfort and privacy

design to give required thermal insulation, sound insulation and fire resistance

adequate lock for security

architrave

skirting

colour and design to blend with internal decor

INTERNAL DOOR

External Doors ~ these are available in a wide variety of types and styles in timber, aluminium alloy or steel. The majority of external doors are however made from timber, the metal doors being mainly confined to fully glazed doors such as 'patio doors'.

Typical Examples of External Doors ~

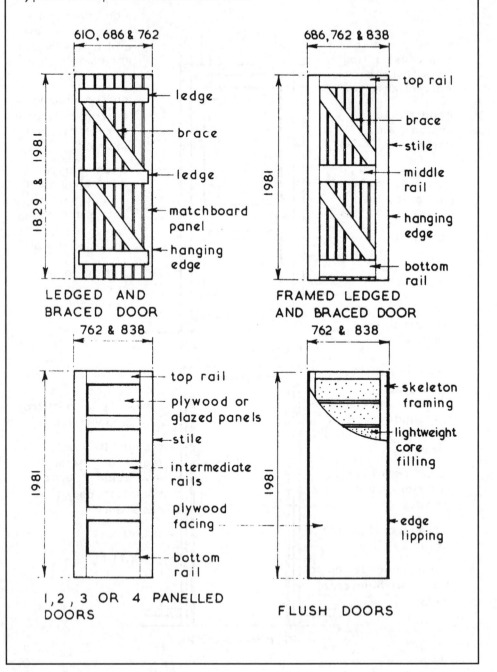

610, 686 & 762

1829 & 1981

- ledge
- brace
- ledge
- matchboard panel
- hanging edge

LEDGED AND BRACED DOOR

686, 762 & 838

1981

- top rail
- brace
- stile
- middle rail
- hanging edge
- bottom rail

FRAMED LEDGED AND BRACED DOOR

762 & 838

1981

- top rail
- plywood or glazed panels
- stile
- intermediate rails
- plywood facing
- bottom rail

1, 2, 3 OR 4 PANELLED DOORS

762 & 838

1981

- skeleton framing
- lightweight core filling
- edge lipping

FLUSH DOORS

Typical examples of purpose made and non-standard external doors ~

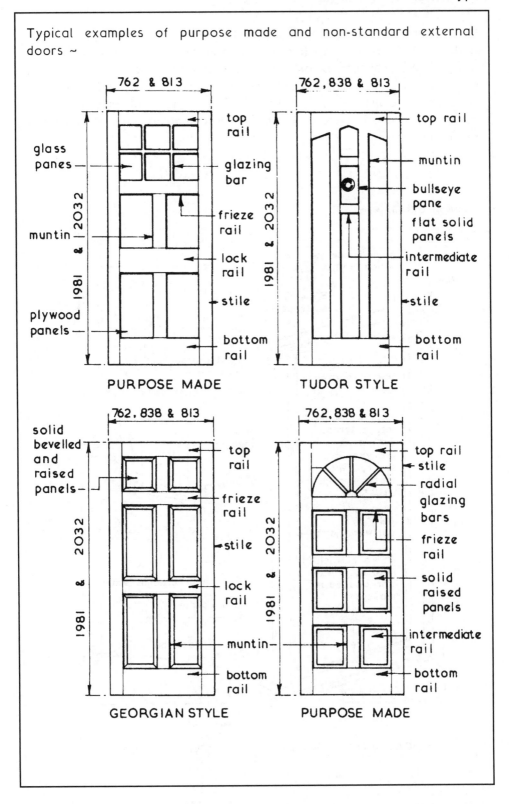

PURPOSE MADE

TUDOR STYLE

GEORGIAN STYLE

PURPOSE MADE

Door Frames ~ these are available for all standard external doors and can be obtained with a fixed solid or glazed panel above a door height transom. Door frames are available for doors opening inwards or outwards. Most door frames are made to the recommendations set out in BS 4787: Internal and external wood doorsets, door leaves and frames. Specification for dimensional requirements.

Typical Example ~

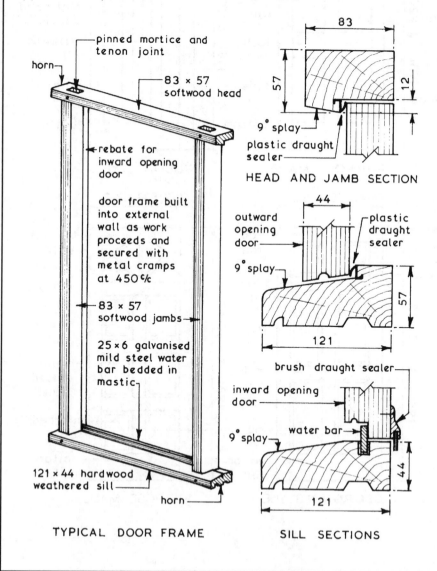

TYPICAL DOOR FRAME

HEAD AND JAMB SECTION

SILL SECTIONS

Door Ironmongery ~ available in a wide variety of materials, styles and finishers but will consist of essentially the same components:- Hinges or Butts - these are used to fix the door to its frame or lining and to enable it to pivot about its hanging edge.

Locks, Latches and Bolts ~ the means of keeping the door in its closed position and providing the required degree of security. The handles and cover plates used in conjunction with locks and latches are collectively called door furniture.

Letter Plates - fitted in external doors to enable letters etc., to be deposited through the door.

Other items include Finger and Kicking Plates which are used to protect the door fabric where there is high usage.

Draught Excluders to seal the clearance gap around the edges of the door and Security Chains to enable the door to be partially opened and thus retain some security.

Typical Examples ~

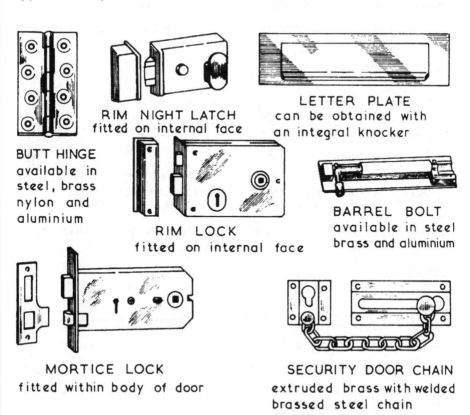

BUTT HINGE
available in
steel, brass
nylon and
aluminium

RIM NIGHT LATCH
fitted on internal face

LETTER PLATE
can be obtained with
an integral knocker

RIM LOCK
fitted on internal face

BARREL BOLT
available in steel
brass and aluminium

MORTICE LOCK
fitted within body of door

SECURITY DOOR CHAIN
extruded brass with welded
brassed steel chain

Industrial Doors ~ these doors are usually classified by their method of operation and construction. There is a very wide range of doors available and the choice should be based on the following considerations:-

1. Movement - vertical or horizontal.
2. Size of opening.
3. Position and purpose of door(s).
4. Frequency of opening and closing door(s).
5. Manual or mechanical operation.
6. Thermal and/or sound insulation requirements.
7. Fire resistance requirements.

Typical Industrial Door Types ~

1. **Straight Sliding -**

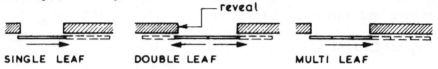

SINGLE LEAF   DOUBLE LEAF   MULTI LEAF

These types can be top hung with a bottom guide roller or hung with bottom rollers and top guides – see page 397

2 **Sliding / Folding -**

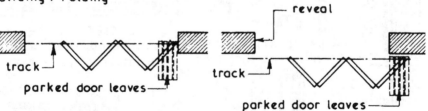

MULTI LEAF END FOLDING   MULTI LEAF END FOLDING
HUNG BETWEEN REVEALS   HUNG BEHIND OPENING

These types can be top hung with a bottom guide roller or hung with bottom rollers and top guides – see page 398

3. **Shutters -**

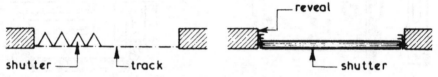

HORIZONTAL FOLDING SHUTTER   ROLLER SHUTTER

Shutters can be installed between, behind or in front of the reveals – see page 399

Straight Sliding Doors ~ these doors are easy to operate, economic to maintain and present no problems for the inclusion of a wicket gate. They do however take up wall space to enable the leaves to be parked in the open position. The floor guide channel associated with top hung doors can become blocked with dirt causing a malfunction of the sliding movement whereas the rollers in bottom track doors can seize up unless regularly lubricated and kept clean. Straight sliding doors are available with either manual or mechanical operation.

Typical Example ~

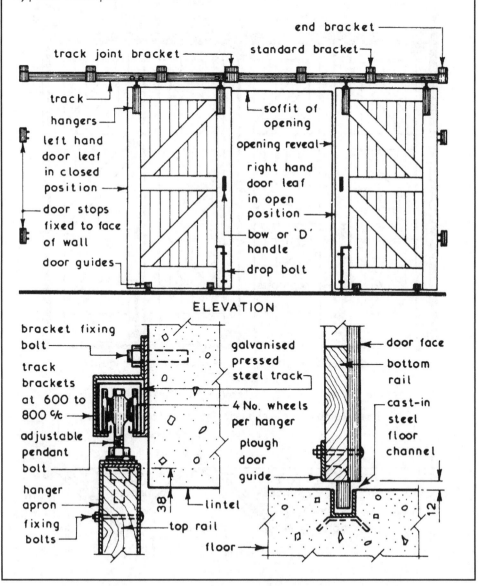

ELEVATION

Sliding/Folding Doors ~ these doors are an alternative format to the straight sliding door types and have the same advantages and disadvantages except that the parking space required for the opened door is less than that for straight sliding doors. Sliding/folding are usually manually operated and can be arranged in groups of 2 to 8 leaves.

Typical Example ~

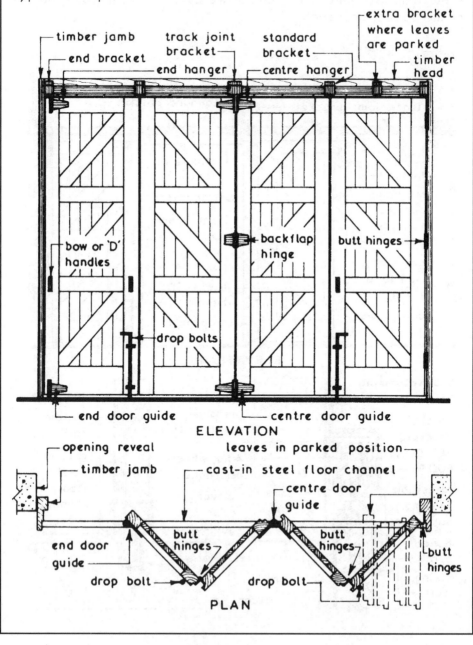

ELEVATION

PLAN

Shutters ~ horizontal folding shutters are similar in operation to sliding/folding doors but are composed of smaller leaves and present the same problems. Roller shutters however do not occupy any wall space but usually have to be fully opened for access. They can be manually operated by means of a pole when the shutters are self coiling, operated by means of an endless chain winding gear or mechanically raised and lowered by an electric motor but in all cases they are slow to open and close. Vision panels cannot be incorporated in the roller shutter but it is possible to include a small wicket gate or door in the design.

Typical Details ~

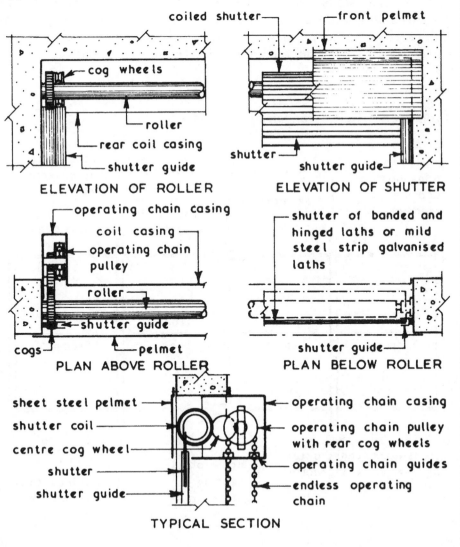

coiled shutter
front pelmet
cog wheels
roller
rear coil casing
shutter guide
ELEVATION OF ROLLER

shutter
shutter guide
ELEVATION OF SHUTTER

operating chain casing
coil casing
operating chain pulley
roller
shutter guide
cogs
pelmet
PLAN ABOVE ROLLER

shutter of banded and hinged laths or mild steel strip galvanised laths

shutter guide
PLAN BELOW ROLLER

sheet steel pelmet
shutter coil
centre cog wheel
shutter
shutter guide

operating chain casing
operating chain pulley with rear cog wheels
operating chain guides
endless operating chain

TYPICAL SECTION

Crosswall Construction ~ this is a form of construction where load bearing walls are placed at right angles to the lateral axis of the building, the front and rear walls being essentially non-load bearing cladding. Crosswall construction is suitable for buildings up to 5 storeys high where the floors are similar and where internal separating or party walls are required such as in blocks of flats or maisonettes. The intermediate floors span longitudinally between the crosswalls providing the necessary lateral restraint and if both walls and floors are of cast in-situ reinforced concrete the series of 'boxes' so formed is sometimes called box frame construction. Great care must be taken in both design and construction to ensure that the junctions between the non-load bearing claddings and the crosswalls are weathertight. If a pitched roof is to be employed with the ridge parallel to the lateral axis an edge beam will be required to provide a seating for the trussed or common rafters and to transmit the roof loads to the crosswalls.

Typical Crosswall Arrangement Details ~

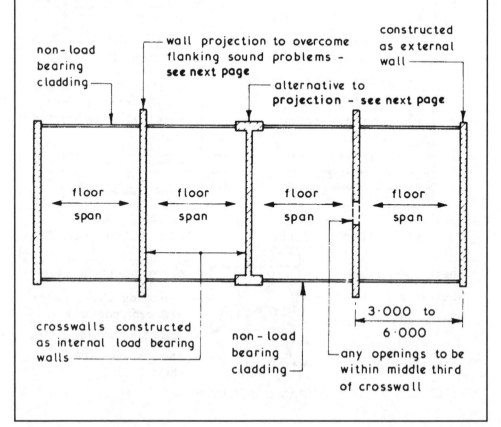

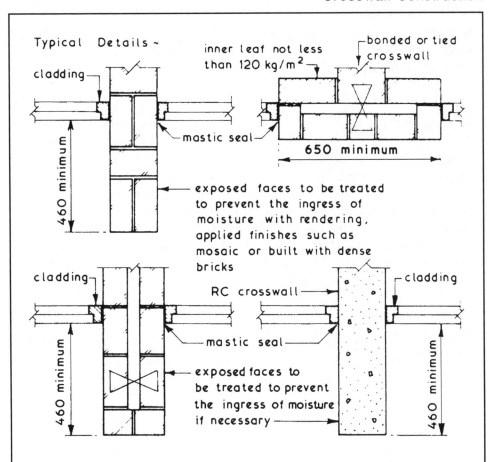

Typical Details ~

cladding

inner leaf not less than 120 kg/m²

bonded or tied crosswall

mastic seal

460 minimum

650 minimum

exposed faces to be treated to prevent the ingress of moisture with rendering, applied finishes such as mosaic or built with dense bricks

cladding

RC crosswall

cladding

mastic seal

460 minimum

460 minimum

exposed faces to be treated to prevent the ingress of moisture if necessary

Advantages of Crosswall Construction:-

1. Load bearing and non-load bearing components can be standardised and in same cases prefabricated giving faster construction times.
2. Fenestration between crosswalls unrestricted structurally.
3. Crosswalls although load bearing need not be weather resistant as is the case with external walls.

Disadvantages of Crosswall Construction:-

1. Limitations of possible plans.
2. Need for adequate lateral ties between crosswalls.
3. Need to weather adequately projecting crosswalls.

Floors:-

An in-situ solid reinforced concrete floor will provide the greatest rigidity, all other form must be adequately tied to walls.

System ~ comprises quality controlled factory produced components of plain reinforced concrete walls and prestressed concrete hollow or solid core plank floors.

Site Assembly ~ components are crane lifted and stacked manually with the floor panel edges bearing on surrounding walls. Temporary support will be necessary until the units are "stitched" together with horizontal and vertical steel reinforcing ties located through reinforcement loops projecting from adjacent panels. In-situ concrete completes the structural connection to provide full transfer of all forces and loads through the joint. Precast concrete stair flights and landings are located and connected to support panels by steel angle bracketing and in-situ concrete joints.

Typical "stitched" joint between precast concrete crosswall components ~

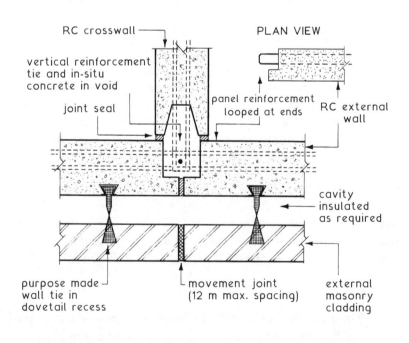

RC crosswall

PLAN VIEW

vertical reinforcement tie and in-situ concrete in void

joint seal

panel reinforcement looped at ends

RC external wall

cavity insulated as required

purpose made wall tie in dovetail recess

movement joint (12 m max. spacing)

external masonry cladding

Concept ~ a cost effective simple and fast site assembly system using load-bearing partitions and external walls to transfer vertical loads from floor panels. The floor provides lateral stability by diaphragm action between the walls.

Application ~ precast reinforced concrete crosswall construction systems may be used to construct multi-storey buildings, particularly where the diaphragm floor load distribution is transferred to lift or stair well cores. Typical applications include schools, hotels, hostels apartment blocks and hospitals. External appearance can be enhanced by a variety of cladding possibilities, including the traditional look of face brickwork secured to the structure by in-built ties. Internal finishing may be with paint or plaster, but it is usually dry lined with plasterboard.

Location of ''stitched'' in-situ reinforced concrete ties ~

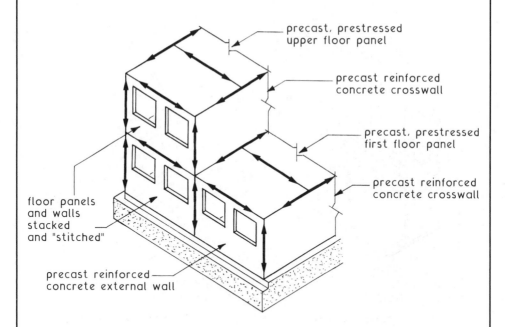

precast, prestressed upper floor panel

precast reinforced concrete crosswall

precast, prestressed first floor panel

precast reinforced concrete crosswall

floor panels and walls stacked and "stitched"

precast reinforced concrete external wall

Fire resistance and sound insulation are achieved by density and quality of concrete. The thermal mass of concrete can be enhanced by applying insulation in between the external precast panel and the masonry or other cladding.

Framing ~ an industry based pre-fabricated house manufacturing process permitting rapid site construction, with considerably fewer site operatives than traditional construction. This technique has a long history of conventional practice in Scandinavia and North America, but has only gained credibility in the UK since the 1960s. Factory-made panels are based on a stud framework of timber, normally ex. 100 × 50 mm, an outer sheathing of plywood, particleboard or similar sheet material, insulation between the framing members and an internal lining of plasterboard. An outer cladding of brickwork weatherproofs the building and provides a traditional appearance.

Assembly techniques are derived from two systems:-

1. Balloon frame

2. Platform frame

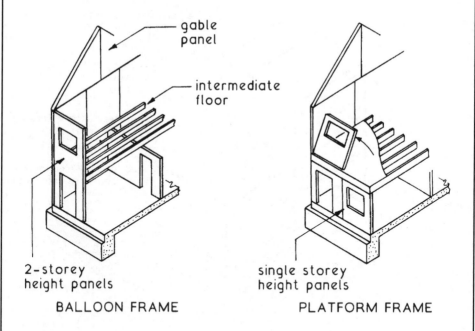

gable panel

intermediate floor

2-storey height panels

BALLOON FRAME

single storey height panels

PLATFORM FRAME

A balloon frame consists of two-storey height panels with an intermediate floor suspended from the framework. In the UK, the platform frame is preferred with intermediate floor support directly on the lower panel. It is also easier to transport, easier to handle on site and has fewer shrinkage and movement problems.

Typical Details ~

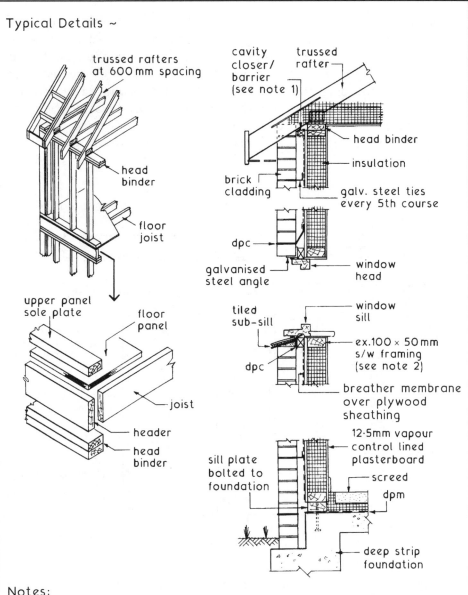

trussed rafters at 600 mm spacing

head binder

floor joist

upper panel sole plate

floor panel

joist

header

head binder

cavity closer/ barrier (see note 1)

trussed rafter

head binder

insulation

brick cladding

galv. steel ties every 5th course

dpc

galvanised steel angle

window head

tiled sub-sill

window sill

ex.100 × 50 mm s/w framing (see note 2)

dpc

breather membrane over plywood sheathing

12·5mm vapour control lined plasterboard

sill plate bolted to foundation

screed

dpm

deep strip foundation

Notes:

1. Cavity barriers prevent fire spread. The principal locations are between elements and compartments of construction (see B. Regs. A.D. B3).

2. Thermal bridging through solid framing may be reduced by using rigid EPS insulation and lighter 'I' section members of plywood or oriented strand board (OSB).

Framing ~ comprising inner leaf wall panels of standard cold-formed galvanised steel channel sections as structural support, with a lined inner face of vapour check layer under plasterboard. These panels can be site assembled, but it is more realistic to order them factory made. Panels are usually produced in 600mm wide modules and bolted together on site. Roof trusses are made up from steel channel or sigma sections. See page 532 for examples of standard steel sections and BS EN 10162: Cold rolled steel sections.

Standard channel and panel.

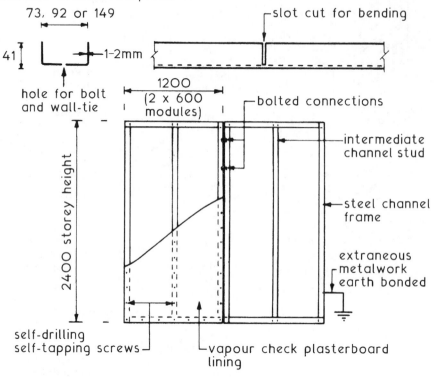

Background/history ~ the concept of steel framing for house construction evolved in the early 1920s, but development of the lightweight concrete "breeze" block soon took preference. Due to a shortage of traditional building materials, a resurgence of interest occurred again during the early post-war building boom of the late 1940s. Thereafter, steel became relatively costly and uncompetitive as a viable alternative to concrete block or timber frame construction techniques. Since the 1990s more efficient factory production processes, use of semi-skilled site labour and availability of economic cold-formed sections have revived an interest in this alternative means of house construction.

Typical Details ~

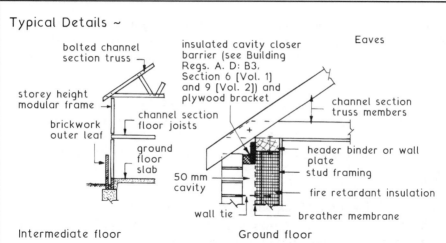

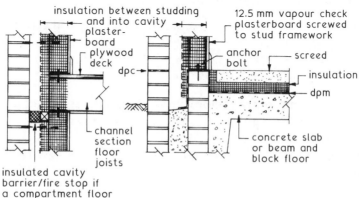

Advantages ~

- Factory made, therefore produced to quality controlled standards and tolerances.
- Relatively simple to assemble on site – bolted connections in pre-formed holes.
- Dimensionally stable, consistent composition, insignificant movement.
- Unaffected by moisture, therefore will not rot.
- Does not burn.
- Inedible by insects.
- Roof spans potentially long relative to weight.

Disadvantages ~

- Possibility of corrosion if galvanised protective layer is damaged.
- Deforms at high temperature, therefore unpredictable in fire.
- Electricity conductor – must be earthed.

Render ~ a mix of binder (cement) and fine aggregate (sand) with the addition of water and lime or a plasticiser to make the mix workable. Applied to walls as a decorative and/or waterproofing treatment.

Mix ratios ~ for general use, mix ratios are between $1:0.5:4{-}4.5$ and $1:1:5{-}6$ of cement, lime and sand. Equivalent using masonry cement and sand is $1:2.5{-}3.5$ and $1:4{-}5$. Unless a fine finish is required, coarse textured sharp sand is preferred for stability.

Background ~

Masonry – brick and block-work joints raked out 12 to 15mm to provide a key for the first bonding coat. Metal mesh can be nailed to the surface as supplementary support and reinforcing.

Wood or similar sheeting – metal lathing, wire mesh or expanded metal of galvanised (zinc coated) or stainless steel secured every 300mm. A purpose made lathing is produced for timber-framed walls.

Concrete – and other smooth, dense surfaces can be hacked to provide a key or spatter-dashed. Spatter-dash is a strong mix of cement and sand $(1:2)$ mixed into a slurry, trowelled roughly or thrown on to leave an irregular surface as a key to subsequent applications.

Three coat application to a masonry background

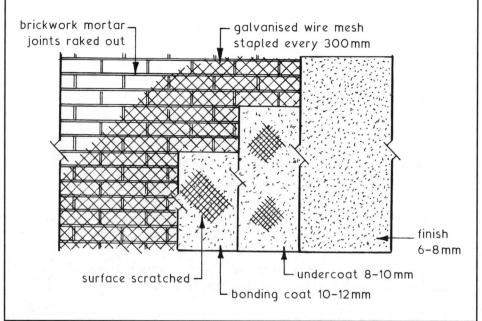

brickwork mortar joints raked out

galvanised wire mesh stapled every 300mm

finish 6-8mm

surface scratched

undercoat 8-10mm

bonding coat 10-12mm

Number of coats (layers) and composition ~ in sheltered locations, one 10mm layer is adequate for regular backgrounds. Elsewhere, two or possibly three separate applications are required to adequately weatherproof the wall and to prevent the brick or block-work joints from "grinning" through. Render mixes should become slightly weaker towards the outer layer to allow for greater flexure at the surface, ie. less opportunity for movement and shrinkage cracking.

Finishes ~ smooth, textured, rough-cast and pebble-dashed.

Smooth – fine sand and cement finished with a steel trowel (6 to 8 mm).

Textured – final layer finished with a coarse brush, toothed implement or a fabric roller (10 to 12 mm with 3 mm surface treated).

Rough-cast – irregular finish resulting from throwing the final coat onto the wall (6 to 10 mm).

Pebble or dry dash – small stones thrown onto a strong mortar finishing coat (10 to 12 mm).

Render application to a timber framed background

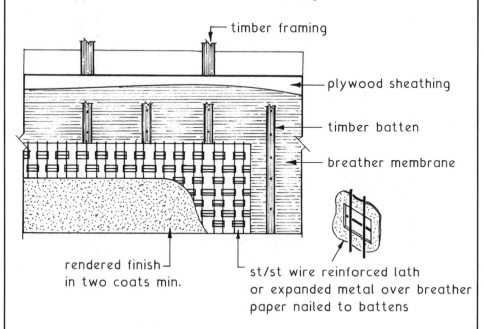

timber framing

plywood sheathing

timber batten

breather membrane

rendered finish in two coats min.

st/st wire reinforced lath or expanded metal over breather paper nailed to battens

Ref. BS EN 13914 – 1: Design, preparation and application of external rendering and internal plastering. External rendering.

Claddings to External Walls ~ external walls of block or timber frame construction can be clad with tiles, timber boards or plastic board sections. The tiles used are plain roofing tiles with either a straight or patterned bottom edge. They are applied to the vertical surface in the same manner as tiles laid on a sloping surface (see pages 430 and 433) except that the gauge can be wider and each tile is twice nailed. External and internal angles can be formed using special tiles or they can be mitred. Timber boards such as matchboarding and shiplap can be fixed vertically to horizontal battens or horizontally to vertical battens. Plastic moulded board claddings can be applied in a similar manner. The battens to which the claddings are fixed should be treated with a preservative against fungi and beetle attack and should be fixed with corrosion resistant nails.

Typical Details ~

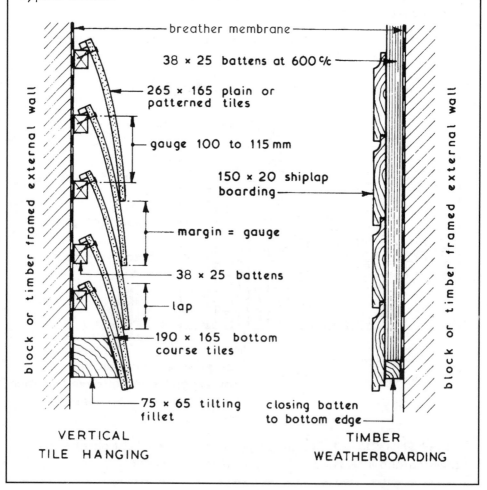

VERTICAL
TILE HANGING

TIMBER
WEATHERBOARDING

External corner tiles are made to order as special fittings to standard plain tiles. In effect they are tile and a halfs turned through 90° (other angles can be made) and handed left or right, fixed alternately to suit the overlapping pattern.

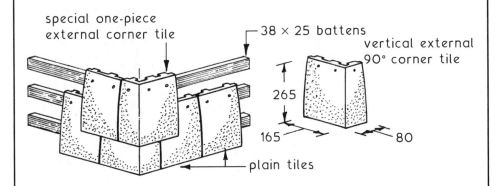

special one-piece external corner tile

38 × 25 battens

vertical external 90° corner tile

265

165

80

plain tiles

An alternative is to accurately mitre cut the meeting sides of tiles. This requires pairs of tile and a halfs in alternate courses to maintain the bond. Lead soakers (1.75mm) of at least 225 × 200mm are applied to every course to weather the mitred cut edges. The top of each soaker is turned over the tiles and the bottom finished flush or slightly above the lower edge of tiles.

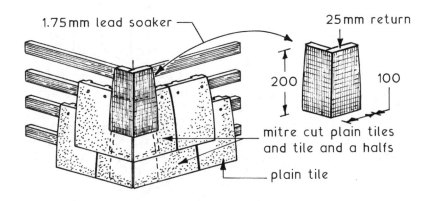

1.75mm lead soaker

25mm return

200

100

mitre cut plain tiles and tile and a halfs

plain tile

411

Internal corners are treated similarly to external corners by using special tiles of approximately tile and a half overall dimensions, turned through 90° (or other specified angle) in the opposing direction to external specials. These tiles are left and right handed and fixed alternately to vertical courses to maintain the overlapping bond.

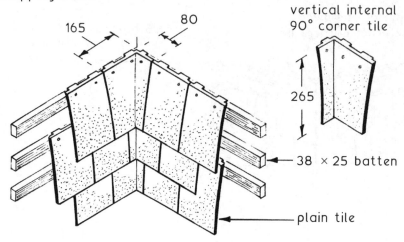

165   80

vertical internal
90° corner tile

265

38 × 25 batten

plain tile

Internal angles can also be formed with mitre cut tiles, with tile and a half tiles in alternate courses. Lead soakers (1.75 mm) of 175 × 175 mm minimum dimensions are placed under each pair of corner tiles to weather the cut edges. As with external soakers, the lead is discretely hidden by accurate mitre cutting and shaping of tiles.

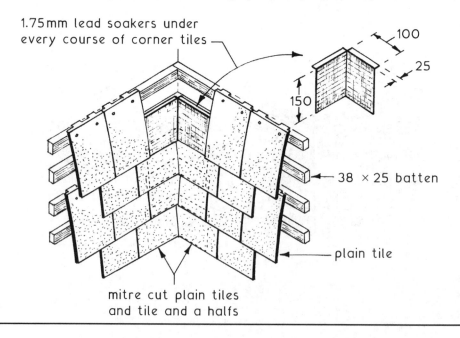

1.75 mm lead soakers under
every course of corner tiles

100

25

150

38 × 25 batten

plain tile

mitre cut plain tiles
and tile and a halfs

Some standard patterned bottom edge tiles ~

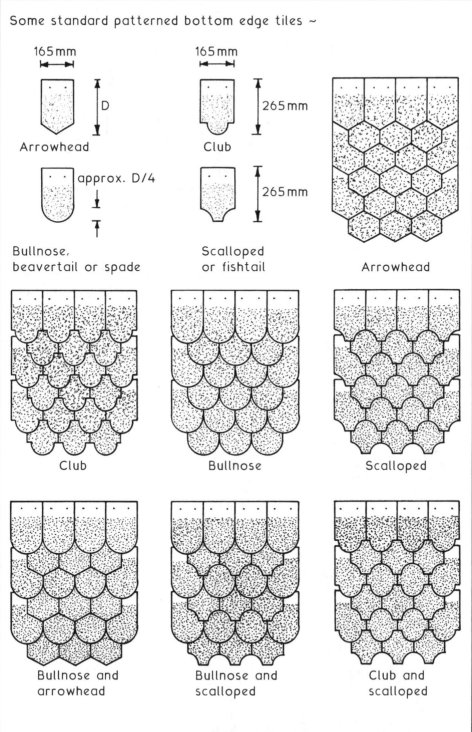

165 mm

Arrowhead

D

Bullnose,
beavertail or spade

approx. D/4

165 mm

265 mm

Club

265 mm

Scalloped
or fishtail

Arrowhead

Club

Bullnose

Scalloped

Bullnose and
arrowhead

Bullnose and
scalloped

Club and
scalloped

Note: May also be used as a roof-tiling feature, generally in two
to three courses at 1.5 to 2.0 m intervals to relieve plain tiling.

Ordinary splay cutting to roof verge ~

Tile and a half tiles at the end of each course, cut to the undercloak course of roof tiling. A second nail hole can be drilled at the head of each cut tile for secure fixing to the parallel batten.

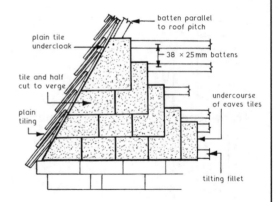

batten parallel to roof pitch

plain tile undercloak

38 × 25mm battens

tile and half cut to verge

undercourse of eaves tiles

plain tiling

tilting fillet

Winchester cutting to roof verge ~

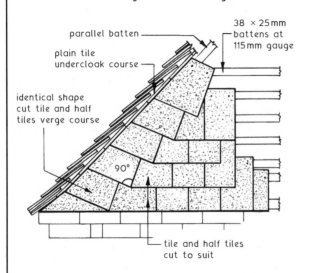

parallel batten

plain tile undercloak course

38 × 25mm battens at 115mm gauge

identical shape cut tile and half tiles verge course

90°

tile and half tiles cut to suit

More attractive than ordinary splay cutting. Two cut tiles are required at the end of each course, varying in size depending on the roof pitch. Tile and half tiles should be used for these to avoid narrow cuts.

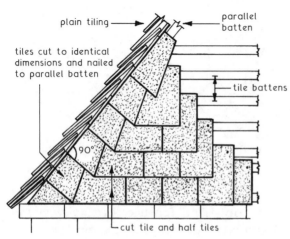

plain tiling

parallel batten

tiles cut to identical dimensions and nailed to parallel batten

tile battens

90°

cut tile and half tiles

A variation that also has two cut tiles at the end of each course, has the square (90°) edge of the first course of tiles next to the roof tiling undercloak course.

Appearance and concept ~ a type of fake brickwork made up of clay tiles side and head lapped over each other to create the impression of brickwork, but without the expense. Joints/pointing can be in lime mortar or left dry.

History ~ originated during the 18th century, when they were used quite frequently on timber framed buildings notably in Kent and Sussex. Possibly this was to update and improve deteriorated weather boarding or to avoid the brick tax of 1784. This tax was repealed in 1835.

Application ~ restoration to mathematically tiled older structures and as a lightweight cladding to modern timber framed construction where the appearance of a brickwork façade is required.

Application shown in a Flemish bond ~

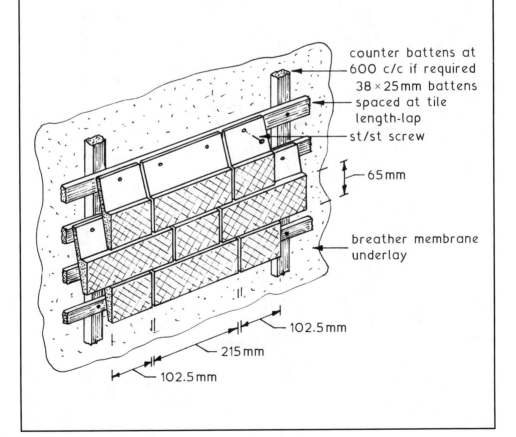

counter battens at 600 c/c if required
38 × 25mm battens spaced at tile length-lap
st/st screw

65mm

breather membrane underlay

102.5mm

215mm

102.5mm

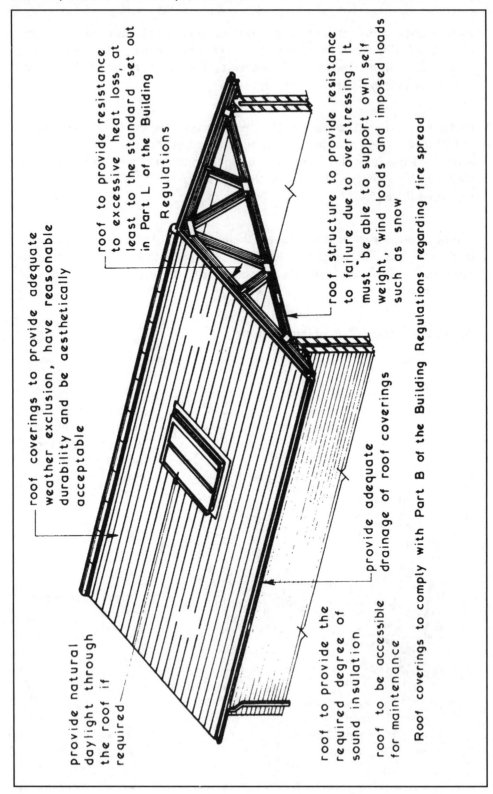

roof coverings to provide adequate weather exclusion, have reasonable durability and be aesthetically acceptable

roof to provide resistance to excessive heat loss, at least to the standard set out in Part L of the Building Regulations

roof structure to provide resistance to failure due to overstressing. It must be able to support own self weight, wind loads and imposed loads such as snow

provide adequate drainage of roof coverings

provide natural daylight through the roof if required

roof to provide the required degree of sound insulation

roof to be accessible for maintenance

Roof coverings to comply with Part B of the Building Regulations regarding fire spread

Roofs ~ these can be classified as either:-
     Flat – pitch from 0° to 10°
     Pitched – pitch over 10°

It is worth noting that for design purposes roof pitches over 70° are classified as walls.

Roofs can be designed in many different forms and in combinations of these forms some of which would not be suitable and/or economic for domestic properties.

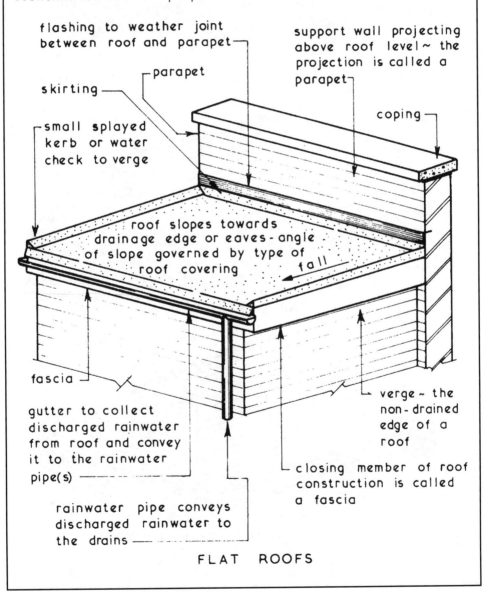

flashing to weather joint between roof and parapet

support wall projecting above roof level ~ the projection is called a parapet

parapet

skirting

coping

small splayed kerb or water check to verge

roof slopes towards drainage edge or eaves-angle of slope governed by type of roof covering

fall

fascia

verge ~ the non-drained edge of a roof

gutter to collect discharged rainwater from roof and convey it to the rainwater pipe(s)

closing member of roof construction is called a fascia

rainwater pipe conveys discharged rainwater to the drains

FLAT ROOFS

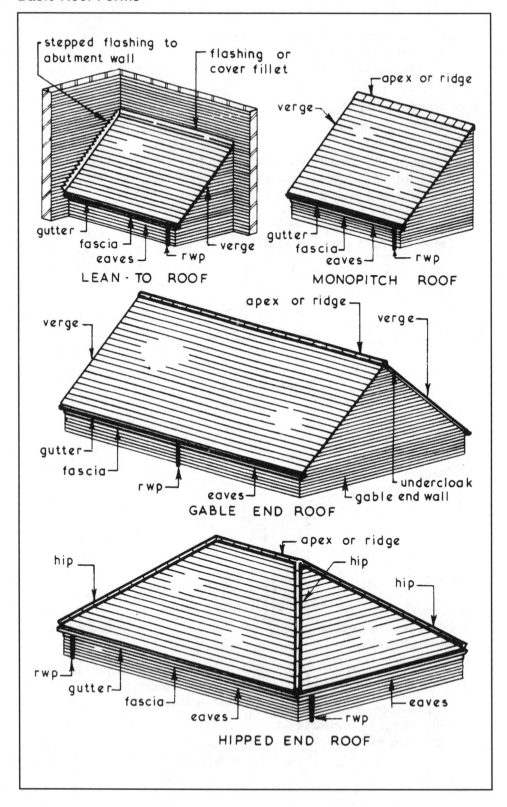

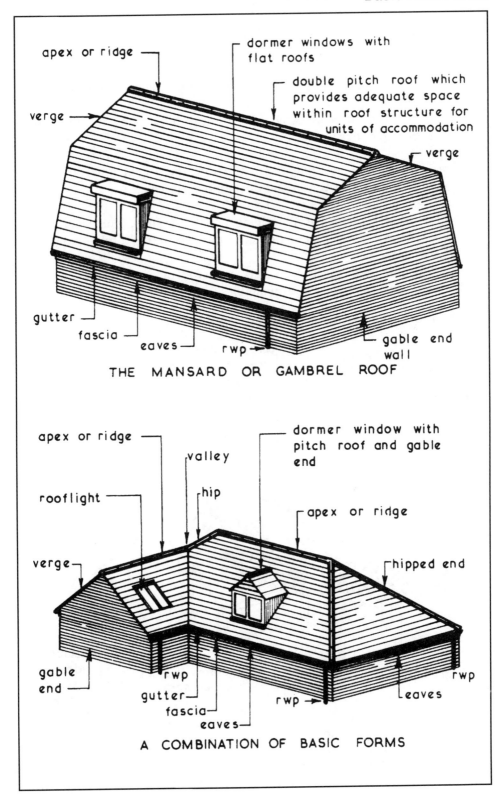

THE MANSARD OR GAMBREL ROOF

A COMBINATION OF BASIC FORMS

Pitched Roofs ~ the primary functions of any domestic roof are to:-

1. Provide an adequate barrier to the penetration of the elements.
2. Maintain the internal environment by providing an adequate resistance to heat loss.

A roof is in a very exposed situation and must therefore be designed and constructed in such a manner as to:-

1. Safely resist all imposed loadings such as snow and wind.
2. Be capable of accommodating thermal and moisture movements.
3. Be durable so as to give a satisfactory performance and reduce maintenance to a minimum.

Component Parts of a Pitched Roof ~

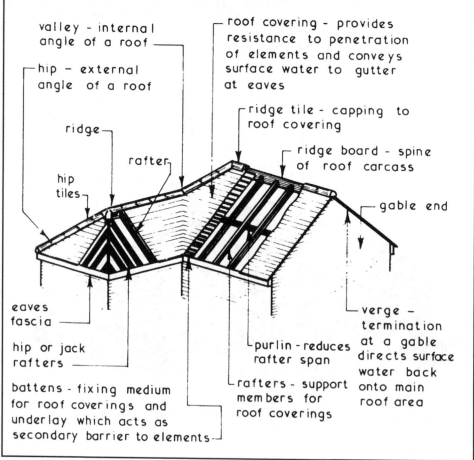

valley - internal angle of a roof

roof covering - provides resistance to penetration of elements and conveys surface water to gutter at eaves

hip - external angle of a roof

ridge tile - capping to roof covering

ridge

ridge board - spine of roof carcass

rafter

hip tiles

gable end

eaves fascia

verge - termination at a gable directs surface water back onto main roof area

hip or jack rafters

purlin - reduces rafter span

battens - fixing medium for roof coverings and underlay which acts as secondary barrier to elements

rafters - support members for roof coverings

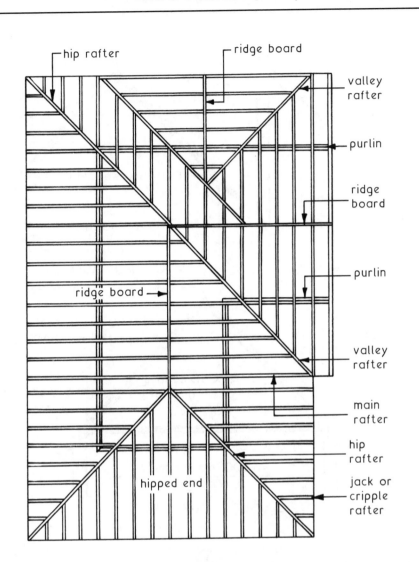

Purlins ~ guide to minimum size (mm) relative to span and spacing:

Span (m)	Spacing (m)		
	1.75	2.25	2.75
2.0	125 × 75	150 × 75	150 × 100
2.5	150 × 75	175 × 75	175 × 100
3.0	175 × 100	200 × 100	200 × 125
3.5	225 × 100	225 × 100	225 × 125
4.0	225 × 125	250 × 125	250 × 125

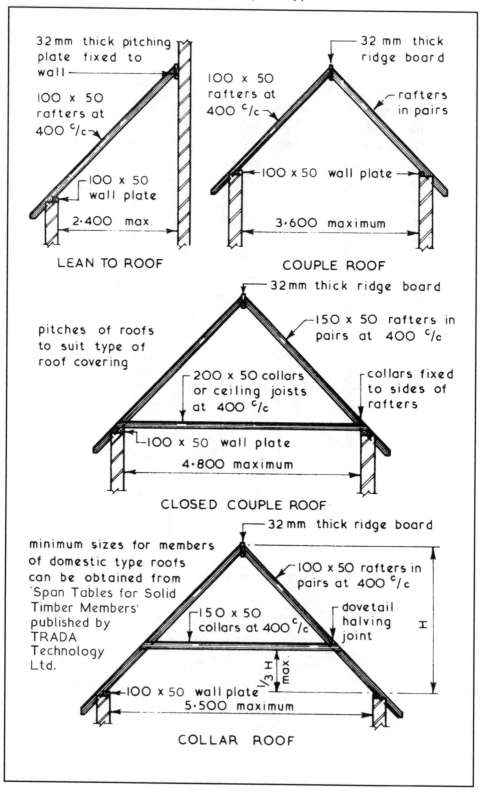

32 mm thick pitching plate fixed to wall

100 x 50 rafters at 400 c/c

100 x 50 wall plate

2·400 max.

LEAN TO ROOF

32 mm thick ridge board

100 x 50 rafters at 400 c/c

rafters in pairs

100 x 50 wall plate

3·600 maximum

COUPLE ROOF

32 mm thick ridge board

pitches of roofs to suit type of roof covering

150 x 50 rafters in pairs at 400 c/c

200 x 50 collars or ceiling joists at 400 c/c

collars fixed to sides of rafters

100 x 50 wall plate

4·800 maximum

CLOSED COUPLE ROOF

32 mm thick ridge board

minimum sizes for members of domestic type roofs can be obtained from 'Span Tables for Solid Timber Members' published by TRADA Technology Ltd.

100 x 50 rafters in pairs at 400 c/c

150 x 50 collars at 400 c/c

dovetail halving joint

H

1/3 H max.

100 x 50 wall plate

5·500 maximum

COLLAR ROOF

422

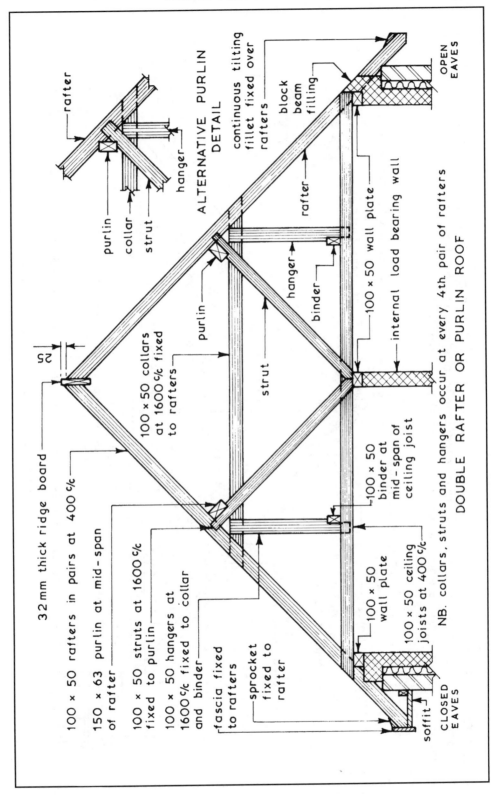

ALTERNATIVE PURLIN DETAIL

rafter

purlin
collar
strut
hanger

continuous tilting fillet fixed over rafters

block
beam
filling

rafter

hanger

binder

100 × 50 wall plate

internal load bearing wall

OPEN EAVES

32 mm thick ridge board

100 × 50 rafters in pairs at 400 c/c

150 × 63 purlin at mid-span of rafter

100 × 50 struts at 1600 c/c fixed to purlin

100 × 50 hangers at 1600 c/c fixed to collar and binder

fascia fixed to rafters

sprocket fixed to rafter

25

100 × 50 collars at 1600 c/c fixed to rafters

purlin

strut

100 × 50 binder at mid-span of ceiling joist

100 × 50 wall plate

100 × 50 ceiling joists at 400 c/c

CLOSED EAVES

soffit

NB. collars, struts and hangers occur at every 4th. pair of rafters

DOUBLE RAFTER OR PURLIN ROOF

423

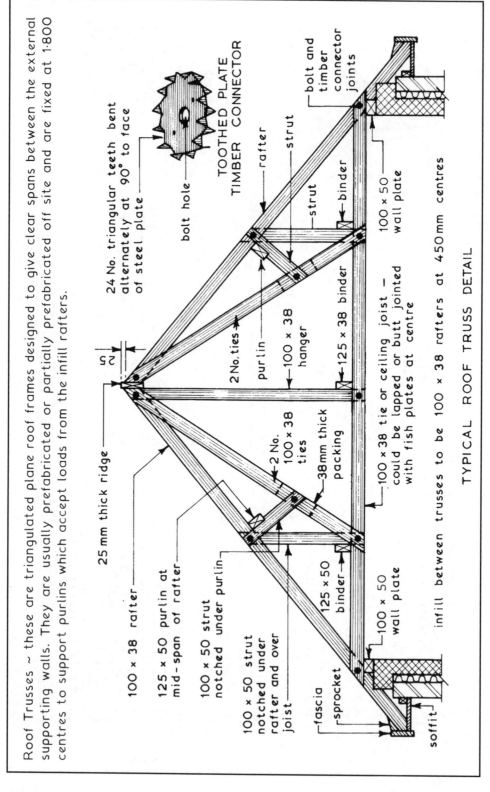

Roof Trusses ~ these are triangulated plane roof frames designed to give clear spans between the external supporting walls. They are usually prefabricated or partially prefabricated off site and are fixed at 1·800 centres to support purlins which accept loads from the infill rafters.

24 No. triangular teeth bent alternately at 90° to face of steel plate

bolt hole

TOOTHED PLATE TIMBER CONNECTOR

25 mm thick ridge

100 × 38 rafter

125 × 50 purlin at mid-span of rafter

100 × 50 strut notched under purlin

100 × 50 strut notched under rafter and over joist

fascia

sprocket

125 × 50 binder

100 × 50 wall plate

2 No. 100 × 38 ties

38 mm thick packing

2 No. ties

purlin

100 × 38 hanger

125 × 38 binder

rafter

strut

strut

binder

bolt and timber connector joints

100 × 50 wall plate

100 × 38 tie or ceiling joist — could be lapped or butt jointed with fish plates at centre

infill between trusses to be 100 × 38 rafters at 450 mm centres

soffit

TYPICAL ROOF TRUSS DETAIL

Trussed Rafters ~ these are triangulated plane roof frames designed to give clear spans between the external supporting walls. They are delivered to site as a prefabricated component where they are fixed to the wall plates at 600mm centres. Trussed rafters do not require any ridge board or purlins since they receive their lateral stability by using larger tiling battens (50 × 25mm) than those used on traditional roofs.

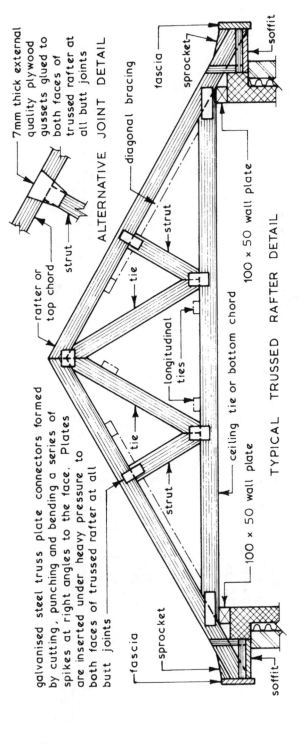

7mm thick external quality plywood gussets glued to both faces of trussed rafter at all butt joints

ALTERNATIVE JOINT DETAIL

rafter or top chord

strut

diagonal bracing

fascia

sprocket

soffit

tie

strut

longitudinal ties

tie

ceiling tie or bottom chord

100 × 50 wall plate

100 × 50 wall plate

strut

100 × 50 wall plate

fascia

sprocket

soffit

TYPICAL TRUSSED RAFTER DETAIL

galvanised steel truss plate connectors formed by cutting, punching and bending a series of spikes at right angles to the face. Plates are inserted under heavy pressure to both faces of trussed rafter at all butt joints

Longitudinal ties (75 × 38) fixed over ceiling ties and under internal ties near to roof apex and rafter diagonal bracing (75 × 38) fixed under rafters at gable ends from eaves to apex may be required to provide stability bracing – actual requirements specified by manufacturer. Lateral restraint to gable walls at top and bottom chord levels in the form of mild steel straps at 2·000 maximum centres over 2 No. trussed rafters may also be required.

425

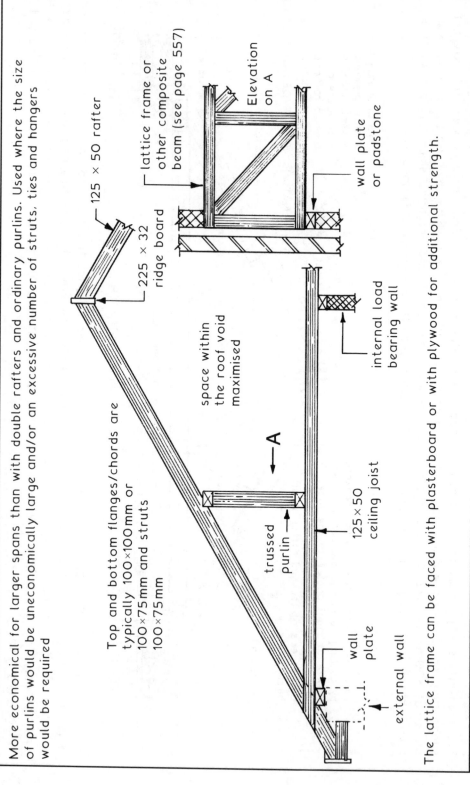

More economical for larger spans than with double rafters and ordinary purlins. Used where the size of purlins would be uneconomically large and/or an excessive number of struts, ties and hangers would be required

125 × 50 rafter

lattice frame or other composite beam (see page 557)

Elevation on A

wall plate or padstone

225 × 32 ridge board

Top and bottom flanges/chords are typically 100×100mm or 100×75mm and struts 100×75mm

space within the roof void maximised

A

trussed purlin

125×50 ceiling joist

internal load bearing wall

wall plate

external wall

The lattice frame can be faced with plasterboard or with plywood for additional strength.

Gambrel roofs are double pitched with a break in the roof slope. The pitch angle above the break is less than 45° relative to the horizontal, whilst the pitch angle below the break is greater. Generally, these angles are 30° and 60°.

Gambrels are useful in providing more attic headroom and frequently incorporate dormers and rooflights. They have a variety of constructional forms.

Typically —

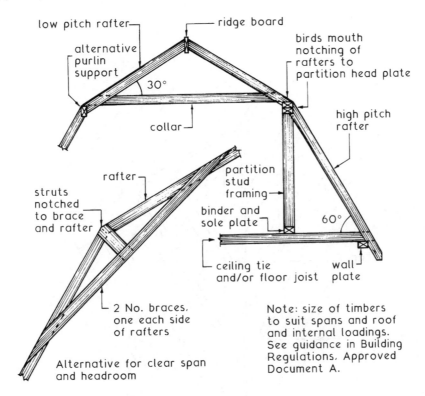

Intermediate support can be provided in various ways as shown above. To create headroom for accommodation in what would otherwise be attic space, a double head plate and partition studding is usual. The collar beam and rafters can conveniently locate on the head plates or prefabricated trusses can span between partitions.

Valley construction and associated pitched roofing is used:

* to visually enhance an otherwise plain roof structure.
* where the roof plan turns through an angle (usually 90°) to follow the building layout or a later extension.
* at the intersection of main and projecting roofs above a bay window or a dormer window.

Construction may be by forming a framework of cut rafters trimmed to valley rafters as shown in the roof plan on page 421. Alternatively, and as favoured with building extensions, by locating a valley or lay board over the main rafters to provide a fixing for each of the jack rafters.

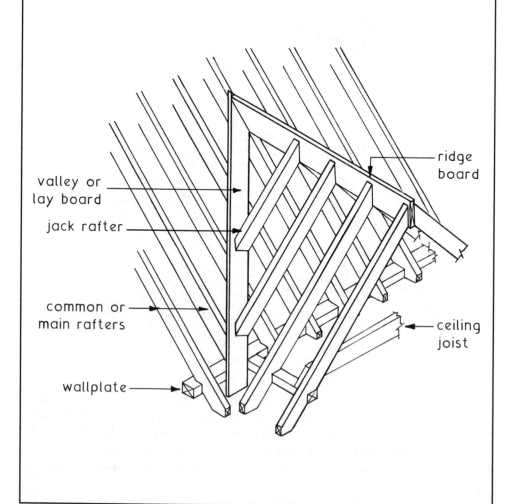

Sprockets may be provided at the eaves to reduce the slope of a pitched roof. Sprockets are generally most suitable for use on wide steeply pitched roofs to:

• enhance the roof profile by creating a feature.
• to slow the velocity of rainwater running off the roof and prevent it over-shooting the gutter.

Where the rafters overhang the external wall, taper cut timber sprockets can be attached to the top of the rafters. Alternatively, the ends of rafters can be birds-mouthed onto the wall plate and short lengths of timber the same size as the rafters secured to the rafter feet. In reducing the pitch angle, albeit for only a short distance, it should not be less than the minimum angle recommended for specific roof coverings.

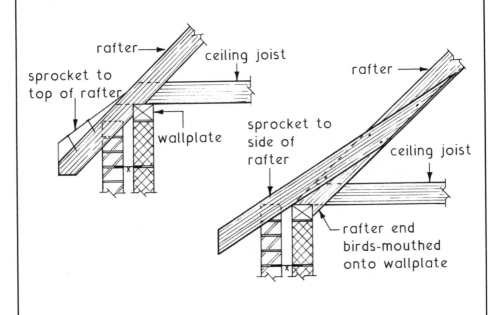

Types of sprocketed eaves

Roof Underlays ~ sometimes called sarking or roofing felt provides the barrier to the entry of snow, wind and rain blown between the tiles or slates. It also prevents the entry of water from capillary action.

Suitable Materials ~

Bitumen fibre based felts – supplied in rolls 1m wide and up to 25m long. Traditionally used in house construction with a cold ventilated roof.

Breather or vapour permeable underlay – typically produced from HDPE fibre or extruded polypropylene fibre, bonded by heat and pressure. Materials permeable to water vapour are preferred as these do not need to be perforated to ventilate the roof space. Also, subject to manufacturer's guidelines, traditional eaves ventilation may not be necessary. Underlay of this type should be installed taut across the rafters with counter battens support to the tile battens. Where counter battens are not used, underlay should sag slightly between rafters to allow rain penetration to flow under tile battens.

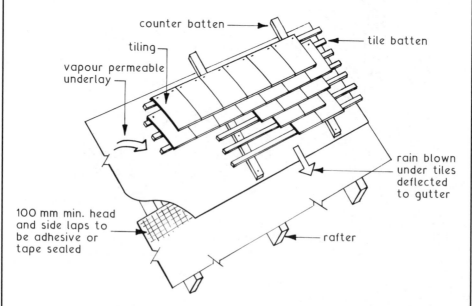

counter batten

tiling

tile batten

vapour permeable underlay

rain blown under tiles deflected to gutter

100 mm min. head and side laps to be adhesive or tape sealed

rafter

Underlays are fixed initially with galvanised clout nails or st/st staples but are finally secured with the tiling or slating batten fixings

Double Lap Tiles ~ these are the traditional tile covering for pitched roofs and are available made from clay and concrete and are usually called plain tiles. Plain tiles have a slight camber in their length to ensure that the tail of the tile will bed and not ride on the tile below. There is always at least two layers of tiles covering any part of the roof. Each tile has at least two nibs on the underside of its head so that it can be hung on support battens nailed over the rafters. Two nail holes provide the means of fixing the tile to the batten, in practice only every 4th course of tiles is nailed unless the roof exposure is high. Double lap tiles are laid to a bond so that the edge joints between the tiles are in the centre of the tiles immediately below and above the course under consideration.

Minimum pitch 35° machine made, 45° hand made.

Typical Plain Tile Details ~

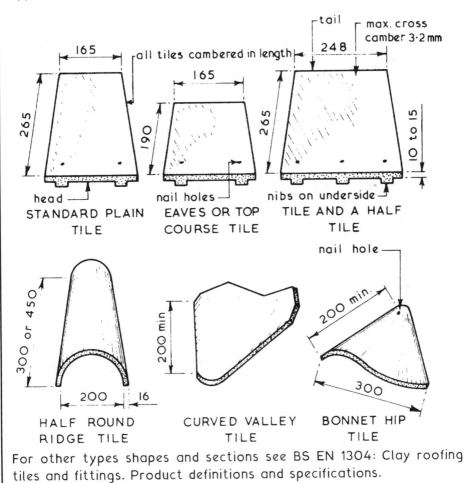

For other types shapes and sections see BS EN 1304: Clay roofing tiles and fittings. Product definitions and specifications.

Hand made from extracted clay sub-strata. Sources of suitable clay in the UK are the brick making areas of Kent, Sussex and Leicestershire.

Tiles are shaped in a timber frame or clamp before being kiln fired at about 1000°C. Early examples of these tiles have been attributed to the Romans, but after they left the UK manufacture all but ceased until about the 12th Century. Historically and today, tile dimensions vary quite significantly, especially those from different regions and makers. In 1477 a Royal Charter attempted to standardise tiles to $10\frac{1}{2}'' \times 6\frac{1}{2}'' \times \frac{1}{2}''$ thick (265 × 165 × 12 mm) and this remains as the BS dimensions shown on the previous page. However, peg tile makers were set in their ways and retained their established local dimensions. This means that replacements have to be specifically produced to match existing.

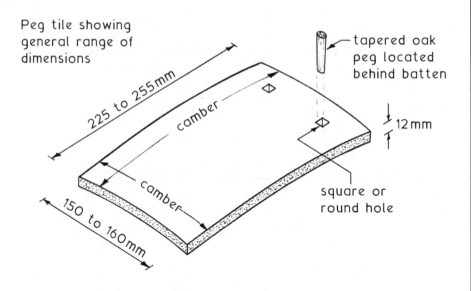

Peg tile showing general range of dimensions

225 to 255 mm

camber

camber

150 to 160 mm

tapered oak peg located behind batten

12 mm

square or round hole

Typical regional sizes ~

Sussex	$9\frac{1}{2}'' \times 6\frac{1}{4}''$	(240 × 160 mm)
Kent	$10'' \times 6''$	(255 × 150 mm)
Leicestershire	$11'' \times 7''$	(280 × 180 mm)

Typical details (cold roof) ~

purpose made in-line tile ventilators spaced to provide equivalent of 5mm continuous gap postitioned at high level

half round ridge tiles bedded in cm. mt. (1:3) butt jointed in length with ends of ridge tiles filled with mortar and tile slip inserts

margin

lap

under ridge
top course tile
38 x 25 timber battens
(see Note 2)

plain tiles

airflow

rafters

ridge

gauge

underlay, see page 430

RIDGE DETAIL

margin = gauge

$$= \frac{tile\ length - lap}{2}$$

$$= \frac{265 - 65}{2}$$

$$= 100\ mm$$

plain tiles nailed to battens every 4th course

underlay

timber battens

rafters

ceiling joists

ventilation spacer

eaves tile

insulation between and over joists

vapour-check plasterboard ceiling (see Note 1)

50 mm deep wall plate

external wall with insulated cavity

cavity insulation

gutter

fascia

soffit board

10 mm wide continuous ventilation gap

EAVES DETAIL

Note 1: Through ventilation is necessary to prevent condensation occurring in the roof space. A vapour check can also help limit the amount of moisture entering the roof void.

Note 2: 50 × 25 where rafter spacing is 600mm.

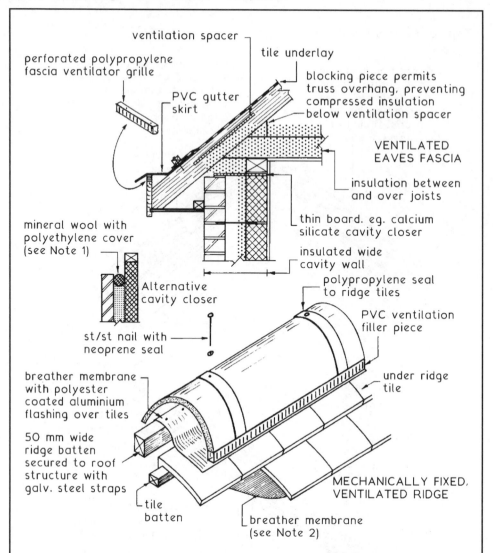

ventilation spacer

tile underlay

perforated polypropylene
fascia ventilator grille

blocking piece permits
truss overhang, preventing
compressed insulation
below ventilation spacer

PVC gutter
skirt

VENTILATED
EAVES FASCIA

insulation between
and over joists

mineral wool with
polyethylene cover
(see Note 1)

thin board. eg. calcium
silicate cavity closer

insulated wide
cavity wall

Alternative
cavity closer

polypropylene seal
to ridge tiles

PVC ventilation
filler piece

st/st nail with
neoprene seal

under ridge
tile

breather membrane
with polyester
coated aluminium
flashing over tiles

50 mm wide
ridge batten
secured to roof
structure with
galv. steel straps

MECHANICALLY FIXED,
VENTILATED RIDGE

tile
batten

breather membrane
(see Note 2)

Note 1. If a cavity closer is also required to function as a cavity barrier to prevent fire spread, it should provide at least 30 minutes fire resistance, (B. Reg. A.D. B3 Section 6 [Vol. 1] and 9 [Vol. 2]).

Note 2. A breather membrane is an alternative to conventional bituminous felt as an under-tiling layer. It has the benefit of restricting liquid water penetration whilst allowing water vapour transfer from within the roof space. This permits air circulation without perforating the under-tiling layer.

Typical detail (warm roof) ~

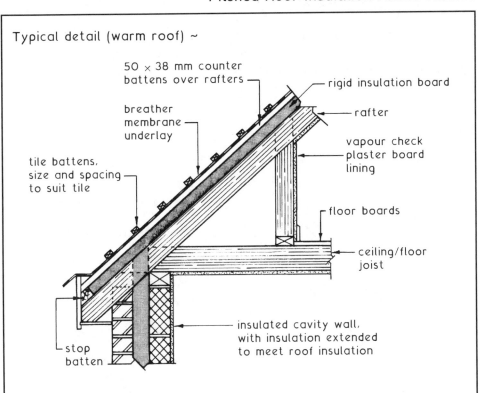

50 × 38 mm counter battens over rafters

rigid insulation board

breather membrane underlay

rafter

tile battens, size and spacing to suit tile

vapour check plaster board lining

floor boards

ceiling/floor joist

insulated cavity wall, with insulation extended to meet roof insulation

stop batten

Where a roof space is used for habitable space, insulation must be provided within the roof slope. Insulation above the rafters (as shown) creates a 'warm roof', eliminating the need for continuous ventilation. Insulation placed between the rafters creates a 'cold roof', where a continuous 50mm ventilation void above the insulation will assist in the control of condensation, (see next page).

Suitable rigid insulants include; low density polyisocyanurate (PIR) foam, reinforced with long strand glass fibres, both faces bonded to aluminium foil with joints aluminium foil taped on the upper surface; high density mineral wool slabs over rafters with less dense mineral wool between rafters.

An alternative location for the breather membrane is under the counter battens. This is often preferred as the insulation board will provide uniform support for the underlay. Otherwise, extra insulation could be provided between the counter battens, retaining sufficient space for the underlay to sag between rafter positions to permit any rainwater penetration to drain to eaves.

Insulation in between rafters is an alternative to placing it above. The following details show two possibilities, where if required supplementary insulation can be secured to the underside of rafters.

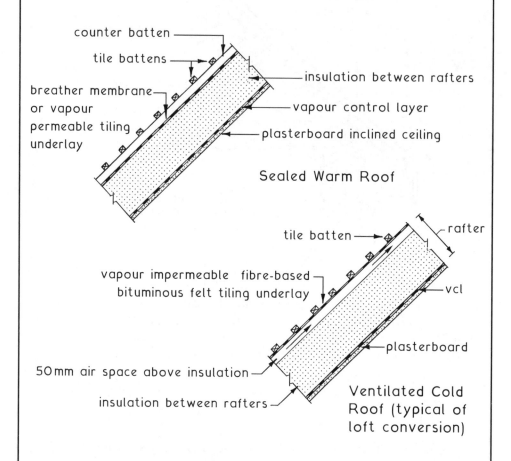

counter batten

tile battens

breather membrane or vapour permeable tiling underlay

insulation between rafters

vapour control layer

plasterboard inclined ceiling

Sealed Warm Roof

tile batten

rafter

vapour impermeable fibre-based bituminous felt tiling underlay

vcl

plasterboard

50mm air space above insulation

insulation between rafters

Ventilated Cold Roof (typical of loft conversion)

Condensation can be controlled where a well sealed vapour control layer (for instance, foil [metallised polyester] backed plasterboard) is incorporated in the ceiling lining and used with a vapour permeable (breather membrane) underlay to the tiling. Joints and openings in the vcl ceiling (eg. cable or pipe penetrations) should be sealed, but if this is impractical ventilation should be provided to the underside of the tile underlay.

Typical Details ~

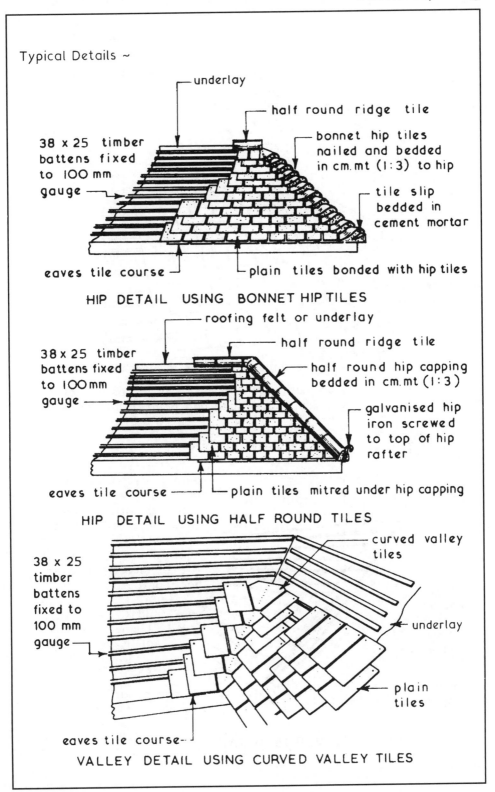

38 x 25 timber battens fixed to 100 mm gauge

underlay

half round ridge tile

bonnet hip tiles nailed and bedded in cm.mt (1:3) to hip

tile slip bedded in cement mortar

eaves tile course

plain tiles bonded with hip tiles

HIP DETAIL USING BONNET HIP TILES

roofing felt or underlay

38 x 25 timber battens fixed to 100mm gauge

half round ridge tile

half round hip capping bedded in cm.mt (1:3)

galvanised hip iron screwed to top of hip rafter

eaves tile course

plain tiles mitred under hip capping

HIP DETAIL USING HALF ROUND TILES

curved valley tiles

38 x 25 timber battens fixed to 100 mm gauge

underlay

plain tiles

eaves tile course

VALLEY DETAIL USING CURVED VALLEY TILES

Typical Details ~

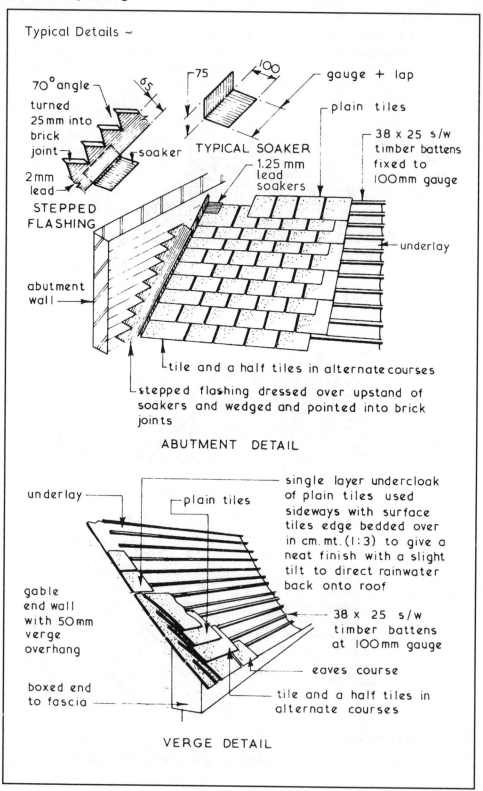

70° angle
turned
25 mm into
brick
joint

2 mm
lead

STEPPED
FLASHING

65

soaker

75

100

TYPICAL SOAKER

gauge + lap

plain tiles

38 x 25 s/w
timber battens
fixed to
100mm gauge

1.25 mm
lead
soakers

underlay

abutment
wall

tile and a half tiles in alternate courses

stepped flashing dressed over upstand of
soakers and wedged and pointed into brick
joints

ABUTMENT DETAIL

underlay

plain tiles

single layer undercloak
of plain tiles used
sideways with surface
tiles edge bedded over
in cm. mt.(1:3) to give a
neat finish with a slight
tilt to direct rainwater
back onto roof

gable
end wall
with 50 mm
verge
overhang

38 x 25 s/w
timber battens
at 100mm gauge

eaves course

boxed end
to fascia

tile and a half tiles in
alternate courses

VERGE DETAIL

Single Lap Tiling ~ so called because the single lap of one tile over another provides the weather tightness as opposed to the two layers of tiles used in double lap tiling. Most of the single lap tiles produced in clay and concrete have a tongue and groove joint along their side edges and in some patterns on all four edges which forms a series of interlocking joints and therefore these tiles are called single lap interlocking tiles. Generally there will be an overall reduction in the weight of the roof covering when compared with double lap tiling but the batten size is larger than that used for plain tiles and as a minimum every tile in alternate courses should be twice nailed, although a good specification will require every tile to be twice nailed. The gauge or batten spacing for single lap tiling is found by subtracting the end lap from the length of the tile.

Typical Single Lap Tiles ~

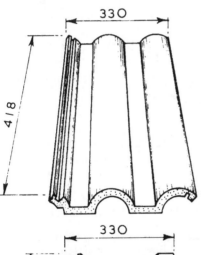

ROLL TYPE TILE

minimum pitch 30°

head lap 75mm

side lap 30mm

gauge 343mm

linear coverage 300mm

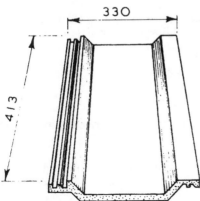

TROUGH TYPE TILE

minimum pitch 15°

head lap 75mm

side lap 38mm

gauge 338mm

linear coverage 292mm

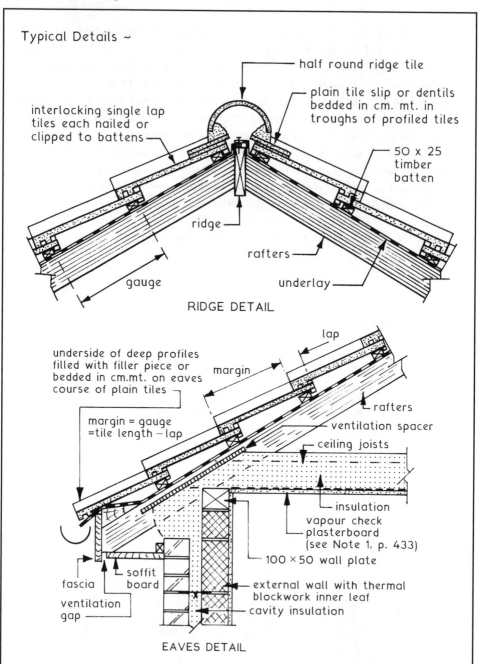

Typical Details ~

half round ridge tile

plain tile slip or dentils bedded in cm. mt. in troughs of profiled tiles

interlocking single lap tiles each nailed or clipped to battens

50 x 25 timber batten

ridge

rafters

gauge

underlay

RIDGE DETAIL

lap

underside of deep profiles filled with filler piece or bedded in cm.mt. on eaves course of plain tiles

margin

margin = gauge =tile length −lap

rafters

ventilation spacer

ceiling joists

insulation

vapour check plasterboard (see Note 1, p. 433)

100 × 50 wall plate

fascia

soffit board

external wall with thermal blockwork inner leaf

ventilation gap

cavity insulation

EAVES DETAIL

Hips – can be finished with a half round tile as a capping as shown for double lap tiling on page 437.

Valleys – these can be finished by using special valley trough tiles or with a lead lined gutter – see manufacturer's data.

Slates ~ slate is a natural dense material which can be split into thin sheets and cut to form a small unit covering suitable for pitched roofs in excess of 25° pitch. Slates are graded according to thickness and texture, the thinnest being known as 'Bests'. These are of 4mm nominal thickness. Slates are laid to the same double lap principles as plain tiles. Ridges and hips are normally covered with half round or angular tiles whereas valley junctions are usually of mitred slates over soakers. Unlike plain tiles every course is fixed to the battens by head or centre nailing, the latter being used on long slates and on pitches below 35° to overcome the problem of vibration caused by the wind which can break head nailed long slates.

Typical Details ~

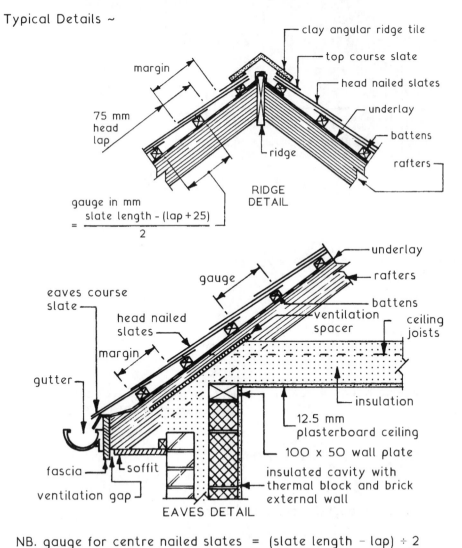

$$\text{gauge in mm} = \frac{\text{slate length} - (\text{lap} + 25)}{2}$$

RIDGE DETAIL

EAVES DETAIL

NB. gauge for centre nailed slates = (slate length − lap) ÷ 2

The UK has been supplied with its own slate resources from quarries in Wales, Cornwall and Westermorland. Imported slate is also available from Spain, Argentina and parts of the Far East.

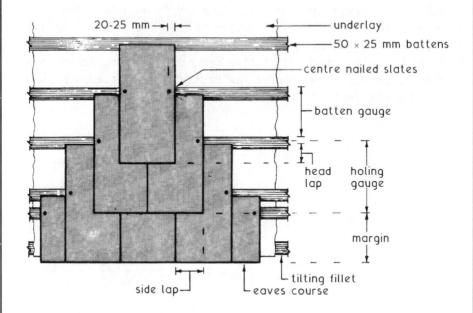

e.g. Countess slate, 510 × 255 mm laid to a 30° pitch with 75 mm head lap.

Batten gauge = (slate length – lap) ÷ 2
= (510 – 75) ÷ 2 = 218 mm.

Holing gauge = batten gauge + head lap + 8 to 15 mm,
= 218 + 75 + (8 to 15 mm) = 301 to 308 mm.

Side lap = 255 ÷ 2 = 127 mm.

Margin = batten gauge of 218 mm.

Eaves course length = head lap + margin = 293 mm.

Traditional slate names and sizes (mm) –

Empress	650 × 400	Wide Viscountess	460 × 255
Princess	610 × 355	Viscountess	460 × 230
Duchess	610 × 305	Wide Ladies	405 × 255
Small Duchess	560 × 305	Broad Ladies	405 × 230
Marchioness	560 × 280	Ladies	405 × 205
Wide Countess	510 × 305	Wide Headers	355 × 305
Countess	510 × 255	Headers	355 × 255
.. ..	510 × 230	Small Ladies	355 × 203
.. ..	460 × 305	Narrow Ladies	355 × 180

Sizes can also be cut to special order.

Generally, the larger the slate, the lower the roof may be pitched. Also, the lower the roof pitch, the greater the head lap.

Slate quality	Thickness (mm)	Weight at 75 mm headlap (kg/m²)
Best	4	26
Medium strong	5	Thereafter in propotion
Heavy	6	to thickness
Extra heavy	9	

Roof pitch (degrees)	Min. head lap (mm)
20	115
25	85
35	75
45	65

See also:

1. BS EN 12326-1: Slate and stone products for discontinuous roofing and cladding. Product specification.
2. Slate producers' catalogues.
3. BS 5534: Code of practice for slating and tiling.

Roof hip examples –

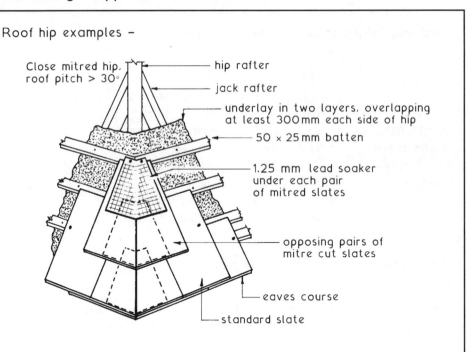

Close mitred hip, roof pitch > 30° — hip rafter

— jack rafter

— underlay in two layers, overlapping at least 300 mm each side of hip

— 50 × 25 mm batten

— 1.25 mm lead soaker under each pair of mitred slates

— opposing pairs of mitre cut slates

— eaves course

— standard slate

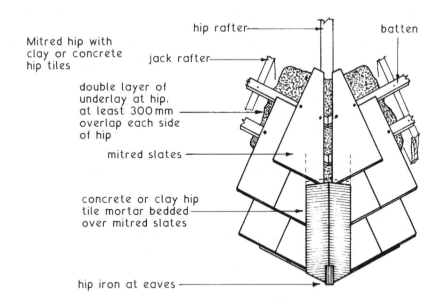

Mitred hip with clay or concrete hip tiles

hip rafter — batten

jack rafter —

double layer of underlay at hip, at least 300 mm overlap each side of hip —

mitred slates —

concrete or clay hip tile mortar bedded over mitred slates —

hip iron at eaves —

Roof valley examples –

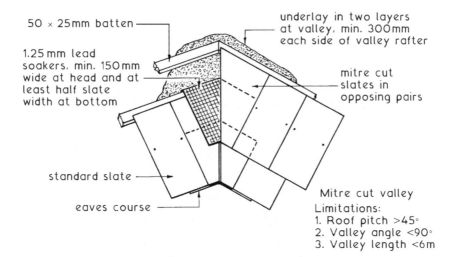

50 × 25mm batten

1.25 mm lead soakers, min. 150mm wide at head and at least half slate width at bottom

underlay in two layers at valley, min. 300mm each side of valley rafter

mitre cut slates in opposing pairs

standard slate

eaves course

Mitre cut valley
Limitations:
1. Roof pitch >45°
2. Valley angle <90°
3. Valley length <6m

Alternatives

valley rafter

wide lay boards in valley to support taper cut slates

two supplementary layers of underlay over lay boards to overlap normal underlay

valley slates tapered to a smooth curve

Swept valley

225mm min. lay board on valley rafter, usually with additional board either side

valley rafter

two layers of underlay at valley

jack rafter

Laced valley

Materials – water reed (Norfolk reed), wheat straw (Spring or Winter), Winter being the most suitable. Wheat for thatch is often known as wheat reed, long straw or Devon reed. Other thatches include rye and oat straws, and sedge. Sedge is harvested every fourth year to provide long growth, making it most suitable as a ridging material.

There are various patterns and styles of thatching, relating to the skill of the thatcher and local traditions.

Typical details –

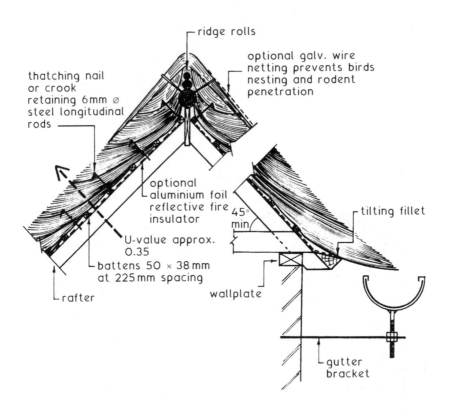

The material composition of thatch with its natural voids and surface irregularities provides excellent insulation when dry and compact. However, when worn with possible accumulation of moss and rainwater, the U-value is less reliable. Thatch is also very vulnerable to fire. Therefore in addition to imposing a premium, insurers may require application of a surface fire retardant and a fire insulant underlay.

Flat Roofs ~ these roofs are very seldom flat with a pitch of 0° but are considered to be flat if the pitch does not exceed 10°. The actual pitch chosen can be governed by the roof covering selected and/or by the required rate of rainwater discharge off the roof. As a general rule the minimum pitch for smooth surfaces such as asphalt should be 1:80 or 0°-43′ and for sheet coverings with laps 1:60 or 0°-57′.

Methods of Obtaining Falls ~

1. Joists cut to falls

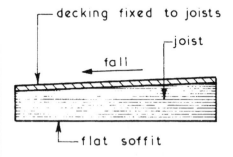

Simple to fix but could be wasteful in terms of timber unless two joists are cut from one piece of timber

2. Joists laid to falls

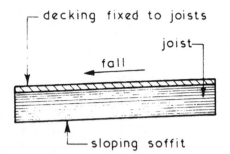

Economic and simple but sloping soffit may not be acceptable but this could be hidden by a flat suspended ceiling

3. Firrings with joist run

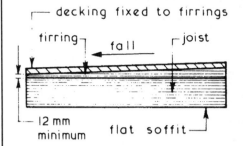

Simple and effective but does not provide a means of natural cross ventilation. Usual method employed.

4. Firrings against joist run

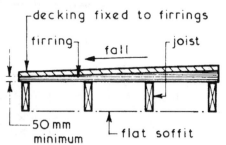

Simple and effective but uses more timber than 3 but does provide a means of natural cross ventilation

Wherever possible joists should span the shortest distance of the roof plan.

Timber Roof Joists ~ the spacing and sizes of joists is related to the loadings and span, actual dimensions for domestic loadings can be taken direct from recommendations in Approved Document A or they can be calculated as shown for timber beam designs. Strutting between joists should be used if the span exceeds 2·400 to restrict joist movements and twisting.

Typical Eaves Details ~

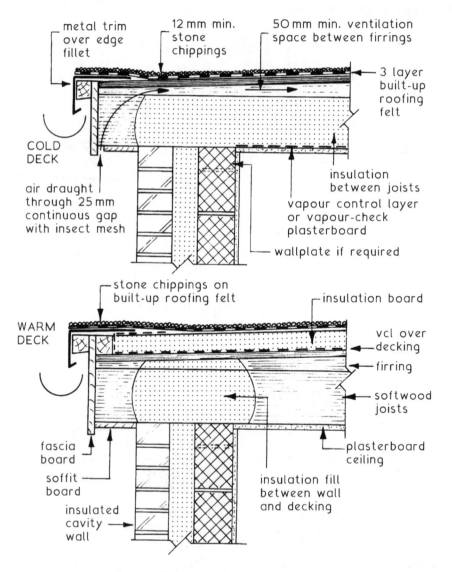

Ref. BS EN 13707: Flexible sheets for waterproofing. Reinforced bitumen sheets for roof waterproofing. Definitions and characteristics.

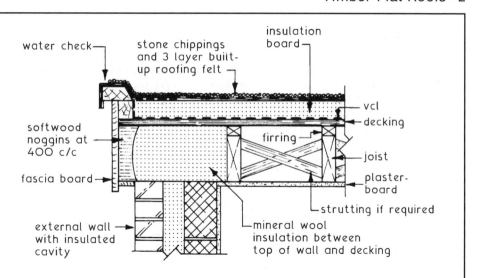

water check

stone chippings
and 3 layer built-
up roofing felt

insulation
board

vcl

decking

firring

joist

softwood
noggins at
400 c/c

plaster-
board

fascia board

strutting if required

external wall
with insulated
cavity

mineral wool
insulation between
top of wall and decking

TYPICAL VERGE DETAILS – WARM DECK

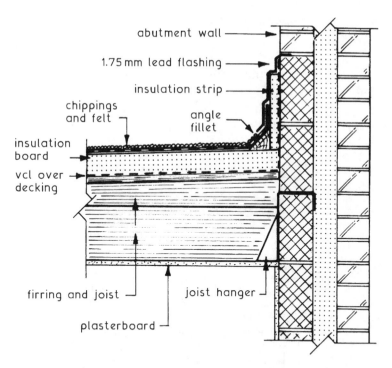

abutment wall

1.75 mm lead flashing

insulation strip

chippings
and felt

angle
fillet

insulation
board

vcl over
decking

firring and joist

joist hanger

plasterboard

TYPICAL ABUTMENT DETAILS – WARM DECK

Ref. BS 8217: Reinforced bitumen membranes for roofing. Code of practice.

449

Conservation of Energy ~ this can be achieved in two ways:

1. Cold Deck – insulation is placed on the ceiling lining, between joists. See page 448 for details. A metallized polyester lined plasterboard ceiling functions as a vapour control layer, with a minimum 50 mm air circulation space between insulation and decking. The air space corresponds with eaves vents and both provisions will prevent moisture build-up, condensation and possible decay of timber.

2. (a) Warm Deck – rigid* insulation is placed below the waterproof covering and above the roof decking. The insulation must be sufficient to maintain the vapour control layer and roof members at a temperature above dewpoint, as this type of roof does not require ventilation.

   (b) Inverted Warm Deck – rigid* insulation is positioned above the waterproof covering. The insulation must be unaffected by water and capable of receiving a stone dressing or ceramic pavings.

* Resin bonded mineral fibre roof boards, expanded polystyrene or polyurethane slabs.

Typical Warm Deck Details ~

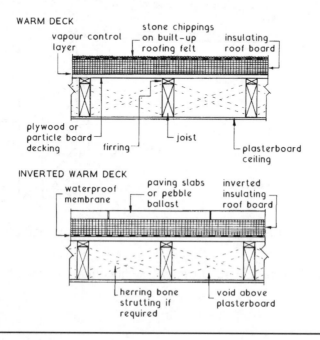

WARM DECK

vapour control layer — stone chippings on built-up roofing felt — insulating roof board

plywood or particle board decking — firring — joist — plasterboard ceiling

INVERTED WARM DECK

waterproof membrane — paving slabs or pebble ballast — inverted insulating roof board

herring bone strutting if required — void above plasterboard

Built-up Roofing Felt ~ this consists of three layers of bitumen roofing felt to BS EN 13707, and should be laid to the recommendations of BS 8217. The layers of felt are bonded together with hot bitumen and should have staggered laps of 50 mm minimum for side laps and 75 mm minimum for end laps – for typical details and references see pages 448 & 449.

Other felt materials which could be used are the two layer polyester based roofing felts which use a non-woven polyester base instead of the woven rag fibre base used in traditional felts.

Mastic Asphalt ~ this consists of two layers of mastic asphalt laid breaking joints and built up to a minimum thickness of 20 mm and should be laid to the recommendations of BS 8218. The mastic asphalt is laid over an isolating membrane of black sheathing felt which should be laid loose with 50 mm minimum overlaps.

Typical Details ~

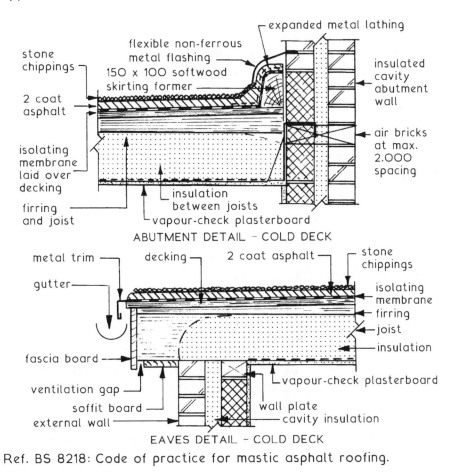

ABUTMENT DETAIL – COLD DECK

EAVES DETAIL – COLD DECK

Ref. BS 8218: Code of practice for mastic asphalt roofing.

Milled Lead Sheet ~ produced from refined lead to an initial thickness of about 125mm. Thereafter it is rolled and cut to 12.000m lengths, 2.400m wide into the following thicknesses and categories:

BS 1178* Code No.	BS EN 12588/standard milled thickness (mm)	Weight (kg/m^2) BS EN/milled	Colour marking
3	1.25/1.32	14.17/14.97	Green
-	1.50/1.59	17.00/18.03	Yellow
4	1.75/1.80	19.84/20.41	Blue
5	2.00/2.24	22.67/25.40	Red
6	2.50/2.65	28.34/30.05	Black
7	3.00/3.15	34.02/35.72	White
8	3.50/3.55	39.69/40.26	Orange

*BS 1178: Specification for milled sheet lead and strip for building purposes. This BS has been superseded by BS EN 12588: Lead and lead alloys. Rolled lead sheet for building purposes. The former BS codes are replaced with lead sheet thicknesses between 1.25 and 3.50 millimetres. They are included here, as these codes remain common industry reference. Codes originated before metrication as the approximate weight of lead sheet in pounds per square foot (lb/ft^2). eg. 3lb/ft^2 became Code 3.

Other Dimensions ~ cut widths between 75mm and 600mm in coils.

Density ~ approximately 11,325kg/m^3.

Application (colour marking) ~

Green and yellow – soakers.

Blue, red and black – flat roof covering in small, medium and large areas respectively (see table on page 455).

White and orange – lead lining to walls as protection from X-rays or for sound insulation, but can be used for relatively large areas of roof covering.

Thermal Movement ~ the coefficient of linear expansion for lead is 0.0000297 (2.97×10^{-6}) for every degree Kelvin.

Eg. If the exposure temperature range throughout a year is from $-10°C$ to $35°C$ (45 K), then a 2.000 m length of sheet lead could increase by: $0.0000297 \times 45 \times 2 = 0.00267$ m, or 2.67 mm.

Over time this movement will cause fatigue stress, manifesting in cracking. To prevent fracture, a smooth surface underlay should be used and the areas of lead sheet limited with provision of joints designed to accommodate movement.

Underlay ~ placed over plywood or similar smooth surface decking, or over rigid insulation boards. Bitumen impregnated felt or waterproof building paper have been the established underlay, but for new work a non-woven, needle punched polyester textile is now generally preferred.

Fixings ~ clips, screws and nails of copper, brass or stainless steel.

Jointing ~ for small areas such as door canopies and dormers where there is little opportunity for thermal movement, a simply formed welt can be used if the depth of rainwater is unlikely to exceed the welt depth.

Welted joint

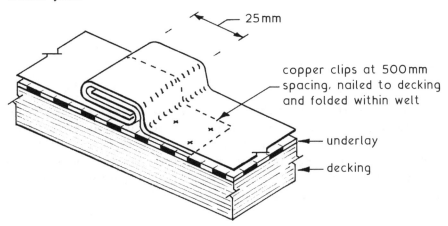

25 mm

copper clips at 500 mm spacing, nailed to decking and folded within welt

underlay

decking

Jointing to absorb movement ~

• Wood cored rolls in the direction of the roof slope (see next page).

• Drips at right angles to and across the roof slope (see next page).

453

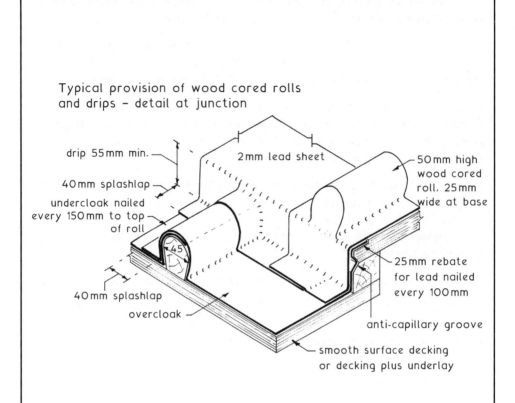

Typical provision of wood cored rolls
and drips – detail at junction

drip 55mm min.

40mm splashlap

undercloak nailed
every 150mm to top
of roll

40mm splashlap

overcloak

2mm lead sheet

50mm high
wood cored
roll, 25mm
wide at base

25mm rebate
for lead nailed
every 100mm

anti-capillary groove

smooth surface decking
or decking plus underlay

45

Lead is a soft and malleable material. A skilled craftsman
(traditionally a plumber) can manipulate lead sheet with hand tools
originally made from dense timber such as boxwood, but now
produced from high-density polythene. This practice is known as
'bossing' the lead to the profiles shown. Alternatively, the lead
sheet can be cut and welded to shape.

Spacing of wood cored rolls and drips varies with the thickness specification of lead sheet. The following is a guide ~

BS EN 12588 thickness (mm)	Maximum distance between drips (mm) [A]	Maximum distance between rolls (mm) [B]
1.25 and 1.50	Use for soakers only	
1.75	1500	500
2.00	2000	600
2.50	2250	675
3.00	2500	675
3.50	3000	750

Typical flat roof plan (page 417)

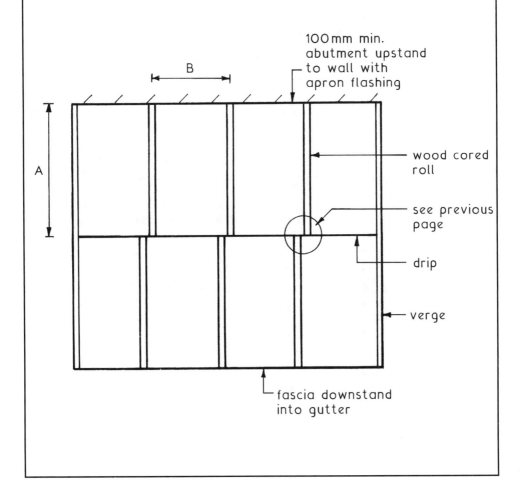

100 mm min. abutment upstand to wall with apron flashing

wood cored roll

see previous page

drip

verge

fascia downstand into gutter

A dormer is the framework for a vertical window constructed from the roof slope. It may be used as a feature, but is more likely as an economical and practical means for accessing light and ventilation to an attic room. Dormers are normally external with the option of a flat or pitched roof. Frame construction is typical of the following illustrations, with connections made by traditional housed and tenoned joints or simpler galvanized steel brackets and hangers.

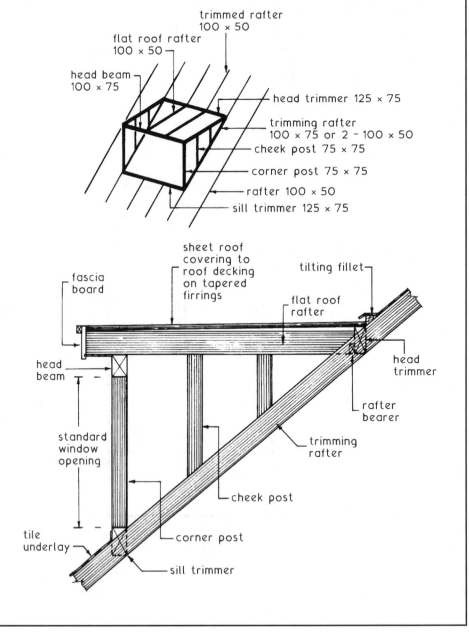

trimmed rafter
100 × 50

flat roof rafter
100 × 50

head beam
100 × 75

head trimmer 125 × 75

trimming rafter
100 × 75 or 2 – 100 × 50

cheek post 75 × 75

corner post 75 × 75

rafter 100 × 50

sill trimmer 125 × 75

sheet roof
covering to
roof decking
on tapered
firrings

tilting fillet

fascia
board

flat roof
rafter

head
trimmer

head
beam

rafter
bearer

standard
window
opening

trimming
rafter

cheek post

tile
underlay

corner post

sill trimmer

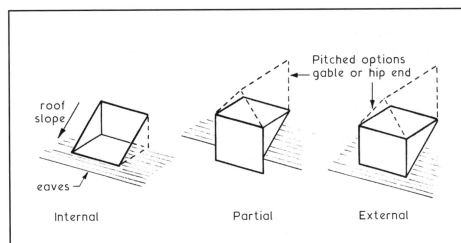

Internal Partial External

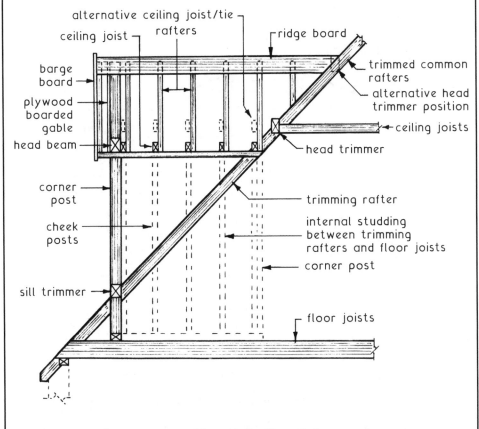

Section through gable ended external dormer

A graceful interruption to the routine of a pitched roof, derived from thatched roofs where the thatch is swept over window openings. Other suitable coverings are timber shingles, plain tiles and small slates.

Main roof pitch $\geq 50°$. Eyebrow pitch $\geq 35°$.

Transition curve should be smooth with span to height ratio $> 8:1$. Less is possible, but may prove impractical and disproportionate.

Possible profile ~

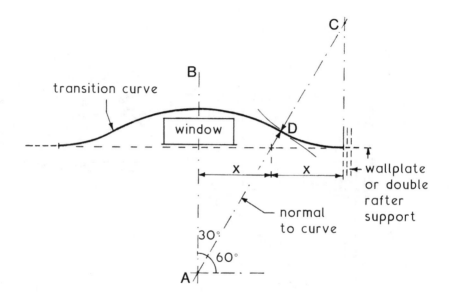

A – B is a line through the centre of the window opening.

A – D radius is positioned to clear the window head.

C – D radius is established near window base in continuity to A – D.

A purpose made gluelam beam can be used to create the transition curve, effectively extending the wallplate to receive the eyebrow rafters. The curved beam for an intermediate eyebrow may be supported on joist hangers to double trimming rafters each side.

Air carries water vapour, the amount increasing proportionally with the air temperature. As the water vapour increases so does the pressure and this causes the vapour to migrate from warmer to cooler parts of a building. As the air temperature reduces, so does its ability to hold water and this manifests as condensation on cold surfaces. Insulation between living areas and roof spaces increases the temperature differential and potential for condensation in the roof void.

Condensation can be prevented by either of the following:

* Providing a vapour control layer on the warm side of any insulation.
* Removing the damp air by ventilating the colder area.

The most convenient form of vapour layer is vapour check plasterboard which has a moisture resistant lining bonded to the back of the board. A typical patented product is a foil or metallised polyester backed plasterboard in 9·5 and 12·5mm standard thicknesses. This is most suitable where there are rooms in roofs and for cold deck flat roofs. Ventilation is appropriate to larger roof spaces.

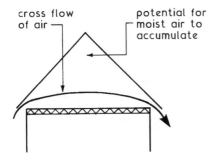

cross flow of air

potential for moist air to accumulate

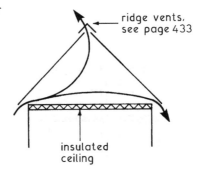

ridge vents,
see page 433

insulated
ceiling

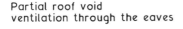

Partial roof void
ventilation through the eaves

Total roof void ventilation
through eaves and high level vents

Roof ventilation – provision of eaves ventilation alone should allow adequate air circulation in most situations. However, in some climatic conditions and where the air movement is not directly at right angles to the building, moist air can be trapped in the roof apex. Therefore, supplementary ridge ventilation is recommended.

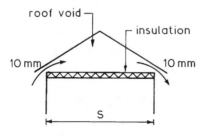

Insulation at ceiling level (1)
S = span <10m for
roof pitches 15°–35°

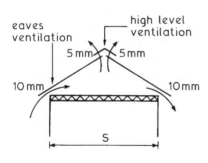

Insulation at ceiling level (2)
S = span >10m for
roof pitches 15°–35°
Any span for roof
pitches >35°

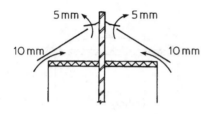

Insulation at ceiling level and
central dividing wall
Roof pitches >15°
for any span

Note: ventilation dimensions shown relate to a continuous strip (or equivalent) of at least the given gap.

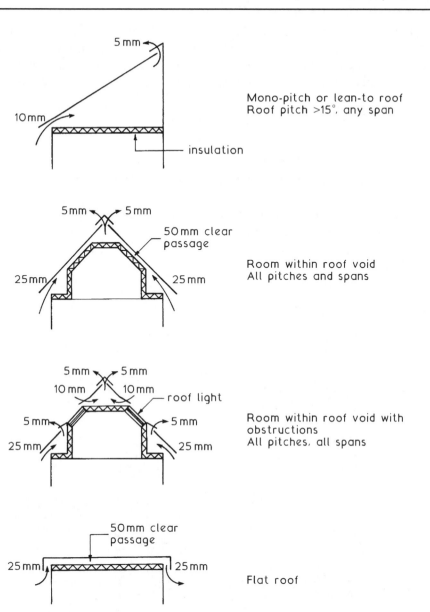

Mono-pitch or lean-to roof
Roof pitch >15°, any span

insulation

5mm

10mm

5mm   5mm

50mm clear passage

25mm   25mm

Room within roof void
All pitches and spans

5mm   5mm

10mm   10mm

roof light

5mm   5mm

25mm   25mm

Room within roof void with obstructions
All pitches, all spans

50mm clear passage

25mm   25mm

Flat roof

Refs. Building Regulations, Approved Document C – Site
preparation and resistance to contaminants and moisture.
Section 6 – Roofs.
BS 5250: Code of practice for control of condensation in
buildings.

BRE report – Thermal Insulation: avoiding risks (3rd. ed.).

Lateral Restraint – stability of gable walls and construction at the eaves, plus integrity of the roof structure during excessive wind forces, requires complementary restraint and continuity through 30 × 5mm cross sectional area galvanised steel straps.

Exceptions may occur if the roof:-

1. exceeds 15° pitch, and
2. is tiled or slated, and
3. has the type of construction known locally to resist gusts, and
4. has ceiling joists and rafters bearing onto support walls at not more than 1·2m centres.

Application ~

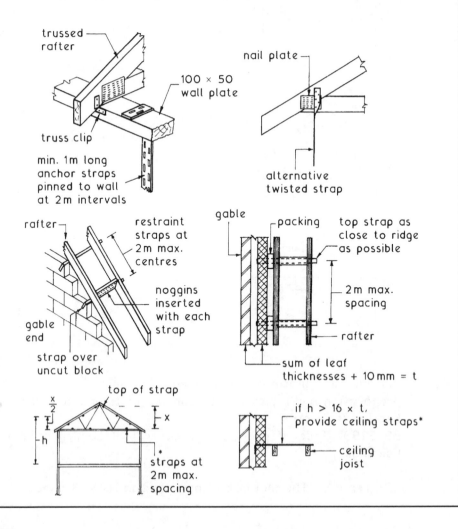

Preservation ~ ref. Building Regulations: Materials and Workmanship. Approved Document to support Regulation 7.

Woodworm infestation of untreated structural timbers is common. However, the smaller woodborers such as the abundant Furniture beetle are controllable. It is the threat of considerable damage potential from the House Longhorn beetle that has forced many local authorities in Surrey and the fringe areas of adjacent counties to seek timber preservation listing in the Building Regulations (see Table 1 in the above reference). Prior to the introduction of pretreated timber (c. 1960s), the House Longhorn beetle was once prolific in housing in the south of England, establishing a reputation for destroying structural roof timbers, particularly in the Camberley area.

House Longhorn beetle data:-

Latin name – Hylotrupes bajulus

Life cycle – Mature beetle lays up to 200 eggs on rough surface of untreated timber.

After 2-3 weeks, larvae emerge and bore into wood, preferring sapwood to denser growth areas. Up to 10 years in the damaging larval stage. In 3 weeks, larvae change to chrysalis to emerge as mature beetles in summer to reproduce.

Timber appearance – powdery deposits (frass) on the surface and the obvious mature beetle flight holes.

Beetle appearance –

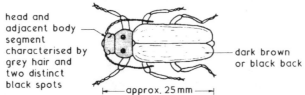

head and adjacent body segment characterised by grey hair and two distinct black spots

dark brown or black back

approx. 25mm

Other woodborers:-

Furniture beetle – dark brown, 6-8mm long, lays 20-50 eggs on soft or hardwoods. Bore holes only 1-2mm diameter.

Lyctus powder post beetle – reddish brown, 10-15mm long, lays 70-200 eggs on sapwood of new hardwood. Bore holes only 1-2mm in diameter.

Death Watch beetle – dark brown, sometimes speckled in lighter shades. Lays 40-80 eggs on hardwood. Known for preferring the oak timbers used in old churches and similar buildings.

Bore holes about 3mm diameter.

Preservation ~ treatment of timber to prevent damage from House Longhorn beetle.

In the areas specified (see previous page), all softwood used in roof structures including ceiling joists and any other softwood fixings should be treated with insecticide prior to installation. Specific chemicals and processes have not been listed in the Building Regulations since the 1976 issue. Timber treatment then was either:

• Vacuum/pressure impregnation with a blend of copper, chromium and arsenic (CCA), known commercially as 'tanalising'.
• Diffusion with sodium borate (boron salts).
• Steeping (min. 10 mins.) in organic solvent wood preservative.
• Steeping or soaking in tar oil (creosote). This has limitations due to staining of adjacent surfaces.

The current edition of Approved Document A (Structure) to the Building Regulations refers to guidance on preservative treatments in the British Wood Preserving and Damp-Proofing Association's Manual. Other guidance is provided in:

BS 1282: Wood preservatives. Guidance on choices, use and application.

BS 5707: Specification for preparation for wood preservatives in organic solvents.

BS 8417: Preservation of timber. Recommendations.

Insect treatment adds about 10% to the cost of timber and also enhances its resistance to moisture. Other parts of the structure, e.g. floors and partitions are less exposed to woodworm damage as they are enclosed. Also, there is a suggestion that if these areas received treated timber, the toxic fumes could be harmful to the health of building occupants. Current requirements for through ventilation in roofs has the added benefit of discouraging wood boring insects, as they prefer draught-free damp areas.

Note: EU directive CEN/TC 38 prohibits the use of CCA preservatives for domestic applications and in places where the public may be in contact with it.

Green roof ~ green with reference to the general appearance of plant growths and for being environmentally acceptable. Part of the measures for constructing sustainable and ecologically friendly buildings.

Categories ~
• Extensive ~ a relatively shallow soil base (typically 50 mm) and lightweight construction. Maximum roof pitch is 40° and slopes greater than 20° will require a system of baffles to prevent the soil moving. Plant life is limited by the shallow soil base to grasses, mosses, herbs and sedum (succulents, generally with fleshy leaves producing pink or white flowers).
• Intensive ~ otherwise known as a roof garden. This category has a deeper soil base (typically 400 mm) that will provide for landscaping features, small ponds, occasional shrubs and small trees. A substantial building structure is required for support and it is only feasible to use a flat roof.

Advantages ~
• Absorbs and controls water run-off.
• Integral thermal insulation.
• Integral sound insulation.
• Absorbs air pollutants, dust and CO_2.
• Passive heat storage potential.

Disadvantages ~
• Weight.
• Maintenance.

Construction ~ the following build-up will be necessary to fulfil the objectives and to create stability:
• Vapour control layer above the roof structure.
• Rigid slab insulation.
• Root resilient waterproof under-layer.
• Drainage layer.
• Filter.
• Growing medium (soil).
• Vegetation (grass, etc.)

Examples of both extensive and intensive green roof construction are shown on the next page.

Typical extensive roof build up ~

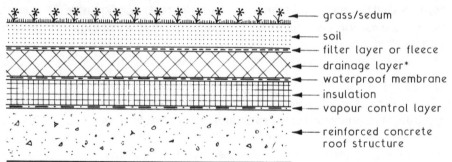

- grass/sedum
- soil
- filter layer or fleece
- drainage layer*
- waterproof membrane
- insulation
- vapour control layer
- reinforced concrete roof structure

* typically, expanded polystyrene with slots

Component	Weight (kg/m²)	Thickness (mm)
vcl	3	3
insulation	3	50
membrane	5	5
drainage layer	3	50
filter	3	3
soil	90	50
turf	40	20
	-------------	---------
	147 kg/m²	181 mm

147 kg/m² saturated weight x 9.81 = 1442 N/m² or 1.44 kN/m²

Typical intensive roof build up ~

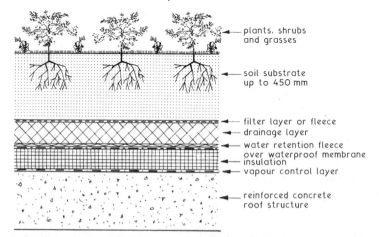

- plants, shrubs and grasses
- soil substrate up to 450 mm
- filter layer or fleece
- drainage layer
- water retention fleece over waterproof membrane
- insulation
- vapour control layer
- reinforced concrete roof structure

Depth to vcl, approximately 560 mm at about 750 kg/m² saturated weight. 750 kg/m² × 9.81 = 7358 N/m² or 7.36 kN/m².

Thermal insulation of external elements of construction is measured in terms of thermal transmittance rate, otherwise known as the U-value. It is the amount of heat energy in watts transmitted through one square metre of construction for every one degree Kelvin between external and internal air temperature, i.e. W/m²K.

U-values are unlikely to be entirely accurate, due to:

* the varying effects of solar radiation, atmospheric dampness and prevailing winds.

* inconsistencies in construction, even with the best of supervision.

* 'bridging' where different structural components meet, e.g. dense mortar in lightweight blockwork.

Nevertheless, calculation of the U-value for a particular element of construction will provide guidance as to whether the structure is thermally acceptable. The Building Regulations, Approved Document L, Conservation of fuel and power, determines acceptable energy efficiency standards for modern buildings, with the objective of limiting the emission of carbon dioxide and other burnt gases into the atmosphere.

The U-value is calculated by taking the reciprocal of the summed thermal resistances (R) of the component parts of an element of construction:

$$U = \frac{1}{\Sigma R} = W/m^2K$$

R is expressed in m²K/W. The higher the value, the better a component's insulation. Conversely, the lower the value of U, the better the insulative properties of the structure.

Building Regulations, Approved Document references:
L1A, Work in new dwellings.
L1B, Work in existing dwellings.
L2A, Work in new buildings other than dwellings.
L2B, Work in existing buildings other than dwellings.

Thermal resistances (R) are a combination of the different structural, surface and air space components which make up an element of construction. Typically:

$$U = \frac{1}{R_{so} + R_1 + R_2 + R_a + R_3 + R_4 \, etc \cdots + R_{si}(m^2K/W)}$$

Where: R_{so} = Outside or external surface resistance.

R_1, R_2, etc. = Thermal resistance of structural components.

R_a = Air space resistance, eg. wall cavity.

R_{si} = Internal surface resistance.

The thermal resistance of a structural component (R_1, R_2, etc.) is calculated by dividing its thickness (L) by its thermal conductivity (λ), i.e.

$$R(m^2K/W) = \frac{L(m)}{\lambda(W/mK)}$$

eg. 1. A 102 mm brick with a conductivity of 0·84 W/mK has a thermal resistance (R) of: 0·102 ÷ 0·84 = 0·121 m²K/W.

eg. 2.

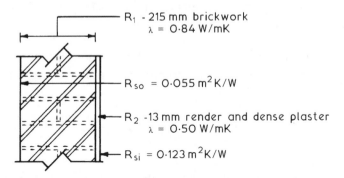

R_1 - 215 mm brickwork
λ = 0·84 W/mK

R_{so} = 0·055 m²K/W

R_2 -13 mm render and dense plaster
λ = 0·50 W/mK

R_{si} = 0·123 m²K/W

Note: the effect of mortar joints in the brickwork can be ignored, as both components have similar density and insulative properties.

$$U = \frac{1}{R_{so} + R_1 + R_2 + R_{si}}$$

$R_1 = 0·215 ÷ 0·84 = 0·256$
$R_2 = 0·013 ÷ 0·50 = 0·026$

$$U = \frac{1}{0.055 + 0.256 + 0.026 + 0.123} = 2.17W/m^2K$$

Typical values in: m^2K/W

Internal surface resistances (R_{si}):

Walls – 0·123
Floors or ceilings for upward heat flow – 0·104
Floors or ceilings for downward heat flow – 0·148
Roofs (flat or pitched) – 0·104

External surface resistances (R_{so}):

Surface	Exposure		
	Sheltered	Normal	Severe
Wall – high emissivity	0·080	0·055	0·030
Wall – low emissivity	0·110	0·070	0·030
Roof – high emissivity	0·070	0·045	0·020
Roof – low emissivity	0·090	0·050	0·020
Floor – high emissivity	0·070	0·040	0·020

Sheltered – town buildings to 3 storeys.
Normal – town buildings 4 to 8 storeys and most suburban premises.

Severe – > 9 storeys in towns.
> 5 storeys elsewhere and any buildings on exposed coasts and hills.

Air space resistances (R_a):

Pitched or flat roof space – 0·180
Behind vertical tile hanging – 0·120
Cavity wall void – 0·180
Between high and low emissivity surfaces – 0·300
Unventilated/sealed – 0·180

Emissivity relates to the heat transfer across and from surfaces by radiant heat emission and absorption effects. The amount will depend on the surface texture, the quantity and temperature of air movement across it, the surface position or orientation and the temperature of adjacent bodies or materials. High surface emissivity is appropriate for most building materials. An example of low emissivity would be bright aluminium foil on one or both sides of an air space.

Typical values –

Material	Density (kg/m³)	Conductivity (λ) (W/mK)
WALLS:		
Boarding (hardwood)	700	0·18
......(softwood)	500	0·13
Brick outer leaf	1700	0·84
....inner leaf	1700	0·62
Calcium silicate board	875	0·17
Ceramic tiles	2300	1·30
Concrete	2400	1·93
....	2200	1·59
....	2000	1·33
....	1800	1·13
.... (lightweight)	1200	0·38
.... (reinforced)	2400	2·50
Concrete block (lightweight)	600	0·18
.... (mediumweight)	1400	0·53
Cement mortar (protected)	1750	0·88
.... (exposed)	1750	0·94
Fibreboard	350	0·08
Gypsum plaster (dense)	1300	0·57
Gypsum plaster (lightweight)	600	0·16
Plasterboard	950	0·16
Tile hanging	1900	0·84
Rendering	1300	0·57
Sandstone	2600	2·30
Wall ties (st/st)	7900	17·00
ROOFS:		
Aerated concrete slab	500	0·16
Asphalt	1900	0·60
Bituminous felt in 3 layers	1700	0·50
Sarking felt	1700	0·50
Stone chippings	1800	0·96
Tiles (clay)	2000	1·00
......(concrete)	2100	1·50
Wood wool slab	500	0·10

Typical values –

Material	Density (kg/m³)	Conductivity (λ) (W/mK)
FLOORS:		
Cast concrete	2000	1·33
Hardwood block/strip	700	0·18
Plywood/particle board	650	0·14
Screed	1200	0·41
Softwood board	500	0·13
Steel tray	7800	50·00
INSULATION:		
Expanded polystyrene board	20	0·035
Mineral wool batt/slab	25	0·038
Mineral wool quilt	12	0·042
Phenolic foam board	30	0·025
Polyurethane board	30	0·025
Urea formaldehyde foam	10	0·040
GROUND:		
Clay/silt	1250	1·50
Sand/gravel	1500	2·00
Homogenous rock	3000	3·50

Notes:

1. For purposes of calculating U-values, the effect of mortar in external brickwork is usually ignored as the density and thermal properties of bricks and mortar are similar.

2. Where butterfly wall ties are used at normal spacing in an insulated cavity $\leq$ 75mm, no adjustment is required to calculations. If vertical twist ties are used in insulated cavities >75mm, 0·020 W/m²K should be added to the U-value.

3. Thermal conductivity (λ) is a measure of the rate that heat is conducted through a material under specific conditions (W/mK).

* Tables and charts – Insulation manufacturers' design guides and technical papers (walls, roofs and ground floors).
* Calculation using the Proportional Area Method (walls and roofs).
* Calculation using the Combined Method – BS EN ISO 6946 (walls and roofs).
* Calculation using BS EN ISO 13370 (ground floors and basements).

Tables and charts – these apply where specific U-values are required and standard forms of construction are adopted. The values contain appropriate allowances for variable heat transfer due to different components in the construction, e.g. twisted pattern wall-ties and non-uniformity of insulation with the interruption by ceiling joists. The example below shows the tabulated data for a solid ground floor with embedded insulation of λ = 0.03 W/mK

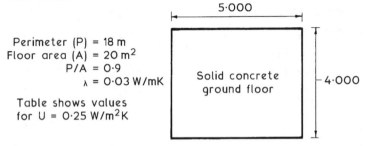

Perimeter (P) = 18 m
Floor area (A) = 20 m^2
P/A = 0·9
λ = 0·03 W/mK

Table shows values
for U = 0·25 W/m^{2}K

Typical table for floor insulation:

P/A	0·020	0·025	0·030*	0·035	0·040	0·045	W/mK
1·0	61	76	91	107	122	137	mm ins.
0·9*	60	75	<u>90</u>	105	120	135	.. ..
0·8	58	73	88	102	117	132	.. ..
0·7	57	71	85	99	113	128	.. ..
0·6	54	68	82	95	109	122	.. ..
0·5	51	64	77	90	103	115	.. ..

90 mm of insulation required.

Refs. BS EN ISO 6946: Building components and building elements. Thermal resistance and thermal transmittance. Calculation method.
BS EN ISO 13370: Thermal performance of buildings. Heat transfer via the ground. Calculation methods.

Various applications to different ground floor situations are considered in BS EN ISO 13370. The following is an example for a solid concrete slab in direct contact with the ground. The data used is from the previous page.

Floor section

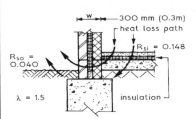

Perimeter = 18 m (exposed)
Floor area = 20 m²
λ for 90 mm insulation = 0.03 W/mK
Characteristic floor dimension = B^1
B^1 = Floor area ÷ (1/2 exp. perimeter)
B^1 = 20 ÷ 9 = 2.222 m

Formula to calculate total equivalent floor thickness for uninsulated and insulated all over floor:

$$dt = w + \lambda \, (R_{si} + R_f + R_{so})$$

where: dt = total equivalent floor thickness (m)
w = wall thickness (m)
λ = thermal conductivity of soil (W/mK) [see page 471]
R_{si} = internal surface resistance (m²K/W) [see page 469]
R_f = insulation resistance (0.09 ÷ 0.03 = 3 m²K/W)
R_{so} = external surface resistance (m²K/W) [see page 469]

Uninsulated: dt = 0.3 + 1.5 (0.148 + 0 + 0.04) = 0.582 m
Insulated: dt = 0.3 + 1.5 (0.148 + 3 + 0.04) = 5.082 m

Formulae to calculate U-values ~
Uninsulated or poorly insulated floor, $dt < B^1$:

$$U = (2\lambda) \div [(\pi B^1) + dt] \times \ln [(\pi B^1 \div dt) + 1]$$

Well insulated floor, $dt \geq B^1$:

$$U = \lambda \div [(0.457 \times B^1) + dt]$$

where: U = thermal transmittance coefficient (W/m²/K)
λ = thermal conductivity of soil (W/mK)
B^1 = characteristic floor dimension (m)
dt = total equivalent floor thickness (m)
ln = natural logarithm

Uninsulated floor ~
U = (2 × 1.5) ÷ [(3.142 × 2.222) + 0.582] × ln [(3.142 × 2.222) ÷ 0.582 + 1]
U = 0.397 × ln 12.996 = 1.02 W/m²K

Insulated floor ~
   U = 1.5 ÷ [(0.457 × 2.222) + 5.082] = 1.5 ÷ 6.097 = 0.246 W/m²K

Compares with the tabulated figure of 0.250 W/m²K on the previous page.

Proportional Area Method (Wall)

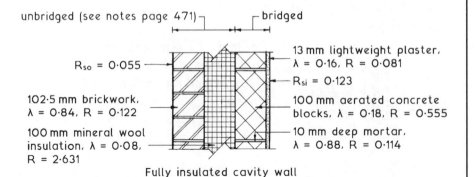

unbridged (see notes page 471) ⌐ bridged

R_{so} = 0·055

102·5 mm brickwork, λ = 0·84, R = 0·122

100 mm mineral wool insulation, λ = 0·08, R = 2·631

13 mm lightweight plaster, λ = 0·16, R = 0·081

R_{si} = 0·123

100 mm aerated concrete blocks, λ = 0·18, R = 0·555

10 mm deep mortar, λ = 0·88, R = 0·114

Fully insulated cavity wall

A standard block with mortar is 450 × 225 mm = 101250 mm²

A standard block format of 440 × 215 mm   = 94600 mm²

The area of mortar per block   =   6650 mm²

Proportional area of mortar = $\dfrac{6650}{101250} \times \dfrac{100}{1}$ = 6·57% (0.066)

Therefore the proportional area of blocks = 93·43% (0·934)

Thermal resistances (R):

Outer leaf + insulation (unbridged)
R_{so} = 0·055
brickwork = 0·122
insulation = 2·631
            2·808
        × 100% = 2·808

Inner leaf (unbridged)
blocks = 0·555
plaster = 0·081
R_{si} = 0·123
            0·759
        × 93·43% = 0·709

Inner leaf (bridged)
mortar   = 0·114
plaster   = 0·081
R_{si} = 0·123
            = 0·318
    × 6·57% = 0·021

$$U = \frac{1}{\Sigma R} = \frac{1}{2·808 + 0·709 + 0·021} = 0·283\,W/m^2K$$

Combined Method (Wall)

This method considers the upper and lower thermal resistance (R) limits of an element of structure. The average of these is reciprocated to provide the U-value.

Formula for upper and lower resistances $= \dfrac{1}{\Sigma(F_x \div R_x)}$

Where: F_x = Fractional area of a section
R_x = Total thermal resistance of a section

Using the wall example from the previous page:

Upper limit of resistance (R) through section containing blocks – (R_{so}, 0·055) + (brkwk, 0·122) + (ins, 2·631) + (blocks, 0·555) + (plstr, 0·081) + (R_{si}, 0·123) = 3·567 m²K/W

Fractional area of section (F) = 93·43% or 0·934

Upper limit of resistance (R) through section containing mortar – (R_{so} 0·055) + (brkwk, 0·122) + (ins, 2·631) + (mortar, 0·114) + (plstr, 0·081) + (R_{si}, 0·123) = 3·126 m²K/W

Fractional area of section (F) = 6·57% or 0·066

The upper limit of resistance =

$$\frac{1}{\Sigma(0·943 \div 3·567) + (0·066 \div 3·126)} = 3·533 \, m^2K/W$$

Lower limit of resistance (R) is obtained by summating the resistance of all the layers –
(R_{so}, 0·055) + (brkwk, 0·122) + (ins, 2·631) + (bridged layer, 1 ÷ [0·934 ÷ 0·555] + [0·066 ÷ 0·114] = 0·442) + (plstr, 0·081) + (R_{si}, 0·123) = 3·454 m²K/W

Total resistance (R) of wall is the average of upper and lower limits = (3·533 + 3·454) ÷ 2 = 3·493 m²K/W

$$U\text{-value} = \frac{1}{R} = \frac{1}{3·493} = 0·286 \, W/m^2K$$

Note: Both proportional area and combined method calculations require an addition of 0·020 W/m²K to the calculated U-value. This is for vertical twist type wall ties in the wide cavity. See page 338 and note 2 on page 471.

Proportional Area Method (Roof)

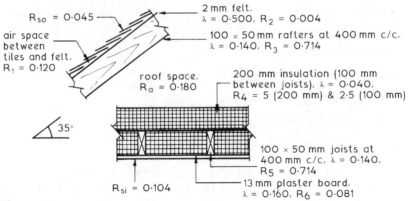

$R_{so} = 0.045$

air space between tiles and felt, $R_1 = 0.120$

2 mm felt, $\lambda = 0.500$, $R_2 = 0.004$

100×50 mm rafters at 400 mm c/c, $\lambda = 0.140$, $R_3 = 0.714$

roof space, $R_a = 0.180$

200 mm insulation (100 mm between joists), $\lambda = 0.040$, $R_4 = 5$ (200 mm) & 2.5 (100 mm)

35°

100×50 mm joists at 400 mm c/c, $\lambda = 0.140$, $R_5 = 0.714$

$R_{si} = 0.104$

13 mm plaster board, $\lambda = 0.160$, $R_6 = 0.081$

Notes:

1. The air space in the loft area is divided between pitched and ceiling components, ie. $R_a = 0.180 \div 2 = 0.090 \, m^2K/W$.

2. The U-value is calculated perpendicular to the insulation, therefore the pitched component resistance is adjusted by multiplying by the cosine of the pitch angle, ie. 0.819.

3. Proportional area of bridging parts (rafters and joists) is $50 \div 400 = 0.125$ or 12.5%.

4. With an air space resistance value (R1) of $0.120 \, m^2K/W$ between tiles and felt, the resistance of the tiling may be ignored.

Thermal resistance (R) of the pitched component:

Raftered part
$R_{so} = 0.045$
$R_1 = 0.120$
$R_2 = 0.004$
$R_3 = 0.714$
$R_a = \underline{0.090}$
$ 0.973 \times 12.5\% = 0.122$

Non-raftered part
$R_{so} = 0.045$
$R_1 = 0.120$
$R_2 = 0.004$
$R_a = \underline{0.090}$
$ 0.259 \times 87.5\%$
$ = 0.227$

Total resistance of pitched components =
$ (0.122 + 0.227) \times 0.819 = 0.286 \, m^2K/W$

Thermal resistance (R) of the ceiling component:

Joisted part
$R_{si} = 0.104$
$R_6 = 0.081$
$R_5 = 0.714$
$R_4 = 2.500$ (100 mm)
$R_a = \underline{0.090}$
$ 3.489 \times 12.5\% = 0.436$

Fully insulated part
$R_{si} = 0.104$
$R_6 = 0.081$
$R_4 = 5.000$ (200mm)
$R_a = \underline{0.090}$
$ 5.275 \times 87.5\%$
$ = 4.615$

Total resistance of ceiling components = $0.436 + 4.615$
$ = 5.051 \, m^2K/W.$

$$U = \frac{1}{\Sigma R} = \frac{1}{0.286 + 5.051} = 0.187 \, W/m^2K$$

Standard Assessment Procedure ~ the Approved Document to Part L of the Building Regulations emphasises the importance of quantifying the energy costs of running homes. For this purpose it uses the Government's Standard Assessment Procedure (SAP). SAP has a numerical scale of 1 to 100, although it can exceed 100 if a dwelling is a net energy exporter. It takes into account the effectiveness of a building's fabric relative to insulation and standard of construction. It also appraises the energy efficiency of fuel consuming installations such as ventilation, hot water, heating and lighting. Incidentals like solar gain also feature in the calculations.

As part of the Building Regulations approval procedure, energy rating (SAP) calculations are submitted to the local building control authority. SAP ratings are also required to provide prospective home purchasers or tenants with an indication of the expected fuel costs for hot water and heating. This information is documented and included with the property conveyance. The SAP calculation involves combining data from tables, work sheets and formulae. Guidance is found in Approved Document L, or by application of certified SAP computer software programmes.

SAP rating average for all homes is about 50. A modernised 1930s house about 70, that built to 1995 energy standards about 80 and a 2002 house about 90. Current quality construction standards should rate dwellings close to 100.

Ref. Standard Assessment Procedure for Energy Rating of Dwellings. The Stationery Office.

Air Permeability ~ air tightness in the construction of dwellings is an important quality control objective. Compliance is achieved by attention to detail at construction interfaces, e.g. by silicone sealing built-in joists to blockwork inner leafs and door and window frames to masonry surrounds; draft proofing sashes, doors and loft hatches. Guidance for compliance is provided in, Limiting thermal bridging and air leakage: Robust construction details for dwellings and similar buildings, published by The Stationery Office. Dwellings failing to comply with these measures are penalised in SAP calculations. Alternatively, a certificate must be obtained to show pre-completion testing satisfying air permeability of less than 10 m^3/h per m^2 envelope area at 50 Pascals (Pa or N/m^2) pressure.

Domestic buildings (England and Wales) ~

Element of construction	Area weighted ave. U-value (W/m^{2}K)	Limiting individual component U-value
Roof	0.25	0.35
Wall	0.35	0.70
Floor	0.25	0.70
Windows, doors, rooflights and roof windows	2.20	3.30

The area weighted average U-value for an element of construction depends on the individual U-values of all components and the area they occupy within that element. E.g. The part of a wall with a meter cupboard built in will have less resistance to thermal transmittance than the rest of the wall (max. U-value at cupboard, 0.45).

Element of construction	U-value targets (W/m^{2}K)
Pitched roof (insulation between rafters)	0.15
Pitched roof (insulation between joists)	0.15
Flat roof	0.15
Wall	0.28
Floor	0.20
Windows, doors, rooflights and roof windows	1.80 (area weighted ave.)

Note: Maximum area of windows, doors, rooflights and roof windows, are not specifically defined.

An alternative to the area weighted average U-value for windows, etc., may be a window energy rating of not less than minus 30 (see page 480).

Energy source ~ gas or oil fired central heating boiler with a minimum SEDBUK efficiency rating of 86% (band rating A or B, A only from Oct. 2010). There are transitional and exceptional circumstances that permit lower band rated boilers. Where this occurs, the construction of the building envelope should compensate with very low U-values.

SEDBUK = Seasonal Efficiency of a Domestic Boiler in the United Kingdom. SEDBUK values are defined in the Government's Standard Assessment Procedure for Energy Rating of Dwellings. There is also a SEDBUK website, www.sedbuk.com.

Note: SEDBUK band A = >90% efficiency
band B = 86–90%   ..
band C = 82–86%   ..
band D = 78–82%   ..

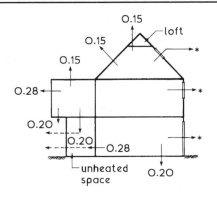

*area weighted average U-value of rooflights, roof windows, windows and doors = 1.80 or a window energy rating of not less than −30.

Further Quality Procedures (Structure) ~

* Provision of insulation to be continuous. Gaps are unacceptable and if allowed to occur will invalidate the insulation value by thermal bridging.
* Junctions at elements of construction (wall/floor, wall/roof) to receive particular attention with regard to continuity of insulation.
* Openings in walls for windows and doors to be adequately treated with insulating cavity closers.

Further Quality Procedures (Energy Consumption) ~

* Hot water and heating systems to be fully commissioned on completion and controls set with regard for comfort, health and economic use.
* As part of the commissioning process, the sealed heating system should be flushed out and filled with a proprietary additive diluted in accordance with the manufacturer's guidance.

This is necessary to enhance system performance by resisting corrosion, scaling and freezing.

* A certificate confirming system commissioning and water treatment should be available for the dwelling occupant. This document should be accompanied with component manufacturer's operating and maintenance instructions.

Note: Commissioning of heating installations and the issue of certificates is by a qualified "competent person" as recognised by the appropriate body, i.e. CAPITA GROUP, OFTEC or HETAS.
CAPITA GROUP ~ 'Gas Safe Register' of Installers (formerly CORGI).
OFTEC ~ Oil Firing Technical Association for the Petroleum Industry.
HETAS ~ Solid Fuel. Heating Equipment Testing and Approval Scheme.

European Window Energy Rating Scheme (EWERS) ~ an alternative to U-values for measuring the thermal efficiency of windows. U-values form part of the assessment, in addition to factors for solar heat gain and air leakage. In the UK, testing and labelling of window manufacturer's products is promoted by the British Fenestration Rating Council (BFRC). The scheme uses a computer to simulate energy movement over a year through a standard window of 1.480 × 1.230 m containing a central mullion and opening sash to one side.

Data is expressed on a scale from A–G in units of kWh/m^2/year.

A > zero
B –10 to 0
C –20 to –10
D –30 to –20
E –50 to –30
F –70 to –50
G < –70

By formula, rating = (218.6 × g value) – 68.5 (U-value × L value)
Where: g value = factor measuring effectiveness of solar heat block expressed between 0 and 1. For comparison:

0.48 (no curtains)
0.43 (curtains open)
0.17 (curtains closed)

U value = weighted average transmittance coefficient
L value = air leakage factor

From the label shown opposite:
Rating = (218·6 × 0·5)
        – 68·5 (1·8 + 0·10)
      = 109·3 – 130·15
      = –20·85 i.e. –21

Typical format of a window energy rating label ~

ABC Joinery Ltd.
Window ref. XYZ 123

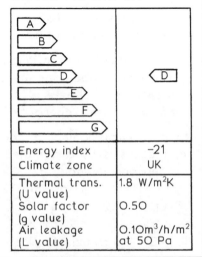

Energy index	–21
Climate zone	UK
Thermal trans. (U value)	1.8 W/m^{2}K
Solar factor (g value)	0.50
Air leakage (L value)	0.10m^3/h/m^2 at 50 Pa

The UK Government's Dept. of Energy and Climate Change (DECC) Standard Assessment Procedure (SAP) for energy rating dwellings, includes a facility to calculate carbon dioxide (CO_2) emissions in kilograms or tonnes per year. The established carbon index method allows for adjustment to dwelling floor area to obtain a carbon factor (CF):

$$CF = CO_2 \div (\text{total floor area} + 45)$$

The carbon index (CI) = $17 \cdot 7 - (9 \log. CF)$

Note: log. = logarithm to the base 10.

e.g. A dwelling of total floor area $125m^2$, with CO_2 emissions of 2000 kg/yr.

$$CF = 2000 \div (125 + 45) = 11 \cdot 76$$
$$CI = 17 \cdot 7 - (9 \log. 11 \cdot 76) = 8 \cdot 06$$

The carbon index (CI) is expressed on a scale of 0 to 10. The higher the number the better. Every new dwelling should have a CI value of a least 8.

Approved Document L to the Building Regulations refers to the Dwelling Carbon Emissions Rate (DER) as another means for assessing carbon discharge. The DER is compared by calculation to a Target Carbon Emissions Rate (TER), based on data for type of lighting, floor area, building shape and choice of fuel.

The DER is derived primarily by appraising the potential CO_2 emission from a dwelling relative to the consumption of fuel (directly or indirectly) in hot water, heating, lighting, cooling (if fitted), fans and pumps.

$$DER \leq TER$$

Buildings account for about half of the UK's carbon emissions. Therefore, there are considerable possibilities for energy savings and reductions in atmospheric pollution.

- Basis for improvement ~ total annual CO_2 emissions are around 150 million tonnes (MtC).
- CO_2 represents about 85% of all greenhouse gases produced by burning fossil fuels (methane 6%, nitrous oxide 5%, industrial trace gases the remainder).
- 25 million homes produce about 27% (41 MtC) of carbon emissions, representing a significant target for improvement (non-domestic buildings about 18%, 27 MtC).
- The table below shows the disposition of domestic carbon emissions.

Source	1990 (%)	2003 (%) (Note 2)
Cooking	8	5
Lighting and appliances	33	22
Hot water	18	20
Heating (see Note 1)	41	53

Source: Climate change – The UK Programme, TSO.

Note 1: Expectations for comfort standards in dwellings are rising. Domestic air conditioning is on the increase, partly in response to climatic change and global warming. Energy expended could increase to include a factor for cooling.
Note 2: Carbon emissions for 2003 are about 5% lower than in 1990.

The energy efficiency of new homes is about 70% higher than those built in 1990. However, many older homes have been improved to include some of the following provisions:

Application	Potential reduction, CO_2 per annum (kg)
Loft insulation	1000
Double glazing	700
Draft proofing (doors, windows, floors)	300
Wall cavity insulation	750
Condensing boiler	875
Insulated hot water storage cylinder	160
Energy saving light bulb	45 (each)

In new buildings and those subject to alterations, the objective is to optimise the use of fuel and power to minimise emission of carbon dioxide and other burnt fuel gases into the atmosphere. This applies principally to the installation of hot water, heating, lighting, ventilation and air conditioning systems. Pipes, ducting, storage vessels and other energy consuming plant should be insulated to limit heat losses. The fabric or external envelope of a building is constructed with regard to limiting heat losses through the structure and to regulate solar gains.

Approved Document L2 of the Building Regulations is not prescriptive. It sets out a series of objectives relating to achievement of a satisfactory carbon emission standard. A number of other technical references and approvals are cross referenced in the Approved Document and these provide a significant degree of design flexibility in achieving the objectives.

Energy efficiency of buildings other than dwellings is determined by applying a series of procedures modelled on a notional building of the same size and shape as the proposed building. The performance standards used for the notional building are similar to the 2002 edition of Approved Document L2. Therefore the proposed or actual building must be seen to be a significant improvement in terms of reduced carbon emissions by calculation. Improvements can be achieved in a number of ways, including the following:

- Limit the area or number of rooflights, windows and other openings.
- Improve the U-values of the external envelope. The limiting values are shown on the next page.
- Improve the airtightness of the building from the poorest acceptable air permeability of $10 \, m^3/hour/m^2$ of external envelope at 50 Pa pressure.
- Improve the heating system efficiency by installing thermostatic controls, zone controls, optimum time controls, etc. Fully insulate pipes and equipment.
- Use of high efficacy lighting fittings, automated controls, low voltage equipment, etc.
- Apply heat recovery systems to ventilation and air conditioning systems. Insulate ducting.
- Install a building energy management system to monitor and regulate use of heating and air conditioning plant.
- Limit overheating of the building with solar controls and appropriate glazing systems.
- Ensure that the quality of construction provides for continuity of insulation in the external envelope.
- Establish a commissioning and plant maintenance procedure. Provide a log-book to document all repairs, replacements and routine inspections.

Buildings Other Than Dwellings (England and Wales) ~

Element of construction	Limiting area weighted ave. U-value (W/m²K)	Limiting individual component U-value
Roof	0.25	0.35
Wall	0.35	0.70
Floor	0.25	0.70
Windows, doors, roof-lights, roof windows and curtain walling	2.20	3.30
High use entrances and roof vents	6.00	6.00
Large and vehicle access doors	1.50	4.00

Notes:
- For display windows separate consideration applies. See Section 5 in A.D., L2A.
- The poorest acceptable thermal transmittance values provide some flexibility for design, allowing a trade off against other thermally beneficial features such as energy recovery systems.
- The minimum U-value standard is set with regard to minimise the risk of condensation.
- The concept of area weighted values is explained on page 443.
- Elements will normally be expected to have much better insulation than the limiting U-values. Suitable objectives or targets could be as shown for domestic buildings.

Further requirements for the building fabric ~

Insulation continuity ~ this requirement is for a fully insulated external envelope with no air gaps in the fabric. Vulnerable places are at junctions between elements of construction, e.g. wall to roof, and around openings such as door and window reveals. Conformity can be shown by producing evidence in the form of a report produced for the local authority building control department by an accredited surveyor. The report must indicate that:

* the approved design specification and construction practice are to an acceptable standard of conformity, OR
* a thermographic survey shows continuity of insulation over the external envelope. This is essential when it is impractical to fully inspect the work in progress.

Air tightness ~ requires that there is no air infiltration through gaps in construction and at the intersection of elements. Permeability of air is tested by using portable fans of capacity to suit the building volume. Smoke capsules in conjunction with air pressurisation will provide a visual indication of air leakage paths.

Thermal Insulation ~ this is required within the roof of all dwellings in the UK. It is necessary to create a comfortable internal environment, to reduce the risk of condensation and to economise in fuel consumption costs.

To satisfy these objectives, insulation may be placed between and over the ceiling joists as shown below to produce a *cold roof* void. Alternatively, the insulation can be located above the rafters as shown on page 435. Insulation above the rafters creates a *warm roof* void and space within the roof structure that may be useful for habitable accommodation.

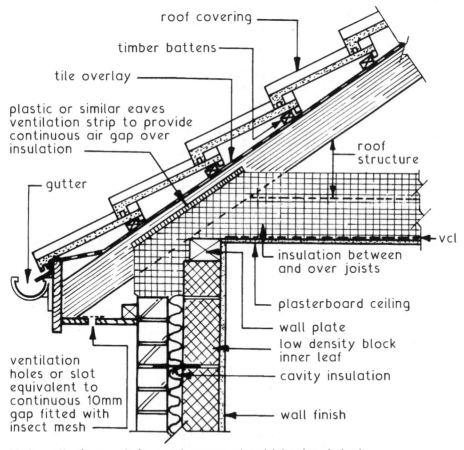

roof covering

timber battens

tile overlay

plastic or similar eaves ventilation strip to provide continuous air gap over insulation

gutter

roof structure

vcl

insulation between and over joists

plasterboard ceiling

wall plate

low density block inner leaf

cavity insulation

ventilation holes or slot equivalent to continuous 10mm gap fitted with insect mesh

wall finish

Note: all pipework in roof space should be insulated to prevent frost attack. The sides and top of cold water storage cisterns should be insulated to prevent freezing.

Thermal insulation to Walls ~ the minimum performance standards for exposed walls set out in Approved Document L to meet the requirements of Part L of the Building Regulations can be achieved in several ways (see pages 478 and 479). The usual methods require careful specification, detail and construction of the wall fabric, insulating material(s) and/or applied finishes.

Typical Examples of existing construction that would require upgrading to satisfy contemporary UK standards ~

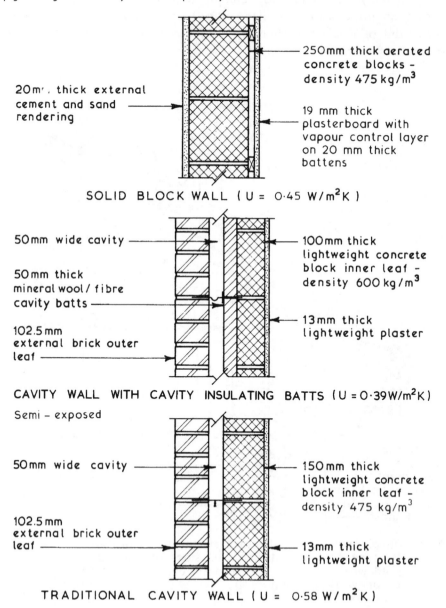

20mm thick external cement and sand rendering

250mm thick aerated concrete blocks – density 475 kg/m^3

19 mm thick plasterboard with vapour control layer on 20 mm thick battens

SOLID BLOCK WALL (U = 0·45 W/m^{2}K)

50mm wide cavity

50mm thick mineral wool/fibre cavity batts

102.5mm external brick outer leaf

100mm thick lightweight concrete block inner leaf – density 600 kg/m^3

13mm thick lightweight plaster

CAVITY WALL WITH CAVITY INSULATING BATTS (U = 0·39W/m^{2}K)

Semi – exposed

50mm wide cavity

102.5mm external brick outer leaf

150mm thick lightweight concrete block inner leaf – density 475 kg/m^3

13mm thick lightweight plaster

TRADITIONAL CAVITY WALL (U = 0·58 W/m^{2}K)

Typical examples of contemporary construction practice that achieve a thermal transmittance or U-value below 0.30 W/m²K ~

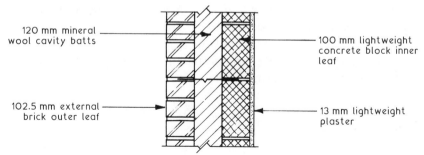

120 mm mineral wool cavity batts

100 mm lightweight concrete block inner leaf

102.5 mm external brick outer leaf

13 mm lightweight plaster

FULL FILL CAVITY WALL. Block density 750 kg/m³   U = 0.25 W/m²K
Block density 600 kg/m³   U = 0.24 W/m²K
Block density 475 kg/m³   U = 0.23 W/m²K

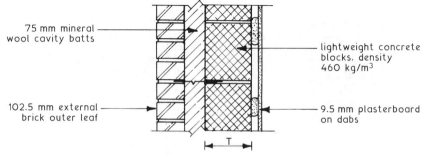

75 mm mineral wool cavity batts

lightweight concrete blocks, density 460 kg/m³

102.5 mm external brick outer leaf

9.5 mm plasterboard on dabs

FULL FILL CAVITY WALL. T = 125 mm   U = 0.28 W/m²K
T = 150 mm   U = 0.26 W/m²K
T = 200 mm   U = 0.24 W/m²K

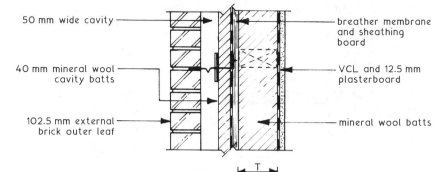

50 mm wide cavity

breather membrane and sheathing board

40 mm mineral wool cavity batts

VCL and 12.5 mm plasterboard

102.5 mm external brick outer leaf

mineral wool batts

TIMBER FRAME PART CAVITY FILL. T = 100 mm   U = 0.26 W/m²K
T = 120 mm   U = 0.24 W/m²K
T = 140 mm   U = 0.21 W/m²K

Note: Mineral wool insulating batts have a typical thermal conductivity (λ) value of 0.038 W/mK.

Thermal or Cold Bridging ~ this is heat loss and possible condensation, occurring mainly around window and door openings and at the junction between ground floor and wall. Other opportunities for thermal bridging occur where uniform construction is interrupted by unspecified components, e.g. occasional use of bricks and/or tile slips to make good gaps in thermal block inner leaf construction.

NB. This practice was quite common, but no longer acceptable by current legislative standards in the UK.

Prime areas for concern ~

WINDOW SILL

WINDOW/DOOR JAMB

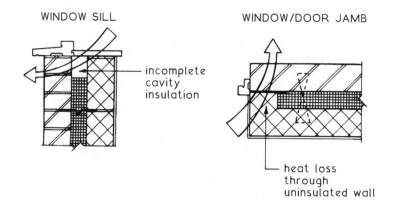

incomplete
cavity
insulation

heat loss
through
uninsulated wall

GROUND FLOOR & WALL

WINDOW/DOOR HEAD

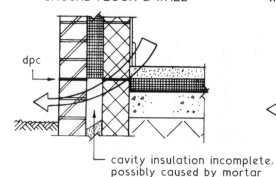

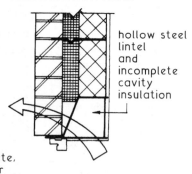

dpc

hollow steel
lintel
and
incomplete
cavity
insulation

cavity insulation incomplete,
possibly caused by mortar
droppings building up and bridging
the lower part of the cavity*

*Note: Cavity should extend down at least 225 mm below the level of the lowest dpc (AD, C: Section 5).

As shown on the preceding page, continuity of insulated construction in the external envelope is necessary to prevent thermal bridging. Nevertheless, some discontinuity is unavoidable where the pattern of construction has to change. For example, windows and doors have significantly higher U-values than elsewhere. Heat loss and condensation risk in these situations is regulated by limiting areas, effectively providing a trade off against very low U-values elsewhere.

The following details should be observed around openings and at ground floor ~

WINDOW SILL

cavity insulation to underside of window board

WINDOW/DOOR JAMB

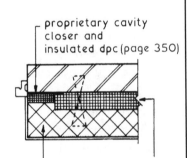

proprietary cavity closer and insulated dpc (page 350)

lightweight insulation blocks

full or part full cavity insulation

GROUND FLOOR & WALL

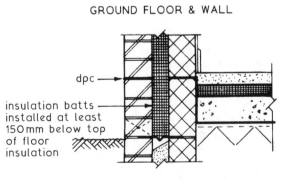

dpc

insulation batts installed at least 150mm below top of floor insulation

WINDOW/DOOR HEAD

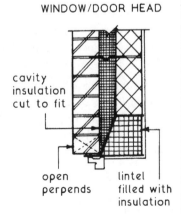

cavity insulation cut to fit

open perpends

lintel filled with insulation

The possibility of a thermal or cold bridge occurring in a specific location can be appraised by calculation. Alternatively, the calculations can be used to determine how much insulation will be required to prevent a cold bridge. The composite lintel of concrete and steel shown below will serve as an example ~

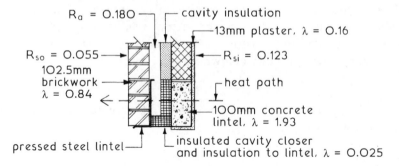

Wall components, less insulation (steel in lintel is insignificant):

102.5 mm brickwork outer leaf,	$\lambda = 0.84$ W/mK
100 mm dense concrete lintel,	$\lambda = 1.93$ ..
13 mm lightweight plaster,	$\lambda = 0.16$ ..

Resistances of above components:

Brickwork,	$0.1025 \div 0.84 = 0.122$ m²K/W
Concrete lintel,	$0.100 \div 1.93 = 0.052$ ..
Lightweight plaster,	$0.013 \div 0.16 = 0.081$ ..

Resistances of surfaces:

$$\text{Internal } (R_{si}) = 0.123 ..$$
$$\text{Cavity } (R_a) = 0.180 ..$$
$$\text{External } (R_{so}) = 0.055 ..$$
$$\text{Summary of resistances} = 0.613 ..$$

To achieve a U-value of say 0.27 W/m²K, total resistance required = $1 \div 0.27 = 3.703$ m²K/W

The insulation in the cavity at the lintel position is required to have a resistance of $3.703 - 0.613 = 3.09$ m²K/W

Using a urethane insulation with a thermal conductivity (λ) of 0.025 W/mK, $0.025 \times 3.09 = 0.077$ m or 77 mm minimum thickness.

If the cavity closer has the same thermal conductivity, then:
Summary of resistance = $0.613 - 0.180$ (Ra) = 0.433 m²K/W
Total resistance required = 3.703 m²K/W, therefore the cavity closer is required to have a resistance of: $3.703 - 0.433 = 3.270$ m²K/W
Min. cavity closer width = 0.025 W/mK × 3.270 m²K/W = 0.082 m or 82 mm.

In practice, the cavity width and the lintel insulation would exceed 82 mm.

Note: data for resistances and λ values taken from pages 469 to 471.

Air Infiltration ~ heating costs will increase if cold air is allowed to penetrate peripheral gaps and breaks in the continuity of construction. Furthermore, heat energy will escape through structural breaks and the following are prime situations for treatment:-

1. Loft hatch
2. Services penetrating the structure
3. Opening components in windows, doors and rooflights
4. Gaps between dry lining and masonry walls

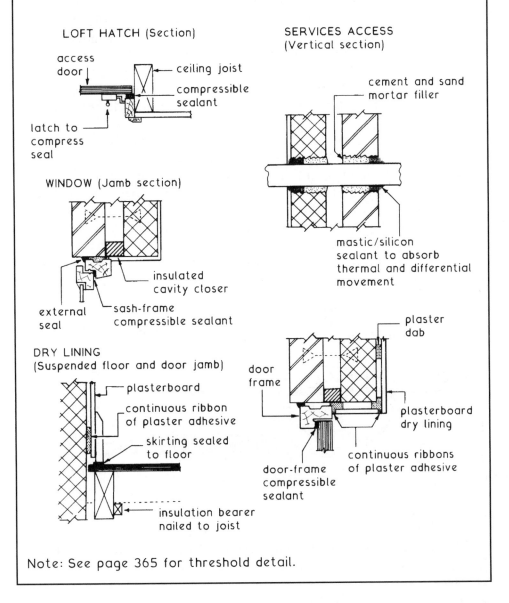

Note: See page 365 for threshold detail.

Main features of Approved Document (A.D.) M: Access to and use of buildings, and other associated guidance –

* Site entrance or car parking space to building entrance to be firm and level. Building approach width 900mm min. A gentle slope is acceptable with a gradient up to 1 in 20 and up to 1 in 40 in cross falls. A slightly steeper ramped access or easy steps should satisfy A.D. Sections 6.14 & 6.15, and 6.16 & 6.17 respectively.
* An accessible threshold for wheelchairs is required at the principal entrance – see illustration.
* Entrance door – minimum clear opening width of 775mm.
* Corridors, passageways and internal doors of adequate width for wheelchair circulation. Minimum 750mm – see also table 1 in A.D. Section 7.
* Stair minimum clear width of 900mm, with provision of handrails both sides. Other requirements as A.D. K for private stairs.
* Accessible light switches, power, telephone and aerial sockets between 450 and 1200mm above floor level.
* WC provision in the entrance storey or first habitable storey. Door to open outwards. Clear wheelchair space of at least 750mm in front of WC and a preferred dimension of 500mm either side of the WC as measured from its centre.
* Special provisions are required for passenger lifts and stairs in blocks of flats, to enable disabled people to access other storeys. See A.D. Section 9 for details.

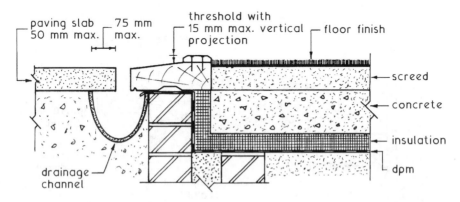

Refs. Accessible thresholds in new housing – Guidance for house builders and designers. The Stationery Office.
BS 8300: Design of buildings and their approaches to meet the needs of disabled people. Code of practice.

Main features ~

* Site entrance, or car parking space to building entrance to be firm and level, ie. maximum gradient 1 in 20 with a minimum car access zone of 1200mm. Ramped and easy stepped approaches are also acceptable.

    * Access to include tactile warnings, ie. profiled (blistered or ribbed) pavings over a width of at least 1200mm, for the benefit of people with impaired vision. Dropped kerbs are required to ease wheelchair use.

    * Special provision for handrails is necessary for those who may have difficulty in negotiating changes in level.

    * Guarding and warning to be provided where projections or obstructions occur, eg. tactile paving could be used around window opening areas.

* Sufficient space for wheelchair manoeuvrability in entrances.

Minimum entrance width of 800mm. Unobstructed space of at least 300mm to the leading (opening) edge of door. Glazed panel in the door to provide visibility from 500 to 1500mm above floor level. Entrance lobby space should be sufficient for a wheelchair user to clear one door before opening another.

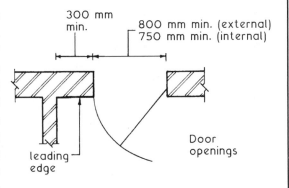

300 mm min.

800 mm min. (external)
750 mm min. (internal)

leading edge

Door openings

* Internal door openings, minimum width 750mm. Unobstructed space of at least 300mm to the leading edge. Visibility panel as above.

continued......

* Main access and internal fire doors that self-close should have a maximum operating force of 20 Newtons at the leading edge. If this is not possible, a power operated door opening and closing system is required.

* Corridors and passageways, minimum unobstructed width 1200 mm. Internal lobbies as described on the previous page for external lobbies.

* Lift dimensions and capacities to suit the building size. Ref. BS EN 81 series: *Safety rules for the construction and installation of lifts*. Alternative vertical access may be by wheelchair stairlift – BS 5776: *Specification for powered stairlifts*, or a platform lift – BS 6440: *Powered lifting platforms for use by disabled persons. Code of practice.*

* Stair minimum width 1200 mm, with step nosings brightly distinguished. Rise maximum 12 risers external, 16 risers internal between landings. Landings to have 1200 mm of clear space from any door swings. Step rise, maximum 170 mm and uniform throughout. Step going, minimum 250 mm (internal), 280 mm (external) and uniform throughout. No open risers. Handrail to each side of the stair.

* Number and location of WCs to reflect ease of access for wheelchair users. In no case should a wheelchair user have to travel more than one storey. Provision may be 'unisex' which is generally more suitable, or 'integral' with specific sex conveniences. Particular provision is outlined in Section 5 of the Approved Document.

* Section 4 of the Approved Document should be consulted for special provisions in restaurants, bars and hotel bedrooms, and for special provisions for spectator seating in theatres, stadia and conference facilities.

Refs. Building Regulations, Approved Document M: Access to and use of buildings.
Disability Discrimination Act.
BS 9999: Code of practice for fire safety in the design, management and use of buildings.
PD 6523: Information on access to and movement within and around buildings and on certain facilities for disabled people.
BS 8300: Design of buildings and their approaches to meet the needs of disabled people. Code of practice.

6 SUPERSTRUCTURE – 2

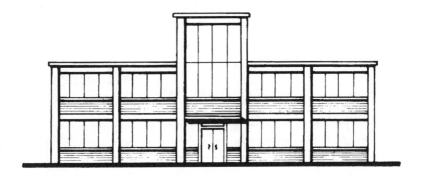

REINFORCED CONCRETE SLABS
REINFORCED CONCRETE FRAMED STRUCTURES
STRUCTURAL CONCRETE FIRE PROTECTION
FORMWORK
PRECAST CONCRETE FRAMES
PRESTRESSED CONCRETE
STRUCTURAL STEELWORK ASSEMBLY
STRUCTURAL STEELWORK CONNECTIONS
STRUCTURAL FIRE PROTECTION
COMPOSITE TIMBER BEAMS
ROOF SHEET COVERINGS
LONG SPAN ROOFS
SHELL ROOF CONSTRUCTION
ROOFLIGHTS
MEMBRANE ROOFS
RAINSCREEN CLADDING
PANEL WALLS AND CURTAIN WALLING
CONCRETE CLADDINGS
CONCRETE SURFACE FINISHES AND DEFECTS

Simply Supported Slabs ~ these are slabs which rest on a bearing and for design purposes are not considered to be fixed to the support and are therefore, in theory, free to lift. In practice however they are restrained from unacceptable lifting by their own self weight plus any loadings.

Concrete Slabs ~ concrete is a material which is strong in compression and weak in tension and if the member is overloaded its tensile resistance may be exceeded leading to structural failure.

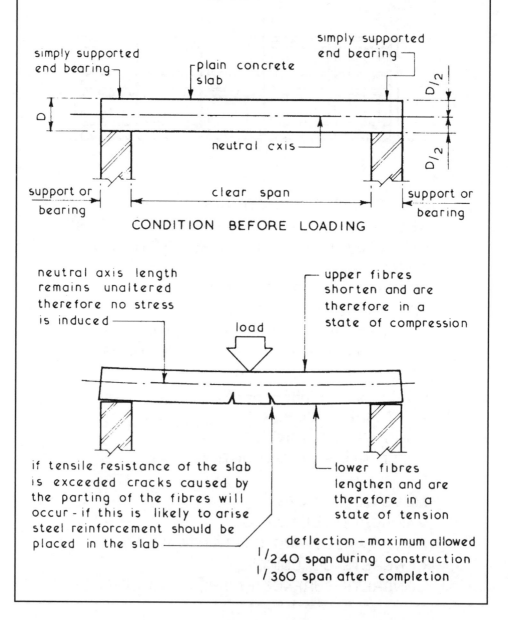

simply supported
end bearing

plain concrete
slab

simply supported
end bearing

D/2

D

neutral axis

D/2

support or
bearing

clear span

support or
bearing

CONDITION BEFORE LOADING

neutral axis length
remains unaltered
therefore no stress
is induced

upper fibres
shorten and are
therefore in a
state of compression

load

if tensile resistance of the slab
is exceeded cracks caused by
the parting of the fibres will
occur - if this is likely to arise
steel reinforcement should be
placed in the slab

lower fibres
lengthen and are
therefore in a
state of tension

deflection – maximum allowed
$^1/240$ span during construction
$^1/360$ span after completion

Reinforcement ~ generally in the form of steel bars which are used to provide the tensile strength which plain concrete lacks. The number, diameter, spacing, shape and type of bars to be used have to be designed; a basic guide is shown on pages 501 and 502. Reinforcement is placed as near to the outside as practicable, with sufficient cover of concrete over the reinforcement to protect the steel bars from corrosion and to provide a degree of fire resistance. Slabs which are square in plan are considered to be spanning in two directions and therefore main reinforcing bars are used both ways whereas slabs which are rectangular in plan are considered to span across the shortest distance and main bars are used in this direction only with smaller diameter distribution bars placed at right angles forming a mat or grid.

Typical Details ~

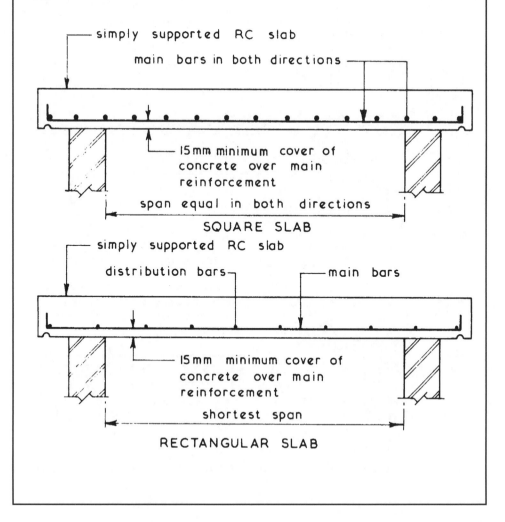

Construction ~ whatever method of construction is used the construction sequence will follow the same pattern-

1. Assemble and erect formwork.
2. Prepare and place reinforcement.
3. Pour and compact or vibrate concrete.
4. Strike and remove formwork in stages as curing proceeds.

Typical Example ~

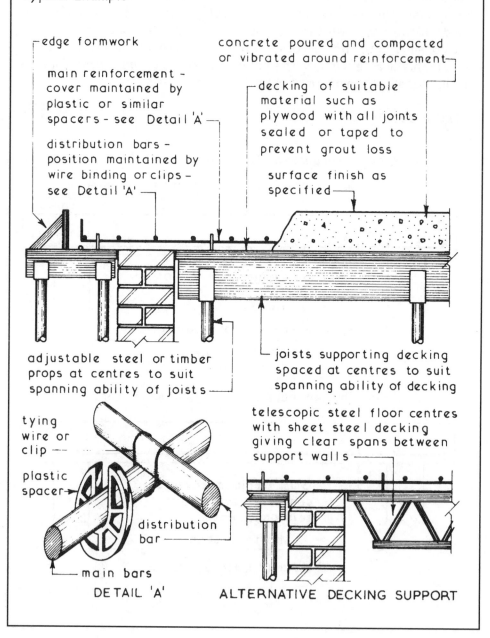

edge formwork

main reinforcement - cover maintained by plastic or similar spacers - see Detail 'A'

distribution bars - position maintained by wire binding or clips - see Detail 'A'

concrete poured and compacted or vibrated around reinforcement

decking of suitable material such as plywood with all joints sealed or taped to prevent grout loss

surface finish as specified

adjustable steel or timber props at centres to suit spanning ability of joists

joists supporting decking spaced at centres to suit spanning ability of decking

tying wire or clip

plastic spacer

distribution bar

main bars

DETAIL 'A'

telescopic steel floor centres with sheet steel decking giving clear spans between support walls

ALTERNATIVE DECKING SUPPORT

Profiled galvanised steel decking is a permanent formwork system for construction of composite floor slabs. The steel sheet has surface indentations and deformities to effect a bond with the concrete topping. The concrete will still require reinforcing with steel rods or mesh, even though the metal section will contribute considerably to the tensile strength of the finished slab.

Typical detail –

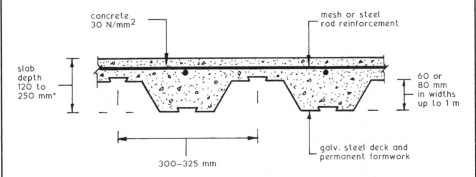

concrete 30 N/mm^2

mesh or steel rod reinforcement

slab depth 120 to 250 mm*

60 or 80 mm in widths up to 1 m

galv. steel deck and permanent formwork

300–325 mm

* For slab depth and span potential, see BS 5950-4: Code of practice for design of composite slabs with profiled steel sheeting.

Where structural support framing is located at the ends of a section and at intermediate points, studs are through-deck welded to provide resistance to shear –

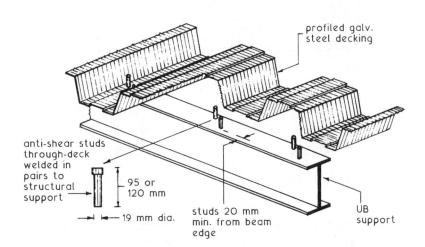

profiled galv. steel decking

anti-shear studs through-deck welded in pairs to structural support

95 or 120 mm

19 mm dia.

studs 20 mm min. from beam edge

UB support

There are considerable savings in concrete volume compared with standard in-situ reinforced concrete floor slabs. This reduction in concrete also reduces structural load on foundations.

Beams ~ these are horizontal load bearing members which are classified as either main beams which transmit floor and secondary beam loads to the columns or secondary beams which transmit floor loads to the main beams.

Concrete being a material which has little tensile strength needs to be reinforced to resist the induced tensile stresses which can be in the form of ordinary tension or diagonal tension (shear). The calculation of the area, diameter, type, position and number of reinforcing bars required is one of the functions of a structural engineer.

Typical RC Beam Details ~

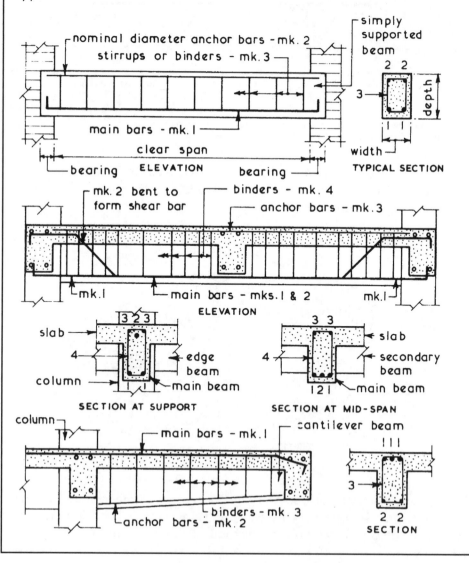

Mild Steel Reinforcement – located in areas where tension occurs in a beam or slab. Concrete specification is normally 25 or 30 N/mm² in this situation.

Simple beam or slab

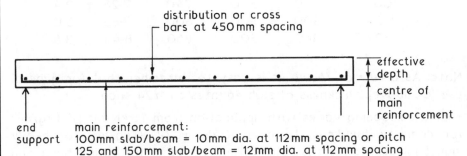

main reinforcement:
100mm slab/beam = 10mm dia. at 112mm spacing or pitch
125 and 150 mm slab/beam = 12mm dia. at 112mm spacing

Continuous beam or slab

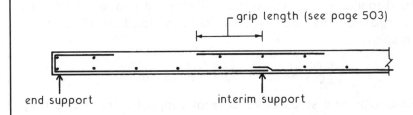

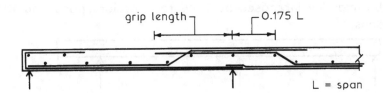

Note: Distribution or cross bars function as lateral reinforcement and supplement the units strength in tensile areas. They also provide resistance to cracking in the concrete as the unit contracts during setting and drying.

Pitch of main bars ≤3 × effective depth.

Pitch of distribution bars ≤5 × effective depth.

Guidance – simply supported slabs are capable of the following loading relative to their thickness:

Thickness (mm)	Self weight (kg/m²)	Imposed load* (kg/m²)	Total load		Span (m)
			(kg/m²)	(kN/m²)	
100	240	500	740	7·26	2·4
125	300	500	800	7·85	3·0
150	360	500	860	8·44	3·6

Note: As a *rule of thumb*, it is easy to remember that for general use (as above), thickness of slab equates to 1/24 span.

* Imposed loading varies with application from 1·5 kN/m² (153 kg/m²) for domestic buildings, to over 10 kN/m² (1020 kg/m²) for heavy industrial storage areas. 500 kg/m² is typical for office filing and storage space. See BS 6399-1: Loading for buildings. Code of practice for dead and imposed loads.

For larger spans – thickness can be increased proportionally to the span, eg. 6 m span will require a 250 mm thickness.

For greater loading – slab thickness is increased proportionally to the square root of the load, eg. for a total load of 1500 kg/m² over a 3 m span:

$$\sqrt{\frac{1500}{800}} \times 125 = 171 \cdot 2 \quad \text{i.e. 175 mm}$$

Continuous beams and slabs have several supports, therefore they are stronger than simple beams and slabs. The spans given in the above table may be increased by 20% for interior spans and 10% for end spans.

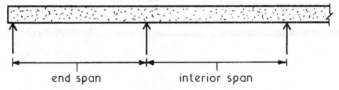

end span   interior span

Deflection limit on reinforced concrete beams and slabs is 1/250 span. Refs. BS 8110-1: Structural use of concrete. Code of practice for design and construction.

BS EN 1992-1-1: Design of concrete structures. General rules and rules for buildings.

See page 546 for deflection formulae.

Bond Between Concrete and Steel – permissible stress for the bond between concrete and steel can be taken as one tenth of the compressive concrete stress, plus $0.175 \, N/mm^2$*. Given the stresses in concrete and steel, it is possible to calculate sufficient grip length.

e.g. concrete working stress of $5 \, N/mm^2$
  steel working stress of $125 \, N/mm^2$
  sectional area of reinf. bar = $3.142 \, r^2$ or $0.7854 \, d^2$
  tensile strength of bar = $125 \times 0.7854 \, d^2$
  circumference of bar = $3.142 \, d$
  area of bar in contact = $3.142 \times d \times L$

Key: r = radius of steel bar
  d = diameter of steel bar
  L = Length of bar in contact

* Conc. bond stress = $(0.10 \times 5 \, N/mm^2) + 0.175 = 0.675 \, N/mm^2$

  Total bond stress = $3.142 \, d \times L \times 0.675 \, N/mm^2$

Thus, developing the tensile strength of the bar:

$$125 \times 0.7854 \, d^2 = 3.142 \, d \times L \times 0.675$$

$$98.175 \, d = 2.120 \, L$$

$$L = 46 \, d$$

As a guide to good practice, a margin of $14 \, d$ should be added to L. Therefore the bar bond or grip length in this example is equivalent to 60 times the bar diameter.

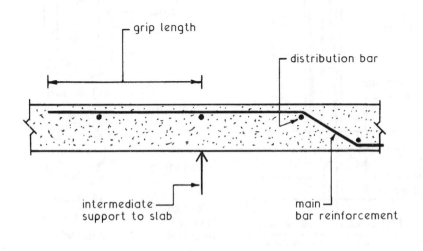

Columns ~ these are the vertical load bearing members of the structural frame which transmits the beam loads down to the foundations. They are usually constructed in storey heights and therefore the reinforcement must be lapped to provide structural continuity.

upper column main bars →

crank - length 300mm minimum

minimum lap – 20 x main bar diameter + 150 mm

75 mm high kicker

main bars – minimum 4 No.

binders in pairs

TYPICAL SECTION

upper floor slab and beams

binders in pairs – minimum diameter 0·25 main bar diameter spacing not more than 12 x main bar diameter

main bars

main bars – minimum 6 No.

helical binding

pitch

ELEVATION

crank

lap

75 mm high kicker

ground floor

starter bars

reinforced concrete foundation

main bars

helical binding

SECTION

CIRCULAR COLUMNS

With the exception of where bars are spliced ~

BEAMS

The distance between any two parallel bars in the horizontal should be not less than the greater of:

* 25 mm
* the bar diameter where they are equal
* the diameter of the larger bar if they are unequal
* 6 mm greater than the largest size of aggregate in the concrete

The distance between successive layers of bars should be not less than the greater of:

* 15 mm (25 mm if bars > 25 mm dia.)
* the maximum aggregate size

An exception is where the bars transverse each other, e.g. mesh reinforcement.

COLUMNS

Established design guides allow for reinforcement of between 0.8% and 8% of column gross cross sectional area. A lesser figure of 0.6% may be acceptable. A relatively high percentage of steel may save on concrete volume, but consideration must be given to the practicalities of placing and compacting wet concrete. If the design justifies a large proportion of steel, it may be preferable to consider using a concrete clad rolled steel I section.

Transverse reinforcement ~ otherwise known as binders or links. These have the purpose of retaining the main longitudinal reinforcement during construction and restraining each reinforcing bar against buckling. Diameter, take the greater of:

* 6 mm
* 0.25 × main longitudinal reinforcement

Spacing or pitch, not more than the lesser of:

* least lateral column dimension
* 12 × diameter of smallest longitudinal reinforcement
* 300 mm

Helical binding ~ normally, spacing or pitch as above, unless the binding has the additional function of restraining the concrete core from lateral expansion, thereby increasing its load carrying potential. This increased load must be allowed for with a pitch:

* not greater than 75 mm
* not greater than 0.166 × core diameter of the column
* not less than 25 mm
* not less than 3 × diameter of the binding steel

Note: Core diameter is measured across the area of concrete enclosed within the centre line of the binding.

Typical RC Column Details ~

Steel Reinforced Concrete – a modular ratio represents the amount of load that a square unit of steel can safely transmit relative to that of concrete. A figure of 18 is normal, with some variation depending on materials specification and quality.

e.g.

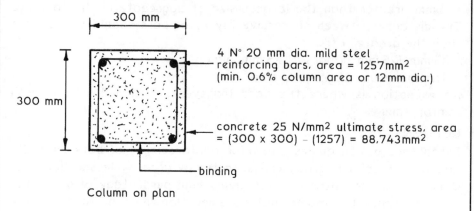

300 mm

300 mm

4 N° 20 mm dia. mild steel reinforcing bars, area = 1257mm² (min. 0.6% column area or 12mm dia.)

concrete 25 N/mm² ultimate stress, area = (300 x 300) – (1257) = 88,743mm²

binding

Column on plan

Area of concrete = 88,743 mm²

Equivalent area of steel = 18 × 1257 mm² = 22626 mm²

Equivalent combined area of concrete and steel:

$$
\begin{array}{r}
88743 \\
+22626 \\
\hline
111369 \ \mathrm{mm^2}
\end{array}
$$

Using concrete with a safe or working stress of 5 N/mm², derived from a factor of safety of 5, i.e.

$$\text{Factory of safety} = \frac{\text{Ultimate stress}}{\text{Working stress}} = \frac{25 \, \text{N/mm}^2}{5 \, \text{N/mm}^2} = 5 \, \text{N/mm}^2$$

5 N/mm² × 111369 mm² = 556845 Newtons
kg × 9·81 (gravity) = Newtons

Therefore: $\dfrac{556845}{9 \cdot 81}$ = 56763 kg or 56·76 tonnes permissible load

Note: This is the safe load calculation for a reinforced concrete column where the load is axial and bending is minimal or nonexistant, due to a very low slenderness ratio (effective length to least lateral dimension). In reality this is unusual and the next example shows how factors for buckling can be incorporated into the calculation.

Buckling or Bending Effect – the previous example assumed total rigidity and made no allowance for column length and attachments such as floor beams.

The working stress unit for concrete may be taken as 0.8 times the maximum working stress of concrete where the effective length of column (see page 548) is less than 15 times its least lateral dimension. Where this exceeds 15, a further factor for buckling can be obtained from the following:

Effective length ÷ Least lateral dimension	Buckling factor
15	1·0
18	0·9
21	0·8
24	0·7
27	0·6
30	0·5
33	0·4
36	0·3
39	0·2
42	0·1
45	0

Using the example from the previous page, with a column effective length of 9 metres and a modular ratio of 18:

Effective length ÷ Least lateral dimension = $9000 \div 300 = 30$

From above table the buckling factor = 0·5

Concrete working stress = $5 \, N/mm^2$

Equivalent combined area of concrete and steel = $111369 \, mm^2$

Therefore: $5 \times 0.8 \times 0.5 \times 111369 = 222738$ Newtons

$$\frac{222738}{9.81} = 22705 \, kg \text{ or } 22.7 \text{ tonnes permissible load}$$

Bar Coding ~ a convenient method for specifying and coordinating the prefabrication of steel reinforcement in the assembly area. It is also useful on site, for checking deliveries and locating materials relative to project requirements. BS EN ISO 3766 provides guidance for a simplified coding system, such that bars can be manufactured and labelled without ambiguity for easy recognition and application on site.

A typical example is the beam shown on page 500, where the lower longitudinal reinforcement (mk·1) could be coded:~

$$2T20\text{-}1\text{-}200B \text{ or, } ①2T\emptyset20\text{-}200\text{-}B\text{-}21$$

$$2 = \text{number of bars}$$
$$T = \text{deformed high yield steel } (460 \text{ N/mm}^2, 8\text{-}40 \text{ mm dia.})$$
$$20 \text{ or, } \emptyset20 = \text{diameter of bar (mm)}$$
$$1 \text{ or } ① = \text{bar mark or ref. no.}$$
$$200 = \text{spacing (mm)}$$
$$B = \text{located in bottom of member}$$
$$21 = \text{shape code}$$

Other common notation:-

$$R = \text{plain round mild steel } (250 \text{ N/mm}^2, 8\text{-}16 \text{ mm dia.})$$
$$S = \text{stainless steel}$$
$$W = \text{wire reinforcement } (4\text{-}12 \text{ mm dia.})$$
$$T \text{ (at the end)} = \text{located in top of member}$$
$$abr = \text{alternate bars reversed (useful for offsets)}$$

Thus, bar mk.2 = 2R10-2-200T or, ②R∅10-200-T-00
and mk.3 = 10R8-3-270 or, ③10R∅8-270-54

All but the most obscure reinforcement shapes are illustrated in the British Standard. For the beam referred to on page 500, the standard listing is:-

BS code	Shape	Total bar length on centre line (mm)
OO	A	A
21	A⌐ r ⌐ C	$A + B + C - r - 2d$ (d = bar diameter) B
54	A ⌐ r ⌐ B	$2(A + B) + 12d$

Ref. BS EN ISO 3766: Construction drawings. Simplified representation of concrete reinforcement.

Bar Schedule ~ this can be derived from the coding explained on the previous page. Assuming 10 No. beams are required:-

Site ref ....   Schedule ref ....

Prepared by ....   Date ....

Member	Bar mark	Type and size	No. of members	No. of bars in each	Total No.	Bar length (mm)	Shape code	A	B	C	D	E/r
Beam	1	T20	10	2	20	3080	21	200	2700	200		
	2	R10	10	2	20	2700	00	2700				
	3	R8	10	10	100	1336	54	400	220			

Note: r = 2 × d for mild steel
3 × d for high yield steel

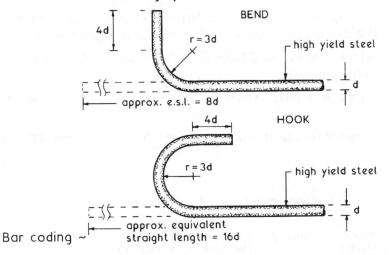

BEND

HOOK

Bar coding ~

Note: 9 is used for special or non-standard shapes

1st character	2nd character
0 No bends	0 Straight bars
1 1 bend	1 90° bends, standard radius, all bends towards same direction
2 2 bends	2 90° bends, non-standard radius, all bends towards same direction
3 3 bends	3 180° bends, non-standard radius, all bends towards same direction
4 4 bends	4 90° bends, standard radius, not all bends towards same direction
5 5 bends	5 Bends <90°, standard radius, all bends towards same direction
6 Arcs of circles	6 Bends <90°, standard radius, not all bends towards same direction
7 Complete helices	7 Arcs or helices

Material ~ Mild steel or high yield steel. Both contain about 99% iron, the remaining constituents are manganese, carbon, sulphur and phosphorus. The proportion of carbon determines the quality and grade of steel; mild steel has 0·25% carbon, high yield steel 0·40%. High yield steel may also be produced by cold working or deforming mild steel until it is strain hardened. Mild steel has the letter R preceding the bar diameter in mm, e.g. R20, and high yield steel the letter T or Y.

Standard bar diameters ~ 6, 8, 10, 12, 16, 20, 25, 32 and 40 mm.

Grade notation ~

• Mild steel – grade 250 or 250 N/mm² characteristic tensile strength (0·25% carbon, 0·06% sulphur and 0·06% phosphorus).
• High yield steel – grade 460/425 (0·40% carbon, 0·05% sulphur and 0·05% phosphorus).

460 N/mm² characteristic tensile strength: 6, 8, 10, 12 and 16 mm diameter.

425 N/mm² characteristic tensile strength: 20, 25, 32 and 40 mm diameter.

Examples of steel reinforcement ~

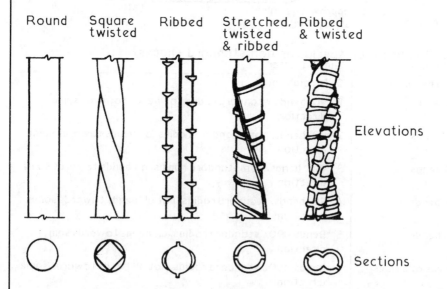

Round	Square twisted	Ribbed	Stretched, twisted & ribbed	Ribbed & twisted	

Elevations

Sections

Ref. BS 4449: Steel for the reinforcement of concrete, weldable reinforcing steel. Bar, coil and decoiled product. Specification.

Steel reinforcement mesh or fabric is produced in four different formats for different applications:

Format	Type	Typical application
A	Square mesh	Floor slabs
B	Rectangular mesh	Floor slabs
C	Long mesh	Roads and pavements
D	Wrapping mesh	Binding wire with concrete fire protection to structural steelwork

Standard sheet size ~ 4·8 m long × 2·4 m wide.
Standard roll size ~ 48 and 72 m long × 2·4 m wide.

Specification ~ format letter plus a reference number. This number equates to the cross sectional area in mm² of the main bars per metre width of mesh.

E.g. B385 is rectangular mesh with 7 mm dia. main bars, i.e. 10 bars of 7 mm dia. @ 100 mm spacing = 385 mm².

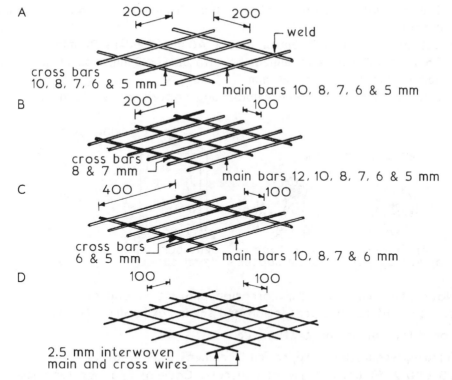

A

200   200

weld

cross bars
10, 8, 7, 6 & 5 mm

main bars 10, 8, 7, 6 & 5 mm

B

200   100

cross bars
8 & 7 mm

main bars 12, 10, 8, 7, 6 & 5 mm

C

400   100

cross bars
6 & 5 mm

main bars 10, 8, 7 & 6 mm

D

100   100

2.5 mm interwoven
main and cross wires

Refs. BS 4483: Steel fabric for the reinforcement of concrete. Specification.

BS 4482: Steel wire for the reinforcement of concrete products. Specification.

Cover to reinforcement in columns, beams, foundations, etc. is required for the following reasons:

- To protect the steel against corrosion.
- To provide sufficient bond or adhesion between steel and concrete.
- To ensure sufficient protection of the steel in a fire (see Note).

If the cover is insufficient, concrete will spall away from the steel.

Minimum cover ~ never less than the maximum size of aggregate in the concrete, or the largest reinforcement bar size (take greater value).
Guidance on minimum cover for particular locations:

Below ground ~

- Foundations, retaining walls, basements, etc., 40 mm, binders 25 mm.
- Marine structures, 65 mm, binders 50 mm.
- Uneven earth and fill 75 mm, blinding 40 mm.

Above ground ~

- Ends of reinforcing bars, not less than 25 mm nor less than 2 × bar diameter.
- Column longitudinal reinforcement 40 mm, binders 20 mm.
- Columns <190 mm min. dimension with bars <12 mm dia., 25 mm.
- Beams 25 mm, binders 20 mm.
- Slabs 20 mm (15 mm where max. aggregate size is <15 mm).

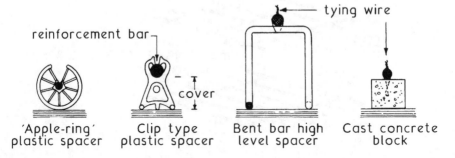

reinforcement bar — tying wire — cover

'Apple-ring' plastic spacer   Clip type plastic spacer   Bent bar high level spacer   Cast concrete block

Note: Minimum cover for corrosion protection and bond may not be sufficient for fire protection and severe exposure situations.

For details of fire protection see ~

Building Regulations, Approved Document B: Fire safety.
BS 8110-2: Structural use of concrete. Code of practice for special circumstances and BS EN 1992-1-2: Design of concrete structures. General rules. Structural fire design.

For general applications, including exposure situations, see next page.

Typical examples using dense concrete of calcareous aggregates (excluding limestone) or siliceous aggregates, eg. flints, quartzites and granites ~

Column fully exposed ~

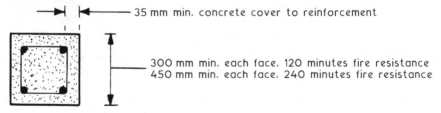

35 mm min. concrete cover to reinforcement

300 mm min. each face, 120 minutes fire resistance
450 mm min. each face, 240 minutes fire resistance

Column, maximum 50% exposed ~

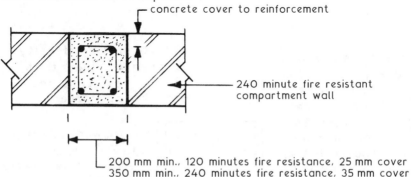

concrete cover to reinforcement

240 minute fire resistant compartment wall

200 mm min., 120 minutes fire resistance, 25 mm cover
350 mm min., 240 minutes fire resistance, 35 mm cover

Column, one face only exposed ~

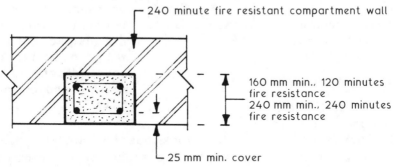

240 minute fire resistant compartment wall

160 mm min., 120 minutes fire resistance
240 mm min., 240 minutes fire resistance

25 mm min. cover

Beam and floor slab ~

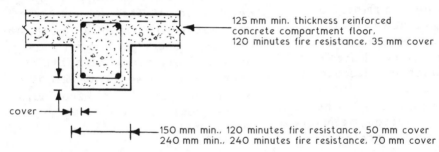

125 mm min. thickness reinforced concrete compartment floor, 120 minutes fire resistance, 35 mm cover

cover

150 mm min., 120 minutes fire resistance, 50 mm cover
240 mm min., 240 minutes fire resistance, 70 mm cover

Formwork ~ concrete when first mixed is a fluid and therefore to form any concrete member the wet concrete must be placed in a suitable mould to retain its shape, size and position as it sets. It is possible with some forms of concrete foundations to use the sides of the excavation as the mould but in most cases when casting concrete members a mould will have to be constructed on site. These moulds are usually called formwork. It is important to appreciate that the actual formwork is the reverse shape of the concrete member which is to be cast.

Falsework ~ the temporary structure which supports the formwork.

Principles ~

formwork sides can be designed to offer all the necessary resistance to the imposed pressures as a single member or alternatively they can be designed to use a thinner material which is adequately strutted — for economic reasons the latter method is usually employed

grout tight joints

formwork soffits can be designed to offer all the necessary resistance to the imposed loads as a single member or alternatively they can be designed to a thinner material which is adequately propped — for economic reasons the latter method is usually employed

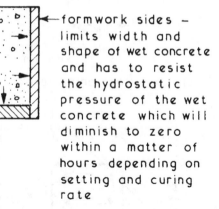

wet concrete — density is greater than that of the resultant set and dry concrete

formwork sides — limits width and shape of wet concrete and has to resist the hydrostatic pressure of the wet concrete which will diminish to zero within a matter of hours depending on setting and curing rate

formwork base or soffit — limits depth and shape of wet concrete and has to resist the initial dead load of the wet concrete and later the dead load of the dry set concrete until it has gained sufficient strength to support its own dead weight which is usually several days after casting depending on curing rate.

Typical Simple Beam Formwork Details ~

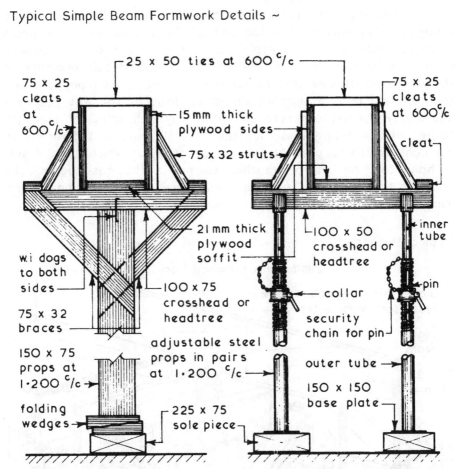

75 × 25 cleats at 600 ᶜ/c

─ 25 × 50 ties at 600 ᶜ/c ─

75 × 25 cleats at 600 ᶜ/c

─ 15mm thick plywood sides

─ 75 × 32 struts ─

cleat

21mm thick plywood soffit

100 × 50 crosshead or headtree

inner tube

w.i dogs to both sides ─

─ collar

─ pin

75 × 32 braces ─

100 × 75 crosshead or headtree

security chain for pin

150 × 75 props at 1·200 ᶜ/c

adjustable steel props in pairs at 1·200 ᶜ/c ─

outer tube ─

150 × 150 base plate ─

folding wedges ─

─ 225 × 75 sole piece ─

SINGLE PROP SUPPORT

DOUBLE PROP SUPPORT

Erecting Formwork

1. Props positioned and levelled through.
2. Soffit placed, levelled and position checked.
3. Side forms placed, their position checked before being fixed.
4. Strutting position and fixed.
5. Final check before casting. Suitable Formwork Materials~ timber, steel and special plastics.

Striking or Removing Formwork

1. Side forms as soon as practicable usually within hours of casting this allows drying air movements to take place around the setting concrete.
2. Soffit formwork as soon as practicable usually within days but as a precaution some props are left in position until concrete member is self supporting.

Beam Formwork ~ this is basically a three sided box supported and propped in the correct position and to the desired level. The beam formwork sides have to retain the wet concrete in the required shape and be able to withstand the initial hydrostatic pressure of the wet concrete whereas the formwork soffit apart from retaining the concrete has to support the initial load of the wet concrete and finally the set concrete until it has gained sufficient strength to be self supporting. It is essential that all joints in the formwork are constructed to prevent the escape of grout which could result in honeycombing and/or feather edging in the cast beam. The removal time for the formwork will vary with air temperature, humidity and consequent curing rate.

Typical Details ~

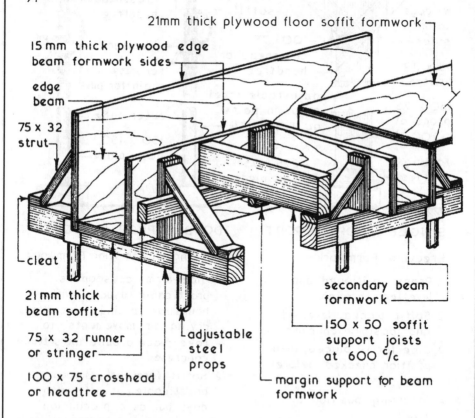

21mm thick plywood floor soffit formwork

15mm thick plywood edge beam formwork sides

edge beam

75 x 32 strut

cleat

21mm thick beam soffit

75 x 32 runner or stringer

100 x 75 crosshead or headtree

adjustable steel props

secondary beam formwork

150 x 50 soffit support joists at 600 c/c

margin support for beam formwork

Typical Formwork Striking Times ~

Beam Sides – 9 to 12 hours
Beam Soffits – 8 to 14 days (props left under) } Using OPC-air temp
Beam Props – 15 to 21 days   } 7 to 16°C

Column Formwork ~ this consists of a vertical mould of the desired shape and size which has to retain the wet concrete and resist the initial hydrostatic pressure caused by the wet concrete. To keep the thickness of the formwork material to a minimum horizontal clamps or yokes are used at equal centres for batch filling and at varying centres for complete filling in one pour. The head of the column formwork can be used to support the incoming beam formwork which gives good top lateral restraint but results in complex formwork. Alternatively the column can be cast to the underside of the beams and at a later stage a collar of formwork can be clamped around the cast column to complete casting and support the incoming beam formwork. Column forms are located at the bottom around a 75 to 100 mm high concrete plinth or kicker which has the dual function of location and preventing grout loss from the bottom of the column formwork.

Typical Details ~

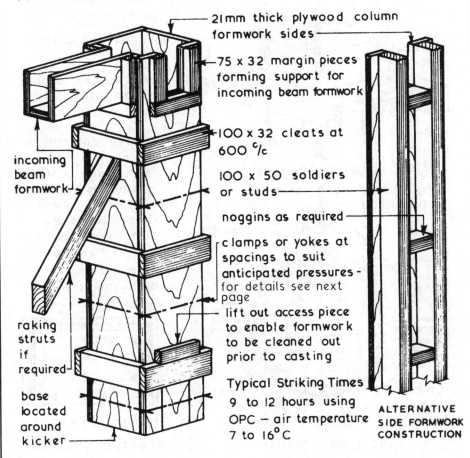

21mm thick plywood column formwork sides

75 x 32 margin pieces forming support for incoming beam formwork

incoming beam formwork

100 x 32 cleats at 600 c/c

100 x 50 soldiers or studs

noggins as required

clamps or yokes at spacings to suit anticipated pressures - for details see next page

lift out access piece to enable formwork to be cleaned out prior to casting

raking struts if required

base located around kicker

Typical Striking Times
9 to 12 hours using OPC — air temperature 7 to 16°C

ALTERNATIVE SIDE FORMWORK CONSTRUCTION

Column Yokes ~ these are obtainable as a metal yoke or clamp or they can be purpose made from timber.

Typical Examples ~

column formwork

security chain or wire

steel blade or arm with 2 rows of 32 x 8 mm slots

metal clamp – available in a range of sizes from 300 to 1400 mm

100 x 32 cleats taken beyond width of panel to form rebate

yoke out of 100 x 75 timber

plate washer to both ends of bolt

steel wedge

hardwood wedges

16 mm diameter bolt

SQUARE COLUMN

gangnail or plywood connecting plates to both faces

12 mm min. gap

shaped timber yokes joined to form half yokes

hardboard or similar lining

25mm thick shaped staves

16 mm diameter bolts

timber yokes out of 200 x 100

CIRCULAR COLUMN

Shaped Columns ~ the basic principles of rectangular or square columns is followed but purpose made shaped yokes are sometimes required. Rebated columns can be formed with blocks or boxing thus —

column formwork

cleat

block or boxing

REBATED COLUMN

518

Wall forms ~ conventionally made up of plywood sheeting that may be steel, plastic or wood faced for specific concrete finishes. Stability is provided by vertical studs and horizontal walings retained in place by adjustable props. Base location is by a kicker of 50 of 75 mm height of width to suit the wall thickness. Spacing of wall forms is shown on the next page.

Wall formwork principles

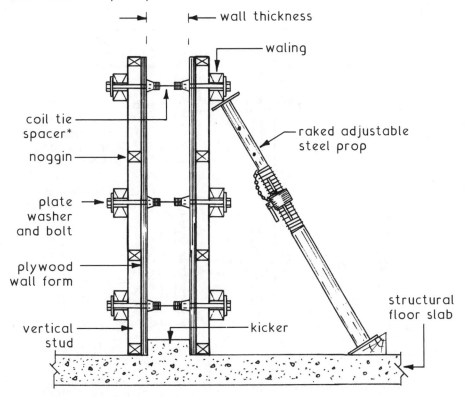

*See next page and page 255

Formwork sides to concrete walls of modest height and load can be positioned with long bolts or threaded dowel bars inserted through the walings on opposing sides. To keep the wall forms apart, tube spacers are placed over the bolts between the forms. For greater load applications, variations include purpose made high tensile steel bolts or dowels. These too are sleeved with plastic tubes and have removable spacer cones inside the forms. Surface voids from the spacers can be made good with strong mortar. Some examples are shown below with the alternative coil tie system. Further applications are shown on pages 255–257.

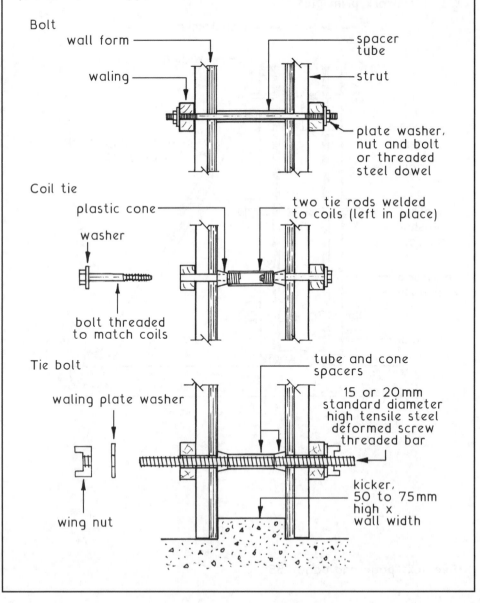

Bolt

wall form

waling

spacer tube

strut

plate washer, nut and bolt or threaded steel dowel

Coil tie

plastic cone

washer

two tie rods welded to coils (left in place)

bolt threaded to match coils

Tie bolt

waling plate washer

tube and cone spacers

15 or 20mm standard diameter high tensile steel deformed screw threaded bar

wing nut

kicker, 50 to 75mm high x wall width

Precast Concrete Frames ~ these frames are suitable for single storey and low rise applications, the former usually in the form of portal frames which are normally studied separately. Precast concrete frames provide the skeleton for the building and can be clad externally and finished internally by all the traditional methods. The frames are usually produced as part of a manufacturer's standard range of designs and are therefore seldom purpose made due mainly to the high cost of the moulds.

Advantages:-

1. Frames are produced under factory controlled conditions resulting in a uniform product of both quality and accuracy.
2. Repetitive casting lowers the cost of individual members.
3. Off site production releases site space for other activities.
4. Frames can be assembled in cold weather and generally by semi-skilled labour.

Disadvantages:-

1. Although a wide choice of frames is available from various manufacturers these systems lack the design flexibility of cast in-situ purpose made frames.
2. Site planning can be limited by manufacturer's delivery and unloading programmes and requirements.
3. Lifting plant of a type and size not normally required by traditional construction methods may be needed.

Typical Site Activities ~

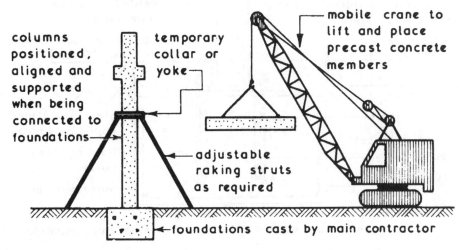

columns positioned, aligned and supported when being connected to foundations

temporary collar or yoke

mobile crane to lift and place precast concrete members

adjustable raking struts as required

foundations cast by main contractor

Foundation Connections ~ the preferred method of connection is to set the column into a pocket cast into a reinforced concrete pad foundation and is suitable for light to medium loadings. Where heavy column loadings are encountered it may be necessary to use a steel base plate secured to the reinforced concrete pad foundation with holding down bolts.

Typical Details ~

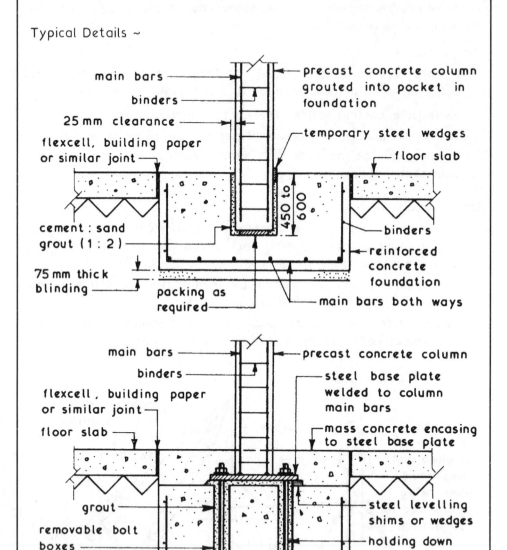

Column to Column Connection ~ precast columns are usually cast in one length and can be up to four storeys in height. They are either reinforced with bar reinforcement or they are prestressed according to the loading conditions. If column to column are required they are usually made at floor levels above the beam to column connections and can range from a simple dowel connection to a complex connection involving in-situ concrete.

Typical Details ~

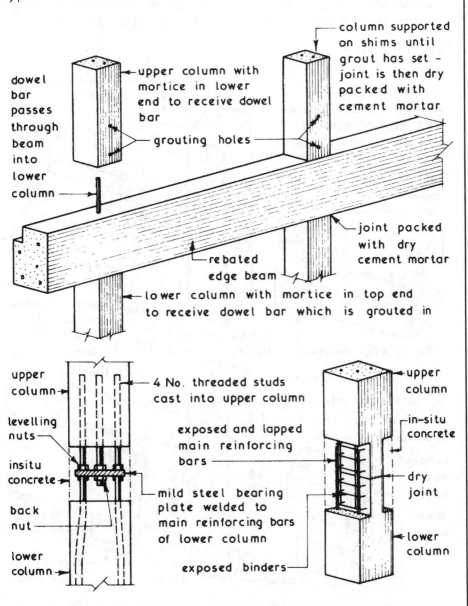

Beam to Column Connections ~ as with the column to column connections (see page 523) the main objective is to provide structural continuity at the junction. This is usually achieved by one of two basic methods:-

1. Projecting bearing haunches cast onto the columns with a projecting dowel or stud bolt to provide both location and fixing.
2. Steel to steel fixings which are usually in the form of a corbel or bracket projecting from the column providing a bolted connection to a steel plate cast into the end of the beam.

Typical Details ~

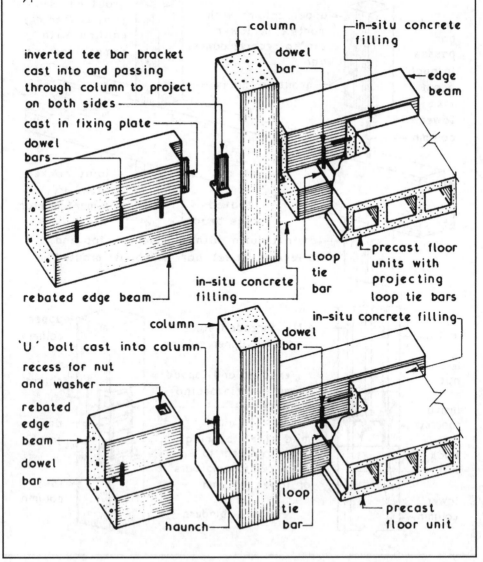

inverted tee bar bracket cast into and passing through column to project on both sides

cast in fixing plate

dowel bars

rebated edge beam

in-situ concrete filling

column

dowel bar

loop tie bar

in-situ concrete filling

edge beam

precast floor units with projecting loop tie bars

'U' bolt cast into column

recess for nut and washer

rebated edge beam

dowel bar

column

dowel bar

haunch

loop tie bar

in-situ concrete filling

precast floor unit

Principles ~ the well known properties of concrete are that it has high compressive strength and low tensile strength. The basic concept of reinforced concrete is to include a designed amount of steel bars in a predetermined pattern to give the concrete a reasonable amount of tensile strength. In prestressed concrete a precompression is induced into the member to make full use of its own inherent compressive strength when loaded. The design aim is to achieve a balance of tensile and compressive forces so that the end result is a concrete member which is resisting only stresses which are compressive. In practice a small amount of tension may be present but providing this does not exceed the tensile strength of the concrete being used tensile failure will not occur.

Comparison of Reinforced and Prestressed Concrete ~

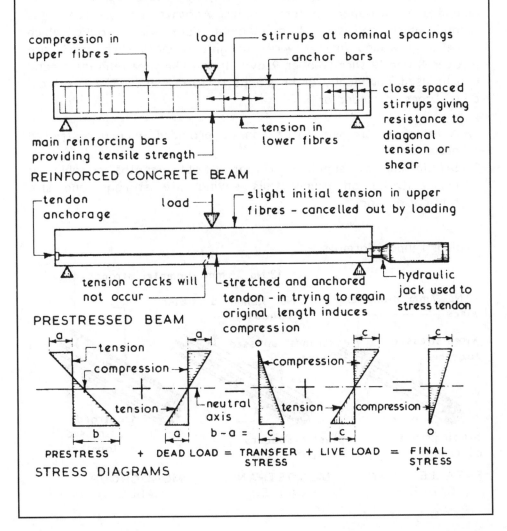

Materials ~ concrete will shrink whilst curing and it can also suffer sectional losses due to creep when subjected to pressure. The amount of shrinkage and creep likely to occur can be controlled by designing the strength and workability of the concrete, high strength and low workability giving the greatest reduction in both shrinkage and creep. Mild steel will suffer from relaxation losses which is where the stresses in steel under load decrease to a minimum value after a period of time and this can be overcome by increasing the initial stress in the steel. If mild steel is used for prestressing the summation of shrinkage, creep and relaxation losses will cancel out any induced compression, therefore special alloy steels must be used to form tendons for prestressed work.

Tendons – these can be of small diameter wires (2 to 7 mm) in a plain round, crimped or indented format, these wires may be individual or grouped to form cables. Another form of tendon is strand which consists of a straight core wire around which is helically wound further wires to give formats such as 7 wire (6 over 1) and 19 wire (9 over 9 over 1) and like wire tendons strand can be used individually or in groups to form cables. The two main advantages of strand are:-

1. A large prestressing force can be provided over a restricted area.
2. Strand can be supplied in long flexible lengths capable of being stored on drums thus saving site storage and site fabrication space.

Typical Tendon Formats ~

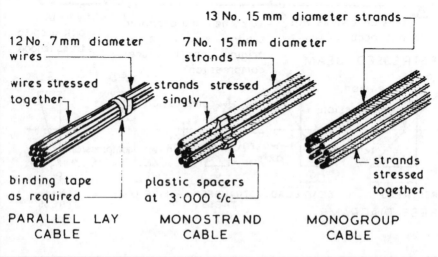

13 No. 15 mm diameter strands

12 No. 7 mm diameter wires

7 No. 15 mm diameter strands

wires stressed together

strands stressed singly

binding tape as required

plastic spacers at 3·000 c/c

strands stressed together

PARALLEL LAY CABLE

MONOSTRAND CABLE

MONOGROUP CABLE

Pre-tensioning ~ this method is used mainly in the factory production of precast concrete components such as lintels, floor units and small beams. Many of these units are formed by the long line method where precision steel moulds up to 120·000 long are used with spacer or dividing plates to form the various lengths required. In pre-tensioning the wires are stressed within the mould before the concrete is placed around them. Steam curing is often used to accelerate this process to achieve a 24 hour characteristic strength of 28 N/mm² with a typical 28 day cube strength of 40 N/mm². Stressing of the wires is carried out by using hydraulic jacks operating from one or both ends of the mould to achieve an initial 10% overstress to counteract expected looses. After curing the wires are released or cut and the bond between the stressed wires and the concrete prevents the tendons from regaining their original length thus maintaining the precompression or prestress.

At the extreme ends of the members the bond between the stressed wires and concrete is not fully developed due to low frictional resistance. This results in a small contraction and swelling at the ends of the wire forming in effect a cone shape anchorage. The distance over which this contraction occurs is called the transfer length and is equal to 80 to 120 times the wire diameter. To achieve a greater total surface contact area it is common practice to use a larger number of small diameter wires rather than a smaller number of large diameter wires giving the same total cross sectional area.

Typical Pre-tensioning Arrangement ~

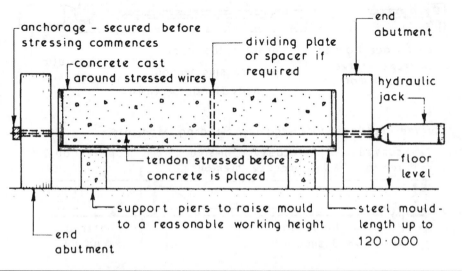

527

Post-tensioning ~ this method is usually employed where stressing is to be carried out on site after casting an in-situ component or where a series of precast concrete units are to be joined together to form the required member. It can also be used where curved tendons are to be used to overcome negative bending moments. In post-tensioning the concrete is cast around ducts or sheathing in which the tendons are to be housed. Stressing is carried out after the concrete has cured by means of hydraulic jacks operating from one or both ends of the member. The anchorages (see next page) which form part of the complete component prevent the stressed tendon from regaining its original length thus maintaining the precompression or prestress. After stressing the annular space in the tendon ducts should be filled with grout to prevent corrosion of the tendons due to any entrapped moisture and to assist in stress distribution. Due to the high local stresses at the anchorage positions it is usual for a reinforcing spiral to be included in the design.

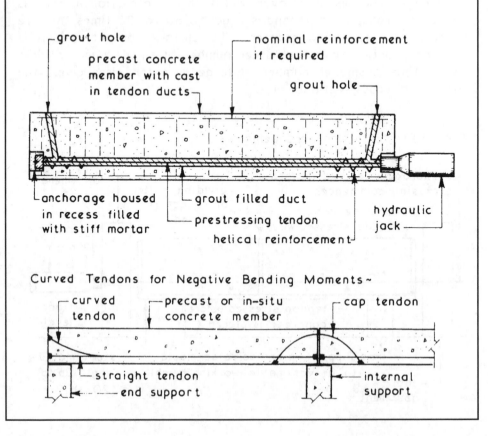

Curved Tendons for Negative Bending Moments ~

Typical Post-tensioning Arrangement ~

Anchorages ~ the formats for anchorages used in conjunction with post-tensioned prestressed concrete works depends mainly on whether the tendons are to be stressed individually or as a group, but most systems use a form of split cone wedges or jaws acting against a form of bearing or pressure plate.

Typical Anchorage Details ~

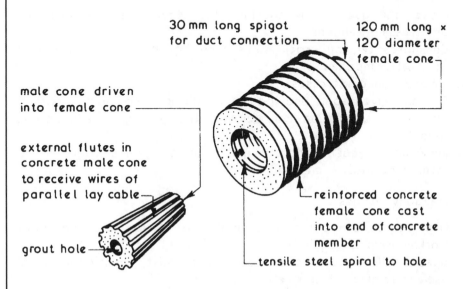

30 mm long spigot for duct connection

120 mm long × 120 diameter female cone

male cone driven into female cone

external flutes in concrete male cone to receive wires of parallel lay cable

grout hole

reinforced concrete female cone cast into end of concrete member

tensile steel spiral to hole

FREYSSINET ANCHORAGE

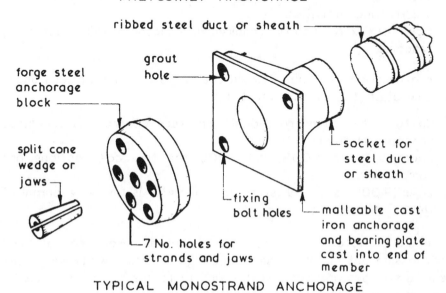

ribbed steel duct or sheath

grout hole

forge steel anchorage block

split cone wedge or jaws

fixing bolt holes

7 No. holes for strands and jaws

socket for steel duct or sheath

malleable cast iron anchorage and bearing plate cast into end of member

TYPICAL MONOSTRAND ANCHORAGE

Comparison with Reinforced Concrete ~ when comparing prestressed concrete with conventional reinforced concrete the main advantages and disadvantages can be enumerated but in the final analysis each structure and/or component must be decided on its own merit.

Main advantages:-

1. Makes full use of the inherent compressive strength of concrete.
2. Makes full use of the special alloy steels used to form the prestressing tendons.
3. Eliminates tension cracks thus reducing the risk of corrosion of steel components.
4. Reduces shear stresses.
5. For any given span and loading condition a component with a smaller cross section can be used thus giving a reduction in weight.
6. Individual precast concrete units can be joined together to form a composite member.

Main Disadvantages:-

1. High degree of control over materials, design and quality of workmanship is required.
2. Special alloy steels are dearer than most traditional steels used in reinforced concrete.
3. Extra cost of special equipment required to carry out the prestressing activities.
4. Cost of extra safety requirements needed whilst stressing tendons.

As a general comparison between the two structural options under consideration it is usually found that:-

1. Up to 6.000 span traditional reinforced concrete is the most economic method.
2. Spans between 6.000 and 9.000 the two cost options are comparable.
3. Over 9.000 span prestressed concrete is more economical than reinforced concrete.

It should be noted that generally columns and walls do not need prestressing but in tall columns and high retaining walls where the bending stresses are high, prestressing techniques can sometimes be economically applied.

Ground Anchors ~ these are a particular application of post-tensioning prestressing techniques and can be used to form ground tie backs to cofferdams, retaining walls and basement walls. They can also be used as vertical tie downs to basement and similar slabs to prevent flotation during and after construction. Ground anchors can be of a solid bar format (rock anchors) or of a wire or cable format for granular and cohesive soils. A lined or unlined bore hole must be drilled into the soil to the design depth and at the required angle to house the ground anchor. In clay soils the bore hole needs to be underreamed over the anchorage length to provide adequate bond. The tail end of the anchor is pressure grouted to form a bond with the surrounding soil, the remaining length being unbonded so that it can be stressed and anchored at head thus inducing the prestress. The void around the unbonded or elastic length is gravity grouted after completion of the stressing operation.

Typical Ground Anchor Details ~

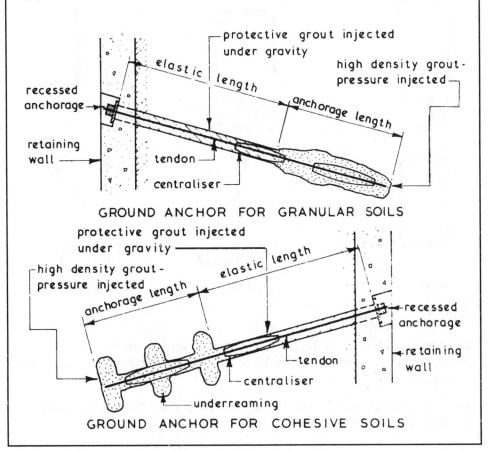

GROUND ANCHOR FOR GRANULAR SOILS

GROUND ANCHOR FOR COHESIVE SOILS

Cold rolled steel sections are a lightweight alternative to the relatively heavy, hot rolled steel sections that have been traditionally used in sub-framing situations, e.g. purlins, joists and sheeting rails. Cold rolled sections are generally only a few millimetres in wall thickness, saving on material and handling costs and building dead load. They are also produced in a wide variety of section profiles, some of which are shown below.

Typical section profiles ~

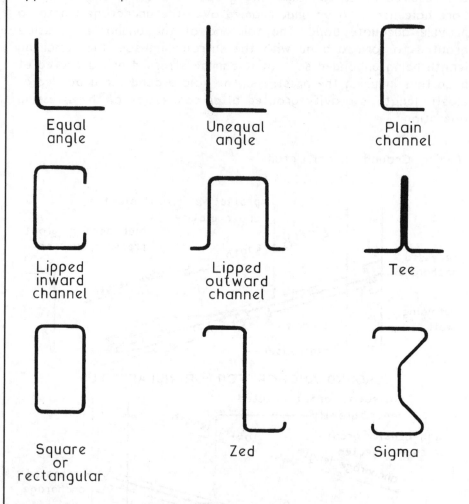

Equal angle

Unequal angle

Plain channel

Lipped inward channel

Lipped outward channel

Tee

Square or rectangular

Zed

Sigma

Dimensions vary considerably and many non-standard profiles are made for particular situations. A range of standard sections are produced to:

BS EN 10162: Cold rolled steel sections. Technical delivery conditions. Dimensional and cross sectional tolerances.

Structural Steelwork ~ standard section references:

BS 4-1: Structural steel sections. Specification for hot rolled sections.

BS EN 10056: Specification for structural steel equal and unequal angles.

BS EN 10210: Hot finished structural hollow sections of non-alloy and fine grain steels.

Typical Standard Steelwork Sections ~

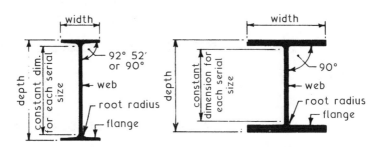

UNIVERSAL BEAMS
127 x 76 x 13 kg/m to
914 x 419 x 388 kg/m

UNIVERSAL COLUMNS
152 x 152 x 23 kg/m to
356 x 406 x 634 kg/m

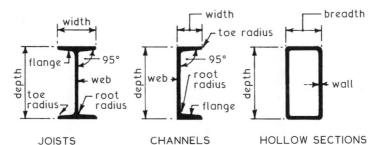

JOISTS
76 x 76 x 13 kg/m to
254 x 203 x 82 kg/m

CHANNELS
100 x 50 x 10 kg/m to
430 x 100 x 64 kg/m

HOLLOW SECTIONS
50 x 30 x 2.89 kg/m to
500 x 300 x 191 kg/m

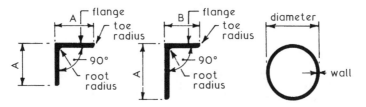

EQUAL ANGLES
25 x 25 x 1.2 kg/m to
200 x 200 x 71.1 kg/m

UNEQUAL ANGLES
40 x 25 x 1.91 kg/m to
200 x 150 x 47.1 kg/m

HOLLOW SECTIONS
21.3 dia. x 1.43 kg/m to
508 dia. x 194 kg/m

NB. Sizes given are serial or nominal, for actual sizes see relevant BS.

Compound Sections – these are produced by welding together standard sections. Various profiles are possible, which can be designed specifically for extreme situations such as very high loads and long spans, where standard sections alone would be insufficient. Some popular combinations of standard sections include:

BEAM or COLUMN

steel plate welded to joists

standard joist, beam or column section

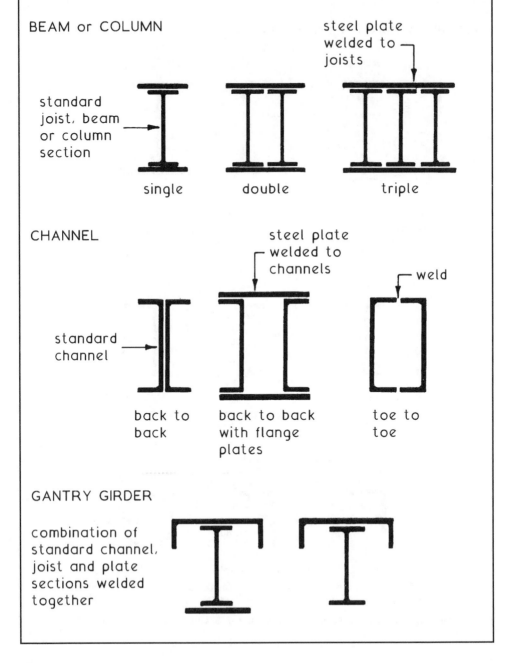

single   double   triple

CHANNEL

steel plate welded to channels

weld

standard channel

back to back

back to back with flange plates

toe to toe

GANTRY GIRDER

combination of standard channel, joist and plate sections welded together

Open Web Beams – these are particularly suited to long spans with light to moderate loading. The relative increase in depth will help resist deflection and voids in the web will reduce structural dead load.

Perforated Beam – a standard beam section with circular voids cut about the neutral axis.

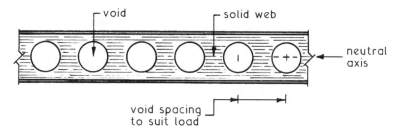

Castellated Beam – a standard beam section web is profile cut into two by oxy-acetylene torch. The projections on each section are welded together to create a new beam 50% deeper than the original.

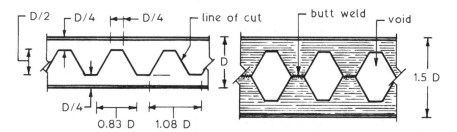

Litzka Beam – a standard beam cut as the castellated beam, but with overall depth increased further by using spacer plates welded to the projections. Minimal increase in weight.

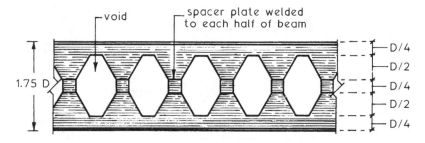

Note: Voids at the end of open web beams should be filled with a welded steel plate, as this is the area of maximum shear stress in a beam.

Lattices – these are an alternative type of open web beam, using standard steel sections to fabricate high depth to weight ratio units capable of spans up to about 15 m. The range of possible components is extensive and some examples are shown below:

PLATE GIRDER

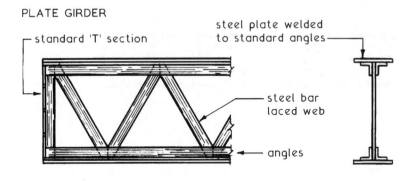

TUBULAR LATTICE

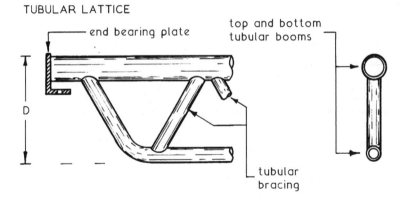

DOWELLED LATTICE

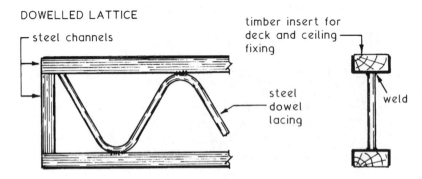

Note: span potential for lattice beams is approximately 24 × D

Structural Steelwork Connections ~ these are either workshop or site connections according to where the fabrication takes place. Most site connections are bolted whereas workshop connections are very often carried out by welding. The design of structural steelwork members and their connections is the province of the structural engineer who selects the type and number of bolts or the size and length of weld to be used according to the connection strength to be achieved.

Typical Connection Examples ~

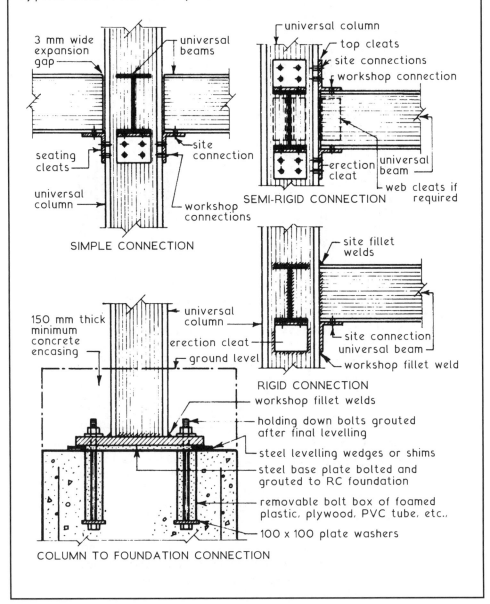

SIMPLE CONNECTION

SEMI-RIGID CONNECTION

RIGID CONNECTION

COLUMN TO FOUNDATION CONNECTION

Typical Connection Examples ~

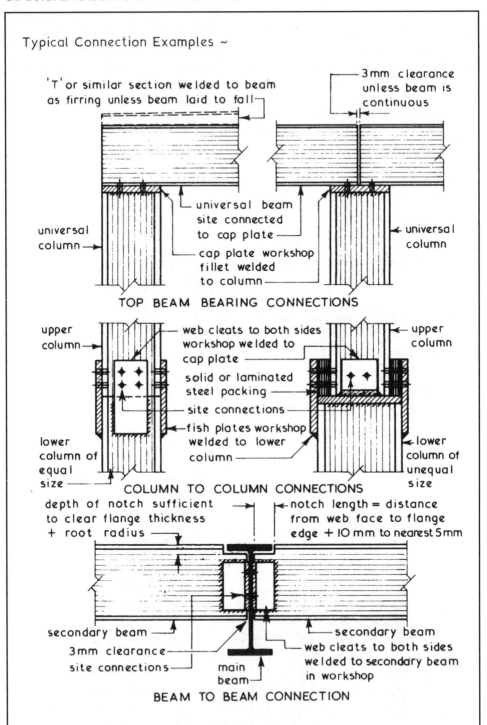

'T' or similar section welded to beam as firring unless beam laid to fall

3mm clearance unless beam is continuous

universal column

universal beam site connected to cap plate

cap plate workshop fillet welded to column

universal column

TOP BEAM BEARING CONNECTIONS

upper column

web cleats to both sides workshop welded to cap plate

solid or laminated steel packing

site connections

fish plates workshop welded to lower column

upper column

lower column of equal size

lower column of unequal size

COLUMN TO COLUMN CONNECTIONS

depth of notch sufficient to clear flange thickness + root radius

notch length = distance from web face to flange edge + 10 mm to nearest 5mm

secondary beam

3mm clearance

site connections

main beam

secondary beam

web cleats to both sides welded to secondary beam in workshop

BEAM TO BEAM CONNECTION

NB. All holes for bolted connections must be made from backmarking the outer surface of the section(s) involved. For actual positions see structural steelwork manuals.

Types ~
Slab or bloom base.
Gusset base.
Steel grillage (see page 225).

The type selected will depend on the load carried by the column and the distribution area of the base plate. The cross sectional area of a UC concentrates the load into a relatively small part of the base plate. Therefore to resist bending and shear, the base must be designed to resist the column loads and transfer them onto the pad foundation below.

SLAB or BLOOM BASE   SLAB BASE WITH ANGLE CLEATS

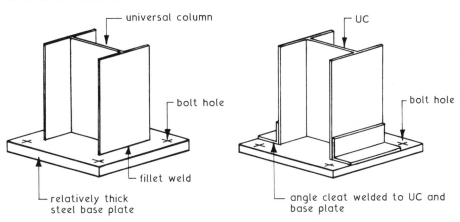

universal column — | — UC

bolt hole | bolt hole

fillet weld

relatively thick steel base plate | angle cleat welded to UC and base plate

GUSSET BASE   BOLT BOX

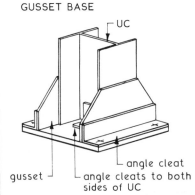
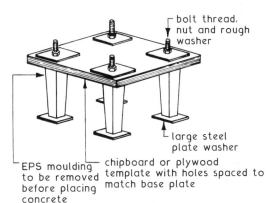

UC

bolt thread, nut and rough washer

angle cleat

gusset | angle cleats to both sides of UC

large steel plate washer

EPS moulding to be removed before placing concrete | chipboard or plywood template with holes spaced to match base plate

Bolt Box ~ a template used to accurately locate column holding down bolts into wet concrete. EPS or plastic tubes provide space around the bolts when the concrete has set. The bolts can then be moved slightly to aid alignment with the column base.

Welding is used to prefabricate the sub-assembly of steel frame components in the workshop, prior to delivery to site where the convenience of bolted joints will be preferred.

Oxygen and acetylene (oxy-acetylene) gas welding equipment may be used to fuse together light steel sections, but otherwise it is limited to cutting profiles of the type shown on page 535. The electric arc process is preferred as it is more effective and efficient. This technique applies an expendable steel electrode to fuse parts together by high amperage current. The current potential and electrode size can be easily changed to suit the thickness of metal.

Overlapping of sections permits the convenience of fillet welds, but if the overlap is obstructive or continuity and direct transfer of loads is necessary, a butt weld will be specified. To ensure adequate weld penetration with a butt weld, the edges of the parent metal should be ground to produce an edge chamfer. For very large sections, both sides of the material should be chamfered to allow for double welds.

BUTT WELD

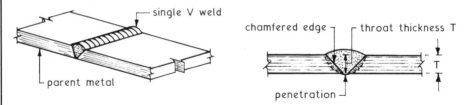

Note: For greater thicknesses of parent metal, both sides are chamfered in preparation for a double weld.

FILLET WELD

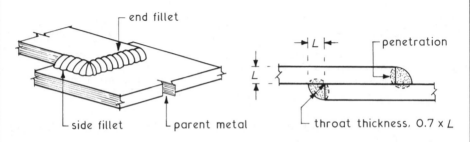

Ref. BS EN 1011-1 and 2: Welding. Recommendations for welding of metallic materials.

Bolts are the preferred method for site assembly of framed building components, although rivets have been popular in the past and will be found when working on existing buildings. Cold driven and 'pop' rivets may be used for factory assembly of light steel frames such as stud walling, but the traditional process of hot riveting structural steel both in the workshop and on site has largely been superseded for safety reasons and the convenience of other practices.

Types of Bolt ~

1. Black Bolts ~ the least expensive and least precise type of bolt, produced by forging with only the bolt and nut threads machined. Clearance between the bolt shank and bolt hole is about 2 mm, a tolerance that provides for ease of assembly. However, this imprecision limits the application of these bolts to direct bearing of components onto support brackets or seating cleats.

2. Bright Bolts ~ also known as turned and fitted bolts. These are machined under the bolt head and along the shank to produce a close fit of 0.5 mm hole clearance. They are specified where accuracy is paramount.

3. High Strength Friction Grip Bolts ~ also known as torque bolts as they are tightened to a predetermined shank tension by a torque controlled wrench. This procedure produces a clamping force that transfers the connection by friction between components and not by shear or bearing on the bolts. These bolts are manufactured from high-yield steel. The number of bolts used to make a connection is less than otherwise required.

Refs.
BS 4190: ISO metric black hexagon bolts, screws and nuts. Specification.

BS 3692: ISO metric precision hexagon bolts, screws and nuts. Specification.

BS 4395 (2 parts): Specification for high strength friction grip bolts and associated nuts and washers for structural engineering.

BS EN 14399 (8 parts): High strength structural bolting assemblies for preloading.

Fire Resistance of Structural Steelwork ~ although steel is a non-combustible material with negligible surface spread of flame properties it does not behave very well under fire conditions. During the initial stages of a fire the steel will actually gain in strength but this reduces to normal at a steel temperature range of 250 to 400°C and continues to decrease until the steel temperature reaches 550°C when it has lost most of its strength. Since the temperature rise during a fire is rapid, most structural steelwork will need protection to give it a specific degree of fire resistance in terms of time. Part B of the Building Regulations sets out the minimum requirements related to building usage and size, BRE Report 128 'Guidelines for the construction of fire resisting structural elements' gives acceptable methods.

Typical Examples for 120 minutes Fire Resistance ~

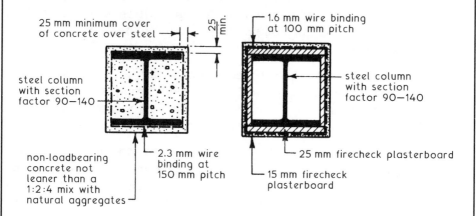

25 mm minimum cover of concrete over steel

25 min.

1.6 mm wire binding at 100 mm pitch

steel column with section factor 90—140

steel column with section factor 90—140

non-loadbearing concrete not leaner than a 1:2:4 mix with natural aggregates

2.3 mm wire binding at 150 mm pitch

25 mm firecheck plasterboard

15 mm firecheck plasterboard

SOLID PROTECTION   HOLLOW PROTECTION

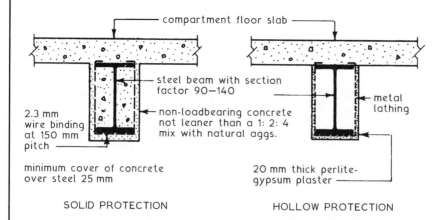

compartment floor slab

steel beam with section factor 90—140

2.3 mm wire binding at 150 mm pitch

non-loadbearing concrete not leaner than a 1: 2: 4 mix with natural aggs.

metal lathing

minimum cover of concrete over steel 25 mm

20 mm thick perlite-gypsum plaster

SOLID PROTECTION   HOLLOW PROTECTION

For section factor calculations see next page.

Section Factors – these are criteria found in tabulated fire protection data such as the Loss Prevention Certification Board's Standards. These factors can be used to establish the minimum thickness or cover of protective material for structural sections. This interpretation is usually preferred by buildings insurance companies, as it often provides a standard in excess of the Building Regulations. Section factors are categorised: <90, 90 – 140 and >140. They can be calculated by the following formula:

Section Factor = Hp/A (m⁻¹)

Hp = Perimeter of section exposed to fire (m)

A = Cross sectional area of steel (m²) [see BS 4-1 or Structural Steel Tables]

Examples:

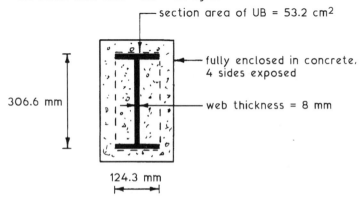

UB serial size, 305 × 127 × 42 kg/m

section area of UB = 53.2 cm²

fully enclosed in concrete, 4 sides exposed

306.6 mm

web thickness = 8 mm

124.3 mm

Hp = (2 × 124·3) + (2 × 306·6) + 2(124·3 – 8) = 1·0944 m
A = 53·2 cm² or 0·00532 m²
Section Factor, Hp/A = 1·0944/0·00532 = 205

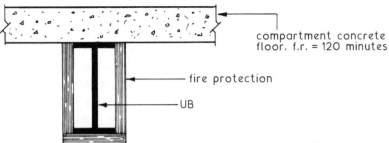

As beam above, but 3 sides only exposed

compartment concrete floor, f.r. = 120 minutes

fire protection

UB

Hp = 124·3 + (2 × 306·6) + 2(124·3 – 8) = 0·9701 m
A = 53·2 cm² or 0·00532 m²
Section Factor, Hp/A = 0·9701/0·00532 = 182

References: BS 4-1: Structural steel sections. Specification for hot-rolled sections.

BS 449-2: Specification for the use of structural steel in building. Metric units.

BS 5950-1: Structural use of steelwork in building. Code of practice for design. Rolled and welded sections.

BS EN 1993-1: Eurocode 3. Design of steel structures.

Simple beam design (Bending)

Formula:

$$Z = \frac{M}{f}$$

where: Z = section or elastic modulus (BS 4-1)

M = moment of resistance > or = max. bending moment

f = fibre stress of the material, (normally 165 N/mm² for rolled steel sections)

In simple situations the bending moment can be calculated:-

(a) Point loads

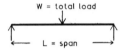

$$BM = \frac{WL}{4}$$

(b) Distributed loads

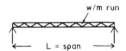

$$BM = \frac{wL^2}{8} \quad or \quad \frac{WL}{8}$$
$$where \ W = w \times L$$

eg.

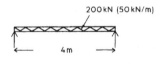

$$BM = \frac{wL^2}{8} = \frac{50 \times 4^2}{8} = 100kNm$$

$$Z = \frac{M}{f} = \frac{100 \times 10^6}{165} = 606cm^3$$

eg.

From structural design tables, e.g. BS 4-1, a Universal Beam 305 × 127 × 48 kg/m with section modulus (Z) of 612·4 cm³ about the x-x axis, can be seen to satisfy the calculated 606 cm³.

Note: Total load in kN can be established by summating the weight of materials – see BS 648: Schedule of Weights of Building Materials, and multiplying by gravity; i.e. kg × 9·81 = Newtons.

This must be added to any imposed loading:-

People and furniture = 1·5 kN/m²

Snow on roofs < 30° pitch = 1·5 kN/m²

Snow on roofs > 30° pitch = 0·75 kN/m²

Simple beam design (Shear)

From the previous example, the section profile is:-

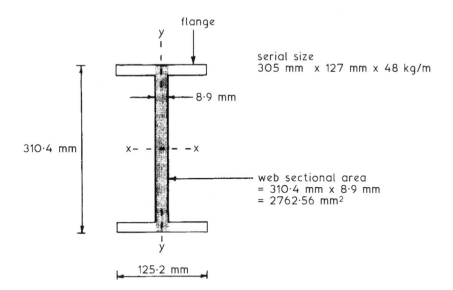

flange

y

serial size
305 mm x 127 mm x 48 kg/m

8·9 mm

310·4 mm

x— — — —x

web sectional area
= 310·4 mm x 8·9 mm
= 2762·56 mm²

y

125·2 mm

Maximum shear force normally occurs at the support points, i.e. near the end of the beam. Calculation is made of the average stress value on the web sectional area.

Using the example of 200 kN load distributed over the beam, the maximum shear force at each end support will be 100 kN.

$$\text{Therefore, the average shear stress} = \frac{\text{shear force}}{\text{web sectional area}}$$

$$= \frac{100 \times 10^3}{2762 \cdot 56}$$

$$= 36 \cdot 20 \text{ N/mm}^2$$

Grade S275 steel has an allowable shear stress in the web of 110 N/mm². Therefore the example section of serial size: 305 mm × 127 mm × 48 kg/m with only 36·20 N/mm² calculated average shear stress is more than capable of resisting the applied forces.

Grade S275 steel has a characteristic yield stress of 275 N/mm² in sections up to 40 mm thickness. This grade is adequate for most applications, but the more expensive grade S355 steel is available for higher stress situations.

Ref. BS EN 10025: Hot rolled products of structural steels.

Simple beam design (Deflection)

The deflection due to loading, other than the weight of the structure, should not exceed 1/360 of the span.

The formula to determine the extent of deflection varies, depending on:-

(a) Point loading

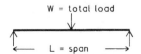

$$\text{Deflection} = \frac{WL^3}{48EI}$$

(b) Uniformly distributed loading

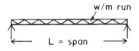

$$\text{Deflection} = \frac{5WL^3}{384EI}$$

where: W = load in kN
L = span in cm
E = Young's modulus of elasticity (typically 21,000 kN/cm² for steel)
I = 2nd moment of area about the x-x axis (see BS 4-1)

Using the example of 200 kN uniformly distributed over a 4 m span:-

$$\text{Deflection} = \frac{5WL^3}{384EI} = \frac{5 \times 200 \times 4^3 \times 100^3}{384 \times 21000 \times 9504} = 0\cdot835\text{cm}$$

Permissible deflection is 1/360 of 4 m = 11·1 mm or 1·11 cm.

Therefore actual deflection of 8·35 mm or 0·835 cm is acceptable.

Ref. BS 5950-1: Structural use of steelwork in building. Code of practice for design. Rolled and welded sections.

Simple column design

Steel columns or stanchions have a tendency to buckle or bend under extreme loading. This can be attributed to:

(a) length
(b) cross sectional area
(c) method of end fixing, and
(d) the shape of section.
(b) and (d) are incorporated into a geometric property of section, known as the radius of gyration (r). It can be calculated:-

$$r = \sqrt{\frac{I}{A}}$$

where: I = 2nd moment of area
A = cross sectional area
Note: r,I and A are all listed in steel design tables, eg. BS 4-1.

The radius of gyration about the y-y axis is used for calculation, as this is normally the weaker axis.

The length of a column will affect its slenderness ratio and potential to buckle. This length is calculated as an effective length relative to the method of fixing each end. Examples of position and direction fixing are shown on the next page. eg. A Universal Column 203 mm × 203 mm × 46 kg/m, 10 m long, position and direction fixed both ends. Determine the maximum axial loading.

Effective length (l) = 0.7 × 10 m = 7 m
(r) from BS 4-1 = 51·1 mm

Slenderness ratio = $\frac{l}{r} = \frac{7 \times 10^3}{51.1} = 137$

From tables in BS 449-2, the maximum allowable stress for grade S275 steel with slenderness ratio of 137 is 48 N/mm²

Cross sectional area of stanchion (UC) = 5880 mm² (BS 4-1)

The total axial load = $\frac{48 \times 5880}{10^3}$ = 283 kN (approx. 28 tonnes)

The tendency for a column to buckle depends on its slenderness as determined by the ratio of its effective length to the radius of gyration about the weaker axis.

Effective lengths of columns and struts in compression ~

End conditions	Effective length relative to actual length
Restrained both ends in position and direction	0.70
Restrained both ends in position with one end in direction	0.85
Restrained both ends in position but not in direction	1.00
Restrained one end in position and direction. The other end restrained in direction only	1.50
Restrained one end in position and direction. The other end unrestrained	2.00

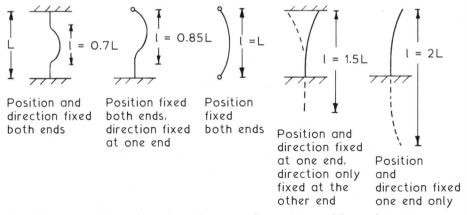

Position and direction fixed both ends

Position fixed both ends, direction fixed at one end

Position fixed both ends

Position and direction fixed at one end, direction only fixed at the other end

Position and direction fixed one end only

Position and direction fixed is location at specific points by beams or other means of retention. Position fixed only means hinged or pinned.

The effective lengths shown apply to the simplest of structures, for example, single storey buildings. Where variations occur such as cross braced support to columns, corrective factors should be applied to the effective length calculations. This data is available in BS 5950-1: Structural use of steelwork in building. Code of practice for design. Rolled and welded sections.

Portal Frames ~ these can be defined as two dimensional rigid frames which have the basic characteristic of a rigid joint between the column and the beam. The main objective of this form of design is to reduce the bending moment in the beam thus allowing the frame to act as one structural unit. The transfer of stresses from the beam to the column can result in a rotational movement at the foundation which can be overcome by the introduction of a pin or hinge joint. The pin or hinge will allow free rotation to take place at the point of fixity whilst transmitting both load and shear from one member to another. In practice a true 'pivot' is not always required but there must be enough movement to ensure that the rigidity at the point of connection is low enough to overcome the tendency of rotational movement.

Typical Single Storey Portal Frame Formats ~

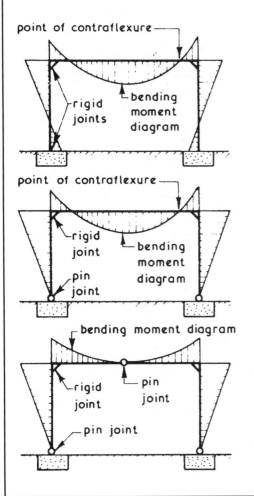

FIXED or RIGID PORTAL FRAME:-

all joints or connections are rigid giving lower bending moments than other formats. Used for small to medium span frames where moments at foundations are not excessive.

TWO PIN PORTAL FRAME:-

pin joints or hinges used at foundation connections to eliminate tendency of base to rotate. Used where high base moments and weak ground are encountered.

THREE PIN PORTAL FRAME:-

pin joints or hinges used at foundation connections and at centre of beam which reduces bending moment in beam but increases deflection. Used as an alternative to a 2 pin frame.

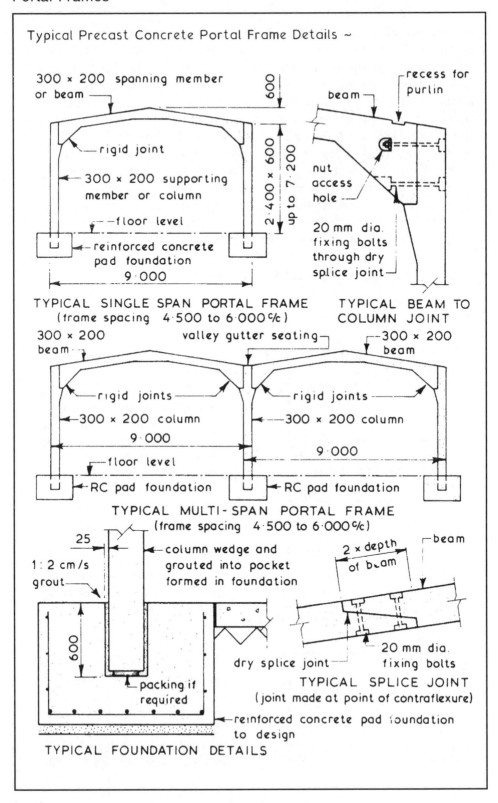

Typical Precast Concrete Portal Frame Details ~

300 × 200 spanning member or beam

rigid joint

300 × 200 supporting member or column

floor level

reinforced concrete pad foundation

9·000

600

2·400 × 600 up to 7·200

TYPICAL SINGLE SPAN PORTAL FRAME
(frame spacing 4·500 to 6·000 c/c)

beam

recess for purlin

nut access hole

20 mm dia. fixing bolts through dry splice joint

TYPICAL BEAM TO COLUMN JOINT

300 × 200 beam

valley gutter seating

300 × 200 beam

rigid joints

rigid joints

300 × 200 column

9·000

300 × 200 column

9·000

floor level

RC pad foundation

RC pad foundation

TYPICAL MULTI-SPAN PORTAL FRAME
(frame spacing 4·500 to 6·000 c/c)

25

1:2 cm/s grout

600

column wedge and grouted into pocket formed in foundation

packing if required

reinforced concrete pad foundation to design

TYPICAL FOUNDATION DETAILS

2 × depth of beam

beam

dry splice joint

20 mm dia. fixing bolts

TYPICAL SPLICE JOINT
(joint made at point of contraflexure)

550

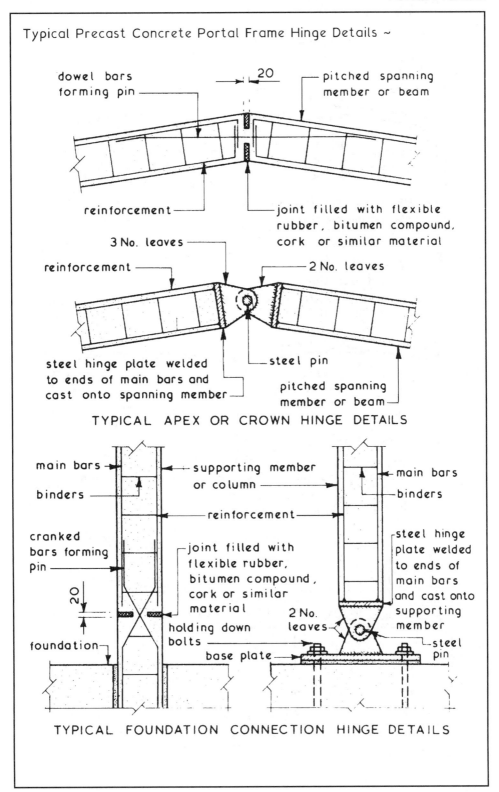

Typical Precast Concrete Portal Frame Hinge Details ~

dowel bars forming pin

20

pitched spanning member or beam

reinforcement

joint filled with flexible rubber, bitumen compound, cork or similar material

3 No. leaves

reinforcement

2 No. leaves

steel hinge plate welded to ends of main bars and cast onto spanning member

steel pin

pitched spanning member or beam

TYPICAL APEX OR CROWN HINGE DETAILS

main bars

binders

supporting member or column

main bars

binders

cranked bars forming pin

reinforcement

steel hinge plate welded to ends of main bars and cast onto supporting member

20

foundation

joint filled with flexible rubber, bitumen compound, cork or similar material

holding down bolts

base plate

2 No. leaves

steel pin

TYPICAL FOUNDATION CONNECTION HINGE DETAILS

551

Typical Steel Portal Frame Details ~

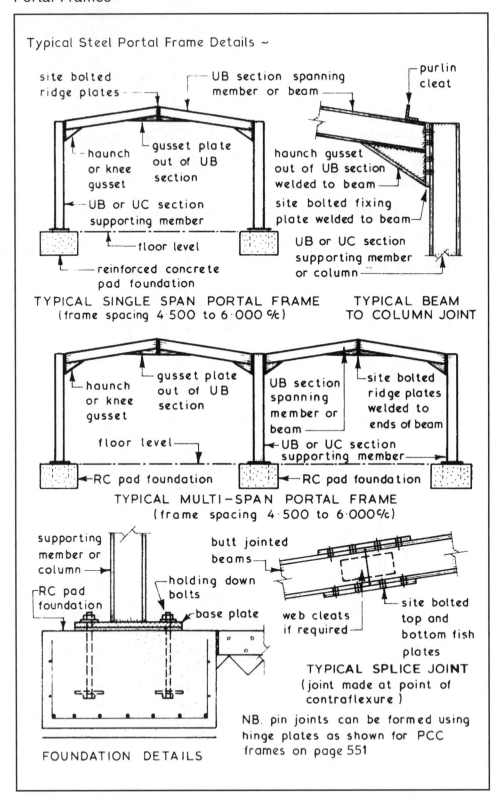

TYPICAL SINGLE SPAN PORTAL FRAME
(frame spacing 4·500 to 6·000 ℅)

TYPICAL BEAM
TO COLUMN JOINT

TYPICAL MULTI−SPAN PORTAL FRAME
(frame spacing 4·500 to 6·000 ℅)

FOUNDATION DETAILS

TYPICAL SPLICE JOINT
(joint made at point of
contraflexure)

NB. pin joints can be formed using
hinge plates as shown for PCC
frames on page 551

Laminated Timber ~ sometimes called 'Gluelam' and is the process of building up beams, ribs, arches, portal frames and other structural units by gluing together layers of timber boards so that the direction of the grain of each board runs parallel with the longitudinal axis of the member being fabricated.

Laminates ~ these are the layers of board and may be jointed in width and length.

Joints ~

Width – joints in consecutive layers should lap twice the board thickness or one quarter of its width whichever is the greater.

Length – scarf and finger joints can be used. Scarf joints should have a minimum slope of 1 in 12 but this can be steeper (say 1 in 6) in the compression edge of a beam:-

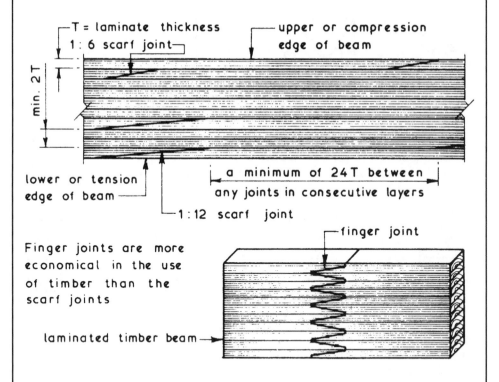

Moisture Content ~ timber should have a moisture content equal to that which the member will reach in service and this is known as its equilibrium moisture content; for most buildings this will be between 11 and 15%. Generally at the time of gluing timber should not exceed 15 ±3% in moisture content.

Vertical Laminations ~ not often used for structural laminated timber members and is unsatisfactory for curved members.

vertical laminates ─────

top edge

beam face

Horizontal Laminations ~ most popular method for all types of laminated timber members. The stress diagrams below show that laminates near the upper edge are subject to a compressive stress whilst those near the lower edge to a tensile stress and those near the neutral axis are subject to shear stress.

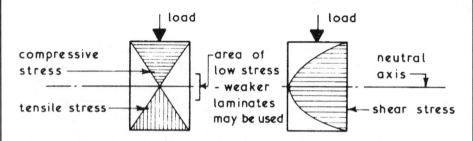

compressive stress ─────

tensile stress ─────

load

area of low stress - weaker laminates may be used

load

neutral axis ─────

shear stress ─────

Flat sawn timber shrinks twice as much as quarter sawn timber therefore flat and quarter sawn timbers should not be mixed in the same member since the different shrinkage rates will cause unacceptable stresses to occur on the glue lines.

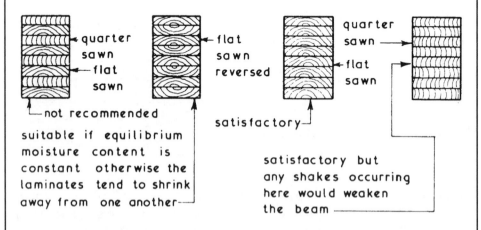

quarter sawn
flat sawn

not recommended
suitable if equilibrium moisture content is constant otherwise the laminates tend to shrink away from one another

flat sawn reversed

satisfactory

quarter sawn
flat sawn

satisfactory but any shakes occurring here would weaken the beam

Planing ~ before gluing, laminates should be planed so that the depth of the planer cutter marks are not greater than 0·025 mm.

Gluing ~ this should be carried out within 48 hours of the planing operation to reduce the risk of the planed surfaces becoming contaminated or case hardened (for suitable adhesives see page 556). Just before gluing up the laminates they should be checked for `cupping.´ The amount of cupping allowed depends upon the thickness and width of the laminates and has a range of 0·75 mm to 1·5 mm.

Laminate Thickness ~ no laminate should be more than 50 mm thick since seasoning up to this thickness can be carried out economically and there is less chance of any individual laminate having excessive cross grain strength.

Straight Members – laminate thickness is determined by the depth of the member, there must be enough layers to allow the end joints (i.e. scarf or finger joints – see page 553) to be properly staggered.

Curved Members – laminate thickness is determined by the radius to which the laminate is to be bent and the species together with the quality of the timber being used. Generally the maximum laminate thickness should be 1/150 of the sharpest curve radius although with some softwoods 1/100 may be used.

Typical Laminated Timber Curved Member ~

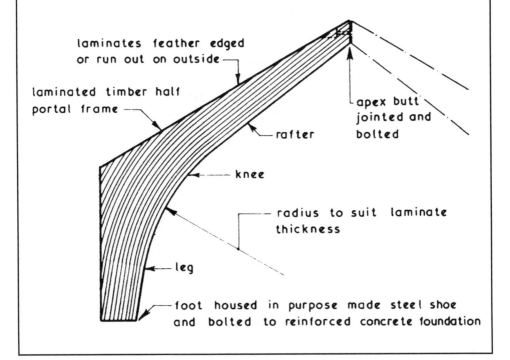

laminates feather edged
or run out on outside

laminated timber half
portal frame

rafter

apex butt
jointed and
bolted

knee

radius to suit laminate
thickness

leg

foot housed in purpose made steel shoe
and bolted to reinforced concrete foundation

Adhesives ~ although timber laminates are carefully machined, the minimum of cupping permitted and efficient cramping methods employed it is not always possible to obtain really tight joints between the laminates. One of the important properties of the adhesive is therefore that it should be gap filling. The maximum permissible gap being 1.25 mm.

There are four adhesives suitable for laminated timber work which have the necessary gap filling property and they are namely:-

1. Casein – the protein in milk, extracted by coagulation and precipitation. It is a cold setting adhesive in the form of a powder which is mixed with water, it has a tendency to stain timber and is only suitable for members used in dry conditions of service.
2. Urea Formaldehyde – this is a cold setting resin glue formulated to MR/GF (moisture resistant/gap filling). Although moisture resistant it is not suitable for prolonged exposure in wet conditions and there is a tendency for the glue to lose its strength in temperatures above 40°C such as when exposed to direct sunlight. The use of this adhesive is usually confined to members used in dry, unexposed conditions of service. This adhesive will set under temperatures down to 10°C.
3. Resorcinol Formaldehyde – this is a cold setting glue formulated to WBP/GF (weather and boilproof/gap filling). It is suitable for members used in external situations but is relatively expensive. This adhesive will set under temperatures down to 15°C and does not lose its strength at high temperatures.
4. Phenol Formaldehyde – this is a similar glue to resorcinol formaldehyde but is a warm setting adhesive requiring a temperature of above 86°C in order to set. Phenol/resorcinol formaldehyde is an alternative, having similar properties to, but less expensive than resorcinol formaldehyde. PRF needs a setting temperature of at least 23°C.

Preservative Treatment – this can be employed if required, provided that the pressure impregnated preservative used is selected with regard to the adhesive being employed. See also page 464.

Ref. BS EN 301: Adhesives, phenolic and aminoplastic, for load-bearing timber structures. Classification and performance requirements.

Composite Beams ~ stock sizes of structural softwood have sectional limitations of about 225 mm and corresponding span potential in the region of 6 m. At this distance, even modest loadings could interpose with the maximum recommended deflection of 0·003 × span.

Fabricated softwood box, lattice and plywood beams are an economic consideration for medium spans. They are produced with adequate depth to resist deflection and with sufficient strength for spans into double figures. The high strength to weight ratio and simple construction provides advantages in many situations otherwise associated with steel or reinforced concrete, e.g. frames, trusses, beams and purlins in gymnasia, workshops, garages, churches, shops, etc. They are also appropriate as purlins in loft conversion.

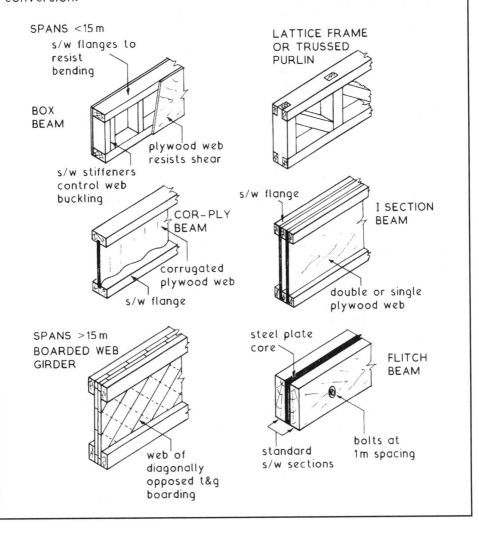

SPANS <15 m
s/w flanges to resist bending

BOX BEAM

plywood web resists shear

s/w stiffeners control web buckling

LATTICE FRAME OR TRUSSED PURLIN

COR-PLY BEAM

corrugated plywood web

s/w flange

s/w flange

I SECTION BEAM

double or single plywood web

SPANS >15 m
BOARDED WEB GIRDER

web of diagonally opposed t&g boarding

steel plate core

FLITCH BEAM

standard s/w sections

bolts at 1m spacing

PSB ~ otherwise known as a parallam beam. Fabricated from long strands of softwood timber bonded with a phenol-formaldehyde adhesive along the length of the beam to produce a structural section of greater strength than natural timber of equivalent section. Used for beams, lintels, structural framing and trimmer sections around floor openings in spans up to 20 m. Can also be used vertically as columns.

Standard sizes ~ range from 200 × 45 mm up to 406 × 178 mm.

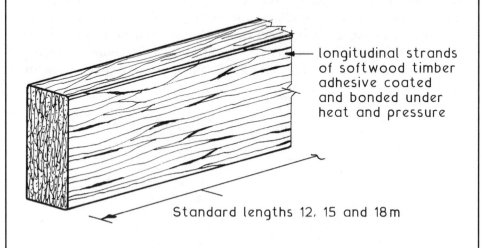

longitudinal strands of softwood timber adhesive coated and bonded under heat and pressure

Standard lengths 12, 15 and 18 m

Typical grade stresses [N/mm^2] (compare with sw timber, page 115) ~

Bending parallel to grain	16.8
Tension parallel to grain	14.8
Compression parallel to grain	15.1
Compression perpendicular to grain	3.6
Shear parallel to grain	2.2
Modulus of elasticity (mean)	12750
Average density	740 kg/m^3

Variation ~ for spans in excess of 20 m or for high loads, a flitch beam as shown on the previous page can be made by bolting a steel plate (typically 10 or 12 mm) between two PSBs.

Ref. BBA Agrément Certificate No. 92/2813.

Composite Joist ~ a type of lattice frame, constructed from a pair of parallel and opposing stress graded softwood timber flanges, separated and jointed with a web of V shaped galvanised steel plate connectors. Manufacture is off-site in a factory quality controlled situation. Here, joists can be made in standard or specific lengths to order. Depending on loading, spans to about 8 m are possible at joist spacing up to 600 mm.

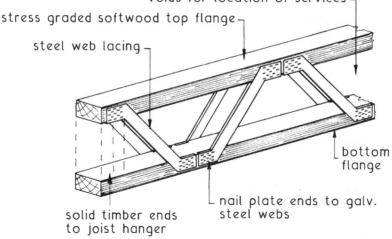

3mm dia. nails at 200 mm c/c, length 2.5 x board thickness

18 mm chipboard floor decking with joists at 450 mm spacing (22 mm @ 600 mm) pva glued in the t & g and onto joists

typically 200, 225, 250 or 300

ex. 100 x 50 or 75 x 50

galv. steel web lacing

12.5 mm plasterboard ceiling

3mm screws, 42 mm long at 150 mm c/c

voids for location of services

stress graded softwood top flange

steel web lacing

bottom flange

solid timber ends to joist hanger

nail plate ends to galv. steel webs

End bearing on inner leaf of cavity wall, silicone sealed to maintain air tightness. Alternatively lateral restraint type joist hanger support or intermedite support from a steel beam.

Advantages over solid timber joists:

Large span to depth ratio.
High strength to weight ratio.
Alternative applications, including roof members, purlins, etc.
Generous space for services without holing or notching.
Minimal movement and shrinkage.
Wide flanges provide large bearing area for decking and ceiling board.

Multi-storey Structures ~ these buildings are usually designed for office, hotel or residential use and contain the means of vertical circulation in the form of stairs and lifts occupying up to 20% of the floor area. These means of circulation can be housed within a core inside the structure and this can be used to provide a degree of restraint to sway due to lateral wind pressures (see next page).

Typical Basic Multi-storey Structure Types ~

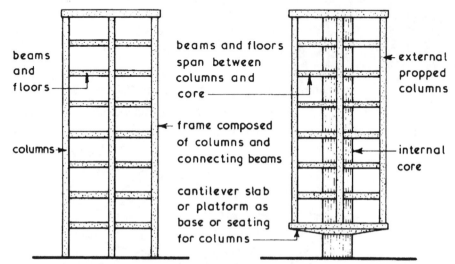

beams and floors

columns

beams and floors span between columns and core

← frame composed of columns and connecting beams

cantilever slab or platform as base or seating for columns

external propped columns

internal core

TRADITIONAL FRAMED STRUCTURES   PROPPED STRUCTURES

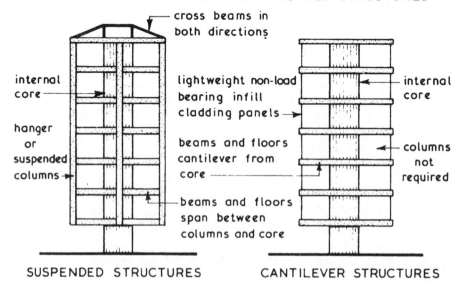

cross beams in both directions

internal core

hanger or suspended columns

lightweight non-load bearing infill cladding panels

beams and floors cantilever from core

beams and floors span between columns and core

internal core

columns not required

SUSPENDED STRUCTURES   CANTILEVER STRUCTURES

Typical Multi-storey Structures ~ the formats shown below are designed to provide lateral restraint against wind pressures.

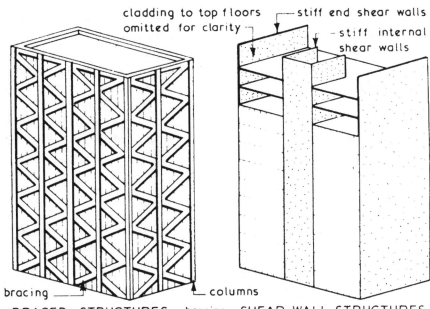

cladding to top floors omitted for clarity —

— stiff end shear walls

— stiff internal shear walls

bracing ——   — columns

BRACED STRUCTURES - bracing used to give stability so that columns can be designed as pure compression members.

SHEAR WALL STRUCTURES - wind pressures transmitted from cladding to shear walls by floors.

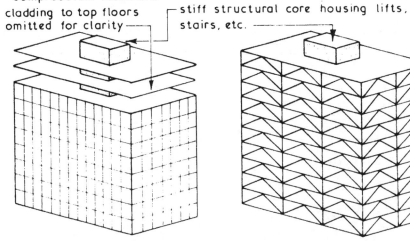
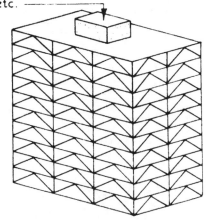

cladding to top floors omitted for clarity ——

— stiff structural core housing lifts, stairs, etc. ——

CORE STRUCTURES - wind pressures transmitted from cladding to core by floors.

HULL CORE STRUCTURES - rigid and braced framework called the hull acts with core through floors to form a rigid structure.

Steel Roof Trusses ~ these are triangulated plane frames which carry purlins to which the roof coverings can be fixed. Steel is stronger than timber and will not spread fire over its surface and for these reasons it is often preferred to timber for medium and long span roofs. The rafters are restrained from spreading by being connected securely at their feet by a tie member. Struts and ties are provided within the basic triangle to give adequate bracing. Angle sections are usually employed for steel truss members since they are economic and accept both tensile and compressive stresses. The members of a steel roof truss are connected together with bolts or by welding to shaped plates called gussets. Steel trusses are usually placed at 3·000 to 4·500 centres which gives an economic purlin size.

Typical Steel Roof Truss Formats ~

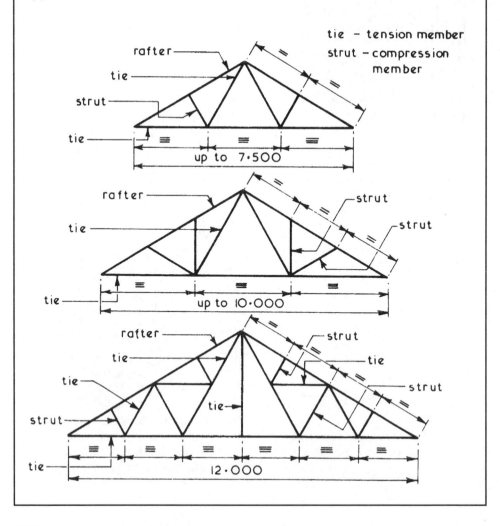

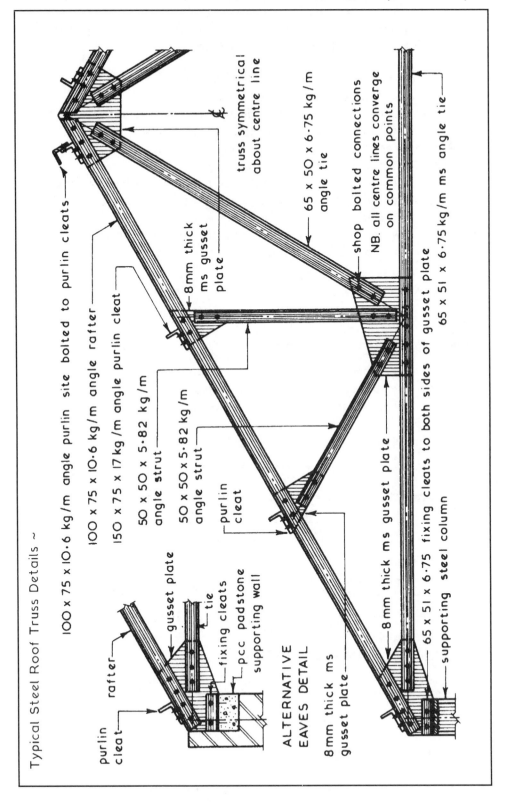

Typical Steel Roof Truss Details ~

100 x 75 x 10·6 kg/m angle purlin site bolted to purlin cleats

100 x 75 x 10·6 kg/m angle rafter

150 x 75 x 17 kg/m angle purlin cleat

50 x 50 x 5·82 kg/m angle strut

50 x 50 x 5·82 kg/m angle strut

purlin cleat

truss symmetrical about centre line

65 x 50 x 6·75 kg/m angle tie

shop bolted connections
NB. all centre lines converge on common points

65 x 51 x 6·75 kg/m ms angle tie

8mm thick ms gusset plate

8mm thick ms gusset plate

8mm thick ms gusset plate

65 x 51 x 6·75 fixing cleats to both sides of gusset plate

65 x 51 x 6·75 fixing cleats to both sides of gusset plate

supporting steel column

purlin cleat

rafter

gusset plate

tie

fixing cleats

pcc padstone

supporting wall

ALTERNATIVE EAVES DETAIL

8mm thick ms gusset plate

563

Sheet Coverings ~ the basic functions of sheet coverings used in conjunction with steel roof trusses are to:-

1. Provide resistance to penetration by the elements.
2. Provide restraint to wind and snow loads.
3. Provide a degree of thermal insulation of not less than that set out in Part L of the Building Regulations.
4. Provide resistance to surface spread of flame as set out in Part B of the Building Regulations.
5. Provide any natural daylight required through the roof in accordance with the maximum permitted areas set out in Part L of the Building Regulations.
6. Be of low self weight to give overall design economy.
7. Be durable to keep maintenance needs to a minimum.

Suitable Materials ~

Hot-dip galvanised corrugated steel sheets – BS 3083

Aluminium profiled sheets – BS 4868.

Asbestos free profiled sheets – various manufacturers whose products are usually based on a mixture of Portland cement, mineral fibres and density modifiers – BS EN 494.

Typical Profiles ~

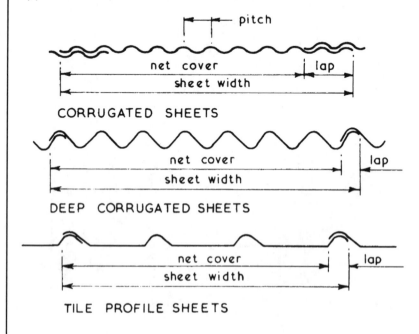

CORRUGATED SHEETS

DEEP CORRUGATED SHEETS

TILE PROFILE SHEETS

Typical Purlin Fixing Details ~

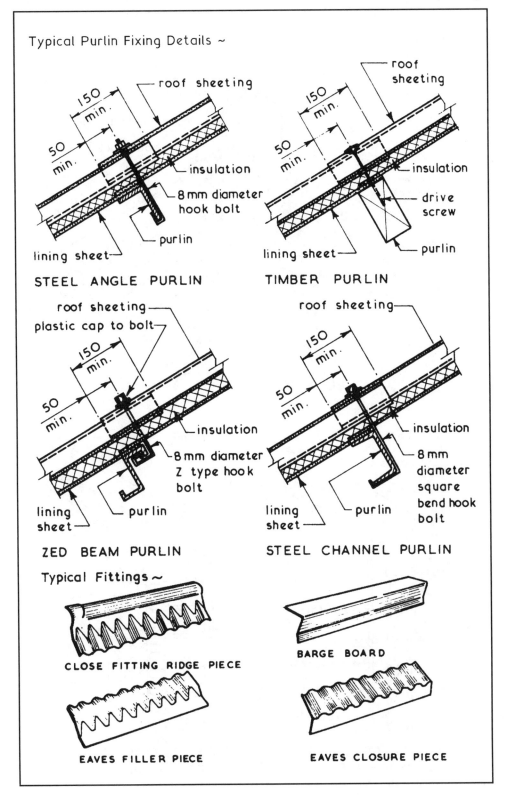

STEEL ANGLE PURLIN

TIMBER PURLIN

ZED BEAM PURLIN

STEEL CHANNEL PURLIN

Typical Fittings ~

CLOSE FITTING RIDGE PIECE

BARGE BOARD

EAVES FILLER PIECE

EAVES CLOSURE PIECE

Typical Details ~

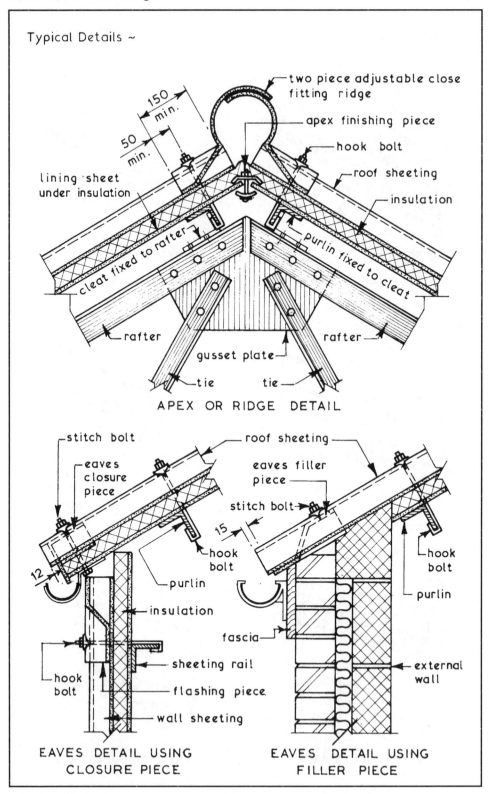

two piece adjustable close fitting ridge

apex finishing piece

hook bolt

roof sheeting

insulation

150 min.

50 min.

lining sheet under insulation

cleat fixed to rafter

purlin fixed to cleat

rafter

rafter

gusset plate

tie

tie

APEX OR RIDGE DETAIL

stitch bolt

roof sheeting

eaves closure piece

eaves filler piece

stitch bolt

15

hook bolt

purlin

12

insulation

hook bolt

purlin

fascia

hook bolt

sheeting rail

flashing piece

wall sheeting

external wall

EAVES DETAIL USING CLOSURE PIECE

EAVES DETAIL USING FILLER PIECE

Double Skin, Energy Roof systems ~ apply to industrial and commercial use buildings. In addition to new projects constructed to current thermal insulation standards, these systems can be specified to upgrade existing sheet profiled roofs with superimposed supplementary insulation and protective decking. Thermal performance with resin bonded mineral wool fibre of up to 250 mm overall depth may provide 'U' values as low as 0·13 W/m²K.

Typical Details ~

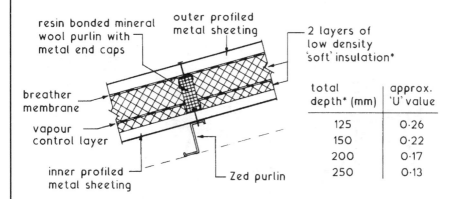

resin bonded mineral wool purlin with metal end caps

outer profiled metal sheeting

2 layers of low density 'soft' insulation*

breather membrane

vapour control layer

inner profiled metal sheeting

Zed purlin

total depth* (mm)	approx. 'U' value
125	0·26
150	0·22
200	0·17
250	0·13

Alternative ~

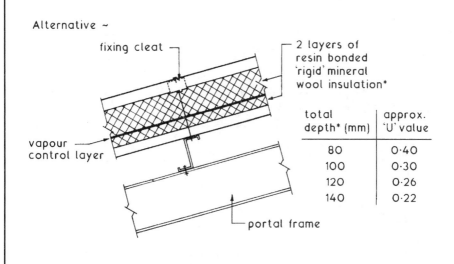

fixing cleat

2 layers of resin bonded 'rigid' mineral wool insulation*

vapour control layer

portal frame

total depth* (mm)	approx. 'U' value
80	0·40
100	0·30
120	0·26
140	0·22

567

Roof Sheet Coverings

Further typical details using profiled galvanised steel or aluminium, colour coated if required ~

RIDGE

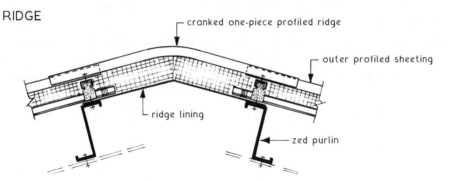

cranked one-piece profiled ridge

outer profiled sheeting

ridge lining

zed purlin

VALLEY GUTTER

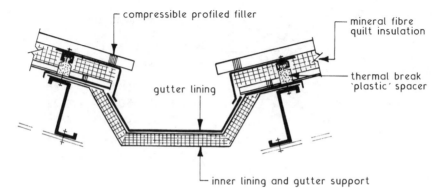

compressible profiled filler

mineral fibre quilt insulation

gutter lining

thermal break 'plastic' spacer

inner lining and gutter support

EAVES GUTTER

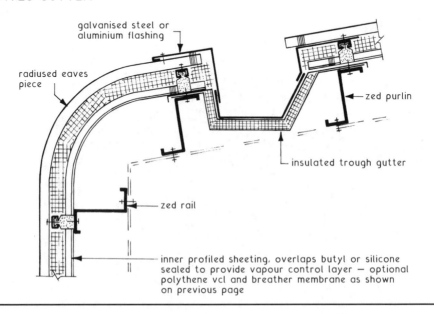

galvanised steel or aluminium flashing

radiused eaves piece

zed purlin

insulated trough gutter

zed rail

inner profiled sheeting, overlaps butyl or silicone sealed to provide vapour control layer — optional polythene vcl and breather membrane as shown on previous page

Long Span Roofs ~ these can be defined as those exceeding 12·000 in span. They can be fabricated in steel, aluminium alloy, timber, reinforced concrete and prestressed concrete. Long span roofs can be used for buildings such as factories. Large public halls and gymnasiums which require a large floor area free of roof support columns. The primary roof functions of providing weather protection, thermal insulation, sound insulation and restricting spread of fire over the roof surface are common to all roof types but these roofs may also have to provide strength sufficient to carry services lifting equipment and provide for natural daylight to the interior by means of rooflights.

Basic Roof Forms ~

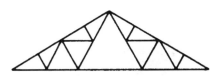

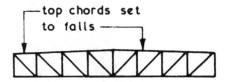

Pitched Trusses - spaced at suitable centres to carry purlins to which the roof coverings are fixed. Good rainwater run off - reasonable daylight spread from rooflights - high roof volume due to the triangulated format - on long spans roof volume can be reduced by using a series of short span trusses.

Flat Top Girders - spaced at suitable centres to carry purlins to which the roof coverings are fixed. Low pitch to give acceptable rainwater run off - reasonable daylight spread from rooflights - can be designed for very long spans but depth and hence roof volume increases with span.

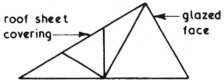

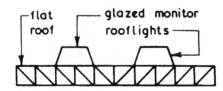

Northlight - spaced at suitable centres to carry purlins to which roof sheeting is fixed. Good rainwater run off - if correctly orientated solar glare is eliminated - long spans can be covered by a series of short span frames

Monitor - girders or cranked beams at centres to suit low pitch decking used. Good even daylight spread from monitor lights which is not affected by orientation of building.

Pitched Trusses ~ these can be constructed with a symmetrical outline (as shown on pages 562 to 563) or with an asymmetrical outline (Northlight – see detail below). They are usually made from standard steel sections with shop welded or bolted connections, alternatively they can be fabricated using timber members joined together with bolts and timber connectors or formed as a precast concrete portal frame.

Typical Multi-span Northlight Roof Details ~

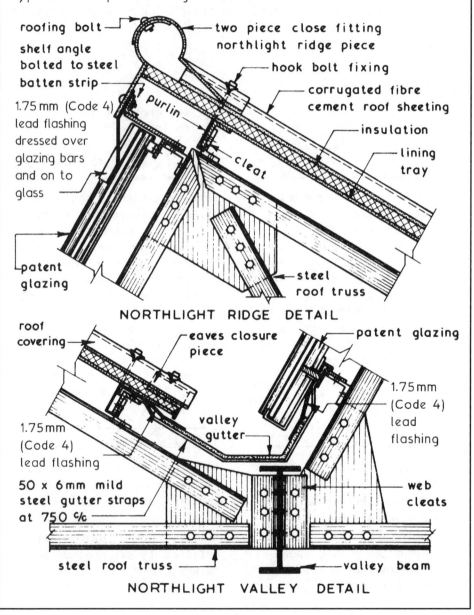

NORTHLIGHT RIDGE DETAIL

NORTHLIGHT VALLEY DETAIL

Monitor Roofs ~ these are basically a flat roof with raised glazed portions called monitors which forms a roof having a uniform distribution of daylight with no solar glare problems irrespective of orientation and a roof with easy access for maintenance. These roofs can be constructed with light long span girders supporting the monitor frames, cranked welded beams following the profile of the roof or they can be of a precast concrete portal frame format.
Typical Monitor Roof Details ~

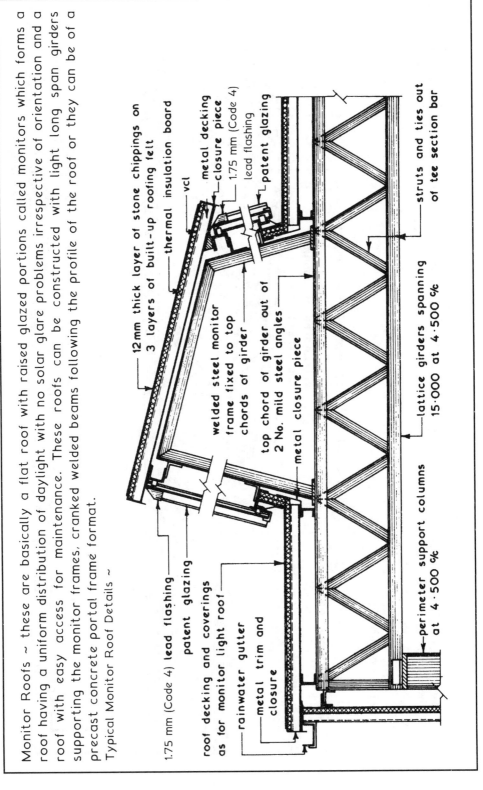

12mm thick layer of stone chippings on
3 layers of built-up roofing felt

thermal insulation board

vcl

metal decking
closure piece

1.75 mm (Code 4)
lead flashing

patent glazing

struts and ties out
of tee section bar

welded steel monitor
frame fixed to top
chords of girder

top chord of girder out of
2 No. mild steel angles

metal closure piece

lattice girders spanning
15·000 at 4·500 c/c

perimeter support columns
at 4·500 c/c

1.75 mm (Code 4) lead flashing

patent glazing

roof decking and coverings
as for monitor light roof

rainwater gutter

metal trim and
closure

Flat Top Girders ~ these are suitable for roof spans ranging from 15·000 to 45·000 and are basically low pitched lattice beams used to carry purlins which support the roof coverings. One of the main advantages of this form of roof is the reduction in roof volume. The usual materials employed in the fabrication of flat top girders are timber and steel.

Typical Flat Top (Pratt Type) Girder Details ~

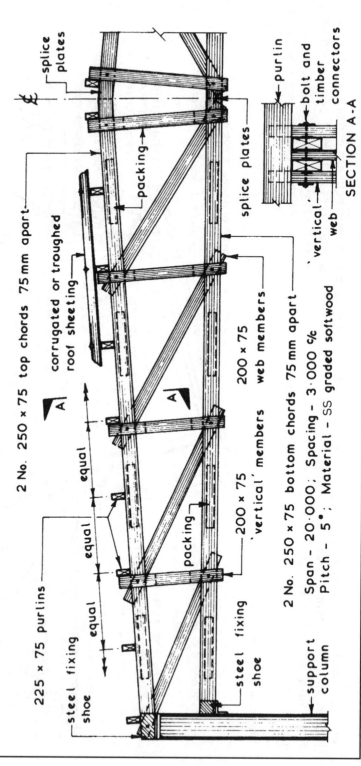

splice plates

℄

2 No. 250 × 75 top chords 75 mm apart

corrugated or troughed roof sheeting

packing

200 × 75 web members

splice plates

200 × 75 'vertical' members

packing

2 No. 250 × 75 bottom chords 75 mm apart

Span - 20·000; Spacing - 3·000 %
Pitch - 5°; Material - SS graded softwood

225 × 75 purlins

steel fixing shoe

steel fixing shoe

support column

purlin

bolt and timber connectors

'vertical' web

SECTION A-A

Bowstring Truss ~ a type of lattice truss formed with a curved upper edge. Bows and strings may be formed in pairs of laminated timber sections that are separated by solid web timber sections of struts and ties.

Spacing ~ 4.000 to 6.000 m apart depending on sizes of timber sections used and span.

Purlins ~ to coincide with web section meeting points and at about 1.000 m interim intervals.

Decking ~ sheet material suitably weathered or profiled metal sheeting. Thermally insulated relative to application.

Top bow radius ~ generally taken as between three-quarters of the span and the whole span.

Application ~ manufacturing assembly areas, factories, aircraft hangers, exhibition centres, sports arenas and other situations requiring a very large open span with featured timbers. Standard steel sections may also be used in this profile where appearance is less important, eg. railway termini.

Variation ~ the Belfast truss that pre-dates the standard bowstring shown. It has much smaller interlaced struts and ties therefore it is more complicated in terms of assembly and for calculation of stress distribution. See next page.

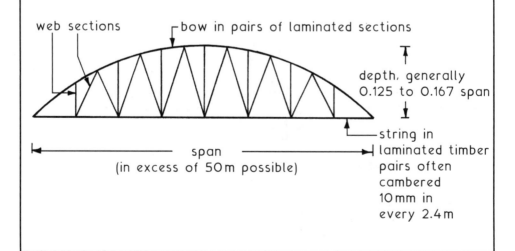

web sections ⌐bow in pairs of laminated sections

depth, generally 0.125 to 0.167 span

string in laminated timber pairs often cambered 10 mm in every 2.4 m

span
(in excess of 50 m possible)

Belfast Truss ~ established as one of the earliest forms of bowstring truss for achieving an efficient and economical roof construction over large spans. It was first used in the latter part of the 19th century and early part of the 20th century for industrial and agricultural buildings in response to the need for space uninterrupted by walls and columns within production, assembly and storage areas. A contribution to the name was a truss design devised by Messrs. D. Anderson and Son, Ltd. of Belfast, for supporting their patent roofing felt.

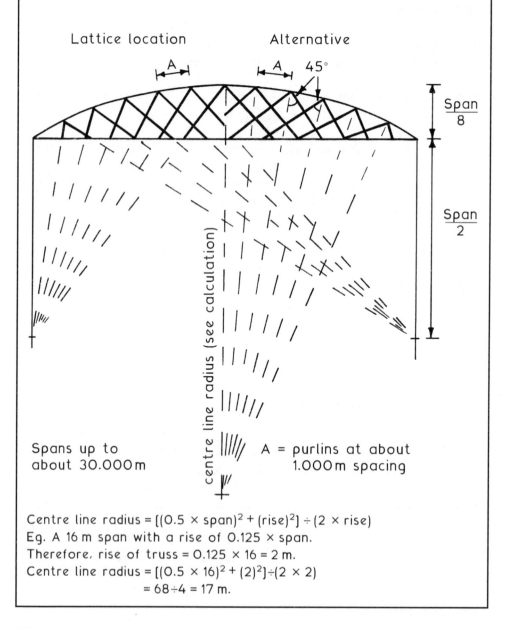

Lattice location

Alternative

$\dfrac{\text{Span}}{8}$

$\dfrac{\text{Span}}{2}$

centre line radius (see calculation)

Spans up to about 30.000 m

A = purlins at about 1.000 m spacing

Centre line radius = [(0.5 × span)² + (rise)²] ÷ (2 × rise)
Eg. A 16 m span with a rise of 0.125 × span.
Therefore, rise of truss = 0.125 × 16 = 2 m.
Centre line radius = [(0.5 × 16)² + (2)²] ÷ (2 × 2)
= 68 ÷ 4 = 17 m.

Bowed Lattice Truss Details ~

Bowstring ~

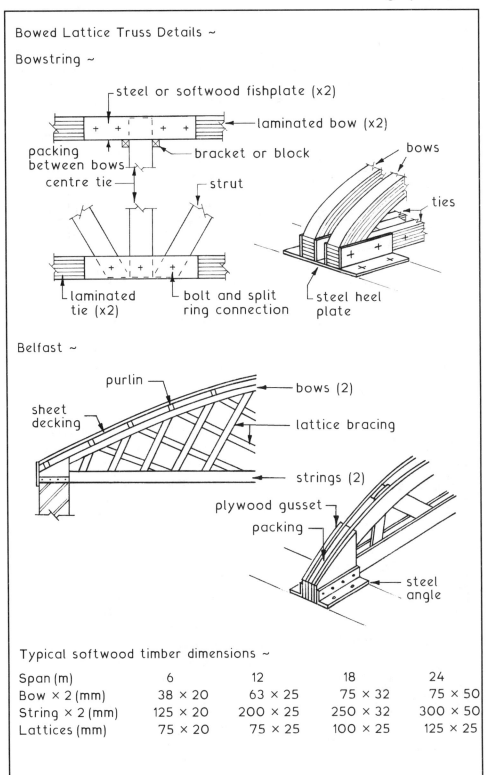

Belfast ~

Typical softwood timber dimensions ~

Span (m)	6	12	18	24
Bow × 2 (mm)	38 × 20	63 × 25	75 × 32	75 × 50
String × 2 (mm)	125 × 20	200 × 25	250 × 32	300 × 50
Lattices (mm)	75 × 20	75 × 25	100 × 25	125 × 25

Connections ~ nails, screws and bolts have their limitations when used to join structural timber members. The low efficiency of joints made with a rigid bar such as a bolt is caused by the usual low shear strength of timber parallel to the grain and the non-uniform distribution of bearing stress along the shank of the bolt –

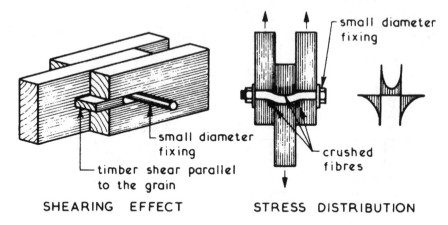

SHEARING EFFECT   STRESS DISTRIBUTION

Timber Connectors ~ these are designed to overcome the problems of structural timber connections outlined above by increasing the effective bearing area of the bolts.

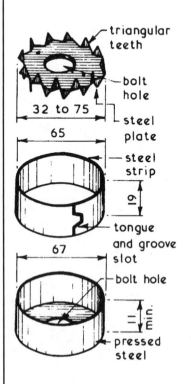

Toothed Plate Connector – provides an efficient joint without special tools or equipment – suitable for all connections especially small sections – bolt holes are drilled 2 mm larger than the bolt diameter, the timbers forming the joint being held together whilst being drilled.

Split Ring Connector – very efficient and develops a high joint strength – suitable for all connections – split ring connectors are inserted into a precut groove formed with a special tool making the connector independent from the bolt.

Shear Plate Connector – counterpart of a split ring connector – housed flush into timber – used for temporary joints.

Space Deck ~ this is a structural roofing system based on a simple repetitive pyramidal unit to give large clear spans of up to 22·000 for single spanning designs and up to 33·000 for two way spanning designs. The steel units are easily transported to site before assembly into beams and the complete space deck at ground level before being hoisted into position on top of the perimeter supports. A roof covering of wood wool slabs with built-up roofing felt could be used, although any suitable structural lightweight decking is appropriate. Rooflights can be mounted directly onto the square top space deck units.

Typical Details ~

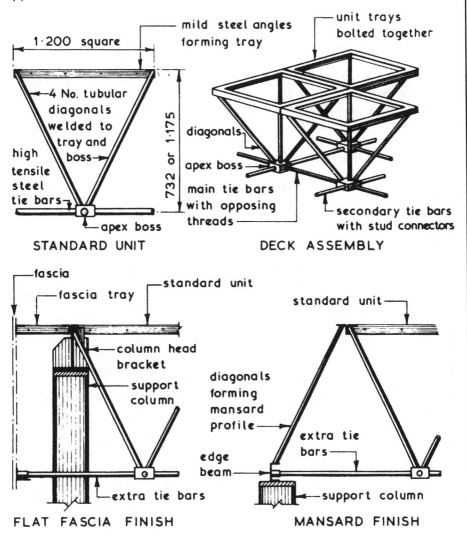

STANDARD UNIT

DECK ASSEMBLY

FLAT FASCIA FINISH

MANSARD FINISH

Space Frames ~ these are roofing systems which consist of a series of connectors which joins together the chords and bracing members of the system. Single or double layer grids are possible, the former usually employed in connection with small domes or curved roofs. Space frames are similar in concept to space decks but they have greater flexibility in design and layout possibilities. Most space frames are fabricated from structural steel tubes or tubes of aluminium alloy although any suitable structural material could be used.

Typical Examples~

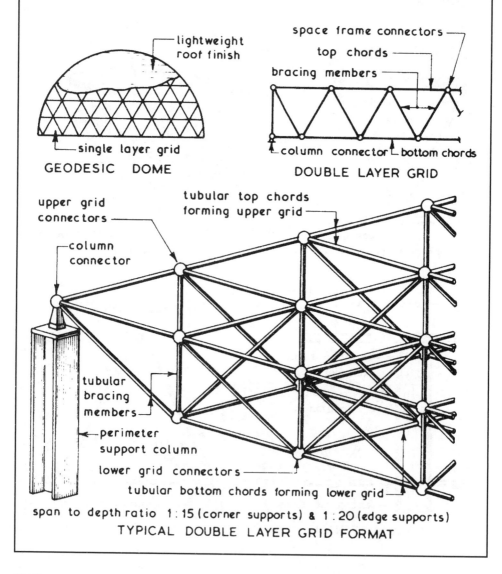

GEODESIC DOME

DOUBLE LAYER GRID

span to depth ratio 1:15 (corner supports) & 1:20 (edge supports)
TYPICAL DOUBLE LAYER GRID FORMAT

Shell Roofs ~ these can be defined as a structural curved skin covering a given plan shape and area where the forces in the shell or membrane are compressive and in the restraining edge beams are tensile. The usual materials employed in shell roof construction are in-situ reinforced concrete and timber. Concrete shell roofs are constructed over formwork which in itself is very often a shell roof making this format expensive since the principle of use and reuse of formwork can not normally be applied. The main factors of shell roofs are:-

1. The entire roof is primarily a structural element.
2. Basic strength of any particular shell is inherent in its geometrical shape and form.
3. Comparatively less material is required for shell roofs than other forms of roof construction.

Domes ~ these are double curvature shells which can be rotationally formed by any curved geometrical plane figure rotating about a central vertical axis. Translation domes are formed by a curved line moving over another curved line whereas pendentive domes are formed by inscribing within the base circle a regular polygon and vertical planes through the true hemispherical dome.

Typical Examples ~

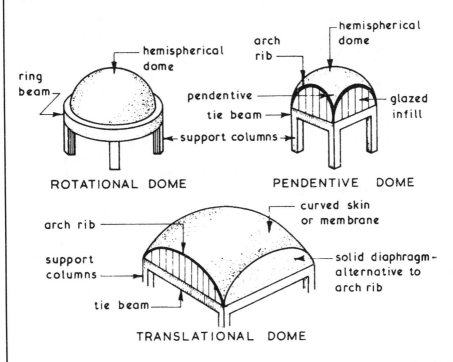

ROTATIONAL DOME

PENDENTIVE DOME

TRANSLATIONAL DOME

Barrel Vaults ~ these are single curvature shells which are essentially a cut cylinder which must be restrained at both ends to overcome the tendency to flatten. A barrel vault acts as a beam whose span is equal to the length of the roof. Long span barrel vaults are those whose span is longer than its width or chord length and conversely short barrel vaults are those whose span is shorter than its width or chord length. In every long span barrel vaults thermal expansion joints will be required at 30·000 centres which will create a series of abutting barrel vault roofs weather sealed together (see next page).

Typical Single Barrel Vault Principles~

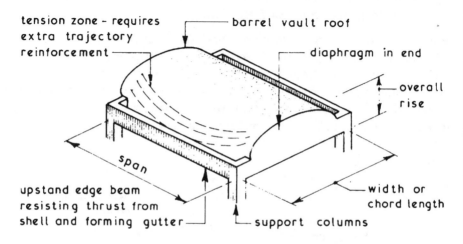

economic design ratios -   width : span   1 : 2 to 1 : 5
                           rise : span 1 : 10 to 1 : 15

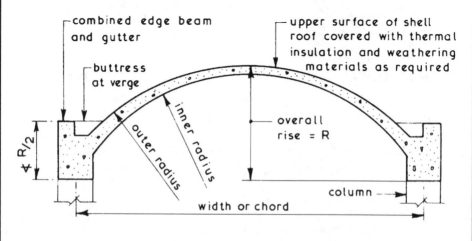

Typical Barrel Vault Expansion Joint Details ~

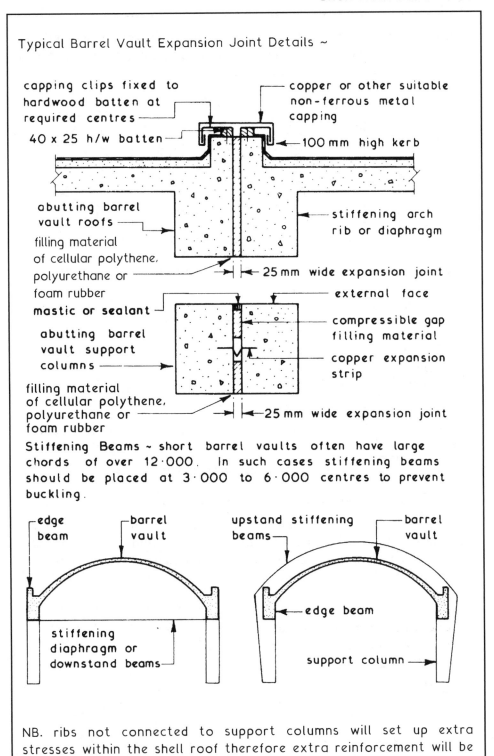

capping clips fixed to
hardwood batten at
required centres

copper or other suitable
non-ferrous metal
capping

40 x 25 h/w batten

100 mm high kerb

abutting barrel
vault roofs

stiffening arch
rib or diaphragm

filling material
of cellular polythene,
polyurethane or
foam rubber

25 mm wide expansion joint

mastic or sealant

external face

abutting barrel
vault support
columns

compressible gap
filling material

copper expansion
strip

filling material
of cellular polythene,
polyurethane or
foam rubber

25 mm wide expansion joint

Stiffening Beams ~ short barrel vaults often have large
chords of over 12·000. In such cases stiffening beams
should be placed at 3·000 to 6·000 centres to prevent
buckling.

edge
beam

barrel
vault

upstand stiffening
beams

barrel
vault

stiffening
diaphragm or
downstand beams

edge beam

support column

NB. ribs not connected to support columns will set up extra
stresses within the shell roof therefore extra reinforcement will be
required at the stiffening rib or beam positions.

Other Forms of Barrel Vault ~ by cutting intersecting and placing at different levels the basic barrel vault roof can be formed into a groin or northlight barrel vault roof:-

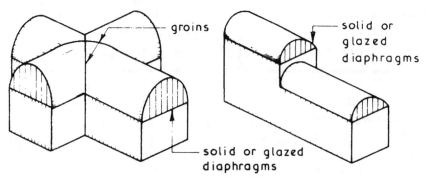

INTERSECTING BARREL VAULTS   STEPPED BARREL VAULTS

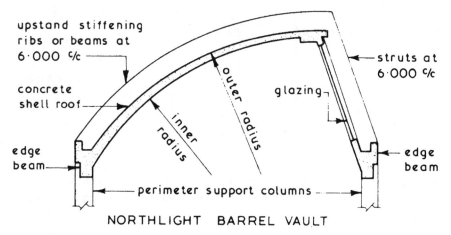

NORTHLIGHT BARREL VAULT

Conoids ~ these are double curvative shell roofs which can be considered as an alternative to barrel vaults. Spans up to 12·000 with chord lengths up to 24·000 are possible. Typical chord to span ratio 2:1.

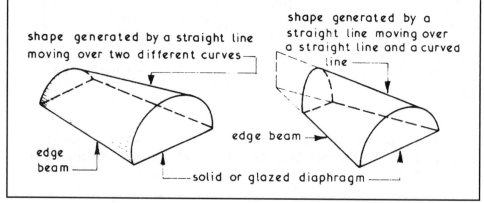

Hyperbolic Paraboloids ~ the true hyperbolic paraboloid shell roof shape is generated by moving a vertical parabola (the generator) over another vertical parabola (the directrix) set at right angles to the moving parabola. This forms a saddle shape where horizontal sections taken through the roof are hyperbolic in format and vertical sections are parabolic. The resultant shape is not very suitable for roofing purposes therefore only part of the saddle shape is used and this is formed by joining the centre points thus:-

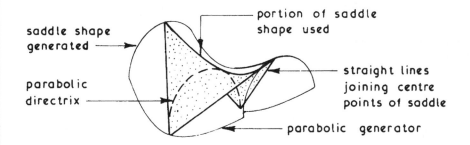

To obtain a more practical shape than the true saddle a straight line limited hyperbolic paraboloid is used. This is formed by raising or lowering one or more corners of a square forming a warped parallelogram thus:-

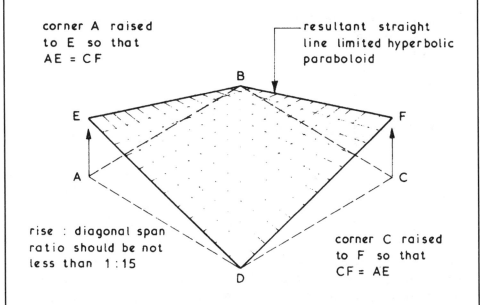

For further examples see next page.

Typical Straight Line Limited Hyperbolic Paraboloid Formats ~

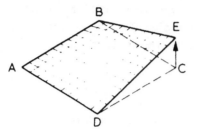

corner C raised to E

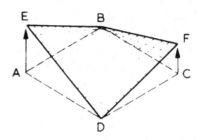

corner A raised to E and
corner C raised to F so that
AE ≠ CF

resultant hyperbolic
paraboloid ─────────

original
square ─────

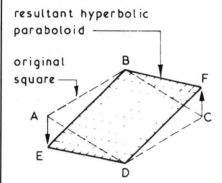

corner A lowered to E and
corner C raised to F so that
AE = CF

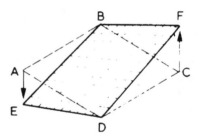

corner A lowered to E and
corner C raised to F so that
AE ≠ CF

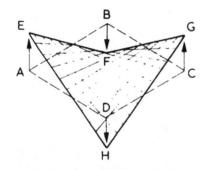

corners A & C raised to E & G
corners B & D lowered to F & H
so that AE = CG & BF = DH

Combination of Hyperbolic
Paraboloid Shell Roofs ~

one corner
raised

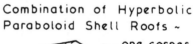
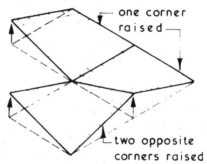

two opposite
corners raised

NB. any combination possible

Concrete Hyperbolic Paraboloid Shell Roofs ~ these can be constructed in reinforced concrete (characteristic strength 25 or 30 N/mm²) with a minimum shell thickness of 50 mm with diagonal spans up to 35·000. These shells are cast over a timber form in the shape of the required hyperbolic paraboloid format. In practice therefore two roofs are constructed and it is one of the reasons for the popularity of timber versions of this form of shell roof.

Timber Hyperbolic Paraboloid Shell Roofs ~ these are usually constructed using laminated edge beams and layers of t & g boarding to form the shell membrane. For roofs with a plan size of up to 6·000 × 6·000 only 2 layers of boards are required and these are laid parallel to the diagonals with both layers running in opposite directions. Roofs with a plan size of over 6·000 × 6·000 require 3 layers of board as shown below. The weather protective cover can be of any suitable flexible material such as built-up roofing felt, copper and lead. During construction the relatively lightweight roof is tied down to a framework of scaffolding until the anchorages and wall infilling have been completed. This is to overcome any negative and positive wind pressures due to the open sides.

Typical Details ~

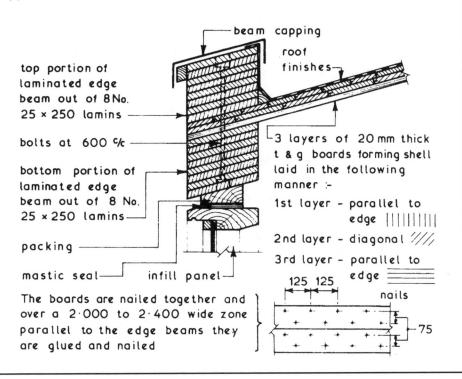

Support Considerations ~ in timber hyperbolic paraboloid shell roofs only two supports are required:-

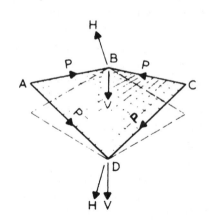

Edge beams are in compression forces P are transmitted to B and D resulting in a vertical force V and a horizontal force H at both positions therefore support columns are required at B and D.

Vertical force V is transmitted directly down the columns to a suitable foundation. The outward or horizontal force H can be accommodated in one of two ways:-

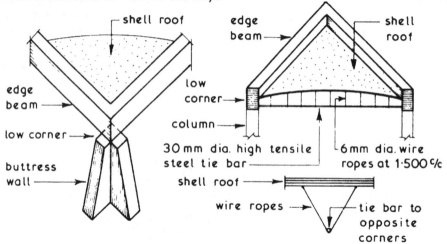

shell roof

edge beam

low corner

buttress wall

edge beam → shell roof

low corner

column

30 mm dia. high tensile steel tie bar

6 mm dia. wire ropes at 1·500 c/c

shell roof

wire ropes → tie bar to opposite corners

If shell roof is to be supported at high corners the edge beams will be in tension and horizontal force will be inwards. This can be resisted by a diagonal strut between the high corners.

Combination Roof Support Example ~

4 No. roof shells of equal loading joined together

Supports required at A; C; G and E. with ties between AC; CE; EG and GA. Forces at J cancel each other therefore no support required at J.

A form of stressed skin reinforced concrete construction also known as folded plate construction. The concept is to profile a flat slab into folds so that the structure behaves as a series of beams spanning parallel with the profile.

Optimum depth to span ratio is between 1:10 to 1:15, or a depth to width of not less than 1:10, whichever is greater.

Roof format ~ pitched, monitor or multi-fold.

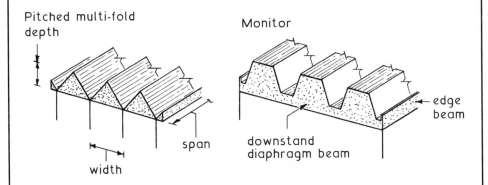

Several slabs of narrow thickness are usually preferred to a few wider slabs as this reduces the material dead loading. Numerous design variations are possible particularly where counter-folds are introduced and the geometry developed to include bowed or arched spans extending to the ground.

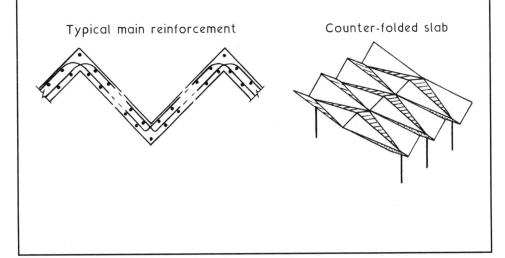

587

Membrane Structure Principles ~ a form of tensioned cable structural support system with a covering of stretched fabric. In principle and origin, this compares to a tent with poles as compression members secured to the ground. The fabric membrane is attached to peripheral stressing cables suspended in a catenary between vertical support members.

Form ~ there are limitless three-dimensional possibilities. The following geometric shapes provide a basis for imagination and elegance in design:

- Hyperbolic paraboloid (Hypar)
- Barrel vault
- Conical or double conical

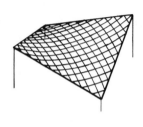  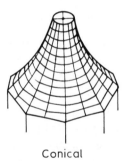

Hyperbolic paraboloid   Barrel vault   Conical

Double conical~

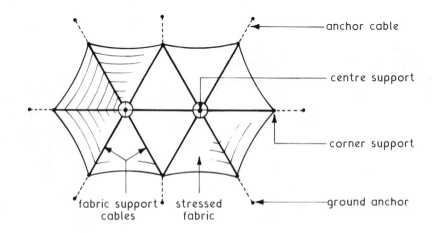

anchor cable

centre support

corner support

fabric support cables   stressed fabric

ground anchor

Simple support structure as viewed from the underside ~

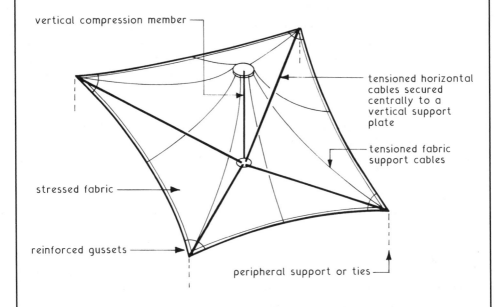

vertical compression member

tensioned horizontal cables secured centrally to a vertical support plate

tensioned fabric support cables

stressed fabric

reinforced gussets

peripheral support or ties

Fabric ~ has the advantages of requiring minimal support, opportunity for architectural expression in colour and geometry and a translucent quality that provides an outside feel inside, whilst combining shaded cover from the sun and shelter from rain. Applications are generally attachments as a feature to entrances and function areas in prominent buildings, notably sports venues, airports and convention centres.

Materials ~ historically, animal hides were the first materials used for tensile fabric structures, but more recently woven fibres of hemp, flax or other natural yarns have evolved as canvas. Contemporary synthetic materials have a plastic coating on a fibrous base. These include polyvinyl chloride (PVC) on polyester fibres, silicone on glass fibres and polytetrafluorethylene (PTFE) on glass fibres. Design life is difficult to estimate, as it will depend very much on type of exposure. Previous use of these materials would indicate that at least 20 years is anticipated, with an excess of 30 years being likely. Jointing can be by fusion welding of plastics, bonding with silicone adhesives and stitching with glass threads.

Rooflights ~ the useful penetration of daylight through the windows in external walls of buildings is from 6·000 to 9·000 depending on the height and size of the window. In buildings with spans over 18·000 side wall daylighting needs to be supplemented by artificial lighting or in the case of top floors or single storey buildings by rooflights. The total maximum area of wall window openings and rooflights for the various purpose groups is set out in the Building Regulations with allowances for increased areas if double or triple glazing is used. In pitched roofs such as northlight and monitor roofs the rooflights are usually in the form of patent glazing (see Long Span Roofs on pages 570 and 571). In flat roof construction natural daylighting can be provided by one or more of the following methods:-

1. Lantern lights – see page 592.

2. Lens lights – see page 592.

3. Dome, pyramid and similar rooflights – see page 593.

Patent Glazing ~ these are systems of steel or aluminium alloy glazing bars which span the distance to be glazed whilst giving continuous edge support to the glass. They can be used in the roof forms noted above as well as in pitched roofs with profiled coverings where the patent glazing bars are fixed above and below the profiled sheets – see page 591.

Typical Patent Glazing Bar Sections ~

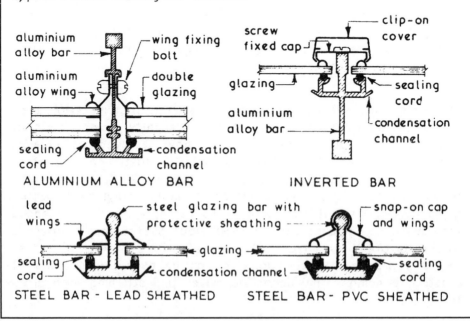

ALUMINIUM ALLOY BAR   INVERTED BAR

STEEL BAR - LEAD SHEATHED   STEEL BAR - PVC SHEATHED

Typical Pitched Roof Patent Glazing Details ~

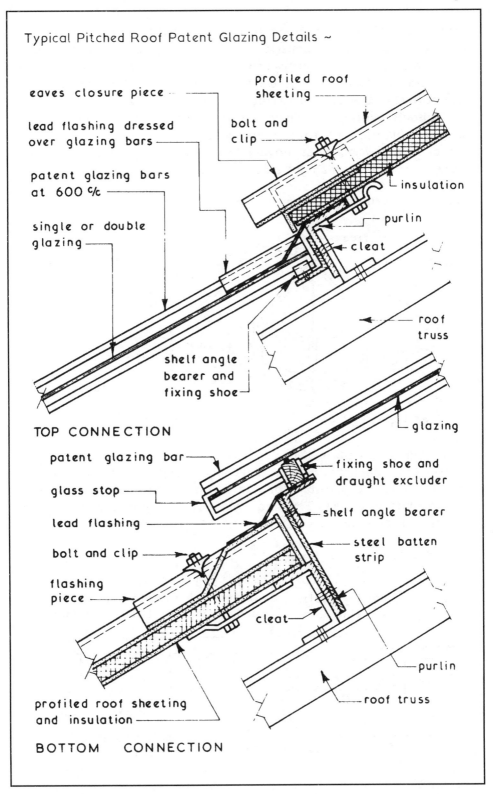

eaves closure piece

profiled roof sheeting

lead flashing dressed over glazing bars

bolt and clip

patent glazing bars at 600 c/c

insulation

single or double glazing

purlin

cleat

roof truss

shelf angle bearer and fixing shoe

TOP CONNECTION

glazing

patent glazing bar

fixing shoe and draught excluder

glass stop

shelf angle bearer

lead flashing

steel batten strip

bolt and clip

flashing piece

cleat

purlin

roof truss

profiled roof sheeting and insulation

BOTTOM CONNECTION

Lantern Lights ~ these are a form of rooflight used in conjuction with flat roofs. They consist of glazed vertical sides and fully glazed pitched roof which is usually hipped at both ends. Part of the glazed upstand sides is usually formed as an opening light or alternatively glazed with louvres to provide a degree of controllable ventilation. They can be constructed of timber, metal or a combination of these two materials. Lantern lights in the context of new buildings have been generally superseded by the various forms of dome light (see next page)

Typical Lantern Light Details ~

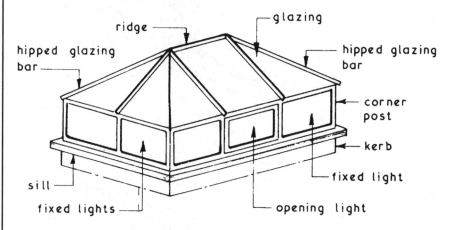

Lens Lights ~ these are small square or round blocks of translucent toughened glass especially designed for casting into concrete and are suitable for use in flat roofs and curved roofs such as barrel vaults. They can also be incorporated in precast concrete frames for inclusion into a cast in-situ roof.

Typical Details ~

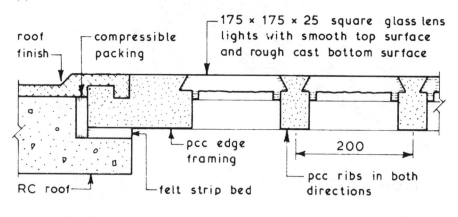

Dome, Pyramid and Similar Rooflights ~ these are used in conjuction with flat roofs and may be framed or unframed. The glazing can be of glass or plastics such as polycarbonate, acrylic, PVC and glass fibre reinforced polyester resin (grp). The whole component is fixed to a kerb and may have a raising piece containing hit and miss ventilators, louvres or flaps for controllable ventilation purposes.

Typical Details ~

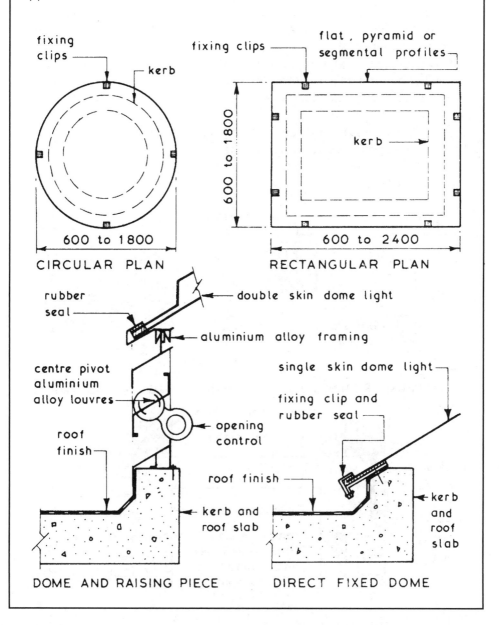

CIRCULAR PLAN

RECTANGULAR PLAN

DOME AND RAISING PIECE

DIRECT FIXED DOME

Non-load Bearing Brick Panel Walls ~ these are used in conjunction with framed structures as an infill between the beams and columns. They are constructed in the same manner as ordinary brick walls with the openings being formed by traditional methods.

Basic Requirements ~

1. To be adequately supported by and tied to the structural frame.
2. Have sufficient strength to support own self weight plus any attached finishes and imposed loads such as wind pressures.
3. Provide the necessary resistance to penetration by the natural elements.
4. Provide the required degree of thermal insulation, sound insulation and fire resistance.
5. Have sufficient durability to reduce maintenance costs to a minimum.
6. Provide for movements due to moisture and thermal expansion of the panel and for contraction of the frame.

Typical Details ~

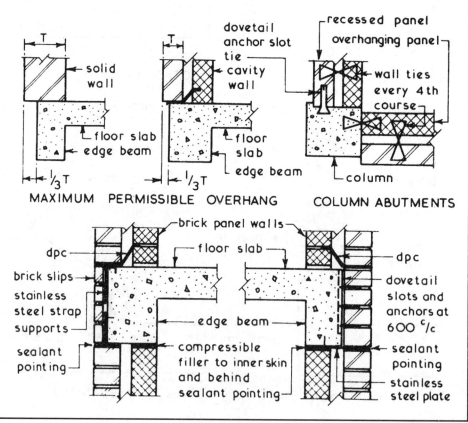

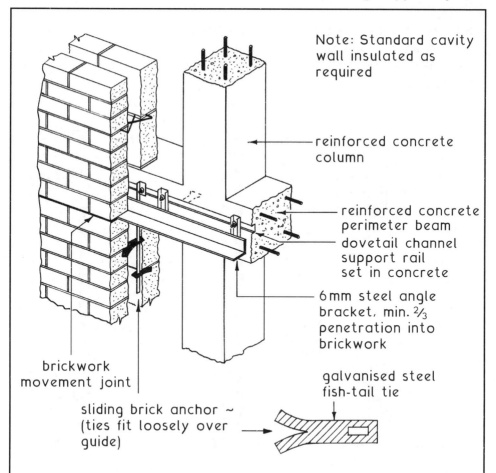

Note: Standard cavity wall insulated as required

reinforced concrete column

reinforced concrete perimeter beam

dovetail channel support rail set in concrete

6 mm steel angle bracket, min. ⅔ penetration into brickwork

galvanised steel fish-tail tie

brickwork movement joint

sliding brick anchor ~ (ties fit loosely over guide)

Application – multi-storey buildings, where a traditional brick façade is required.

Brickwork movement – to allow for climatic changes and differential movement between the cladding and main structure, a 'soft' joint (cellular polyethylene, cellular polyurethane, expanded rubber or sponge rubber with polysulphide or silicon pointing) should be located below the support angle. Vertical movement joints may also be required at a maximum of 12 m spacing.

Lateral restraint – provided by normal wall ties between inner and outer leaf of masonry, plus sliding brick anchors below the support angle.

595

Infill Panel Walls ~ these can be used between the framing members of a building to provide the cladding and division between the internal and external environments and are distinct from claddings and facing:-

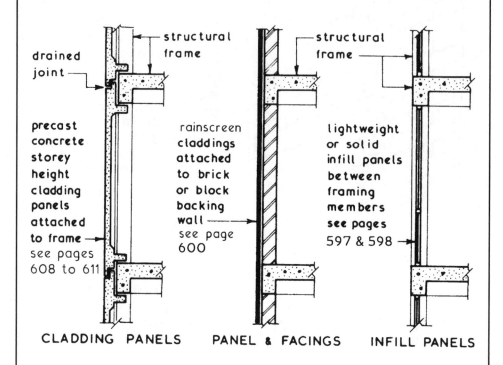

drained joint

structural frame

precast concrete storey height cladding panels attached to frame →
see pages 608 to 611

rainscreen claddings attached to brick or block backing wall
see page 600

structural frame

lightweight or solid infill panels between framing members
see pages 597 & 598 →

CLADDING PANELS   **PANEL & FACINGS**   **INFILL PANELS**

Functional Requirements ~ all forms of infill panel should be designed and constructed to fulfil the following functional requirements:-

1. Self supporting between structural framing members.
2. Provide resistance to the penetration of the elements.
3. Provide resistance to positive and negative wind pressures.
4. Give the required degree of thermal insulation.
5. Give the required degree of sound insulation.
6. Give the required degree of fire resistance.
7. Have sufficient openings to provide the required amount of natural ventilation.
8. Have sufficient glazed area to fulfil the natural daylight and vision out requirements.
9. Be economic in the context of construction and maintenance.
10. Provide for any differential movements between panel and structural frame.

Brick Infill Panels ~ these can be constructed in a solid or cavity format, the latter usually having an inner skin of blockwork to increase the thermal insulation properties of the panel. All the fundamental construction processes and detail of solid and cavity walls (bonding, lintels over openings, wall ties, damp-proof courses etc.,) apply equally to infill panel walls. The infill panel walls can be tied to the columns by means of wall ties cast into the columns at 300 mm centres or located in cast-in dovetail anchor slots. The head of every infill panel should have a compressible joint to allow for any differential movements between the frame and panel.

Typical Details

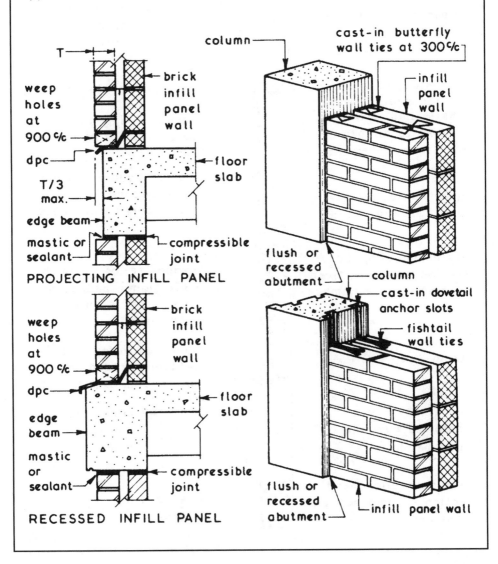

PROJECTING INFILL PANEL

RECESSED INFILL PANEL

Lightweight Infill Panels ~ these can be constructed from a wide variety or combination of materials such as timber, metals and plastics into which single or double glazing can be fitted. If solid panels are to be used below a transom they are usually of a composite or sandwich construction to provide the required sound insulation, thermal insulation and fire resistance properties.

Typical Example ~

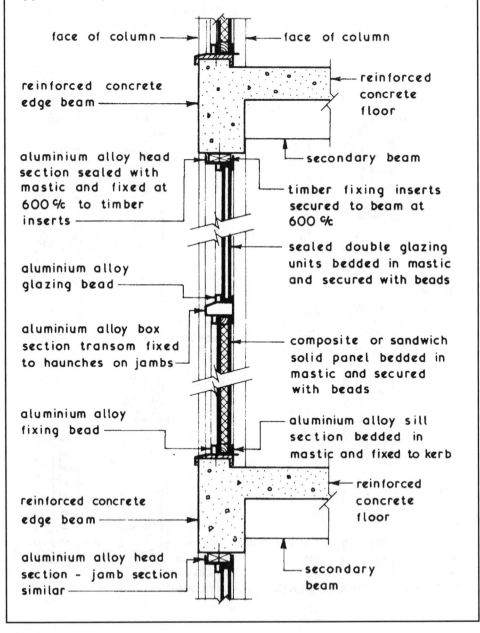

face of column
face of column

reinforced concrete edge beam

reinforced concrete floor

aluminium alloy head section sealed with mastic and fixed at 600 c/c to timber inserts

secondary beam

timber fixing inserts secured to beam at 600 c/c

aluminium alloy glazing bead

sealed double glazing units bedded in mastic and secured with beads

aluminium alloy box section transom fixed to haunches on jambs

composite or sandwich solid panel bedded in mastic and secured with beads

aluminium alloy fixing bead

aluminium alloy sill section bedded in mastic and fixed to kerb

reinforced concrete edge beam

reinforced concrete floor

aluminium alloy head section - jamb section similar

secondary beam

Lightweight Infill Panels ~ these can be fixed between the structural horizontal and vertical members of the frame or fixed to the face of either the columns or beams to give a grid, horizontal or vertical emphasis to the façade thus –

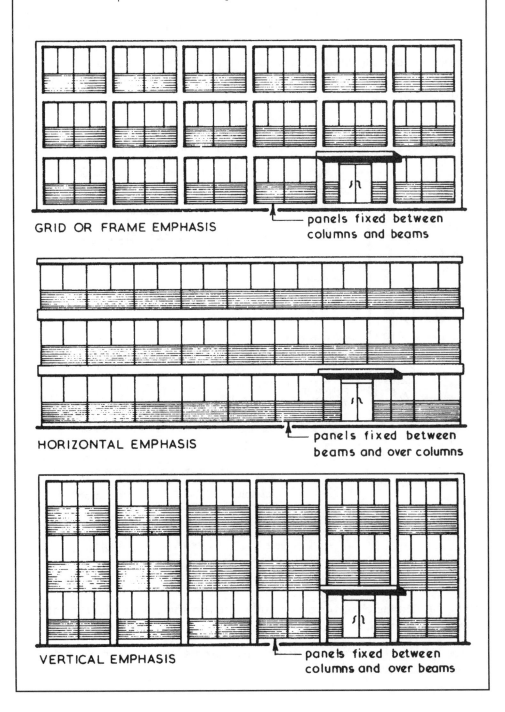

GRID OR FRAME EMPHASIS   panels fixed between columns and beams

HORIZONTAL EMPHASIS   panels fixed between beams and over columns

VERTICAL EMPHASIS   panels fixed between columns and over beams

Overcladding ~ a superficial treatment, applied either as a component of new construction work, or as a façade and insulation enhancement to existing structures. The outer weather resistant decorative panelling is 'loose fit' in concept, which is easily replaced to suit changing tastes, new materials and company image. Panels attach to the main structure with a grid of simple metal framing or vertical timber battens. This allows space for a ventilated and drained cavity, with provision for insulation to be attached to the substructure; a normal requirement in upgrade/ refurbishment work.

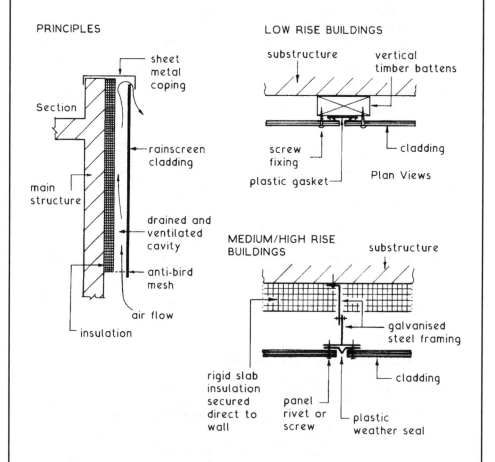

Note (1): Cladding materials include, plastic laminates, fibre cement, ceramics, aluminium, enamelled steel and various stone effects.

Note (2): Anti-bird mesh coated with intumescent material to form a fire stop cavity barrier.

Glazed façades have been associated with hi-tech architecture since the 1970s. The increasing use of this type of cladding is largely due to developments in toughened glass and improved qualities of elastomeric silicone sealants. The properties of the latter must incorporate a resilience to varying atmospheric conditions as well as the facility to absorb structural movement without loss of adhesion.

Systems – two edge and four edge.

The two edge system relies on conventional glazing beads/fixings to the head and sill parts of a frame, with sides silicone bonded to mullions and styles.

The four edge system relies entirely on structural adhesion, using silicone bonding between glazing and support frame – see details.

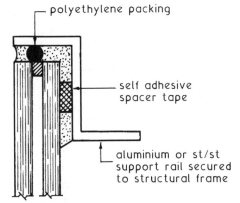

Upper edge or head of support frame

Structural glazing, as shown on this and the next page, is in principle a type of curtain walling. Due to its unique appearance, it is usual to consider full glazing of the building façade as a separate design and construction concept.

BS EN 13830: Curtain walling. Product standard; defines curtain walling as an external vertical building enclosure produced by elements mainly of metal, timber or plastic. Glass as a primary material is excluded.

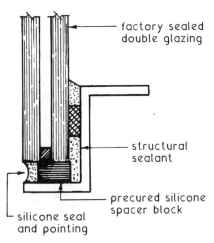

Lower edge of support frame to sill

Note: Sides of frame as head.

Structural glazing is otherwise known as frameless glazing. It is a system of toughened glass cladding without the visual impact of surface fixings and supporting components. Unlike curtain walling, the self-weight of the glass and wind loads are carried by the glass itself and transferred to a subsidiary lightweight support structure behind the glazing.

Assembly principles ~

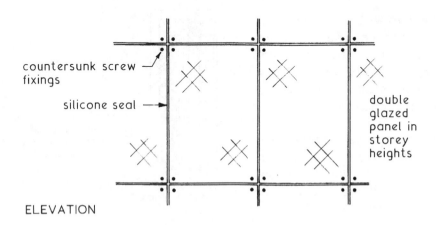

countersunk screw fixings

silicone seal

double glazed panel in storey heights

ELEVATION

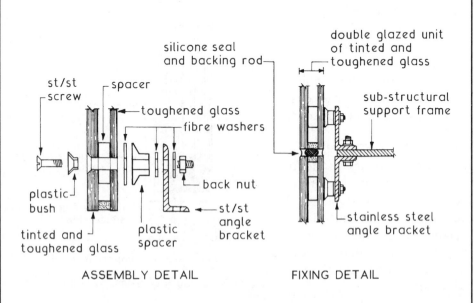

st/st screw

spacer

silicone seal and backing rod

double glazed unit of tinted and toughened glass

toughened glass

sub-structural support frame

fibre washers

plastic bush

back nut

tinted and toughened glass

plastic spacer

st/st angle bracket

stainless steel angle bracket

ASSEMBLY DETAIL

FIXING DETAIL

Curtain Walling ~ this is a form of lightweight non-load bearing external cladding which forms a complete envelope or sheath around the structural frame. In low rise structures the curtain wall framing could be of timber or patent glazing but in the usual high rise context, box or solid members of steel or aluminium alloy are normally employed.

Basic Requirements for Curtain Walls ~

1. Provide the necessary resistance to penetration by the elements.
2. Have sufficient strength to carry own self weight and provide resistance to both positive and negative wind pressures.
3. Provide required degree of fire resistance – glazed areas are classified in the Building Regulations as unprotected areas therefore any required fire resistance must be obtained from the infill or undersill panels and any backing wall or beam.
4. Be easy to assemble, fix and maintain.
5. Provide the required degree of sound and thermal insulation.
6. Provide for thermal and structural movements.

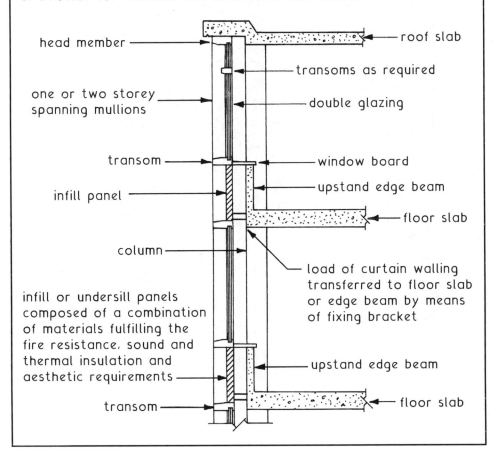

head member — roof slab

transoms as required

one or two storey spanning mullions — double glazing

transom — window board

infill panel — upstand edge beam

floor slab

column — load of curtain walling transferred to floor slab or edge beam by means of fixing bracket

infill or undersill panels composed of a combination of materials fulfilling the fire resistance, sound and thermal insulation and aesthetic requirements — upstand edge beam

transom — floor slab

Typical Curtain Walling Details

Principles ~

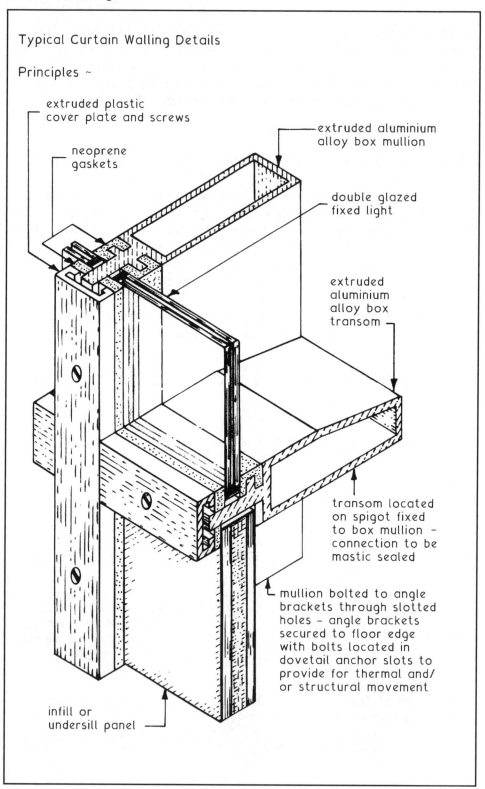

extruded plastic
cover plate and screws

neoprene
gaskets

extruded aluminium
alloy box mullion

double glazed
fixed light

extruded
aluminium
alloy box
transom

transom located
on spigot fixed
to box mullion –
connection to be
mastic sealed

mullion bolted to angle
brackets through slotted
holes – angle brackets
secured to floor edge
with bolts located in
dovetail anchor slots to
provide for thermal and/
or structural movement

infill or
undersill panel

Fixing Curtain Walling to the Structure ~ in curtain walling systems it is the main vertical component or mullion which carries the loads and transfers them to the structural frame at every or alternate floor levels depending on the spanning ability of the mullion. At each fixing point the load must be transferred and an allowance made for thermal expansion and differential movement between the structural frame and curtain walling. The usual method employed is slotted bolt fixings.

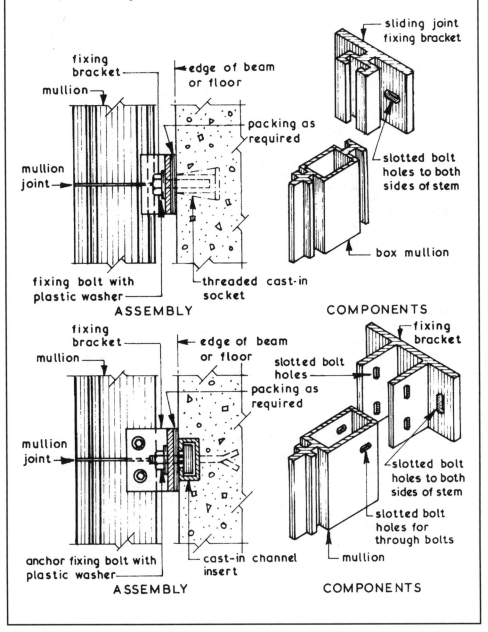

Re-cladding existing framed buildings has become an economical alternative to complete demolition and re-building. This may be justified when a building has a change of use or it is in need of an image upgrade. Current energy conservation measures can also be achieved by the re-dressing of older buildings.

Typical section through an existing structural floor slab with a replacement system attached ~

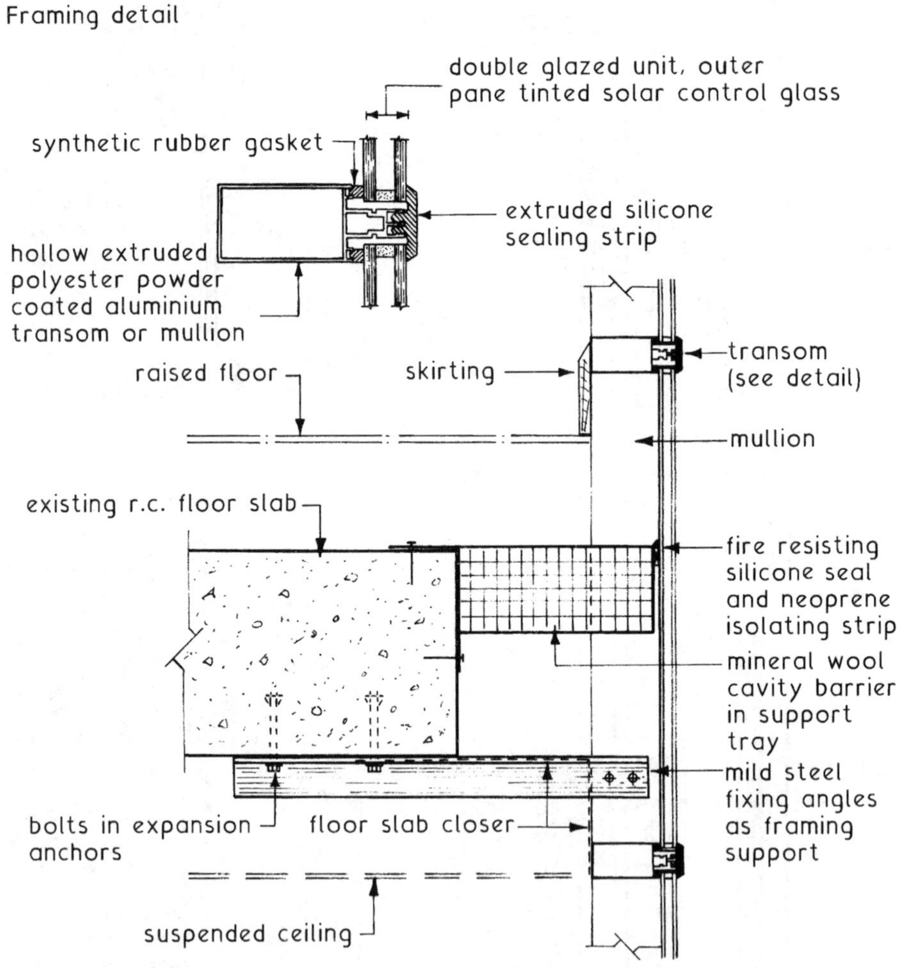

Framing detail

double glazed unit, outer pane tinted solar control glass

synthetic rubber gasket

extruded silicone sealing strip

hollow extruded polyester powder coated aluminium transom or mullion

raised floor

skirting

transom (see detail)

mullion

existing r.c. floor slab

fire resisting silicone seal and neoprene isolating strip

mineral wool cavity barrier in support tray

mild steel fixing angles as framing support

bolts in expansion anchors

floor slab closer

suspended ceiling

VERTICAL SECTION

Load bearing Concrete Panels ~ this form of construction uses storey height load bearing pre-cast reinforced concrete perimeter panels. The width and depth of the panels is governed by the load(s) to be carried, the height and exposure of the building. Panels can be plain or fenestrated providing the latter leaves sufficient concrete to transmit the load(s) around the opening. The cladding panels, being structural, eliminate the need for perimeter columns and beams and provide an internal surface ready to receive insulation, attached services and decorations. In the context of design these structures must be formed in such a manner that should a single member be removed by an internal explosion, wind pressure or similar force, progressive or structural collapse will not occur, the minimum requirements being set out in Part A of the Building Regulations. Load bearing concrete panel construction can be a cost effective method of building.

Typical Details ~

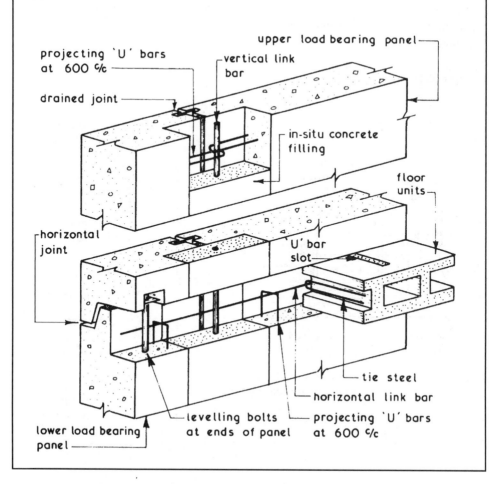

607

Concrete Cladding Panels ~ these are usually of reinforced precast concrete to an undersill or storey height format, the former being sometimes called apron panels. All precast concrete cladding panels should be designed and installed to fulfil the following functions:-

1. Self supporting between framing members.
2. Provide resistance to penetration by the natural elements.
3. Resist both positive and negative wind pressures.
4. Provide required degree of fire resistance.
5. Provide required degree of thermal insulation by having the insulating material incorporated within the body of the cladding or alternatively allow the cladding to act as the outer leaf of cavity wall panel.
6. Provide required degree of sound insulation.

Undersill or Apron Cladding Panels ~ these are designed to span from column to column and provide a seating for the windows located above. Levelling is usually carried out by wedging and packing from the lower edge before being fixed with grouted dowels.

Typical Details ~

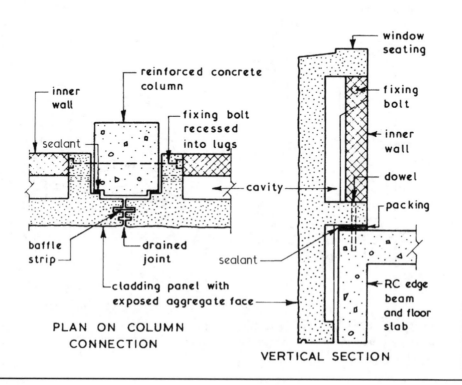

PLAN ON COLUMN
CONNECTION

VERTICAL SECTION

Storey Height Cladding Panels ~ these are designed to span vertically from beam to beam and can be fenestrated if required. Levelling is usually carried out by wedging and packing from floor level before being fixed by bolts or grouted dowels.

Typical Details ~

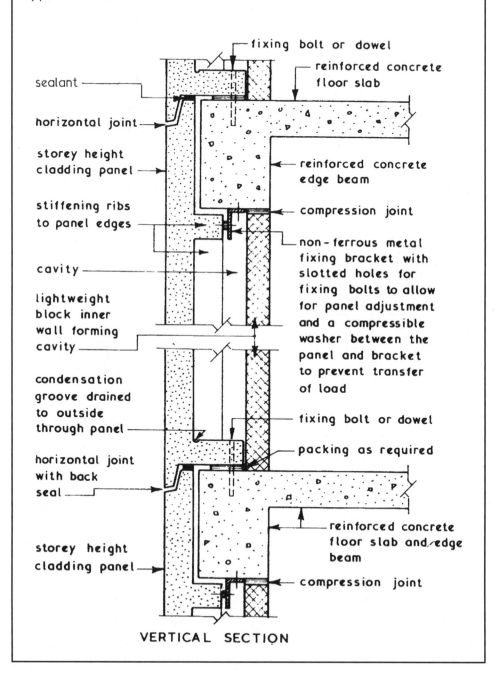

VERTICAL SECTION

fixing bolt or dowel

reinforced concrete floor slab

sealant

horizontal joint

storey height cladding panel

stiffening ribs to panel edges

reinforced concrete edge beam

compression joint

non-ferrous metal fixing bracket with slotted holes for fixing bolts to allow for panel adjustment and a compressible washer between the panel and bracket to prevent transfer of load

cavity

lightweight block inner wall forming cavity

condensation groove drained to outside through panel

fixing bolt or dowel

packing as required

horizontal joint with back seal

reinforced concrete floor slab and edge beam

storey height cladding panel

compression joint

Single-Stage ~ the application of a compressible filling material and a weatherproofing sealant between adjacent cladding panels. This may be adequate for relatively small areas and where exposure to thermal or structural movement is limited. Elsewhere, in order to accommodate extremes of thermal movement between exposed claddings, the use of only a sealant and filler would require an over-frequency of joints or over-wide joints that could slump or fracture.

Two-Stage ~ otherwise known as open drained joints. The preferred choice as there is a greater facility to absorb movement. Drained joints to cladding panels comprise a sealant to the inside or back of the joint and a baffle to the front, both separated by an air seal.

Comparison of single- and two-stage jointing principles ~

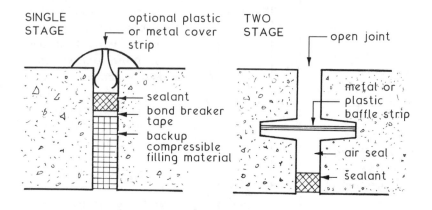

Plan views

Typical coefficients of linear thermal expansion (10^{-6} m/mK) ~
Dense concrete aggregate 14, lightweight aggregate 10.

Ref. BS 6093: Design of joints and jointing in building construction.

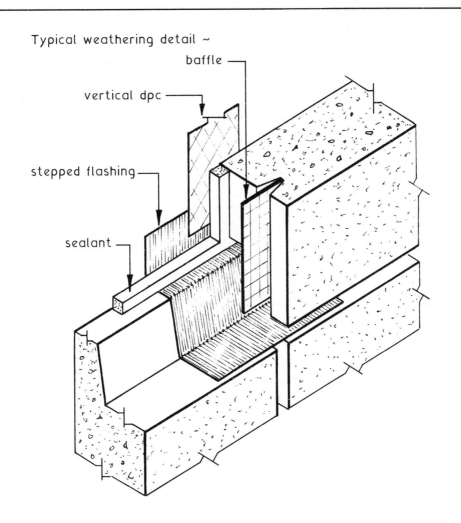

Typical weathering detail ~

baffle

vertical dpc

stepped flashing

sealant

Where the horizontal lapped joint between upper and lower cladding panels coincides with the vertical open drained joint, a stepped apron flashing is required to weather the intersection. Lead is the natural choice for this, but reinforced synthetic rubber or reinforced plastic sheet may be preferred to avoid possible lead oxide staining over the panel surface.

Baffle material is traditionally of non-ferrous metal such as copper, but like lead this can cause staining to the surface. Neoprene, butyl rubber or PVC are alternatives.

Gasket ~ an alternative to using mastic or sealant to close the gap between two cladding panels. They are used specifically where movements or joint widths are greater than could be accommodated by sealants. For this purpose a gasket is defined in BS 6093 as, 'flexible, generally elastic, preformed material that constitutes a seal when compressed'.

Location and fit ~ as shown on the next page, a recess is provided in at least one of the two adjacent claddings. To be effective contact surfaces must be clean and free of imperfections for a gasket to exert pressure on adjacent surfaces and to maintain this during all conditions of exposure. To achieve this, greater dimensional accuracy in manufacture and assembly of components is necessary relative to other sealing systems.

Profile ~ solid or hollow extrusions in a variety of shapes. Generally non-structural but vulcanised polychloroprene rubber can be used if a structural specification is required.

Materials (non-structural) ~ synthetic rubber including neoprene, silicone, ethylene propylene diene monomer (EPDM) and thermoplastic rubber (TPR). These materials are very durable with excellent resistance to compression, heat, water, ultra-violet light, ozone, aging, abrasion and chemical cleaning agents such as formaldehyde. They also have exceptional elastic memory, ie. will resume original shape after stressing. Poly vinyl chloride (PVC) and similar plastics can also be used but they will need protection from the effects of direct sunlight.

Refs.

BS 4255-1: Rubber used in preformed gaskets for weather exclusion from buildings. Specification for non-structural gaskets.

BS 6093: Design of joints and jointing in building construction. Guide.

Non-structural gasket types (shown plan view) ~

HOLLOW/TUBULAR (cast into concrete during manufacture)

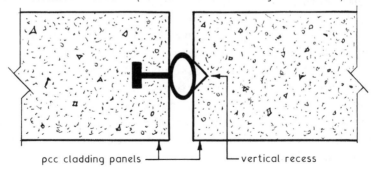

pcc cladding panels ——⸺————— vertical recess

SOLID DOUBLE SPLAY (fitted on site during assembly)

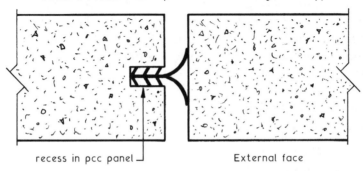

recess in pcc panel ⌐   External face

SOLID CRUCIFORM   shoulder to tapered recess

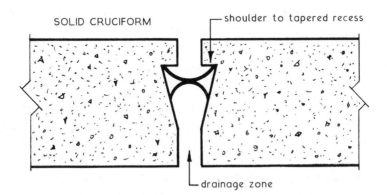

drainage zone

Gaskets have other applications to construction. See Double Glazing, Triple Glazing and Curtain Walling.

Concrete Surface Finishes ~ it is not easy to produce a concrete surface with a smooth finish of uniform colour direct from the mould or formwork since the colour of the concrete can be affected by the cement and fine aggregate used. The concrete surface texture can be affected by the aggregate grading, cement content, water content, degree of compaction, pin holes caused by entrapped air and rough patches caused by adhesion to parts of the formworks. Complete control over the above mentioned causes is difficult under ideal factory conditions and almost impossible under normal site conditions. The use of textured and applied finishes has therefore the primary function of improving the appearance of the concrete surface and in some cases it will help to restrict the amount of water which reaches a vertical joint.

Casting ~ concrete components can usually be cast in-situ or precast in moulds. Obtaining a surface finish to concrete cast in-situ is usually carried out against a vertical face, whereas precast concrete components can be cast horizontally and treated on either upper or lower mould face. Apart from a plain surface concrete the other main options are:-

1. Textured and profiled surfaces.
2. Tooled finishes.
3. Cast-on finishes. (see next page)
4. Exposed aggregate finishes. (see next page)

Textured and Profiled Surfaces ~ these can be produced on the upper surface of a horizontal casting by rolling, tamping, brushing and sawing techniques but variations in colour are difficult to avoid. Textured and profiled surfaces can be produced on the lower face of a horizontal casting by using suitable mould linings.

Tooled Finishes ~ the surface of hardened concrete can be tooled by bush hammering, point tooling and grinding. Bush hammering and point tooling can be carried out by using an electric or pneumatic hammer on concrete which is at least three weeks old provided gravel aggregates have not been used since these tend to shatter leaving surface pits. Tooling up to the arris could cause spalling therefore a 10 mm wide edge margin should be left untooled. Grinding the hardened concrete consists of smoothing the surface with a rotary carborundum disc which may have an integral water feed. Grinding is a suitable treatment for concrete containing the softer aggregates such as limestone.

Cast-on Finishes ~ these finishes include split blocks, bricks, stone, tiles and mosaic. Cast-on finishes to the upper surface of a horizontal casting are not recommended although such finishes could be bedded onto the fresh concrete. Lower face treatment is by laying the materials with sealed or grouted joints onto the base of mould or alternatively the materials to be cast-on may be located in a sand bed spread over the base of the mould.

Exposed Aggregate Finishes ~ attractive effects can be obtained by removing the skin of hardened cement paste or surface matrix, which forms on the surface of concrete, to expose the aggregate. The methods which can be employed differ with the casting position.

Horizontal Casting – treatment to the upper face can consist of spraying with water and brushing some two hours after casting, trowelling aggregate into the fresh concrete surface or by using the felt-float method. This method consists of trowelling 10 mm of dry mix fine concrete onto the fresh concrete surface and using the felt pad to pick up the cement and fine particles from the surface leaving a clean exposed aggregate finish.

Treatment to the lower face can consist of applying a retarder to the base of the mould so that the partially set surface matrix can be removed by water and/or brushing as soon as the castings are removed from the moulds. When special face aggregates are used the sand bed method could be employed.

Vertical Casting – exposed aggregate finishes to the vertical faces can be obtained by tooling the hardened concrete or they can be cast-on by the aggregate transfer process. This consists of sticking the selected aggregate onto the rough side of pegboard sheets with a mixture of water soluble cellulose compounds and sand fillers. The cream like mixture is spread evenly over the surface of the pegboard to a depth of one third the aggregate size and the aggregate sprinkled or placed evenly over the surface before being lightly tamped into the adhesive. The prepared board is then set aside for 36 hours to set before being used as a liner to the formwork or mould. The liner is used in conjunction with a loose plywood or hardboard baffle placed against the face of the aggregate. The baffle board is removed as the concrete is being placed.

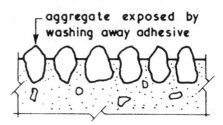

aggregate exposed by washing away adhesive

Discolouration ~ manifests as a patchy surface finish. It is caused where there are differences in hydration or moisture loss during the concrete set, due to concentrations of cement or where aggregates become segregated. Both of these will produce moisture content differences at the surface. Areas with a darker surface indicate the greater loss of moisture, possibly caused by insufficient mixing and/or poorly sealed formwork producing differences in surface absorption.

Crazing ~ surface shrinkage cracks caused by a cement rich surface skin or by too much water in the mix. Out-of-date cement can have the same effect as well as impairing the strength of the concrete.

Lime bloom ~ a chalky surface deposit produced when the calcium present in cement reacts to contamination from moisture in the atmosphere or rainwater during the hydration process. Generally resolved by dry brushing or with a 20:1 water/hydrochloric acid wash.

Scabbing ~ small areas or surface patches of concrete falling away as the formwork is struck. Caused by poor preparation of formwork, ie. insufficient use of mould oil or by formwork having a surface texture that is too rough.

Blow holes ~ otherwise known as surface popping. Possible causes are use of formwork finishes with nil or low absorbency or by insufficient vibration of concrete during placement.

Rust staining ~ if not caused by inadequate concrete cover to reinforcement, this characteristic is quite common where iron rich aggregates or pyrites are used. Rust-brown stains are a feature and there may also be some cracking where the iron reacts with the cement.

Dusting ~ caused by unnaturally rapid hardening of concrete and possibly where out-of-date cement is used. The surface of set concrete is dusty and friable.

7 INTERNAL CONSTRUCTION AND FINISHES

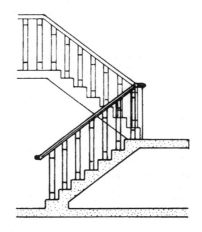

INTERNAL ELEMENTS

INTERNAL WALLS

CONSTRUCTION JOINTS

PARTITIONS AND TIMBER STRUT DESIGN

PLASTERS, PLASTERING AND PLASTERBOARD

DRY LINING TECHNIQUES

WALL TILING

DOMESTIC FLOORS AND FINISHES

LARGE CAST IN-SITU GROUND FLOORS

CONCRETE FLOOR SCREEDS

TIMBER SUSPENDED FLOORS

TIMBER BEAM DESIGN

REINFORCED CONCRETE SUSPENDED FLOORS

PRECAST CONCRETE FLOORS

RAISED ACCESS FLOORS

SOUND INSULATION

TIMBER, CONCRETE AND METAL STAIRS

INTERNAL DOORS

FIRE RESISTING DOORS

PLASTERBOARD CEILINGS

SUSPENDED CEILINGS

PAINTS AND PAINTING

JOINERY PRODUCTION

COMPOSITE BOARDING

PLASTICS IN BUILDING

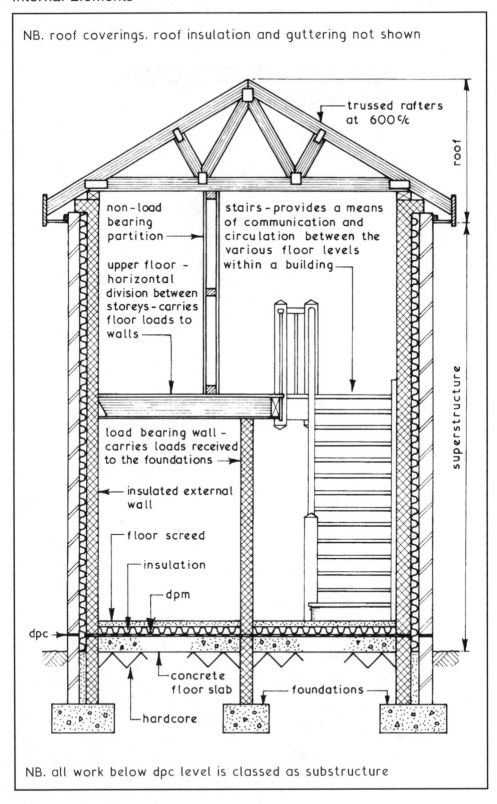

NB. roof coverings, roof insulation and guttering not shown

trussed rafters at 600 c/c

roof

non-load bearing partition

stairs - provides a means of communication and circulation between the various floor levels within a building

upper floor - horizontal division between storeys - carries floor loads to walls

superstructure

load bearing wall - carries loads received to the foundations

insulated external wall

floor screed

insulation

dpm

dpc

concrete floor slab

foundations

hardcore

NB. all work below dpc level is classed as substructure

Internal Walls ~ their primary function is to act as a vertical divider of floor space and in so doing form a storey height enclosing element.

Other Possible Functions:-

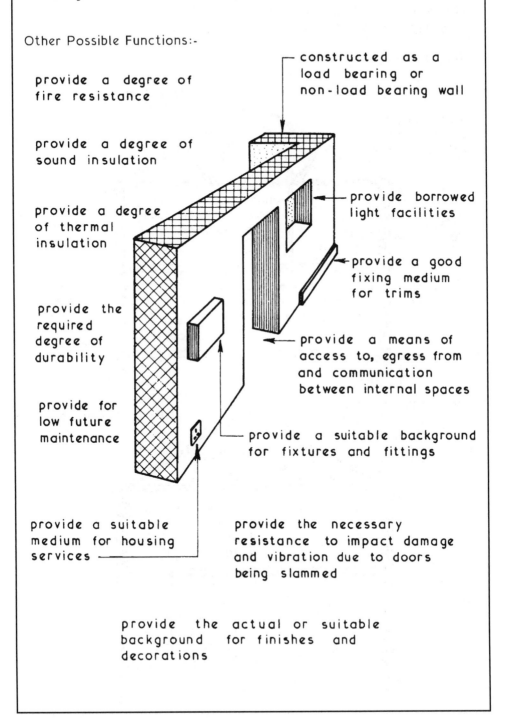

provide a degree of fire resistance

provide a degree of sound insulation

provide a degree of thermal insulation

provide the required degree of durability

provide for low future maintenance

provide a suitable medium for housing services

constructed as a load bearing or non-load bearing wall

provide borrowed light facilities

provide a good fixing medium for trims

provide a means of access to, egress from and communication between internal spaces

provide a suitable background for fixtures and fittings

provide the necessary resistance to impact damage and vibration due to doors being slammed

provide the actual or suitable background for finishes and decorations

Internal Walls ~ there are two basic design concepts for internal walls those which accept and transmit structural loads to the foundations are called Load Bearing Walls and those which support only their own self-weight and do not accept any structural loads are called Non-load Bearing Walls or Partitions.

Typical Examples ~

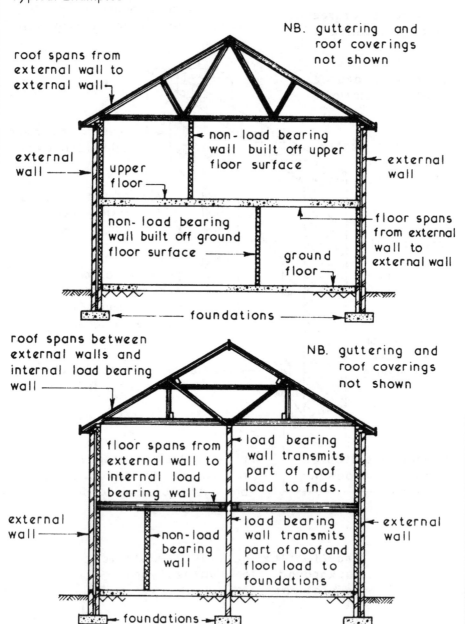

roof spans from external wall to external wall

NB. guttering and roof coverings not shown

external wall

upper floor

non-load bearing wall built off upper floor surface

external wall

non-load bearing wall built off ground floor surface

ground floor

floor spans from external wall to external wall

foundations

roof spans between external walls and internal load bearing wall

NB. guttering and roof coverings not shown

floor spans from external wall to internal load bearing wall

load bearing wall transmits part of roof load to fnds.

external wall

non-load bearing wall

load bearing wall transmits part of roof and floor load to foundations

external wall

foundations

Internal Brick Walls ~ these can be load bearing or non-load bearing (see previous page) and for most two storey buildings are built in half brick thickness in stretcher bond.

Typical Details ~

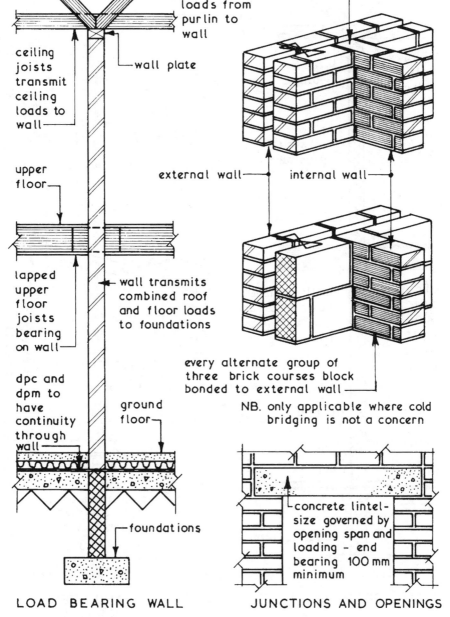

roof struts transmit loads from purlin to wall

wall plate

ceiling joists transmit ceiling loads to wall

upper floor

lapped upper floor joists bearing on wall

wall transmits combined roof and floor loads to foundations

dpc and dpm to have continuity through wall

ground floor

foundations

every alternate course bonded to external wall

external wall

internal wall

every alternate group of three brick courses block bonded to external wall

NB. only applicable where cold bridging is not a concern

concrete lintel- size governed by opening span and loading - end bearing 100 mm minimum

LOAD BEARING WALL

JUNCTIONS AND OPENINGS

Internal Block Walls ~ these can be load bearing or non-load bearing (see page 620) the thickness and type of block to be used will depend upon the loadings it has to carry.

Typical Details ~

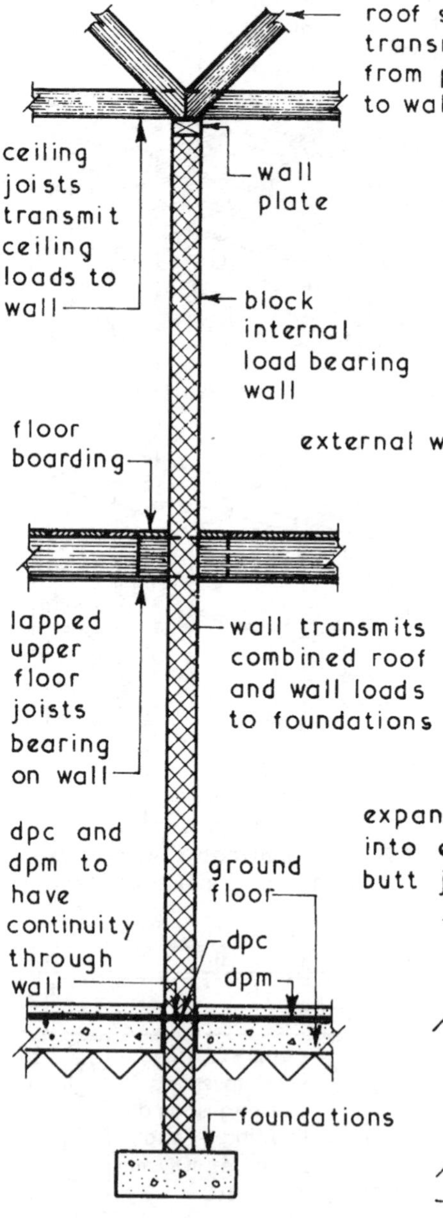

roof struts transmit loads from purlins to wall

every alternate course block bonded to external wall

ceiling joists transmit ceiling loads to wall

wall plate

block internal load bearing wall

floor boarding

external wall

internal wall

lapped upper floor joists bearing on wall

wall transmits combined roof and wall loads to foundations

dpc and dpm to have continuity through wall

ground floor

dpc

dpm

expanded metal strip built into every bed joint of butt jointed internal wall

foundations

concrete lintel - size governed by opening span and loading - end bearing 100 mm min.

LOAD BEARING WALL

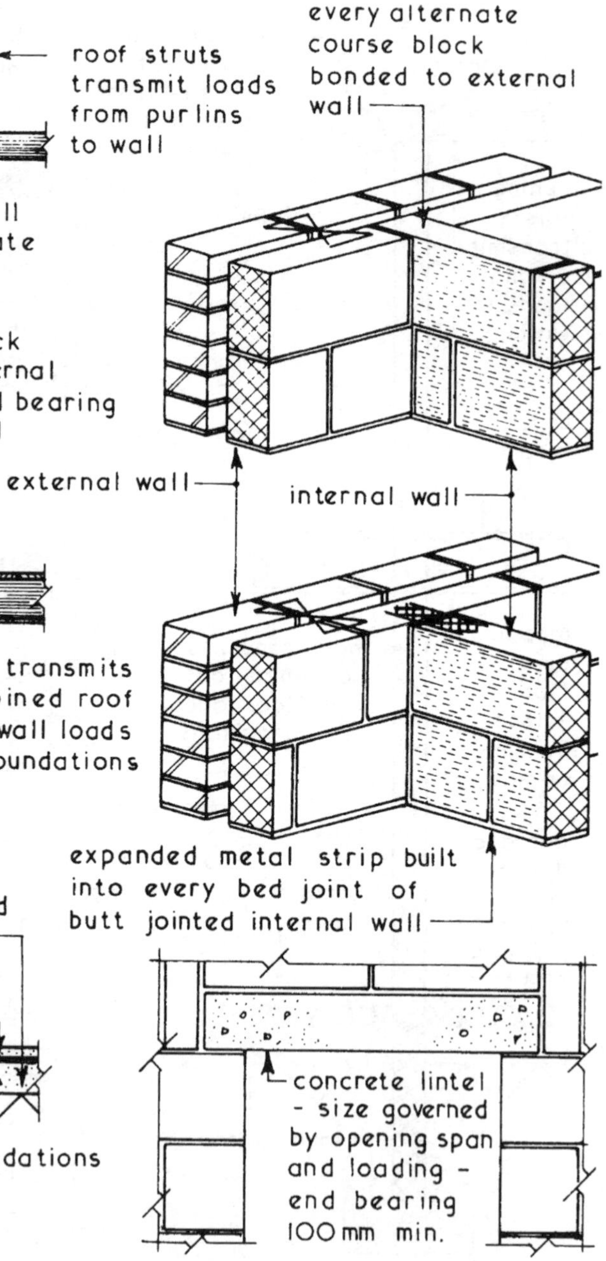

JUNCTIONS AND OPENINGS

Internal Walls ~ an alternative to brick and block bonding shown on the preceding two pages is application of wall profiles. These are quick and simple to install, provide adequate lateral stability, sufficient movement flexibility and will overcome the problem of thermal bridging where a brick partition would otherwise bond into a block inner leaf. They are also useful for attaching extension walls at right angles to existing masonry.

Application ~

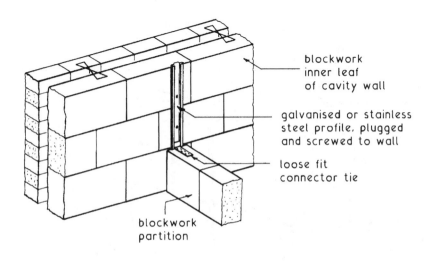

blockwork
inner leaf
of cavity wall

galvanised or stainless
steel profile, plugged
and screwed to wall

loose fit
connector tie

blockwork
partition

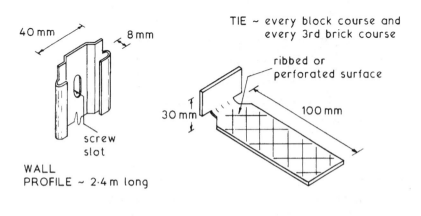

40 mm   8 mm

screw
slot

WALL
PROFILE ~ 2·4 m long

TIE ~ every block course and
every 3rd brick course

ribbed or
perforated surface

30 mm   100 mm

Movement or Construction Joints ~ provide an alternative to ties or mesh reinforcement in masonry bed joints. Even with reinforcement, lightweight concrete block walls are renowned for producing unsightly and possibly unstable shrinkage cracks. Galvanised or stainless steel formers and ties are built in at a maximum of 6m horizontal spacing and within 3m of corners to accommodate initial drying, shrinkage movement and structural settlement. One side of the former is fitted with profiled or perforated ties to bond into bed joints and the other has plastic sleeved ties. The sleeved tie maintains continuity, but restricts bonding to allow for controlled movement.

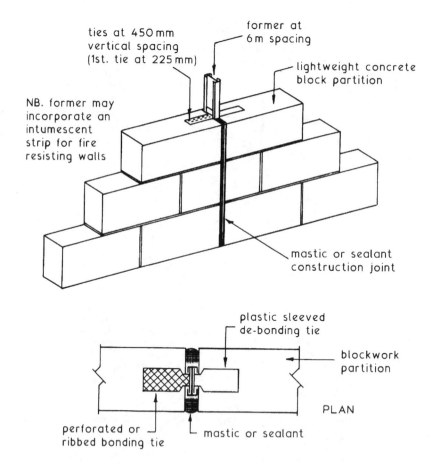

Note: Movement joints in clay brickwork should be provided at 12 m maximum spacing and 7.5 to 9m for calcium silicate.
Ref. BS 5628-3: Code of practice for use of masonry. Materials and components, design and workmanship.

Location ~ specifically in positions of high stress.
Reinforcement ~ expanded metal or wire mesh (see page 328).
Mortar Cover ~ 13 mm minimum thickness, 25 mm to external faces.

Openings~

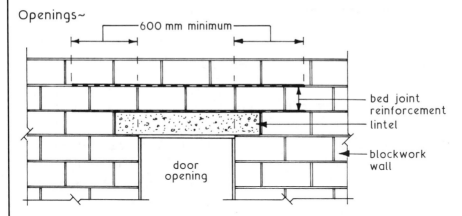

Concentrated Load ~

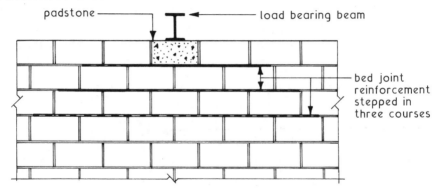

Suspended Floor~

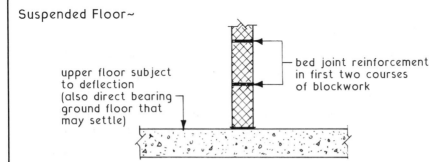

Differential Movement ~ may occur where materials such as steel, brick, timber or dense concrete abut with or bear on lightweight concrete blocks. A smooth separating interface of two layers of non-compressible dpc or similar is suitable in this situation.

Typical examples ~

Solid brickwork

T = thickness

	Fire resistance (minutes)		Material and application
	120	240	
T (mm)	102.5	215	Clay bricks. Load bearing or non-load bearing wall.
T (mm)	102.5	215	Concrete or sand/lime bricks. Load bearing or non-load bearing wall.

Note: For practical reasons a standard one-brick dimension is given for 240 minutes fire resistance. Theoretically a clay brick wall can be 170 mm and a concrete or sand/lime brick wall 200 mm, finishes excluded.

Solid concrete blocks of lightweight aggregate

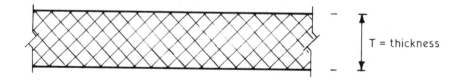

T = thickness

	Fire resistance (minutes)			Material and application
	60	120	240	
T (mm)	100	130	200	Load bearing, 2.8–3.5 N/mm^2 compressive strength.
T (mm)	90	100	190	Load bearing, 4.0–10 N/mm^2 compressive strength.
T (mm)	75	100	140	Non-load bearing, 2.8–3.5 N/mm^2 compressive strength.
T (mm)	75	75	100	Non-load bearing, 4.0–10 N/mm^2 compressive strength.

Note: Finishes excluded

Party Wall ~ a wall separating different owners buildings, ie. a wall that stands astride the boundary line between property of different ownerships. It may also be solely on one owner's land but used to separate two buildings.

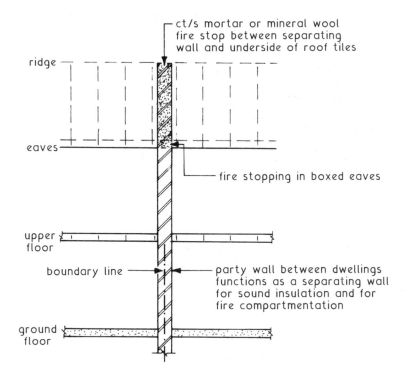

Where an internal separating wall forms a junction with an external cavity wall, the cavity must be fire stopped by using a barrier of fire resisting material. Depending on the application, the material specification is of at least 30 minutes fire resistance. Between terraced and semi-detached dwellings the location is usually limited by the separating elements. For other buildings additional fire stopping will be required in constructional cavities such as suspended ceilings, rainscreen cladding and raised floors. The spacing of these cavity barriers is generally not more than 20 m in any direction, subject to some variation as indicated in Volume 2 of Approved Document B.

Refs.
Party Wall Act 1996.
Building Regulations, A.D. B, Volumes 1 and 2: Fire safety.
Building Regulations, A.D. E: Resistance to the passage of sound.

Requirements for fire and sound resisting construction ~

Typical masonry construction ~

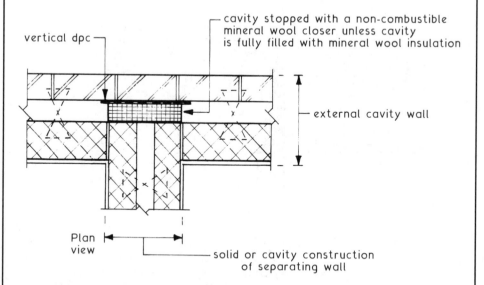

vertical dpc

cavity stopped with a non-combustible mineral wool closer unless cavity is fully filled with mineral wool insulation

external cavity wall

Plan view

solid or cavity construction of separating wall

Typical timber framed construction~

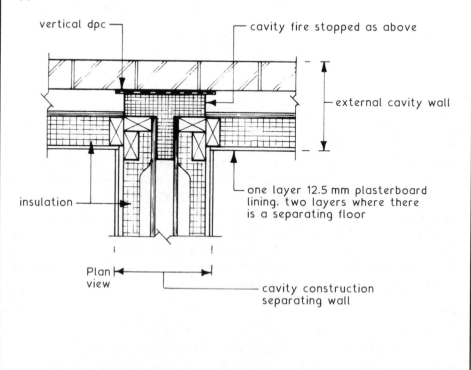

vertical dpc

cavity fire stopped as above

external cavity wall

insulation

one layer 12.5 mm plasterboard lining, two layers where there is a separating floor

Plan view

cavity construction separating wall

Internal Partitions ~ these are vertical dividers which are used to separate the internal space of a building into rooms and circulation areas such as corridors. Partitions which give support to a floor or roof are classified as load bearing whereas those which give no such support are called non-load bearing.

Load Bearing Partitions ~ these walls can be constructed of bricks, blocks or in-situ concrete by traditional methods and have the design advantages of being capable of having good fire resistance and/or high sound insulation. Their main disadvantage is permanence giving rise to an inflexible internal layout.

Non-load Bearing Partitions ~ the wide variety of methods available makes it difficult to classify the form of partition but most can be placed into one of three groups:-

1. Masonry partitions.
2. Stud partitions - see pages 630 to 632.
3. Demountable partitions - see pages 634 & 635.

Masonry Partitions ~ these are usually built with blocks of clay or lightweight concrete which are readily available and easy to construct thus making them popular. These masonry partitions should be adequately tied to the structure or load bearing walls to provide continuity as a sound barrier, provide edge restraint and to reduce the shrinkage cracking which inevitably occurs at abutments. Wherever possible openings for doors should be in the form of storey height frames to provide extra stiffness at these positions.

Typical Details ~

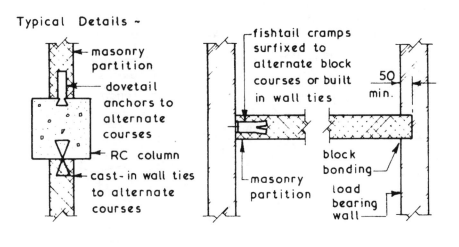

masonry partition

dovetail anchors to alternate courses

RC column

cast-in wall ties to alternate courses

fishtail cramps surfixed to alternate block courses or built in wall ties

masonry partition

50 min.

block bonding

load bearing wall

Timber Stud Partitions ~ these are non-load bearing internal dividing walls which are easy to construct, lightweight, adaptable and can be clad and infilled with various materials to give different finishes and properties. The timber studs should be of prepared or planed material to ensure that the wall is of constant thickness with parallel faces. Stud spacings will be governed by the size and spanning ability of the facing or cladding material.

Typical Details ~

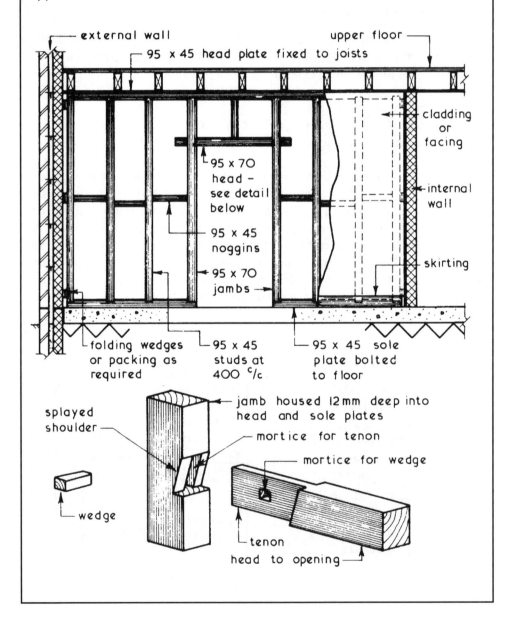

Although generally non-load bearing, timber stud partitions may carry some of the load from the floor and roof structure. In these situations the vertical studs are considered struts.

Example ~ using the stud frame dimensions shown on the previous page, with each stud (strut) supporting a 5kN load.

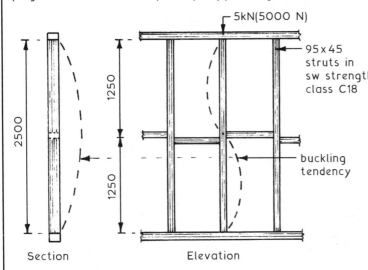

95×45 struts in sw strength class C18

Effective length of struts as shown on page 548. Position fixed at both ends, the effective length is the actual length.

buckling tendency

Section   Elevation

Slenderness ratio (SR) of section = Effective length ÷ breadth
On the partition face = 1250 ÷ 45 = 27.8
At right angles to the face = 2500 ÷ 95 = 26.3
Timber of strength classification C18 (see pages 114 and 115) has the following properties:
       Modulus of elasticity = 6000 N/mm^2
       Grade stress in compression parallel to the grain = 7.1 N/mm^2
       Grade stress ratio = 6000 ÷ 7.1 = 845

See table adapted from BS 5268-2 on page 161. By coordinating the SR of 27.8 (greater value) with a grade stress ratio of 845, a figure of 0.4 is obtained by interpolation.

Allowable applied stress is 7.1 N/mm^2 × 0.4 = 2.84 N/mm^2
Applied stress = axial load ÷ strut section area
          = 5000 N ÷ (95 mm × 45 mm) = 1.17 N/mm^2
1.17 N/mm^2 is well within the allowable stress of 2.84 N/mm^2 therefore 95 mm × 45 mm struts are adequate.

See pages 159 to 161 for an application to dead shoring. Struts in trusses and lattice frames can also be designed using the same principles.

Stud Partitions ~ these non-load bearing partitions consist of a framework of vertical studs to which the facing material can be attached. The void between the studs created by the two faces can be infilled to meet specific design needs. The traditional material for stud partitions is timber (see Timber Stud Partitions on page 630) but a similar arrangement can be constructed using metal studs faced on both sides with plasterboard.

Typical Metal Stud Partition Details ~

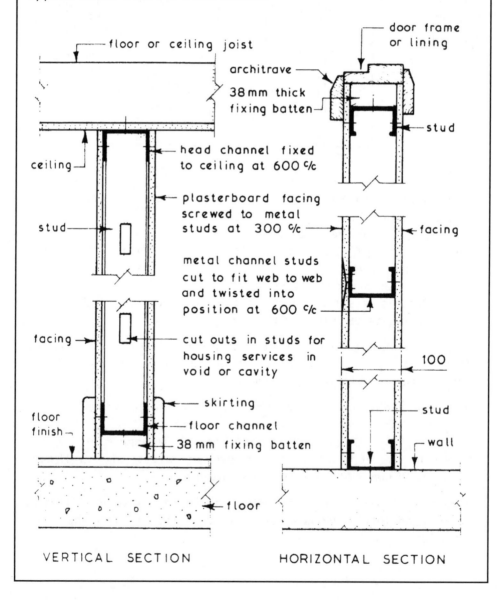

VERTICAL SECTION   HORIZONTAL SECTION

Plasterboard lining to stud framed partition walls satisfies the Building Regulations, Approved Document B – Fire safety, as a material of "limited combustibility" with a Class O rating for surface spread of flame (Class O is better than Classes 1 to 4 as determined by BS 476-7). The plasterboard dry walling should completely protect any combustible timber components such as sole plates. The following shows typical fire resistances as applied to a metal stud frame ~

30 minute fire resistance

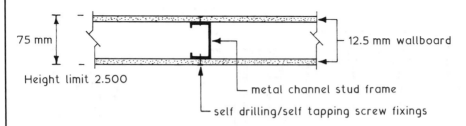

75 mm

Height limit 2.500

12.5 mm wallboard

metal channel stud frame

self drilling/self tapping screw fixings

60(90) minute fire resistance

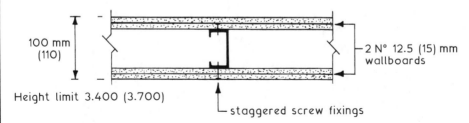

100 mm
(110)

Height limit 3.400 (3.700)

2 N° 12.5 (15) mm wallboards

staggered screw fixings

120 minute fire resistance

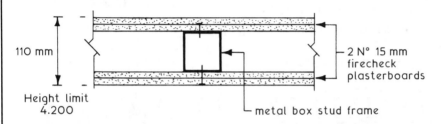

110 mm

Height limit 4.200

2 N° 15 mm firecheck plasterboards

metal box stud frame

For plasterboard types see page 642.

Partitions ~ these can be defined as vertical internal space dividers and are usually non-loadbearing. They can be permanent, constructed of materials such as bricks or blocks or they can be demountable constructed using lightweight materials and capable of being taken down and moved to a new location incurring little or no damage to the structure or finishes. There is a wide range of demountable partitions available constructed from a variety of materials giving a range that will be suitable for most situations. Many of these partitions have a permanent finish which requires no decoration and only periodic cleaning in the context of planned maintenance.

Typical Example ~

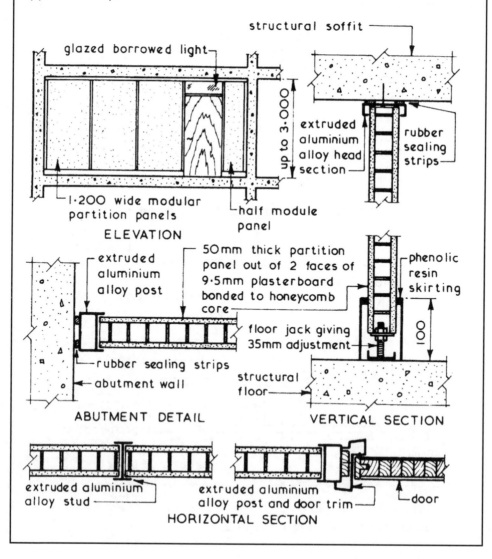

glazed borrowed light

structural soffit

up to 3·000

extruded aluminium alloy head section

rubber sealing strips

1·200 wide modular partition panels

half module panel

ELEVATION

extruded aluminium alloy post

50mm thick partition panel out of 2 faces of 9·5mm plasterboard bonded to honeycomb core

phenolic resin skirting

floor jack giving 35mm adjustment

rubber sealing strips

abutment wall

structural floor

100

ABUTMENT DETAIL

VERTICAL SECTION

extruded aluminium alloy stud

extruded aluminium alloy post and door trim

door

HORIZONTAL SECTION

Demountable Partitions ~ it can be argued that all internal non-load bearing partitions are demountable and therefore the major problem is the amount of demountability required in the context of ease of moving and the possible frequency anticipated. The range of partitions available is very wide including stud partitions, framed panel partitions (see Demountable Partitions on page 634), panel to panel partitions and sliding/folding partitions which are similar in concept to industrial doors (see Industrial Doors on pages 396 to 398) The latter type is often used where movement of the partition is required frequently. The choice is therefore based on the above stated factors taking into account finish and glazing requirements together with any personal preference for a particular system but in all cases the same basic problems will have to be considered:-

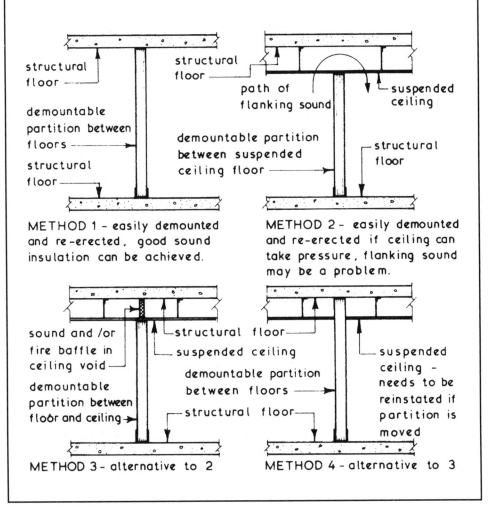

structural floor

demountable partition between floors

structural floor

METHOD 1 - easily demounted and re-erected, good sound insulation can be achieved.

structural floor

path of flanking sound

suspended ceiling

demountable partition between suspended ceiling floor

structural floor

METHOD 2 - easily demounted and re-erected if ceiling can take pressure, flanking sound may be a problem.

sound and /or fire baffle in ceiling void

demountable partition between floor and ceiling

structural floor

suspended ceiling

METHOD 3 - alternative to 2

structural floor

demountable partition between floors

structural floor

suspended ceiling - needs to be reinstated if partition is moved

METHOD 4 - alternative to 3

Plaster ~ this is a wet mixed material applied to internal walls as a finish to fill in any irregularities in the wall surface and to provide a smooth continuous surface suitable for direct decoration. The plaster finish also needs to have a good resistance to impact damage. The material used to fulfil these requirements is gypsum plaster. Gypsum is a crystalline combination of calcium sulphate and water. The raw material is crushed, screened and heated to dehydrate the gypsum and this process together with various additives defines its type as set out in BS EN 13279-1: Gypsum binders and gypsum plasters. Definitions and requirement.

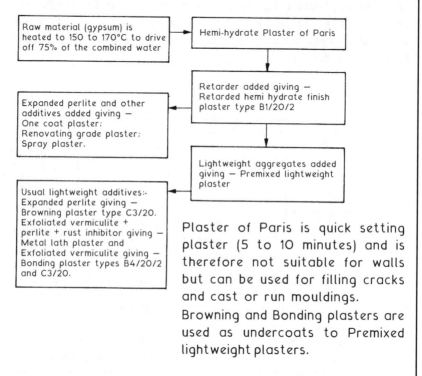

Plaster of Paris is quick setting plaster (5 to 10 minutes) and is therefore not suitable for walls but can be used for filling cracks and cast or run mouldings.

Browning and Bonding plasters are used as undercoats to Premixed lightweight plasters.

All plaster should be stored in dry conditions since any absorption of moisture before mixing may shorten the normal setting time of about one and a half hours which can reduce the strength of the set plaster. Gypsum plasters are not suitable for use in temperatures exceeding 43°C and should not be applied to frozen backgrounds.

A good key to the background and between successive coats is essential for successful plastering. Generally brick and block walls provide the key whereas concrete unless cast against rough formwork will need to be treated to provide the key.

Internal Wall Finishes ~ these can be classified as wet or dry. The traditional wet finish is plaster which is mixed and applied to the wall in layers to achieve a smooth and durable finish suitable for decorative treatments such as paint and wallpaper.

Most plasters are supplied in 25kg paper sacks and require only the addition of clean water or sand and clean water according to the type of plaster being used.

Typical Method of Application ~

1. PREPARATION

surface well brushed with hard broom to remove loose material and dust

chases cut before plastering

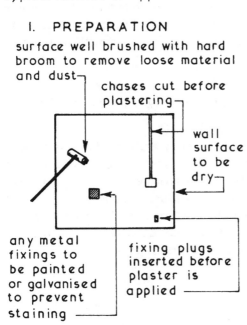

wall surface to be dry

any metal fixings to be painted or galvanised to prevent staining

fixing plugs inserted before plaster is applied

2. UNDERCOATING

thin coats of undercoat plaster applied and built up to required thickness

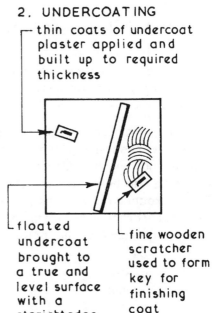

floated undercoat brought to a true and level surface with a straightedge

fine wooden scratcher used to form key for finishing coat

3. FINISHING

finishing coat of plaster applied with steel trowel to give a smooth finish

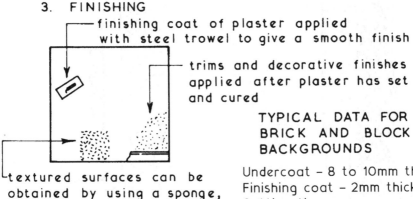

trims and decorative finishes applied after plaster has set and cured

textured surfaces can be obtained by using a sponge, hair brush, felt float or steel combs

TYPICAL DATA FOR BRICK AND BLOCK BACKGROUNDS

Undercoat – 8 to 10mm thick
Finishing coat – 2mm thick
Setting times ~
Undercoat – 2 hours
Finishing coat – 1 hour

Background ~ ideally level and of consistent material. If there are irregularities, three applications may be required; render (ct. and sand) 10–12 mm, undercoat plaster 6–8 mm and finish plaster 2 mm.

Difficult backgrounds such as steel or glazed surfaces require a PVA bonding agent or a cement slurry brushed on to improve plaster adhesion. A wire mesh or expanded metal surface attachment may also be required with metal lathing plaster as the undercoat. This may be mixed with sand in the ratio of 1:1.5.

Soft backgrounds of cork, fibreboard or expanded plastics should have wire mesh or expanded metal stapled to the surface. An undercoat of lightweight bonding plaster with compatible finish is suitable.

Dense regular surfaces of low-medium suction such as plasterboard require only one finishing coat of specially formulated finishing plaster.

Corners ~ reinforced with glass-fibre scrim tape or fine wire mesh to prevent shrinkage cracking at the junction of plasterboard ceiling and wall. Alternatively a preformed gypsum plaster moulding can be nailed or attached with plaster adhesive (see page 725). Expanded metal angle beads are specifically produced for external corner reinforcement. These are attached with plaster dabs or galvanised nails before finishing just below the nosing.

galvanised steel
expanded metal

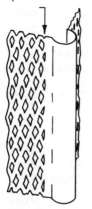

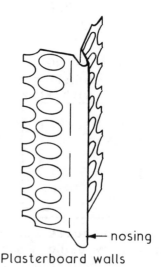

nosing

Plastered walls   Plasterboard walls

External Corner Beads

Plasterboard ~ a board material comprising two outer layers of lining paper with gypsum plaster between – two edge profiles are generally available:-

Tapered Edge – a flush seamless surface is obtained by filling the joint with a special filling plaster, applying a joint tape over the filling and finishing with a thin layer of joint filling plaster, the edge of which is feathered out using a slightly damp jointing sponge or steel trowel.

Square Edge – edges are close butted and finished with a cover fillet or the joint is covered with a glass-fibre scrim tape before being plastered.

Some jointing details are shown on page 644.

Typical Details ~

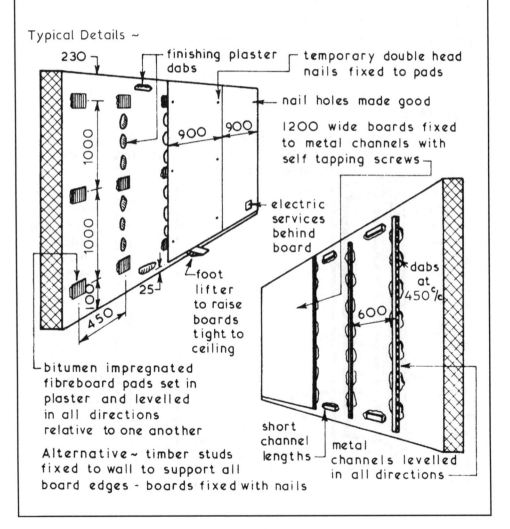

Dry Linings ~ an alternative to wet finishing internal wall surfaces with render and plaster. Dry lining materials can be plasterboard, insulating fibre board, hardboard, timber boards, and plywood, all of which can be supplied with a permanent finish or they can be supplied to accept an applied finish such as paint or wallpaper. For plasterboard a dry wall sealer should be applied before wallpapering to permit easier removal at a later date. The main purpose of lining an internal wall surface is to provide an acceptable but not necessarily an elegant or expensive wall finish. It is also very difficult and expensive to build a brick or block wall which has a fair face to both sides since this would involve the hand selection of bricks and blocks to ensure a constant thickness together with a high degree of skill to construct a satisfactory wall. The main advantage of dry lining walls is that the drying out period required with wet finishes is eliminated. By careful selection and fixing of some dry lining materials it is possible to improve the thermal insulation properties of a wall. Dry linings can be fixed direct to the backing by means of a recommended adhesive or they can be fixed to a suitable arrangement of wall battens.

Typical Example ~

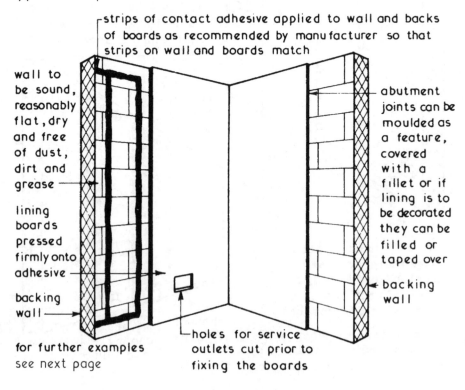

strips of contact adhesive applied to wall and backs of boards as recommended by manufacturer so that strips on wall and boards match

wall to be sound, reasonably flat, dry and free of dust, dirt and grease

lining boards pressed firmly onto adhesive

backing wall

for further examples see next page

holes for service outlets cut prior to fixing the boards

abutment joints can be moulded as a feature, covered with a fillet or if lining is to be decorated they can be filled or taped over

backing wall

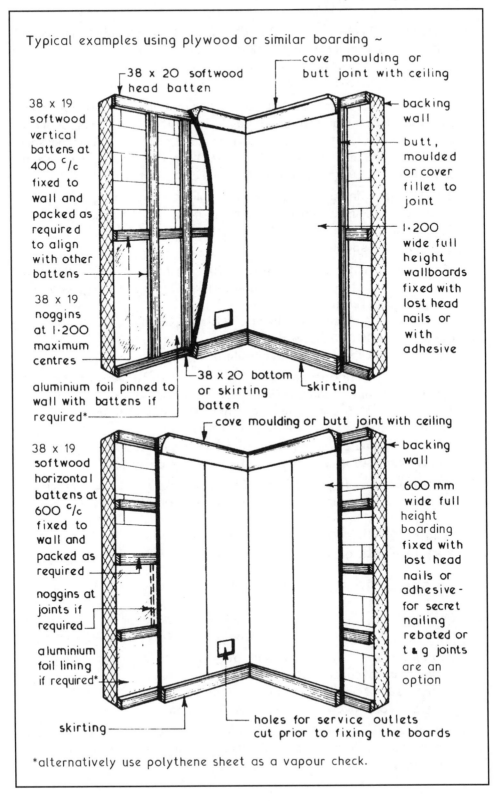

Typical examples using plywood or similar boarding ~

cove moulding or butt joint with ceiling

38 x 20 softwood head batten

backing wall

38 x 19 softwood vertical battens at 400 c/c fixed to wall and packed as required to align with other battens

butt, moulded or cover fillet to joint

38 x 19 noggins at 1.200 maximum centres

1.200 wide full height wallboards fixed with lost head nails or with adhesive

aluminium foil pinned to wall with battens if required*

38 x 20 bottom or skirting batten

skirting

cove moulding or butt joint with ceiling

backing wall

38 x 19 softwood horizontal battens at 600 c/c fixed to wall and packed as required

600 mm wide full height boarding fixed with lost head nails or adhesive - for secret nailing rebated or t & g joints are an option

noggins at joints if required

aluminium foil lining if required*

skirting

holes for service outlets cut prior to fixing the boards

*alternatively use polythene sheet as a vapour check.

641

Plasterboard Types ~ to BS EN 520: Gypsum plasterboards. Definitions, requirements and test methods.

BS PLASTERBOARDS:~

1. Wallboard – ivory faced for taping, jointing and direct decoration; grey faced for finishing plaster or wall adhesion with plaster dabs. General applications, i.e. internal walls, ceilings and partitions. Thicknesses: 9·5, 12·5 and 15 mm. Widths: 900 and 1200 mm. Lengths: vary between 1800 and 3000 mm. Edge profile square or tapered.

2. Baseboard – lining ceilings requiring direct plastering.
Thickness: 9·5 mm. Width: 900 mm. Length: 1220 mm and,
Thickness: 12·5 mm. Width: 600 mm. Length: 1220 mm.
Edge profile square.

3. Moisture Resistant – wallboard for bathrooms and kitchens. Pale green colour to face and back. Ideal base for ceramic tiling or plastering.
Thicknesses: 12·5 mm and 15 mm. Width: 1200 mm.
Lengths: 2400, 2700 and 3000 mm.
Square and taper edges available.

4. Firecheck – wallboard of glass fibre reinforced vermiculite and gypsum for fire cladding. Pink face and grey back.
Thicknesses: 12·5 and 15 mm. Widths: 900 and 1200 mm.
Lengths: 1800, 2400, 2700 and 3000 mm.
A 25 mm thickness is also produced, 600 mm wide × 3000 mm long.
Plaster finished if required. Square or tapered edges.

5. Plank – used as fire protection for structural steel and timber, in addition to sound insulation in wall panels and floating floors.
Thickness: 19 mm. Width: 600 mm.
Lengths: 2350, 2400, 2700 and 3000 mm.
Ivory face with grey back. Tapered edge.

NON – STANDARD PLASTERBOARDS:~

1. Contour – only 6 mm in thickness to adapt to curved featurework. Width: 1200 mm. Lengths: 2400 m and 3000 mm.

2. Vapourcheck – a metallized polyester wallboard lining to provide an integral water vapour control layer.
Thicknesses: 9·5 and 12·5 mm. Widths: 900 and 1200 mm.
Lengths: vary between 1800 and 3000 mm.

3. Thermalcheck – various expanded or foamed insulants are bonded to wallboard. Approximately 25–50 mm overall thickness in board sizes 1200 × 2400 mm.

Fixing ~ the detail below shows board fixing with nails or screws to timber battens. This is an alternative to using plaster dabs and pads shown on page 639.

Typical fixing ~

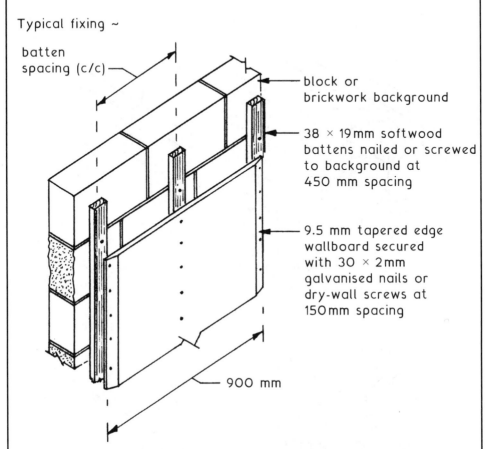

batten spacing (c/c)

block or brickwork background

38 × 19mm softwood battens nailed or screwed to background at 450 mm spacing

9.5 mm tapered edge wallboard secured with 30 × 2mm galvanised nails or dry-wall screws at 150mm spacing

900 mm

Guide to securing plasterboard to battens ~

Board thickness (mm)	Board width (mm)	Batten spacing (mm)	Nail or dry-wall screw length (mm)
9.5	900	450	30
9.5	1200	400	30
12.5	600	600	40
12.5	900	450	40
19.0	600	600	60

Horizontal joints are avoided if possible with cross battens only required at the top and bottom to provide rigidity. Board fixing is with galvanised steel taper-head nails of 2mm diameter (2.6mm dia. for 19mm boards) or with dry-wall screws.

Jointing ~ boards should not directly abut, instead a gap of 3 to 5mm should be provided between adjacent boards for plaster filling and joint reinforcement tape. The illustrations show various applications.

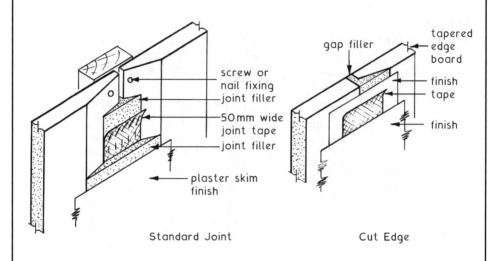

screw or
nail fixing
joint filler
50mm wide
joint tape
joint filler
plaster skim
finish

Standard Joint

gap filler
tapered
edge
board
finish
tape
finish

Cut Edge

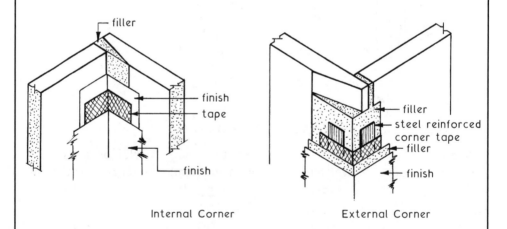

filler
finish
tape
finish

Internal Corner

filler
steel reinforced
corner tape
filler
finish

External Corner

Note: Paper jointing tape is generally specified for dry lining tapered edge boards. External corners are reinforced and strengthened with a particular type of tape that has two strips of thin steel attached.

644

Glazed Wall Tiles ~ internal glazed wall tiles are usually made to the various specifications under BS EN 14411: Ceramic tiles. Definitions, classification, characteristics and marking.

Internal Glazed Wall Tiles ~ the body of the tile can be made from ball-clay, china clay, china stone, flint and limestone. The material is usually mixed with water to the desired consistency, shaped and then fired in a tunnel oven at a high temperature (1150°C) for several days to form the unglazed biscuit tile. The glaze pattern and colour can now be imparted onto to the biscuit tile before the final firing process at a temperature slightly lower than that of the first firing (1050°C) for about two days.

Typical Internal Glazed Wall Tiles and Fittings ~

Sizes – Modular   100 × 100 × 5 mm thick and
                           200 × 100 × 6·5 mm thick.

    Non-modular   152 × 152 × 5 to 8 mm thick and
                           108 × 108 × 4 and 6·5 mm thick.

Other sizes – 200 × 300, 250 × 330, 250 × 400, 300 × 450, 300 × 600 and 330 × 600mm.

Fittings – wide range available particularly in the non-modular format.

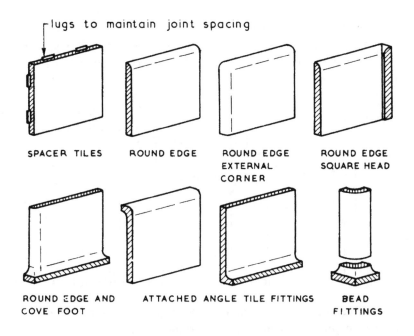

lugs to maintain joint spacing

SPACER TILES   ROUND EDGE   ROUND EDGE EXTERNAL CORNER   ROUND EDGE SQUARE HEAD

ROUND EDGE AND COVE FOOT   ATTACHED ANGLE TILE FITTINGS   BEAD FITTINGS

Bedding of Internal Wall Tiles ~ generally glazed internal wall tiles are considered to be inert in the context of moisture and thermal movement, therefore if movement of the applied wall tile finish is to be avoided attention must be given to the background and the method of fixing the tiles.

Backgrounds ~ these are usually of a cement rendered or plastered surface and should be flat, dry, stable, firmly attached to the substrate and sufficiently established for any initial shrinkage to have taken place. The flatness of the background should be not more than 3 mm in 2·000 for the thin bedding of tiles and not more than 6 mm in 2·000 for thick bedded tiles.

Fixing Wall Tiles ~ two methods are in general use:-

1. Thin Bedding – lightweight internal glazed wall tiles fixed dry using a recommended adhesive which is applied to the wall in small areas 1m² at a time with a notched trowel, the tile being pressed into the adhesive.
2. Thick Bedding – cement mortar within the mix range of 1:3 to 1:4 can be used or a proprietary adhesive, either by buttering the backs of the tiles which are then pressed into position or by rendering the wall surface to a thickness of approximately 10 mm and then applying thin bedded tiles to the rendered wall surface within two hours.

Grouting ~ when the wall tiles have set, the joints can be grouted by rubbing into the joints a grout paste either using a sponge or brush. Most grouting materials are based on cement with inert fillers and are used neat.

Typical Example ~

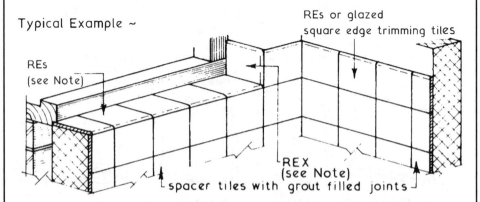

REs or glazed square edge trimming tiles

REs (see Note)

REX (see Note)

spacer tiles with grout filled joints

Note: The alternative treatment at edges is application of a radiused profile plastic trimming to standard spacer tiles.

Primary Functions ~

1. Provide a level surface with sufficient strength to support the imposed loads of people and furniture.
2. Exclude the passage of water and water vapour to the interior of the building.
3. Provide resistance to unacceptable heat loss through the floor.
4. Provide the correct type of surface to receive the chosen finish.

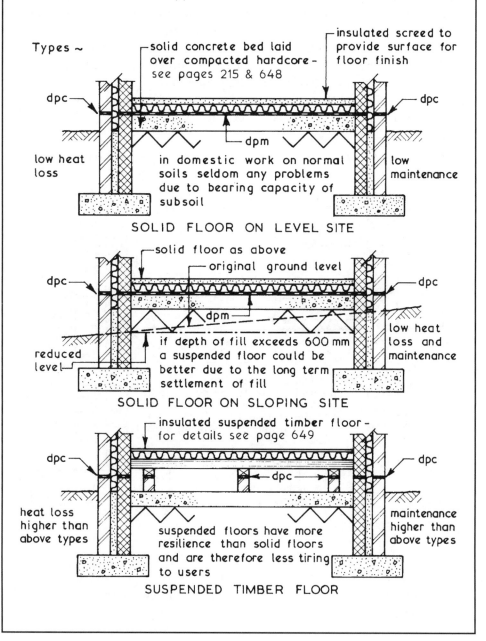

Types ~

SOLID FLOOR ON LEVEL SITE

SOLID FLOOR ON SLOPING SITE

SUSPENDED TIMBER FLOOR

This drawing should be read in conjunction with page 215 – Foundation Beds.

A domestic solid ground floor consists of three components:-

1. Hardcore – a suitable filling material to make up the top soil removal and reduced level excavations. It should have a top surface which can be rolled out to ensure that cement grout is not lost from the concrete. It may be necessary to blind the top surface with a layer of sand especially if the damp-proof membrane is to be placed under the concrete bed.

2. Damp-proof Membrane – an impervious layer such as heavy duty polythene sheeting to prevent moisture passing through the floor to the interior of the building.

3. Concrete Bed – the component providing the solid level surface to which screeds and finishes can be applied.

Typical Details ~

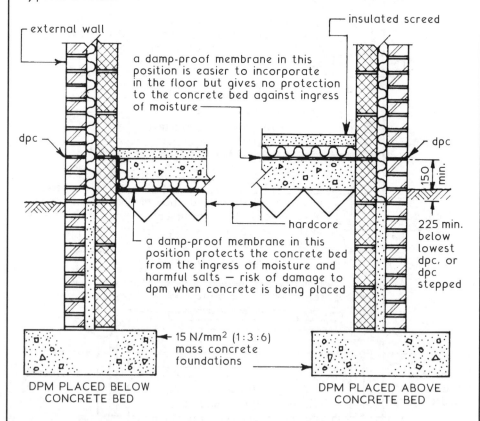

DPM PLACED BELOW
CONCRETE BED

DPM PLACED ABOVE
CONCRETE BED

NB. a compromise to the above methods is to place the dpm in the middle of the concrete bed but this needs two concrete pouring operations.

Suspended Timber Ground Floors ~ these need to have a well ventilated space beneath the floor construction to prevent the moisture content of the timber rising above an unacceptable level (i.e. not more than 20%) which would create the conditions for possible fungal attack.

Typical Details ~

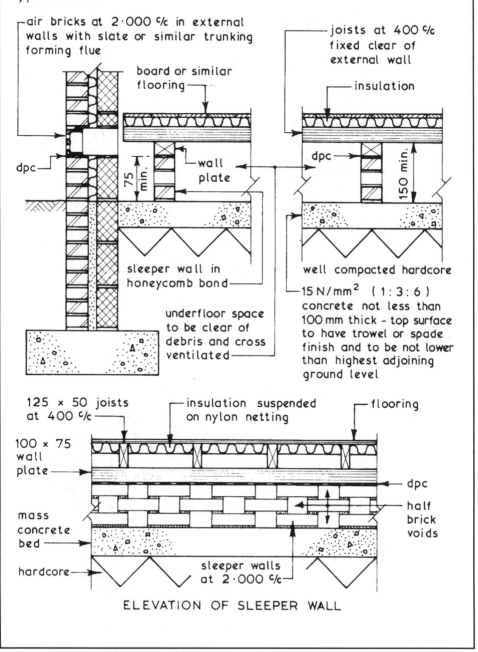

air bricks at 2·000 c/c in external walls with slate or similar trunking forming flue

board or similar flooring

joists at 400 c/c fixed clear of external wall

insulation

dpc

75 min.

wall plate

dpc

150 min.

sleeper wall in honeycomb bond

underfloor space to be clear of debris and cross ventilated

well compacted hardcore

15 N/mm² (1:3:6) concrete not less than 100 mm thick - top surface to have trowel or spade finish and to be not lower than highest adjoining ground level

125 × 50 joists at 400 c/c

insulation suspended on nylon netting

flooring

100 × 75 wall plate

dpc

mass concrete bed

half brick voids

hardcore

sleeper walls at 2·000 c/c

ELEVATION OF SLEEPER WALL

649

Domestic Suspended Concrete Ground Floors

Precast Concrete Floors ~ these have been successfully adapted from commercial building practice (see pages 673 and 674), as an economic alternative construction technique for suspended timber and solid concrete domestic ground (and upper) floors. See also page 352 for special situations.

Typical Details ~

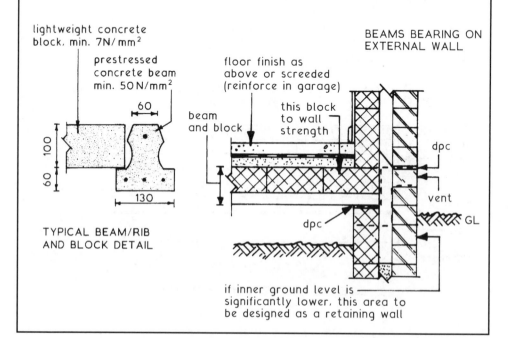

BEAMS PARALLEL WITH EXTERNAL WALL

cavity wall
insulated as required

cavity tray
over vent

18 mm t&g chipboard
over vapour control layer

insulation

dpc
stepped
ventilator

GL

dpc

coursing
slip 150 mm min. block

beam and

organic material stripped;
surface treated with weed
killer: lower level than
adjacent ground if free
draining (not Scotland)

POLYPROPYLENE
VENTILATOR

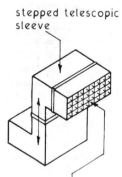

stepped telescopic
sleeve

grill clips to sleeve;
1500 mm² /m run of
wall OR 500 mm² /m²
of floor area (take
greater value)

lightweight concrete
block, min. 7N/mm²

prestressed
concrete beam
min. 50N/mm²

floor finish as
above or screeded
(reinforce in garage)

BEAMS BEARING ON
EXTERNAL WALL

60

100

60

130

beam
and block

this block
to wall
strength

dpc

vent

dpc

GL

TYPICAL BEAM/RIB
AND BLOCK DETAIL

if inner ground level is
significantly lower, this area to
be designed as a retaining wall

650

Precast Reinforced Concrete Beam and Expanded Polystyrene (EPS) Block Floors ~ these have evolved from the principles of beam and block floor systems as shown on the preceding page. The light weight and easy to cut properties of the blocks provide for speed and simplicity in construction. Exceptional thermal insulation is possible, with U-values as low as 0.2 W/m²K.

Typical detail ~

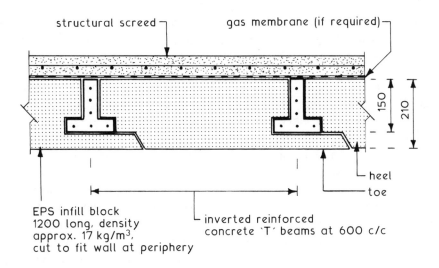

Cold Bridging ~ this is prevented by the EPS "toe" projecting beneath the underside of the concrete beam.

Structural Floor Finish ~ 50 mm structural concrete (C30grade) screed, reinforced with 10 mm steel Type A square mesh or with polypropylene fibres in the mix. A low-density polyethylene (LDPE) methane/radon gas membrane can be incorporated under the screed if local conditions require it.

Floating Floor Finish ~ subject to the system manufacturer's specification and accreditation, 18 mm flooring grade moisture resistant chipboard can be used over a 1000 gauge polythene vapour control layer. All four tongued and grooved edges of the chipboard are glued for continuity.

Floor Finishes ~ these are usually applied to a structural base but may form part of the floor structure as in the case of floor boards. Most finishes are chosen to fulfil a particular function such as:-

1. Appearance – chosen mainly for their aesthetic appeal or effect but should however have reasonable wearing properties. Examples are carpets; carpet tiles and wood blocks.
2. High Resistance – chosen mainly for their wearing and impact resistance properties and for high usage areas such as kitchens. Examples are quarry tiles and granolithic pavings.
3. Hygiene – chosen to provide an impervious easy to clean surface with reasonable aesthetic appeal. Examples are quarry tiles and polyvinyl chloride (PVC) sheets and tiles.

Carpets and Carpet Tiles – made from animal hair, mineral fibres and man made fibres such as nylon and acrylic. They are also available in mixtures of the above. A wide range of patterns; sizes and colours are available. Carpets and carpet tiles can be laid loose, stuck with a suitable adhesive or in the case of carpets edge fixed using special grip strips.

PVC Tiles – made from a blended mix of thermoplastic binders; fillers and pigments in a wide variety of colours and patterns to the recommendations of BS EN 649: Resilient floor coverings. PVC tiles are usually 305 × 305 × 1·6 mm thick and are stuck to a suitable base with special adhesives as recommended by the manufacturer.

Quarry Tiles ~

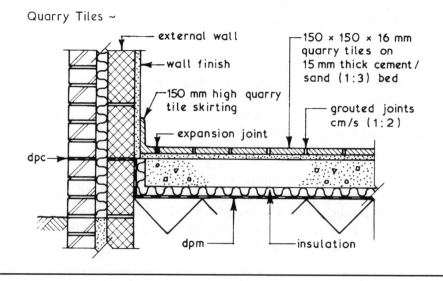

external wall

wall finish

150 mm high quarry tile skirting

expansion joint

dpc

150 × 150 × 16 mm quarry tiles on 15 mm thick cement/sand (1:3) bed

grouted joints cm/s (1:2)

dpm

insulation

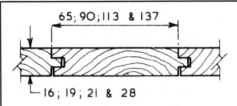

65; 90; 113 & 137

16; 19; 21 & 28

Tongue and Groove Boarding ~ prepared from softwoods to the recommendations of BS 1297. Boards are laid at right angles to the joists and are fixed with 2 No. 65 mm long cut floor brads per joists. The ends of board lengths are butt jointed on the centre line of the supporting joist.

Maximum board spans are:-
16 mm thick – 505 mm
19 mm thick – 600 mm
21 mm thick – 635 mm
28 mm thick – 790 mm

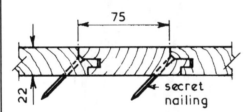

75

22

secret nailing

Timber Strip Flooring ~ strip flooring is usually considered to be boards under 100 mm face width. In good class work hardwoods would be specified the boards being individually laid and secret nailed. Strip flooring can be obtained treated with a spirit-based fungicide. Spacing of supports depends on type of timber used and applied loading. After laying the strip flooring should be finely sanded and treated with a seal or wax. In common with all timber floorings a narrow perimeter gap should be left for moisture movement.

Chipboard ~ sometimes called Particle Board is made from particles of wood bonded with a synthetic resin and/or other organic binders to the recommendations of BS EN 312.

Standard floor boards are in lengths upto 2400 mm, a width of 600 mm × 18 or 22 mm thickness with tongued and grooved joints to all edges. Laid right angles to joists with all cross joints directly supported. May be specified as unfinished or water-proof quality indicated with a dull green dye.

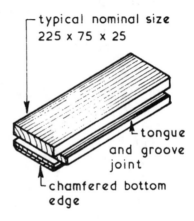

typical nominal size
225 x 75 x 25

tongue and groove joint

chamfered bottom edge

Wood Blocks ~ prepared from hardwoods and softwoods to the recommendations of BS 1187. Wood blocks can be laid to a variety of patterns, also different timbers can be used to create colour and grain effects. Laid blocks should be finely sanded and sealed or polished.

Large Cast-In-situ Ground Floors ~ these are floors designed to carry medium to heavy loadings such as those used in factories, warehouses, shops, garages and similar buildings. Their design and construction is similar to that used for small roads (see pages 132 to 134). Floors of this type are usually laid in alternate 4·500 wide strips running the length of the building or in line with the anticipated traffic flow where applicable. Transverse joints will be required to control the tensile stresses due to the thermal movement and contraction of the slab. The spacing of these joints will be determined by the design and the amount of reinforcement used. Such joints can either be formed by using a crack inducer or by sawing a 20 to 25 mm deep groove into the upper surface of the slab within 20 to 30 hours of casting.

Typical Layout ~

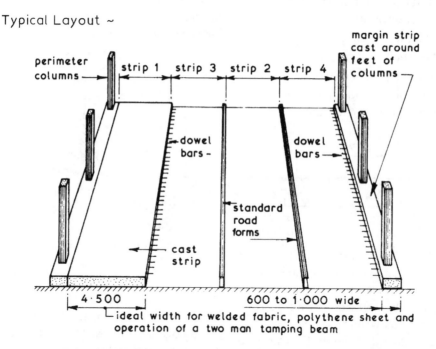

Surface Finishing ~ the surface of the concrete may be finished by power floating or trowelling which is carried out whilst the concrete is still plastic but with sufficient resistance to the weight of machine and operator whose footprint should not leave a depression of more than 3 mm. Power grinding of the surface is an alternative method which is carried out within a few days of the concrete hardening. The wet concrete having been surface finished with a skip float after the initial levelling with a tamping bar has been carried out. Power grinding removes 1 to 2 mm from the surface and is intended to improve surface texture and not to make good deficiencies in levels.

Vacuum Dewatering ~ if the specification calls for a power float surface finish vacuum dewatering could be used to shorten the time delay between tamping the concrete and power floating the surface. This method is suitable for slabs up to 300 mm thick. The vacuum should be applied for approximately 3 minutes for every 25 mm depth of concrete which will allow power floating to take place usually within 20 to 30 minutes of the tamping operation. The applied vacuum forces out the surplus water by compressing the slab and this causes a reduction in slab depth of approximately 2% therefore packing strips should be placed on the side forms before tamping to allow for sufficient surcharge of concrete.

Typical Details~

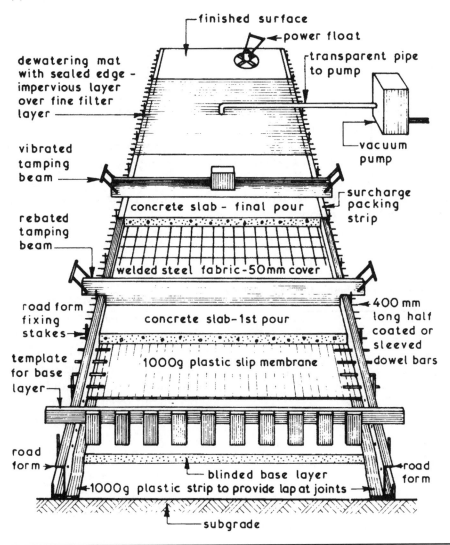

Concrete Floor Screeds ~ these are used to give a concrete floor a finish suitable to receive the floor finish or covering specified. It should be noted that it is not always necessary or desirable to apply a floor screed to receive a floor covering, techniques are available to enable the concrete floor surface to be prepared at the time of casting to receive the coverings at a later stage.

Typical Screed Mixes ~

Screed Thickness	Cement	Dry Fine Aggregate <5 mm	Coarse Aggregate >5 mm <10 mm
up to 40 mm	1	3 to 4 1/2	-
40 to 75 mm	1	3 to 4 1/2	-
	1	1 1/2	3

Laying Floor Screeds ~ floor screeds should not be laid in bays since this can cause curling at the edges, screeds can however be laid in 3·000 wide strips to receive thin coverings. Levelling of screeds is achieved by working to levelled timber screeding batten or alternatively a 75 mm wide band of levelled screed with square edges can be laid to the perimeter of the floor prior to the general screed laying operation.

Screed Types ~

Monolithic Screeds –

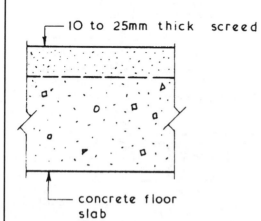

10 to 25mm thick screed

concrete floor slab

screed laid directly on concrete floor slab within three hours of placing concrete – before any screed is placed all surface water should be removed – all screeding work should be carried out from scaffold board runways to avoid walking on the 'green' concrete slab.

Screed Types ⌄

40 mm thick screed

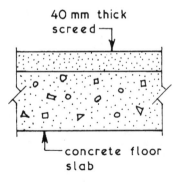

concrete floor slab

50 mm thick screed*

insulation

dpm

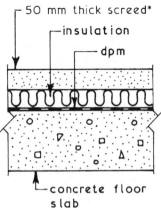

concrete floor slab

65 mm thick screed*

resilient quilt

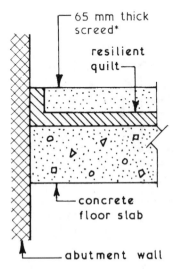

concrete floor slab

abutment wall

Separate Screeds –

screed is laid onto the concrete floor slab after it has cured. The floor surface must be clean and rough enough to ensure an adequate bond unless the floor surface is prepared by applying a suitable bonding agent or by brushing with a cement/water grout of a thick cream like consistency just before laying the screed.

Unbonded Screeds –

screed is laid directly over a damp-proof membrane or over a damp-proof membrane and insulation. A rigid form of floor insulation is required where the concrete floor slab is in contact with the ground. Care must be taken during this operation to ensure that the damp-proof membrane is not damaged.

Floating Screeds –

a resilient quilt of 25 mm thickness is laid with butt joints and turned up at the edges against the abutment walls, the screed being laid directly over the resilient quilt. The main objective of this form of floor screed is to improve the sound insulation properties of the floor.

*preferably wire mesh reinforced

Primary Functions ~

1. Provide a level surface with sufficient strength to support the imposed loads of people and furniture plus the dead loads of flooring and ceiling.
2. Reduce heat loss from lower floor as required.
3. Provide required degree of sound insulation.
4. Provide required degree of fire resistance.

Basic Construction – a timber suspended upper floor consists of a series of beams or joists supported by load bearing walls sized and spaced to carry all the dead and imposed loads.

Joist Sizing – three methods can be used:-

1. Building Regs. Approved Document A – Structure. Refs. *BS 6399-1: Code of practice for dead and imposed loads (min. 1·5 kN/m² distributed, 1.4 kN concentrated). *TRADA publication – Span Tables for Solid Timber Members in Dwellings.	2. Calculation formula:- $$BM = \frac{fbd^2}{6}$$ where BM = bending moment f = fibre stress b = breadth d = depth in mm must be assumed	3. Empirical formula:- $$D = \frac{\text{span in mm}}{24} + 50$$ where D = depth of joist in mm above assumes that joists have a breadth of 50 mm and are at 400 c/c spacing

Support and Restraint ~

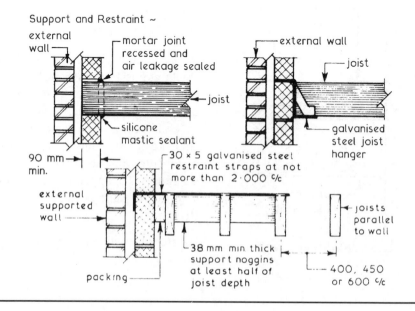

external wall — mortar joint recessed and air leakage sealed — external wall — joist — silicone mastic sealant — joist — galvanised steel joist hanger

90 mm min. — 30 × 5 galvanised steel restraint straps at not more than 2·000 c/c — joists parallel to wall

external supported wall — 38 mm min. thick support noggins at least half of joist depth — 400, 450 or 600 c/c

packing

Strutting ~ used in timber suspended floors to restrict the movements due to twisting and vibration which could damage ceiling finishes. Strutting should be included if the span of the floor joists exceeds 2·5m and is positioned on the centre line of the span. Max. floor span ~ 6m measured centre to centre of bearing (inner leaf centre line in cavity wall).

Typical Details ~

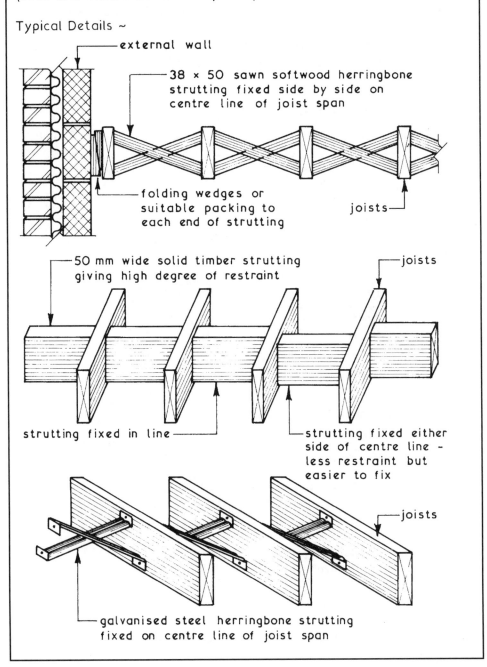

external wall

38 × 50 sawn softwood herringbone strutting fixed side by side on centre line of joist span

folding wedges or suitable packing to each end of strutting

joists

50 mm wide solid timber strutting giving high degree of restraint

joists

strutting fixed in line

strutting fixed either side of centre line - less restraint but easier to fix

joists

galvanised steel herringbone strutting fixed on centre line of joist span

Lateral Restraint ~ external, compartment (fire), separating (party) and internal loadbearing walls must have horizontal support from adjacent floors, to restrict movement. Exceptions occur when the wall is less than 3m long.

Methods:

1. 90 mm end bearing of floor joists, spaced not more than 1·2m apart – see page 658.
2. Galvanised steel straps spaced at intervals not exceeding 2m and fixed square to joists – see page 658.

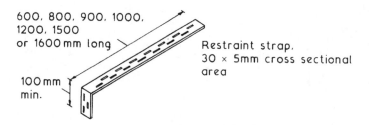

600, 800, 900, 1000, 1200, 1500 or 1600 mm long

100 mm min.

Restraint strap, 30 × 5mm cross sectional area

3. Joists carried by BS 5628-1 approved galvanised steel hangers.

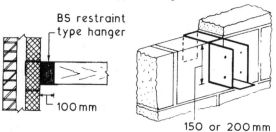

BS restraint type hanger

100 mm

150 or 200 mm

4. Adjacent floors at or about the same level, contacting with the wall at no more than 2 m intervals.

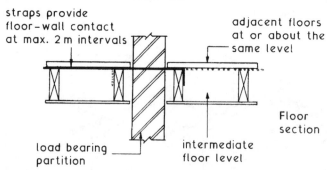

straps provide floor–wall contact at max. 2 m intervals

adjacent floors at or about the same level

load bearing partition

intermediate floor level

Floor section

Ref. BS EN 845-1: Specification for ancillary components for masonry. Ties, tension straps, hangers and brackets.

Wall Stability – at right angles to floor and ceiling joists this is achieved by building the joists into masonry support walls or locating them on approved joist hangers.

Walls parallel to joists are stabilised by lateral restraint straps. Buildings constructed before current stability requirements (see Bldg. Regs. A.D; A – Structure) often show signs of wall bulge due to the effects of eccentric loading and years of thermal movement.

Remedial Measures –

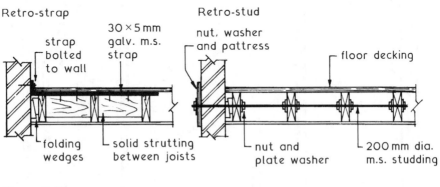

Retro-strap

strap bolted to wall

30×5 mm galv. m.s. strap

folding wedges

solid strutting between joists

Retro-stud

nut, washer and pattress

floor decking

nut and plate washer

200 mm dia. m.s. studding

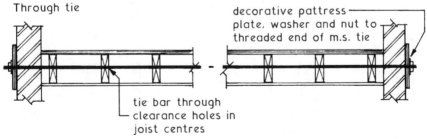

Through tie

decorative pattress plate, washer and nut to threaded end of m.s. tie

tie bar through clearance holes in joist centres

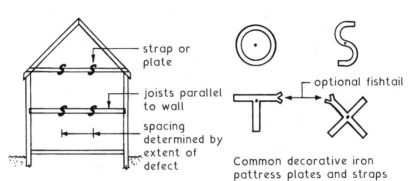

strap or plate

joists parallel to wall

spacing determined by extent of defect

Location of straps and ties

optional fishtail

Common decorative iron pattress plates and straps

Trimming Members ~ these are the edge members of an opening in a floor and are the same depth as common joists but are usually 25 mm wider.

Typical Details ~

200 min.

trimmed joist

200 min.

trimmer

900 min.

trimming joist

2700 min.

minimum recommended dimensions for straight flight stairs

common joist

50 min.

chimney flue

50 min.

wedge

joist hangers

tusk tenon joint

TRIMMING TO STAIRWELL

TRIMMING AROUND FLUE

wedge

mortice for wedge

housing and mortice

$\frac{D}{3}$

housing

$\frac{D}{2}$

galvanised steel joist hanger

W

W

$\frac{D}{4}$ $\frac{D}{4}$

D

150 min.

$\frac{W}{3}$

D

$\frac{W}{4}$

TRADITIONAL TUSK TENON JOINT

TRADITIONAL HOUSED JOINT

MODERN JOIST HANGER

Typical spans and loading for floor joists of general structural grade −

Sawn size (mm × mm)	Dead weight of flooring and ceiling, excluding the self weight of the joists (kg/m²)								
	< 25			25–50			50–125		
	Spacing of joists (mm)								
	400	450	600	400	450	600	400	450	600
	Maximum clear span (m)								
38 × 75	1.22	1.09	0.83	1.14	1.03	0.79	0.98	0.89	0.70
38 × 100	1.91	1.78	1.38	1.80	1.64	1.28	1.49	1.36	1.09
38 × 125	2.54	2.45	2.01	2.43	2.30	1.83	2.01	1.85	1.50
38 × 150	3.05	2.93	2.56	2.91	2.76	2.40	2.50	2.35	1.93
38 × 175	3.55	3.40	2.96	3.37	3.19	2.77	2.89	2.73	2.36
38 × 200	4.04	3.85	3.35	3.82	3.61	3.13	3.27	3.09	2.68
38 × 225	4.53	4.29	3.73	4.25	4.02	3.50	3.65	3.44	2.99
50 × 75	1.45	1.37	1.08	1.39	1.30	1.01	1.22	1.11	0.88
50 × 100	2.18	2.06	1.76	2.06	1.95	1.62	1.82	1.67	1.35
50 × 125	2.79	2.68	2.44	2.67	2.56	2.28	2.40	2.24	1.84
50 × 150	3.33	3.21	2.92	3.19	3.07	2.75	2.86	2.70	2.33
50 × 175	3.88	3.73	3.38	3.71	3.57	3.17	3.30	3.12	2.71
50 × 200	4.42	4.25	3.82	4.23	4.07	3.58	3.74	3.53	3.07
50 × 225	4.88	4.74	4.26	4.72	4.57	3.99	4.16	3.94	3.42
63 × 100	2.41	2.29	2.01	2.28	2.17	1.90	2.01	1.91	1.60
63 × 125	3.00	2.89	2.63	2.88	2.77	2.52	2.59	2.49	2.16
63 × 150	3.59	3.46	3.15	3.44	3.31	3.01	3.10	2.98	2.63
63 × 175	4.17	4.02	3.66	4.00	3.85	3.51	3.61	3.47	3.03
63 × 200	4.73	4.58	4.18	4.56	4.39	4.00	4.11	3.95	3.43
63 × 225	5.15	5.01	4.68	4.99	4.85	4.46	4.62	4.40	3.83
75 × 125	3.18	3.06	2.79	3.04	2.93	2.67	2.74	2.64	2.40
75 × 150	3.79	3.66	3.33	3.64	3.50	3.19	3.28	3.16	2.86
75 × 175	4.41	4.25	3.88	4.23	4.07	3.71	3.82	3.68	3.30
75 × 200	4.92	4.79	4.42	4.77	4.64	4.23	4.35	4.19	3.74
75 × 225	5.36	5.22	4.88	5.20	5.06	4.72	4.82	4.69	4.16

Notes:
1. Where a bath is supported, the joists should be duplicated.
2. See pages 35 and 36 for material dead weights.

Joist and Beam Sizing ~ design tables and formulae have limitations, therefore where loading, span and/or conventional joist spacings are exceeded, calculations are required. BS 5268: Structural Use Of Timber and BS EN 338: Structural Timber – Strength Classes, are both useful resource material for detailed information on a variety of timber species. The following example serves to provide guidance on the design process for determining joist size, measurement of deflection, safe bearing and resistance to shear force:-

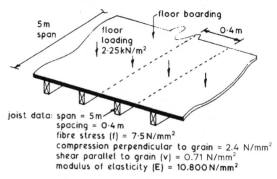

joist data: span = 5 m
spacing = 0·4 m
fibre stress (f) = 7·5 N/mm²
compression perpendicular to grain = 2.4 N/mm²
shear parallel to grain (v) = 0.71 N/mm²
modulus of elasticity (E) = 10.800 N/mm²

Total load (W) per joist = 5 m × 0·4 m × 2·25 kN/m² = 4·5 kN

$$\text{or}: \frac{4 \cdot 5 \, kN}{5 \, m \, span} = 0 \cdot 9 \, kN/m$$

Resistance to bending ~

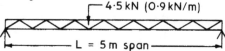

Bending moment formulae are shown on page 544

$$BM = \frac{WL}{8} = \frac{fbd^2}{6}$$

Where: W = total load, 4·5 kN (4500 N)
　　　　L = span, 5 m (5000 mm)
　　　　f = fibre stress of timber, 7·5 N/mm²
　　　　b = breadth of joist, try 50 mm
　　　　d = depth of joist, unknown

Transposing:

$$\frac{WL}{8} = \frac{fbd^2}{6}$$

Becomes:

$$d = \sqrt{\frac{6WL}{8fb}} = \sqrt{\frac{6 \times 4500 \times 5000}{8 \times 7 \cdot 5 \times 50}} = 212 \, mm$$

Nearest commercial size: 50 mm × 225 mm

Joist and Beam Sizing ~ calculating overall dimensions alone is insufficient, checks should also be made to satisfy: resistance to deflection, adequate safe bearing and resistance to shear.

Deflection – should be minimal to prevent damage to plastered ceilings. An allowance of up to 0·003 × span is normally acceptable; for the preceding example this will be:-

0·003 × 5000 mm = 15 mm

The formula for calculating deflection due to a uniformly distributed load (see page 546) is: ~

$$\frac{5WL^3}{384EI} \quad \text{where} \quad I = \frac{bd^3}{12}$$

$$I = \frac{50 \times (225)^3}{12} = 4·75 \times (10)^7$$

So, deflection $= \dfrac{5 \times 4500 \times (5000)^3}{384 \times 10800 \times 4·75 \times (10)^7} = 14·27 \text{ mm}$

NB. This is only just within the calculated allowance of 15 mm, therefore it would be prudent to specify slightly wider or deeper joists to allow for unknown future use.

Safe Bearing ~

$$= \frac{\text{load at the joist end, W/2}}{\text{compression perpendicular to grain} \times \text{ breadth}}$$

$$= \frac{4500/2}{2·4 \times 50} = 19 \text{ mm}.$$

therefore full support from masonry (90 mm min.) or joist hangers will be more than adequate.

Shear Strength ~

$$V = \frac{2bdv}{3}$$

where: V = vertical loading at the joist end, W/2
      v = shear strength parallel to the grain, 0.71 N/mm²
Transposing:-

$$bd = \frac{3V}{2v} = \frac{3 \times 2250}{2 \times 0.71} = 4753 \text{ mm}^2 \text{ minimum}$$

Actual bd = 50 mm × 225 mm = 11,250 mm²

Resistance to shear is satisfied as actual is well above the minimum.

Typical situations ~

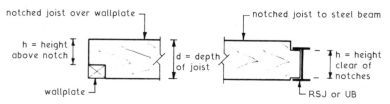

It is necessary to ensure enough timber above and/or below a notch to resist horizontal shear or shear parallel to the grain.

To check whether a joist or beam has adequate horizontal shear strength:

$$v = (V3d) \div (2bh^2)$$

Using the data provided in the previous two pages as applied to the design of a timber joist of 225×50mm cross section, in this instance with a 50mm notch to leave 175mm (h) clear:

$$v = (2250 \times 3 \times 225) \div (2 \times 50 \times 175 \times 175) = 0.496 \, N/mm^2$$

Shear strength parallel to the grain* is given as $0.710 \, N/mm^2$ so sufficient strength is still provided.

*BS 5262-2: Structural use of timber. Code of practice for permissible stress design, materials and workmanship.

Bressummer (or breastsummer) beam ~

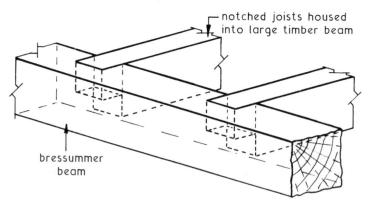

A long, large section timber beam carrying joists directly or notched and housed (as shown). A dated form of construction for supporting a masonry wall and floor/ceiling joists in large openings. Often found during refurbishment work over shop windows. Steel or reinforced concrete now preferred.

For fire protection, floors are categorised depending on their height relative to adjacent ground ~

Height of top floor above ground	Fire resistance (load bearing)
Less than 5 m	30 minutes (60 min. for compartment floors)
More than 5 m	60 minutes (30 min. for a three storey dwelling)

Tests for fire resistance relate to load bearing capacity, integrity and insulation as determined by BS 476 – 21: Fire tests on building materials and structures. Methods for determination of the fire resistence of loadbearing elements of construction.

Integrity ~ the ability of an element to resist fire penetration and capacity to bear load in a fire.

Insulation ~ ability to resist heat penetration so that fire is not spread by radiation and conduction.

Typical applications ~

30 MINUTE FIRE RESISTANCE

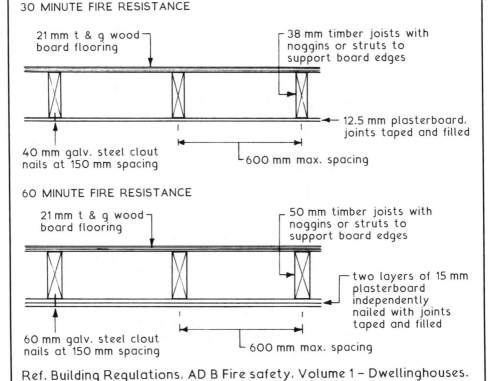

21 mm t & g wood board flooring

38 mm timber joists with noggins or struts to support board edges

12.5 mm plasterboard, joints taped and filled

40 mm galv. steel clout nails at 150 mm spacing

600 mm max. spacing

60 MINUTE FIRE RESISTANCE

21 mm t & g wood board flooring

50 mm timber joists with noggins or struts to support board edges

two layers of 15 mm plasterboard independently nailed with joints taped and filled

60 mm galv. steel clout nails at 150 mm spacing

600 mm max. spacing

Ref. Building Regulations, AD B Fire safety, Volume 1 – Dwellinghouses.

Reinforced Concrete Suspended Floors ~ a simple reinforced concrete flat slab cast to act as a suspended floor is not usually economical for spans over 5·000. To overcome this problem beams can be incorporated into the design to span in one or two directions. Such beams usually span between columns which transfers their loads to the foundations. The disadvantages of introducing beams are the greater overall depth of the floor construction and the increased complexity of the formwork and reinforcement. To reduce the overall depth of the floor construction flat slabs can be used where the beam is incorporated with the depth of the slab. This method usually results in a deeper slab with complex reinforcement especially at the column positions.

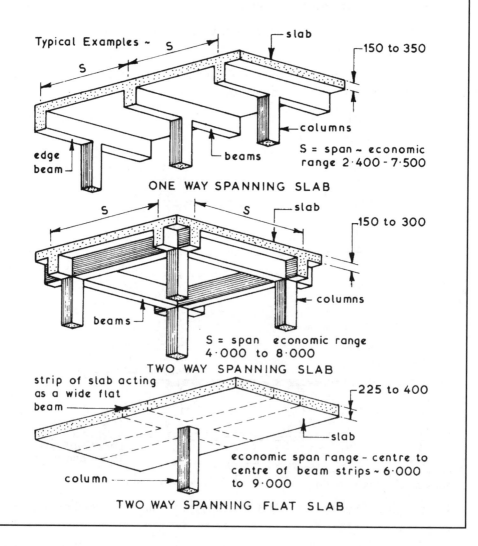

Typical Examples ~

slab — 150 to 350

S = span ~ economic range 2·400 - 7·500

edge beam — columns — beams

ONE WAY SPANNING SLAB

slab — 150 to 300

S = span   economic range 4·000 to 8·000

beams — columns

TWO WAY SPANNING SLAB

strip of slab acting as a wide flat beam — 225 to 400

slab

economic span range - centre to centre of beam strips ~ 6·000 to 9·000

column

TWO WAY SPANNING FLAT SLAB

Ribbed Floors ~ to reduce the overall depth of a traditional cast in-situ reinforced concrete beam and slab suspended floor a ribbed floor could be used. The basic concept is to replace the wide spaced deep beams with narrow spaced shallow beams or ribs which will carry only a small amount of slab loading. These floors can be designed as one or two way spanning floors. One way spanning ribbed floors are sometimes called troughed floors whereas the two way spanning ribbed floors are called coffered or waffle floors. Ribbed floors are usually cast against metal, glass fibre or polypropylene preformed moulds which are temporarily supported on plywood decking, joists and props – see page 498.

Typical Examples ~

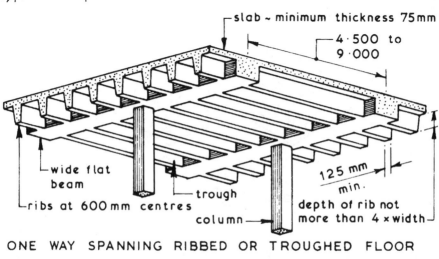

slab ~ minimum thickness 75mm

4·500 to 9·000

wide flat beam

trough

ribs at 600mm centres

column

125 mm min.

depth of rib not more than 4 × width

ONE WAY SPANNING RIBBED OR TROUGHED FLOOR

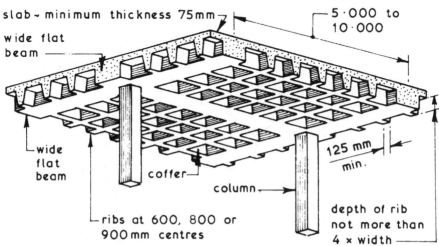

slab ~ minimum thickness 75mm

5·000 to 10·000

wide flat beam

wide flat beam

coffer

column

ribs at 600, 800 or 900mm centres

125 mm min.

depth of rib not more than 4 × width

TWO WAY SPANNING COFFERED OR WAFFLE FLOOR

Ribbed Floors – these have greater span and load potential per unit weight than flat slab construction. This benefits a considerable reduction in dead load, to provide cost economies in other super-structural elements and foundations. The regular pattern of voids created with waffle moulds produces a honeycombed effect, which may be left exposed in utility buildings such as car parks. Elsewhere such as shopping malls, a suspended ceiling would be appropriate. The trough finish is also suitable in various situations and has the advantage of creating a continuous void for accommodation of service cables and pipes. A suspended ceiling can add to this space where air conditioning ducting is required, also providing several options for finishing effect.

Typical mould profile –

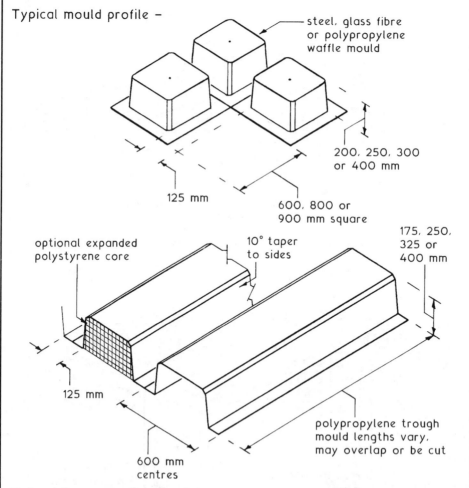

steel, glass fibre or polypropylene waffle mould

200, 250, 300 or 400 mm

125 mm

600, 800 or 900 mm square

optional expanded polystyrene core

10° taper to sides

175, 250, 325 or 400 mm

125 mm

polypropylene trough mould lengths vary, may overlap or be cut

600 mm centres

Note: After removing the temporary support structure, moulds are struck by flexing with a flat tool. A compressed air line is also effective.

Hollow Pot Floors ~ these are in essence a ribbed floor with permanent formwork in the form of hollow clay or concrete pots. The main advantage of this type of cast in-situ floor is that it has a flat soffit which is suitable for the direct application of a plaster finish or an attached dry lining. The voids in the pots can be utilised to house small diameter services within the overall depth of the slab. These floors can be designed as one or two way spanning slabs, the common format being the one way spanning floor.

Typical Example ~

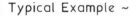

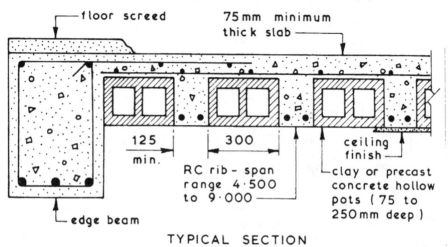

ONE WAY SPANNING HOLLOW POT FLOOR

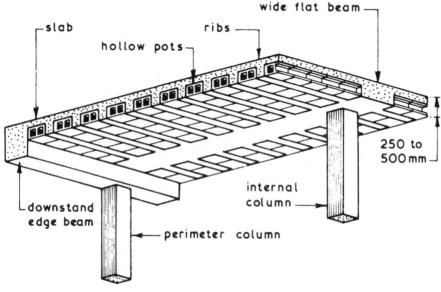

TYPICAL SECTION

Soffit and Beam Fixings ~ concrete suspended floors can be designed to carry loads other than the direct upper surface loadings. Services can be housed within the voids created by the beams or ribs and suspended or attached ceilings can be supported by the floor. Services which run at right angles to the beams or ribs are usually housed in cast-in holes. There are many types of fixings available for use in conjunction with floor slabs, some are designed to be cast-in whilst others are fitted after the concrete has cured. All fixings must be positioned and installed so that they are not detrimental to the structural integrity of the floor.

Typical Examples ~

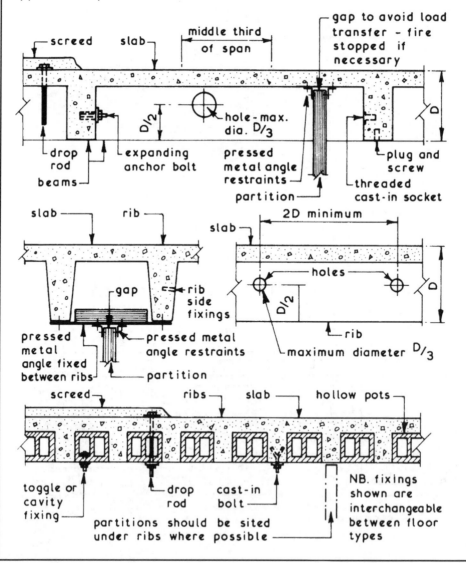

Precast Concrete Floors ~ these are available in several basic formats and provide an alternative form of floor construction to suspended timber floors and in-situ reinforced concrete suspended floors. The main advantages of precast concrete floors are:-

1. Elimination of the need for formwork except for nominal propping which is required with some systems.
2. Curing time of concrete is eliminated therefore the floor is available for use as a working platform at an earlier stage.
3. Superior quality control of product is possible with factory produced components.

The main disadvantages of precast concrete floors when compared with in-situ reinforced concrete floors are:-

1. Less flexible in design terms.
2. Formation of large openings in the floor for ducts, shafts and stairwells usually have to be formed by casting an in-situ reinforced concrete floor strip around the opening position.
3. Higher degree of site accuracy is required to ensure that the precast concrete floor units can be accommodated without any alterations or making good.

Typical Basic Formats ~

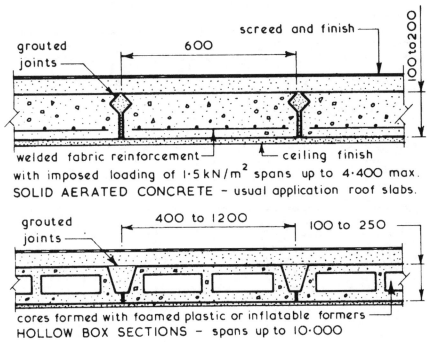

with imposed loading of $1 \cdot 5 \, kN/m^2$ spans up to $4 \cdot 400$ max.
SOLID AERATED CONCRETE – usual application roof slabs.

cores formed with foamed plastic or inflatable formers
HOLLOW BOX SECTIONS – spans up to $10 \cdot 000$

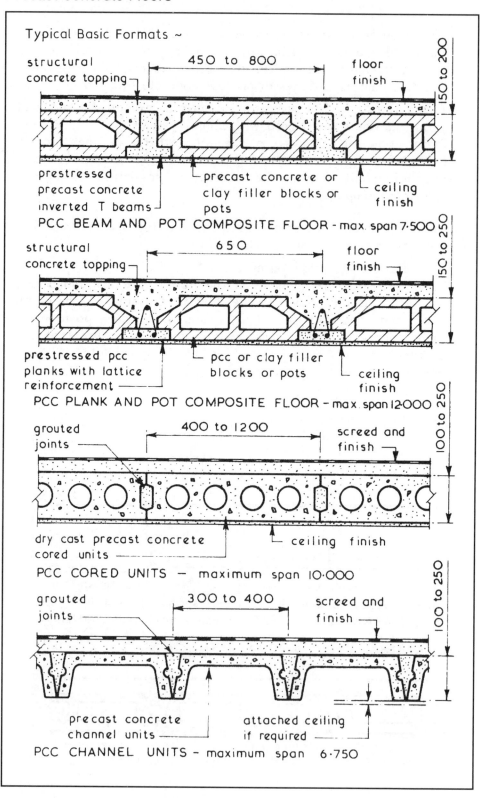

Typical Basic Formats ~

structural concrete topping ⟶   450 to 800   floor finish   150 to 200

prestressed precast concrete inverted T beams ⟶   precast concrete or clay filler blocks or pots   ceiling finish

PCC BEAM AND POT COMPOSITE FLOOR – max. span 7·500

structural concrete topping ⟶   650   floor finish   150 to 250

prestressed pcc planks with lattice reinforcement ⟶   pcc or clay filler blocks or pots   ceiling finish

PCC PLANK AND POT COMPOSITE FLOOR – max. span 12·000

grouted joints ⟶   400 to 1200   screed and finish   100 to 250

dry cast precast concrete cored units ⟶   ceiling finish

PCC CORED UNITS – maximum span 10·000

grouted joints ⟶   300 to 400   screed and finish   100 to 250

precast concrete channel units ⟶   attached ceiling if required

PCC CHANNEL UNITS – maximum span 6·750

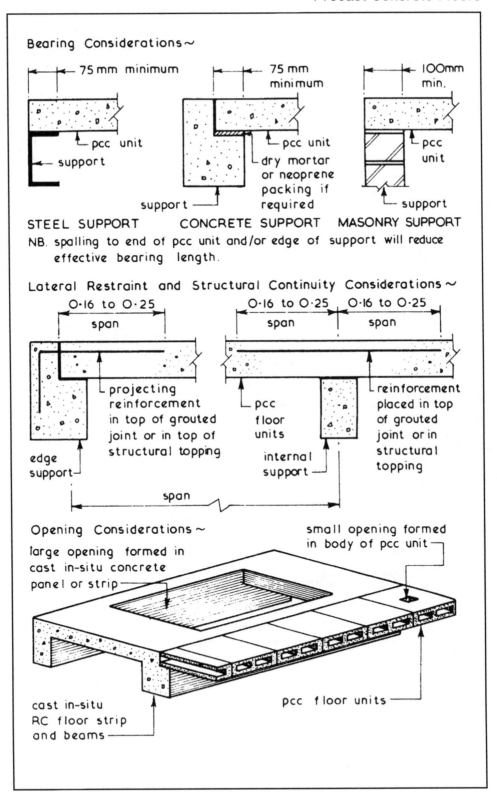

Bearing Considerations ~

75 mm minimum

STEEL SUPPORT

pcc unit

support

75 mm minimum

CONCRETE SUPPORT

pcc unit

dry mortar or neoprene packing if required

support

100mm min.

MASONRY SUPPORT

pcc unit

support

NB. spalling to end of pcc unit and/or edge of support will reduce effective bearing length.

Lateral Restraint and Structural Continuity Considerations ~

0·16 to 0·25

span

projecting reinforcement in top of grouted joint or in top of structural topping

edge support

0·16 to 0·25   0·16 to 0·25

span   span

pcc floor units

internal support

reinforcement placed in top of grouted joint or in structural topping

span

Opening Considerations ~

large opening formed in cast in-situ concrete panel or strip

small opening formed in body of pcc unit

cast in-situ RC floor strip and beams

pcc floor units

Steel fabricated beams can be used as an integral means of support for precast concrete floors. These are an overall depth and space saving alternative compared to down-stand reinforced concrete beams or masonry walls. Only the lower steel flange of the steel beam is exposed.

To attain sufficient strength, a supplementary steel plate is welded to the bottom flange of standard UC sections. This produces a type of compound or plated section that is supported by the main structural frame.

Floor support

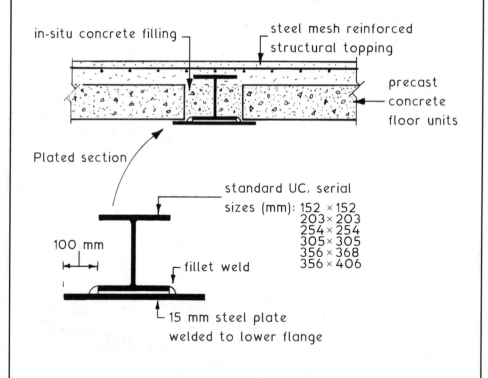

in-situ concrete filling

steel mesh reinforced structural topping

precast concrete floor units

Plated section

standard UC, serial sizes (mm): 152 × 152
203 × 203
254 × 254
305 × 305
356 × 368
356 × 406

100 mm

fillet weld

15 mm steel plate welded to lower flange

A standard manufactured steel beam with similar applications to the plated UC shown on the previous page. This purpose made alternative is used with lightweight flooring units, such as precast concrete hollow core slabs and metal section decking of the type shown on page 499.

Standard production serial size range ~

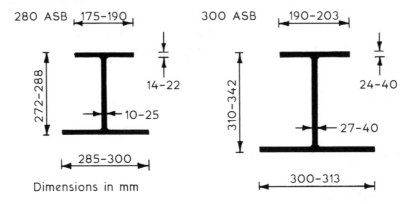

280 ASB

175–190

272–288

14–22

10–25

285–300

Dimensions in mm

300 ASB

190–203

310–342

24–40

27–40

300–313

Application ~

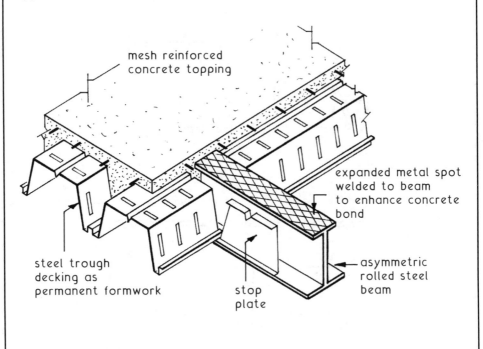

mesh reinforced
concrete topping

expanded metal spot
welded to beam
to enhance concrete
bond

steel trough
decking as
permanent formwork

stop
plate

asymmetric
rolled steel
beam

677

Raised Flooring ~ developed in response to the high-tech boom of the 1970s. It has proved expedient in accommodating computer and communications cabling as well as numerous other established services. The system is a combination of adjustable floor pedestals, supporting a variety of decking materials. Pedestal height ranges from as little as 30 mm up to about 600 mm, although greater heights are possible at the expense of structural floor levels. Decking is usually in loose fit squares of 600 mm, but may be sheet plywood or particleboard screwed direct to closer spaced pedestal support plates on to joists bearing on pedestals.

Cavity fire stops are required between decking and structural floor at appropriate intervals (see Building Regulations, A D B, Volume 2, Section 9).

Application ~

PEDESTAL DETAIL

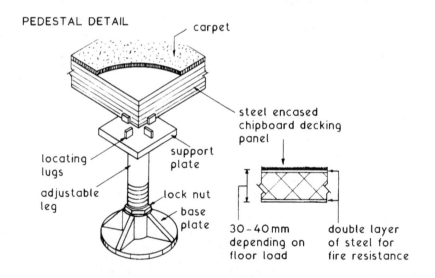

carpet

steel encased chipboard decking panel

locating lugs

support plate

adjustable leg

lock nut

base plate

30 – 40 mm depending on floor load

double layer of steel for fire resistance

FLOOR SECTION

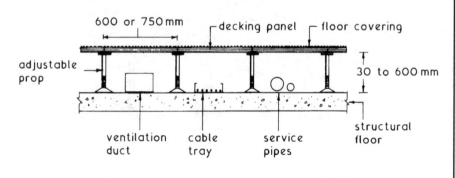

600 or 750 mm

decking panel

floor covering

adjustable prop

30 to 600 mm

ventilation duct

cable tray

service pipes

structural floor

Sound Insulation ~ sound can be defined as vibrations of air which are registered by the human ear. All sounds are produced by a vibrating object which causes tiny particles of air around it to move in unison. These displaced air particles collide with adjacent air particles setting them in motion and in unison with the vibrating object. This continuous chain reaction creates a sound wave which travels through the air until at some distance the air particle movement is so small that it is inaudible to the human ear. Sounds are defined as either impact or airborne sound, the definition being determined by the source producing the sound. Impact sounds are created when the fabric of structure is vibrated by direct contact whereas airborne sound only sets the structural fabric vibrating in unison when the emitted sound wave reaches the enclosing structural fabric. The vibrations set up by the structural fabric can therefore transmit the sound to adjacent rooms which can cause annoyance, disturbance of sleep and of the ability to hold a normal conservation. The objective of sound insulation is to reduce transmitted sound to an acceptable level, the intensity of which is measured in units of decibels (dB).

The Building Regulations, Approved Document E: Resistance to the passage of sound, establishes sound insulation standards as follows

E1: Between dwellings and between dwellings and other buildings.
E2: Within a dwelling, ie. between rooms, particularly WC and habitable rooms, and bedrooms and other rooms.
E3: Control of reverberation noise in common parts (stairwells and corridors) of buildings containing dwellings, ie. flats.
E4: Specific applications to acoustic conditions in schools.
Note: E1 includes, hotels, hostels, student accommodation, nurses' homes and homes for the elderly, but not hospitals and prisons.

Typical Sources and Transmission of Sound ~

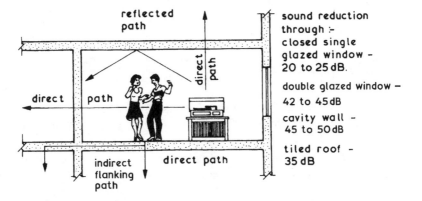

reflected path

direct path

direct   path

indirect flanking path

direct path

sound reduction through :-
closed single glazed window - 20 to 25 dB.

double glazed window - 42 to 45 dB

cavity wall - 45 to 50 dB

tiled roof - 35 dB

The Approved Document to Building Regulation E2 provides for internal walls and floors located between a bedroom or a room containing a WC and other rooms to have a reasonable resistance to airborne sound. Impact sound can be improved by provision of a carpet.

Typical details ~

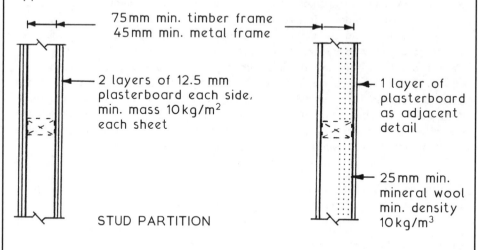

75mm min. timber frame
45mm min. metal frame

2 layers of 12.5 mm plasterboard each side, min. mass 10 kg/m² each sheet

STUD PARTITION

1 layer of plasterboard as adjacent detail

25mm min. mineral wool min. density 10 kg/m³

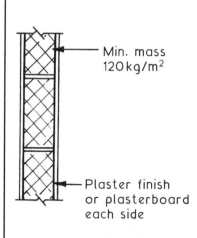

Min. mass 120 kg/m²

Plaster finish or plasterboard each side

CONCRETE BLOCKWORK PARTITION

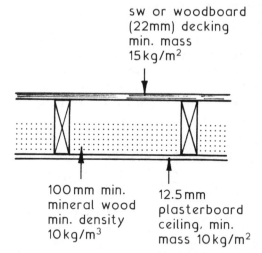

sw or woodboard (22mm) decking min. mass 15 kg/m²

100 mm min. mineral wood min. density 10 kg/m³

12.5 mm plasterboard ceiling, min. mass 10 kg/m²

TRADITIONAL SUSPENDED TIMBER FLOOR

SUSPENDED CONCRETE FLOORS (See page 674)
Cored plank ~ min. mass 180 kg/m² with any top surface and ceiling finish.
Precast beam and filler block (pot) ~ min. mass 220 kg/m² with 40 mm bonded screed and 12.5 mm plasterboard ceiling on battens.

Separating Walls ~ types:-

1. Solid masonry
2. Cavity masonry
3. Masonry between isolating panels
4. Timber frame

Type 1 — relies on mass

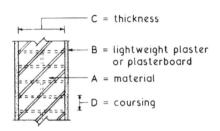

C = thickness
B = lightweight plaster or plasterboard
A = material
D = coursing

Material A	Density of A [Kg/m³]	Finish B	Combined mass A + B (Kg/m²)	Thickness C [mm]	Coursing D [mm]
brickwork	1610	13 mm lwt. pl.	375	215	75
.. ..	.. ..	12·5 mm pl. brd.	.. ..	.. ..	.. ..
Concrete block	1840	13 mm lwt. pl	415	.. ..	110
.. ..	1840	12·5 mm pl. brd	.. ..	.. ..	150
In-situ concrete	2200	Optional	415	190	n/a

Type 2 — relies on mass and isolation

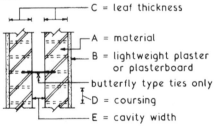

C = leaf thickness
A = material
B = lightweight plaster or plasterboard
butterfly type ties only
D = coursing
E = cavity width

Material A	Density of A [Kg/m³]	Finish B	Mass A + B (Kg/m²)	Thickness C [mm]	Coursing D [mm]	Cavity E [mm]
bkwk.	1970	13 mm lwt. pl.	415	102	75	50
concrete block	1990	..	..	100	225	..
lwt. conc. block	1375	.. or 12.5 mm pl. brd.	300	100	225	75

Type 3 ~ relies on: (a) core material type and mass,
       (b) isolation, and
       (c) mass of isolated panels.

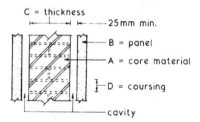

Core material A	Density of A [kg/m³]	Mass A (kg/m²)	Thickness C (mm)	Coursing D (mm)	Cavity (mm)
brickwork	1290	300	215	75	n/a
concrete block	2200	300	140	110	n/a
lwt. conc. block	1400	150	200	225	n/a
Cavity bkwk. or block	any	any	2 × 100	to suit	50

Panel materials – B

(i) Plasterboard with cellular core plus plaster finish, mass 18 kg/m². All joints taped. Fixed floor and ceiling only.

(ii) 2 No. plasterboard sheets, 12·5 mm each, with joints staggered. Frame support or 30 mm overall thickness.

Type 4 — relies on mass, frame separation and absorption of sound.

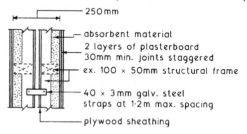

Absorbent material – quilting of unfaced mineral fibre batts with a minimum density of 10 kg/m³, located in the cavity or frames.

Thickness (mm)	Location
25	Suspended in cavity
50	Fixed within one frame
2 × 25	Each quilt fixed within each frame

Separating Floors ~ types:-

1. Concrete with soft covering
2. Concrete with floating layer
3. Timber with floating layer

Type 1. Airborne resistance depends on mass of concrete and ceiling.
Impact resistance depends on softness of covering.

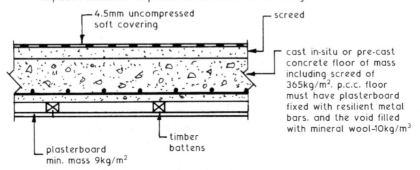

— 4.5mm uncompressed soft covering

— screed

cast in-situ or pre-cast concrete floor of mass including screed of 365kg/m², p.c.c. floor must have plasterboard fixed with resilient metal bars, and the void filled with mineral wool-10kg/m³

— plasterboard min. mass 9kg/m²

— timber battens

Type 2. Airborne resistance depends mainly on concrete mass and partly on mass of floating layer and ceiling.
Impact resistance depends on resilient layer isolating floating layer from base and isolation of ceiling.

Bases: As type 1. but overall mass minimum 300 kg/m².

Floating layers:

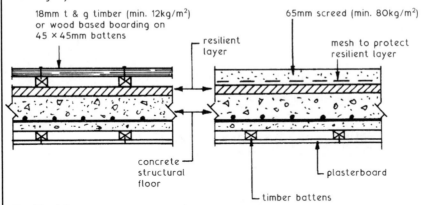

18mm t & g timber (min. 12kg/m²) or wood based boarding on 45 × 45mm battens

65mm screed (min. 80kg/m²)

resilient layer

mesh to protect resilient layer

concrete structural floor

plasterboard

timber battens

Resilient layers:

(a) 25 mm paper faced mineral fibre, density 36 kg/m³.
Timber floor – paper faced underside.
Screeded floor – paper faced upper side to prevent screed entering layer.

(b) Screeded floor only:
13 mm pre-compressed expanded polystyrene (EPS) board, or 5 mm extruded polyethylene foam of density 30–45 kg/m³, laid over a levelling screed for protection.

See BS EN 29052-1: Acoustics. Method for the determination of dynamic stiffness. Materials used under floating floors in dwellings.

Type 3. Airborne resistance varies depending on floor construction, absorbency of materials, extent of pugging and partly on the floating layer. Impact resistance depends mainly on the resilient layer separating floating from structure.

Platform floor ~

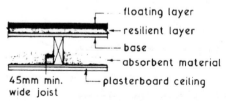

45mm min. wide joist

Note: Minimum mass per unit area = 25 kg/m^2

Floating layer: 18 mm timber or wood based board, t & g joints glued and spot bonded to a sub-strate of 19 mm plasterboard.

Alternatively, cement bonded particle board in 2 thicknesses – 24 mm total, joints staggered, glued and screwed together.

Resilient layer: 25 mm mineral fibre, density 60–100 kg/m^3.

Base: 12 mm timber boarding or wood based board nailed to joists.

Absorbent material: 100 mm unfaced rock fibre, minimum density 10 kg/m^3.

Ceiling: 30 mm plasterboard in 2 layers, joints staggered.

Ribbed floor ~

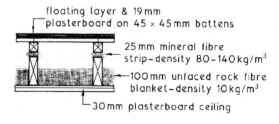

Ribbed floor with dry sand pugging ~

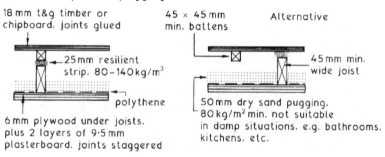

Primary Functions ~

1. Provide a means of circulation between floor levels.
2. Establish a safe means of travel between floor levels.
3. Provide an easy means of travel between floor levels.
4. Provide a means of conveying fittings and furniture between floor levels.

Constituent Parts ~

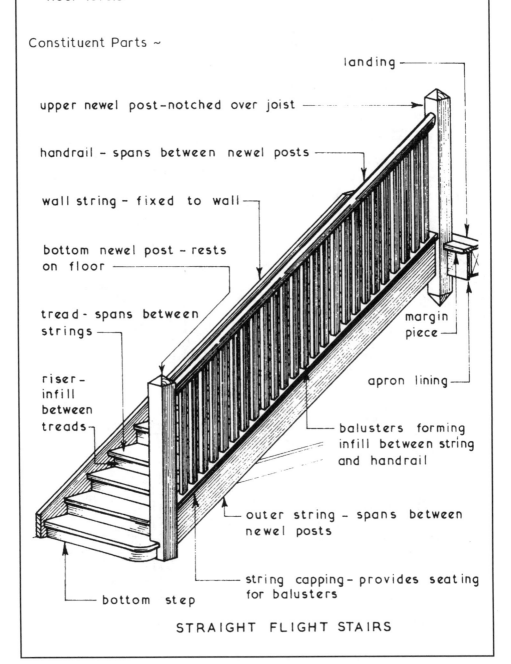

landing

upper newel post-notched over joist

handrail - spans between newel posts

wall string - fixed to wall

bottom newel post - rests on floor

tread - spans between strings

riser - infill between treads

margin piece

apron lining

balusters forming infill between string and handrail

outer string - spans between newel posts

string capping - provides seating for balusters

bottom step

STRAIGHT FLIGHT STAIRS

All dimensions quoted are the minimum required for domestic stairs exclusive to one dwelling as given in Approved Document K unless stated otherwise.

Terminology ~

pitch line - the line joining nosings

going

nosings

= going

rise

tread

riser

riser + tread = step

aggregate of going + twice rise of a step to be 550 min. and 700 max.

handrail - must provide adequate support and is required where total rise exceeds 600mm and to both sides if width exceeds 1·000

upper floor

pitch line

minimum headroom · 2 000

min. going 220 mm

rise height max. 220 mm

900mm min
1000mm max.

1100 mm preferred

900mm min.

recommended floor to floor height 2·600

no openings which will allow a 100mm sphere to pass through

maximum pitch 42°.

* recommended min. width 800 mm
handrail →

← width of string ignored

* AD K does not give a minimum dimension for stair width. See also page 690.

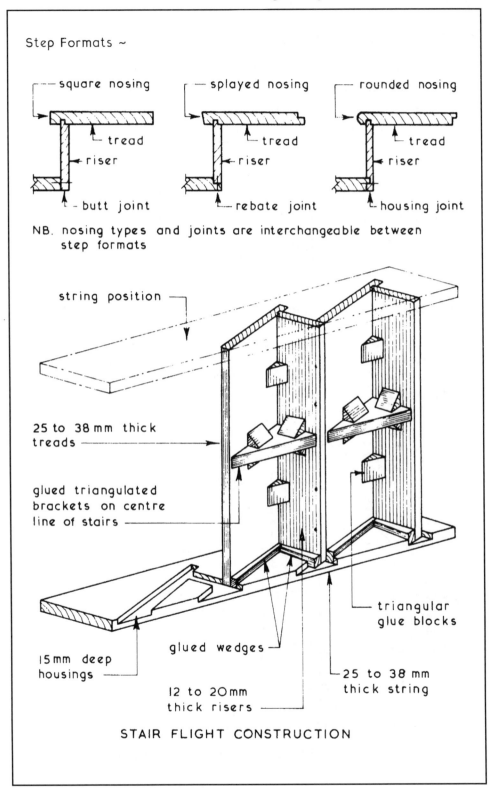

Step Formats ~

square nosing
tread
riser
butt joint

splayed nosing
tread
riser
rebate joint

rounded nosing
tread
riser
housing joint

NB. nosing types and joints are interchangeable between step formats

string position

25 to 38 mm thick treads

glued triangulated brackets on centre line of stairs

triangular glue blocks

15mm deep housings

glued wedges

25 to 38 mm thick string

12 to 20mm thick risers

STAIR FLIGHT CONSTRUCTION

Straight Flight Timber Stair Details

Bottom Step Arrangements ~

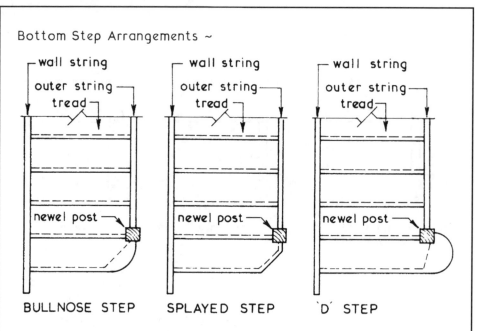

BULLNOSE STEP   SPLAYED STEP   `D` STEP

Projecting bottom steps are usually included to enable the outer string to be securely jointed to the back face of the newel post and to provide an easy line of travel when ascending or descending at the foot of the stairs.

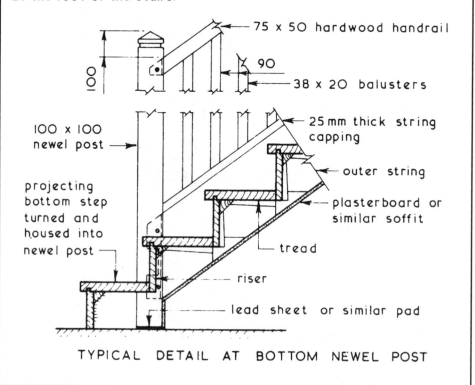

TYPICAL DETAIL AT BOTTOM NEWEL POST

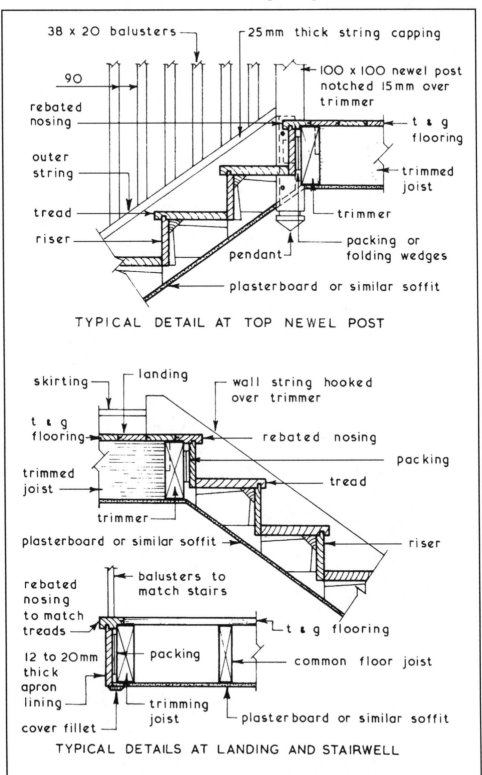

TYPICAL DETAIL AT TOP NEWEL POST

TYPICAL DETAILS AT LANDING AND STAIRWELL

689

Open Riser Timber Stairs ~ these are timber stairs constructed to the same basic principles as standard timber stairs excluding the use of a riser. They have no real advantage over traditional stairs except for the generally accepted aesthetic appeal of elegance. Like the traditional timber stairs they must comply with the minimum requirements set out in Part K of the Building Regulations.

Typical Requirements for Stairs in a Small Residential Building ~

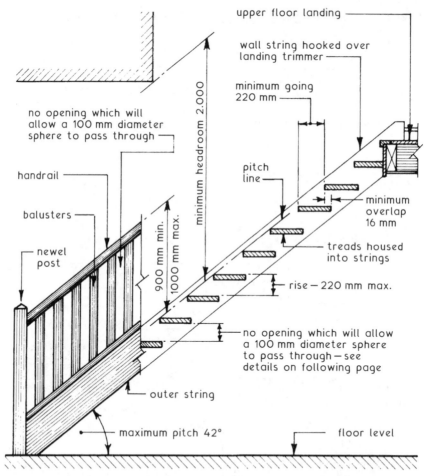

upper floor landing

wall string hooked over landing trimmer

minimum going 220 mm

no opening which will allow a 100 mm diameter sphere to pass through

minimum headroom 2,000

pitch line

handrail

minimum overlap 16 mm

balusters

treads housed into strings

900 mm min.
1000 mm max.

newel post

rise — 220 mm max.

no opening which will allow a 100 mm diameter sphere to pass through — see details on following page

outer string

maximum pitch 42°

floor level

Recommended clear width for all stairs is 800mm minimum, but 900 mm wall to wall or wall to centre of handrail is preferable. Clear width is defined in BS 585-1 as "unobstructed width between handrail and face of newel", but see also page 686. A reduced clear width of 600mm is acceptable for access to limited use space such as a loft.

Aggregate of going plus twice the rise to be 550 mm minimum and 700mm maximum.

Design and Construction ~ because of the legal requirement of not having a gap between any two consecutive treads through which a 100 mm diameter sphere can pass and the limitation relating to the going and rise, as shown on the previous page, it is generally not practicable to have a completely riserless stair for residential buildings since by using minimum dimensions a very low pitch of approximately 27½° would result and by choosing an acceptable pitch a very thick tread would have to be used to restrict the gap to 100 mm.

Possible Solutions ~

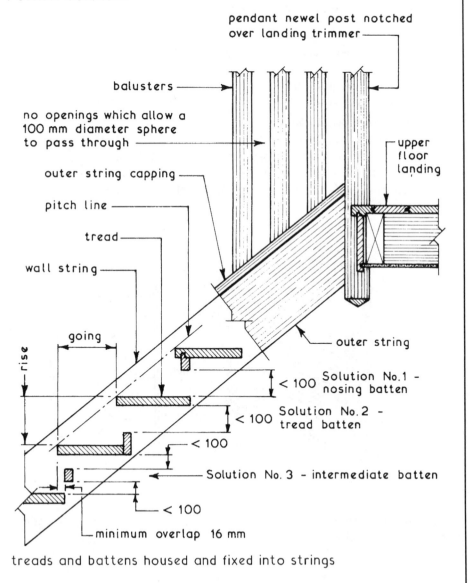

treads and battens housed and fixed into strings

Application – a straight flight for access to a domestic loft conversion only. This can provide one habitable room, plus a bathroom or WC. The WC must not be the only WC in the dwelling.

Practical issues – an economic use of space, achieved by a very steep pitch of about 60° and opposing overlapping treads.

Safety – pitch and tread profile differ considerably from other stairs, but they are acceptable to Building Regulations by virtue of "familiarity and regular use" by the building occupants.

Additional features are:

* a non-slip tread surface.
* handrails to both sides.
* minimum going 220 mm.
* maximum rise 220 mm.
* (2 + rise) + (going) between 550 and 700 mm.
* a stair used by children under 5 years old, must have the tread voids barred to leave a gap not greater than 100 mm.

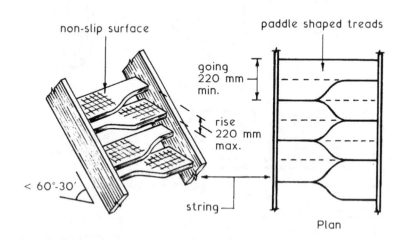

non-slip surface

paddle shaped treads

going 220 mm min.

rise 220 mm max.

< 60°-30'

string

Plan

Ref. Building Regulations, Approved Document K1: Stairs, ladders and ramps: Section 1.29

Timber Stairs ~ these must comply with the minimum requirements set out in Part K of the Building Regulations. Straight flight stairs are simple, easy to construct and install but by the introduction of intermediate landings stairs can be designed to change direction of travel and be more compact in plan than the straight flight stairs.

Landings ~ these are designed and constructed in the same manner as timber upper floors but due to the shorter spans they require smaller joist sections. Landings can be detailed for a 90° change of direction (quarter space landing) or a 180° change of direction (half space landing) and can be introduced at any position between the two floors being served by the stairs.

Typical Layouts ~

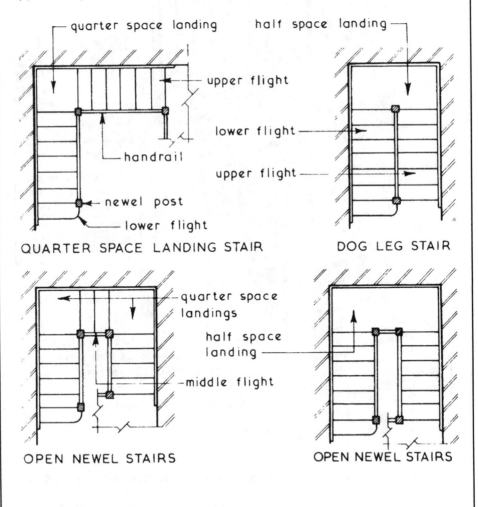

QUARTER SPACE LANDING STAIR

DOG LEG STAIR

OPEN NEWEL STAIRS

OPEN NEWEL STAIRS

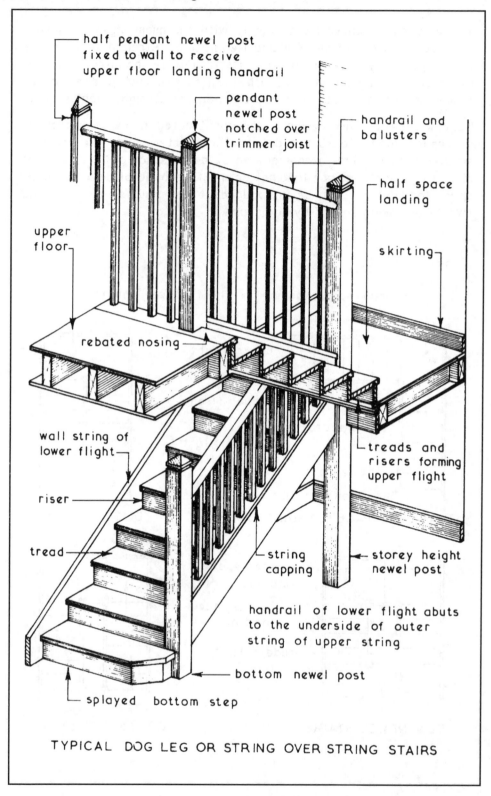

half pendant newel post
fixed to wall to receive
upper floor landing handrail

pendant
newel post
notched over
trimmer joist

handrail and
balusters

half space
landing

upper
floor

skirting

rebated nosing

wall string of
lower flight

treads and
risers forming
upper flight

riser

tread

string
capping

storey height
newel post

handrail of lower flight abuts
to the underside of outer
string of upper string

bottom newel post

splayed bottom step

TYPICAL DOG LEG OR STRING OVER STRING STAIRS

For domestic situations a spiral stair of 800mm clear width can provide an alternative compact, space saving means of access to the upper floor of a private dwelling. With a clear width of only 600mm this type of stair may also be used to access the space available in a roof void. Approved Document K to the Building Regulations refers the design and application of spiral stairs to guidance in BS 5395-2, as summarised on pages 713 and 714.

Typical timber spiral stair components

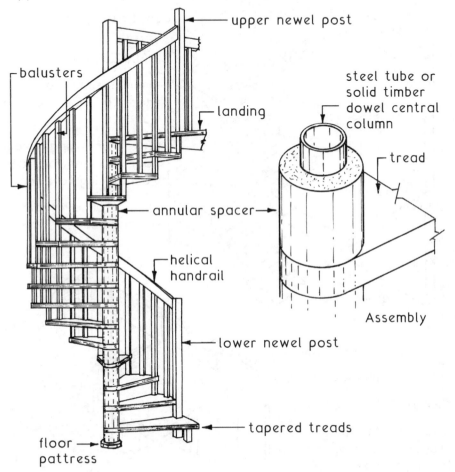

Note: Clear width is measured between handrails or between central column and handrail. If strings are used, measurement is to or between strings. Take greater value.

In-situ Reinforced Concrete Stairs ~ a variety of stair types and arrangements are possible each having its own appearance and design characteristics. In all cases these stairs must comply with the minimum requirements set out in Part K of the Building Regulations in accordance with the purpose group of the building in which the stairs are situated.

Typical Examples ~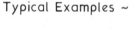

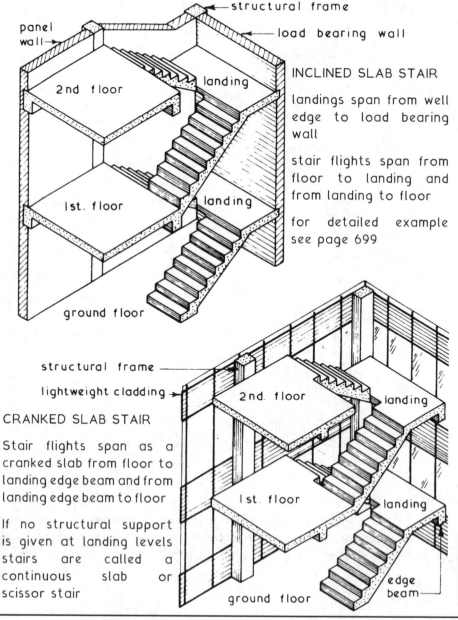

structural frame
panel wall
load bearing wall

INCLINED SLAB STAIR

landings span from well edge to load bearing wall

stair flights span from floor to landing and from landing to floor

for detailed example see page 699

2nd floor

landing

1st. floor

landing

ground floor

structural frame
lightweight cladding

CRANKED SLAB STAIR

Stair flights span as a cranked slab from floor to landing edge beam and from landing edge beam to floor

If no structural support is given at landing levels stairs are called a continuous slab or scissor stair

2nd. floor

landing

1st. floor

landing

ground floor

edge beam

Typical Examples ~

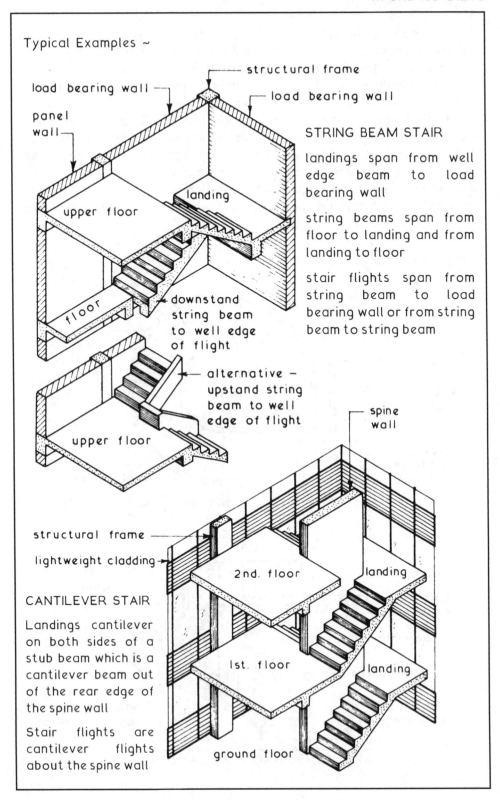

load bearing wall

structural frame

load bearing wall

panel wall

STRING BEAM STAIR

landings span from well edge beam to load bearing wall

upper floor

landing

string beams span from floor to landing and from landing to floor

floor

stair flights span from string beam to load bearing wall or from string beam to string beam

downstand string beam to well edge of flight

alternative – upstand string beam to well edge of flight

upper floor

spine wall

structural frame

lightweight cladding

2nd. floor

landing

CANTILEVER STAIR

Landings cantilever on both sides of a stub beam which is a cantilever beam out of the rear edge of the spine wall

1st. floor

landing

Stair flights are cantilever flights about the spine wall

ground floor

Spiral and Helical Stairs ~ these stairs constructed in in-situ reinforced concrete are considered to be aesthetically pleasing but are expensive to construct. They are therefore mainly confined to prestige buildings usually as accommodation stairs linking floors within the same compartment. Like all other forms of stair they must conform to the requirements of Part K of the Building Regulations and if used as a means of escape in case of fire with the requirements of Part B. Spiral stairs can be defined as those describing a helix around a central column whereas a helical stair has an open well. The open well of a helical stair is usually circular or elliptical in plan and the formwork is built up around a vertical timber core.

Typical Example of a Helical Stair ~

outer handrail

upper floor

balustrade

edge beam under

tapered treads of equal shape and size

circular well

inner handrail

landing - cantilevered from floor

balustrade

PLAN

balustrade

upper floor edge beam

landing slab and edge beam

outer handrail

lower or ground floor

inner handrail

balusters

curved flight spanning from landing to floor

SECTION THROUGH LANDING

698

In-situ RC Inclined Slab Stair — Typical Details ~

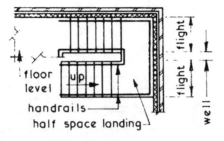

KEY PLAN

floor level

handrails

half space landing

up

flight

flight

well

NB. in plan the risers in the upper flight are not in line with those in the lower flight. This is to ensure that the soffits of the two flights line through at their intersection with the soffit of the half space landing

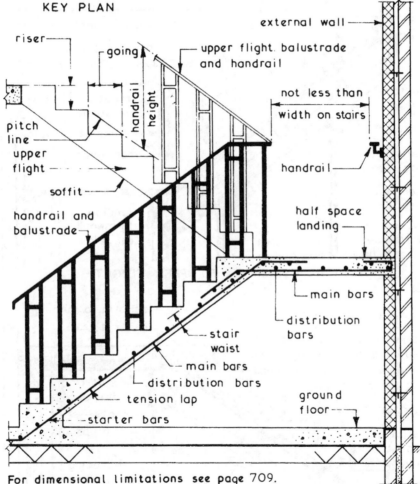

riser

going

pitch line upper flight

soffit

handrail and balustrade

handrail height

upper flight. balustrade and handrail

external wall

not less than

width on stairs

handrail

half space landing

main bars

distribution bars

stair waist

main bars

distribution bars

tension lap

starter bars

ground floor

For dimensional limitations see page 709.

Stair width subject to AD B: Fire safety, and AD M: Access to and use of buildings. Width measured as clear distance between walls or balustrade. Ignore string and handrail if projecting <100 mm (AD B, Volume 2).

In-situ Reinforced Concrete Stair Formwork ~ in specific detail the formwork will vary for the different types of reinforced concrete stair but the basic principles for each format will remain constant.

Typical RC Stair Formwork Details ~ (see page 699 for Key Plan)

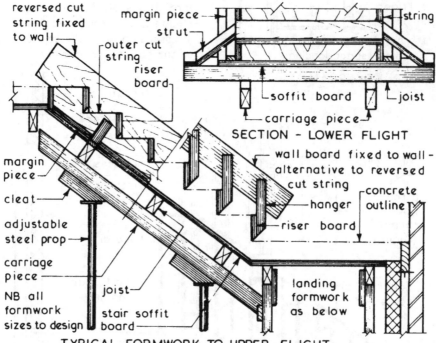

reversed cut string fixed to wall

margin piece

strut

string

outer cut string

riser board

soffit board

joist

carriage piece

SECTION - LOWER FLIGHT

wall board fixed to wall - alternative to reversed cut string

margin piece

cleat

hanger

concrete outline

riser board

adjustable steel prop

carriage piece

NB all formwork sizes to design

joist

stair soffit board

landing formwork as below

TYPICAL FORMWORK TO UPPER FLIGHT

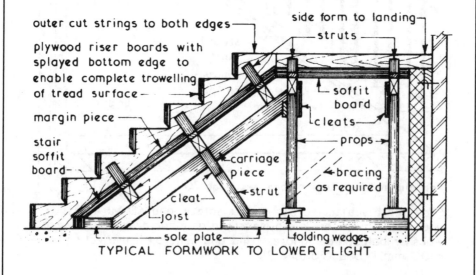

outer cut strings to both edges

side form to landing

struts

plywood riser boards with splayed bottom edge to enable complete trowelling of tread surface

margin piece

stair soffit board

soffit board

cleats

props

carriage piece

cleat

strut

bracing as required

joist

sole plate

folding wedges

TYPICAL FORMWORK TO LOWER FLIGHT

Precast Concrete Stairs ~ these can be produced to most of the formats used for in-situ concrete stairs and like those must comply with the appropriate requirements set out in Part K of the Building Regulations. To be economic the total production run must be sufficient to justify the costs of the moulds and therefore the designers choice may be limited to the stair types which are produced as a manufacturer's standard item.

Precast concrete stairs can have the following advantages:-

1. Good quality control of finished product.
2. Saving in site space since formwork fabrication and storage will not be required.
3. The stairs can be installed at any time after the floors have been completed thus giving full utilisation to the stair shaft as a lifting or hoisting space if required.
4. Hoisting, positioning and fixing can usually be carried out by semi-skilled labour.

Typical Example ~ Straight Flight Stairs

FLOOR JUNCTION DETAIL

Typical Example ~ Cranked Slab Stairs

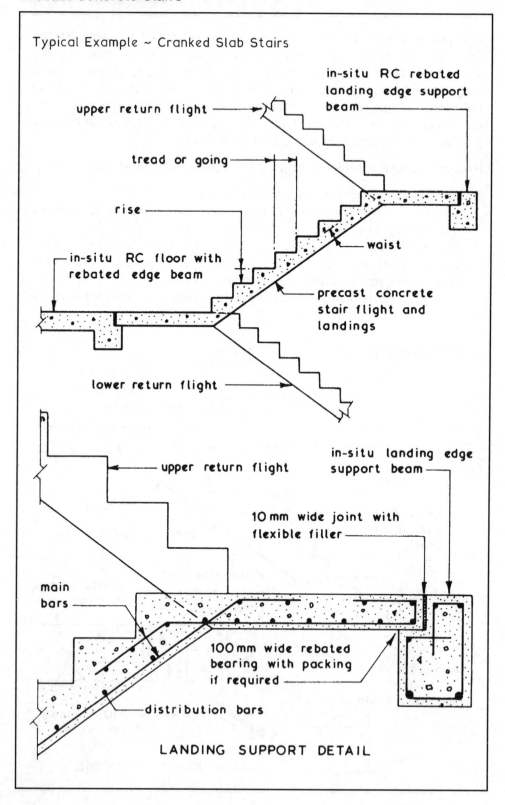

upper return flight

in-situ RC rebated landing edge support beam

tread or going

rise

in-situ RC floor with rebated edge beam

waist

precast concrete stair flight and landings

lower return flight

upper return flight

in-situ landing edge support beam

10 mm wide joint with flexible filler

main bars

100 mm wide rebated bearing with packing if required

distribution bars

LANDING SUPPORT DETAIL

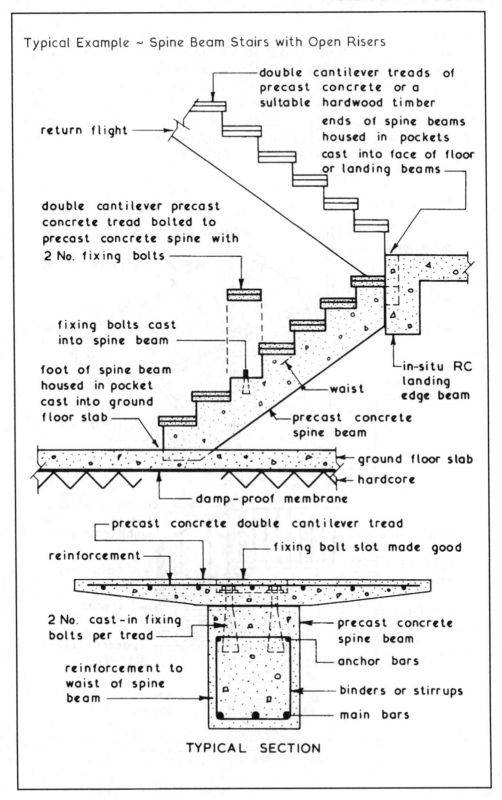

Typical Example ~ Spine Beam Stairs with Open Risers

double cantilever treads of precast concrete or a suitable hardwood timber

return flight

ends of spine beams housed in pockets cast into face of floor or landing beams

double cantilever precast concrete tread bolted to precast concrete spine with 2 No. fixing bolts

fixing bolts cast into spine beam

in-situ RC landing edge beam

foot of spine beam housed in pocket cast into ground floor slab

waist

precast concrete spine beam

ground floor slab

hardcore

damp-proof membrane

precast concrete double cantilever tread

fixing bolt slot made good

reinforcement

2 No. cast-in fixing bolts per tread

precast concrete spine beam

anchor bars

reinforcement to waist of spine beam

binders or stirrups

main bars

TYPICAL SECTION

703

Precast Concrete Spiral Stairs ~ this form of stair is usually constructed with an open riser format using tapered treads which have a keyhole plan shape. Each tread has a hollow cylinder at the narrow end equal to the rise which is fitted over a central steel column usually filled with in-situ concrete. The outer end of the tread has holes through which the balusters pass to be fixed on the underside of the tread below, a hollow spacer being used to maintain the distance between consecutive treads.

Typical Example ~

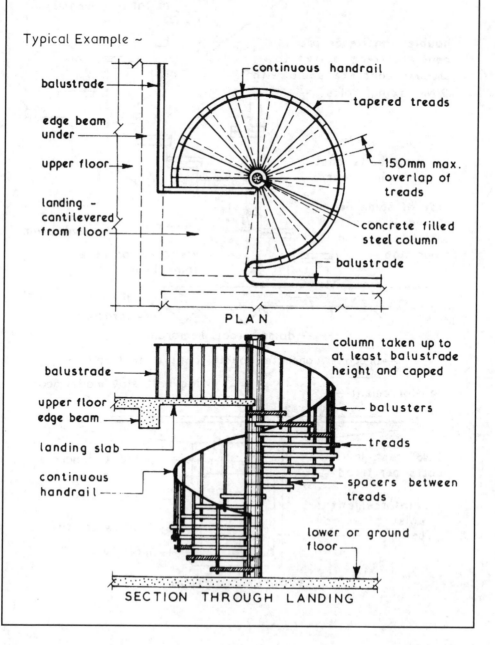

PLAN

SECTION THROUGH LANDING

Metal Stairs ~ these can be produced in cast iron, mild steel or aluminium alloy for use as escape stairs or for internal accommodation stairs. Most escape stairs are fabricated from cast iron or mild steel and must comply with the Building Regulation requirements for stairs in general and fire escape stairs in particular. Most metal stairs are purpose made and therefore tend to cost more than comparable concrete stairs. Their main advantage is the elimination of the need for formwork whilst the main disadvantage is the regular maintenance in the form of painting required for cast iron and mild steel stairs.

Typical Example ~ Straight Flight Steel External Escape Stair

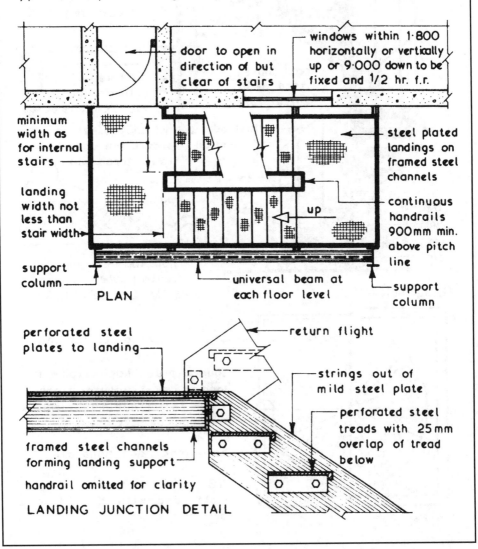

door to open in direction of but clear of stairs

windows within 1·800 horizontally or vertically up or 9·000 down to be fixed and 1/2 hr. f.r.

minimum width as for internal stairs

steel plated landings on framed steel channels

landing width not less than stair width

continuous handrails 900mm min. above pitch line

up

support column

universal beam at each floor level

support column

PLAN

perforated steel plates to landing

return flight

strings out of mild steel plate

framed steel channels forming landing support

perforated steel treads with 25mm overlap of tread below

handrail omitted for clarity

LANDING JUNCTION DETAIL

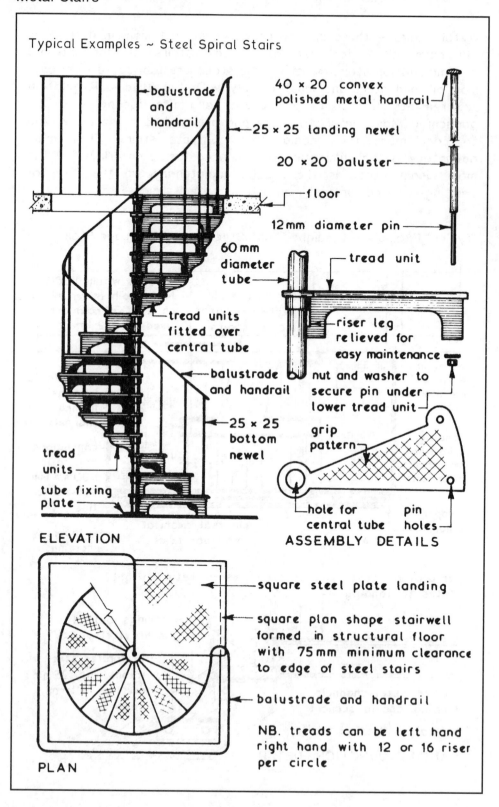

Typical Examples ~ Steel Spiral Stairs

←balustrade and handrail

40 × 20 convex polished metal handrail→

25 × 25 landing newel→

20 × 20 baluster→

←floor

12mm diameter pin→

60 mm diameter tube→

←tread unit

tread units fitted over central tube

riser leg relieved for easy maintenance

←balustrade and handrail

nut and washer to secure pin under lower tread unit

←25 × 25 bottom newel

grip pattern

tread units

tube fixing plate

hole for central tube   pin holes

ELEVATION

ASSEMBLY DETAILS

←square steel plate landing

←square plan shape stairwell formed in structural floor with 75mm minimum clearance to edge of steel stairs

←balustrade and handrail

NB. treads can be left hand right hand with 12 or 16 riser per circle

PLAN

706

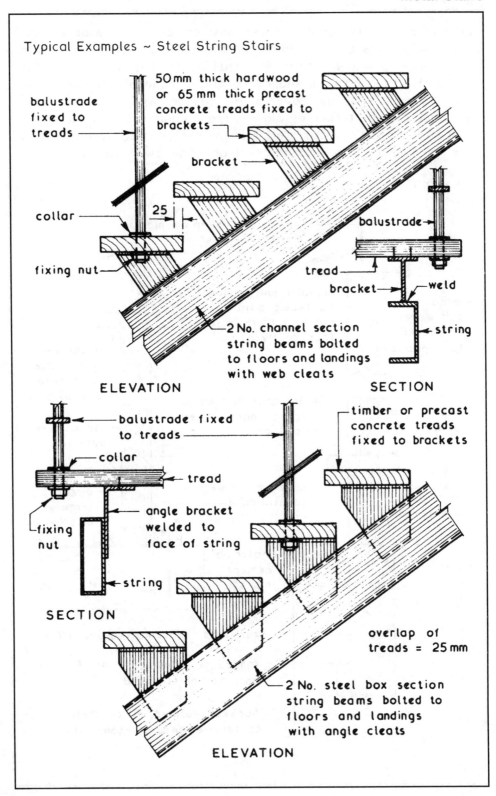

Typical Examples ~ Steel String Stairs

balustrade fixed to treads

50 mm thick hardwood or 65 mm thick precast concrete treads fixed to brackets

bracket

collar

25

fixing nut

balustrade

tread

bracket

weld

string

2 No. channel section string beams bolted to floors and landings with web cleats

ELEVATION

SECTION

balustrade fixed to treads

timber or precast concrete treads fixed to brackets

collar

tread

angle bracket welded to face of string

fixing nut

string

overlap of treads = 25 mm

SECTION

2 No. steel box section string beams bolted to floors and landings with angle cleats

ELEVATION

Balustrades and Handrails ~ these must comply in all respects with the requirements given in Part K of the Building Regulations and in the context of escape stairs are constructed of a non-combustible material with a handrail shaped to give a comfortable hand grip. The handrail may be covered or capped with a combustible material such as timber or plastic. Most balustrades are designed to be fixed after the stairs have been cast or installed by housing the balusters in a preformed pocket or by direct surface fixing.

Typical Details ~

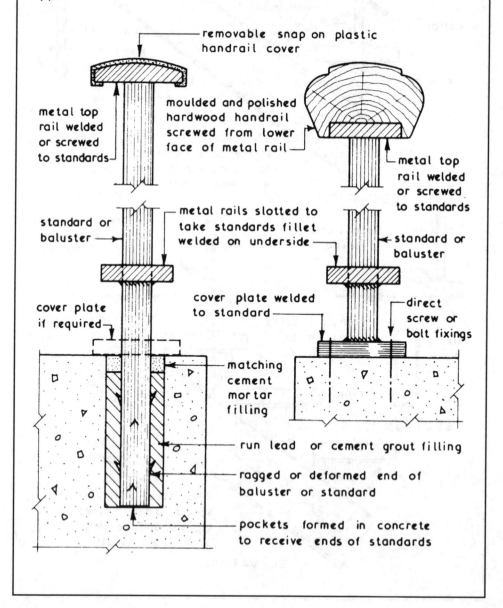

removable snap on plastic handrail cover

metal top rail welded or screwed to standards

moulded and polished hardwood handrail screwed from lower face of metal rail

metal top rail welded or screwed to standards

standard or baluster

metal rails slotted to take standards fillet welded on underside

standard or baluster

cover plate if required

cover plate welded to standard

direct screw or bolt fixings

matching cement mortar filling

run lead or cement grout filling

ragged or deformed end of baluster or standard

pockets formed in concrete to receive ends of standards

Institutional and Assembly Stairs ~ Serving a place where a substantial number of people will gather.

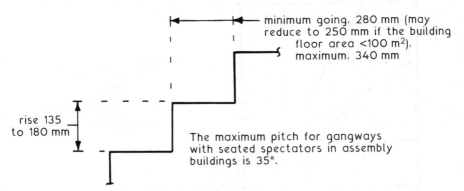

minimum going, 280 mm (may reduce to 250 mm if the building floor area <100 m²), maximum, 340 mm

rise 135 to 180 mm

The maximum pitch for gangways with seated spectators in assembly buildings is 35°.

Other stairs ~ All other buildings.

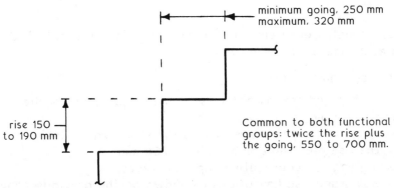

minimum going, 250 mm maximum, 320 mm

rise 150 to 190 mm

Common to both functional groups: twice the rise plus the going, 550 to 700 mm.

Alternative guidance: BS 5395-1: Stairs, ladders and walkways. Code of practice for the design, construction and maintenance of straight stairs and winders.

The rise and going in both situations may be subject to the requirements of Approved Document M: Access to and use of buildings. AD M will take priority and the following will apply:

Going (external steps) ~ 280 mm minimum (300 mm min. preferred).
Going (stairs) ~ 250 mm minimum.
Rise ~ 170 mm maximum.

Other AD M requirements for stairs:

Avoid tapered treads.
Width at least 1200 mm between walls, strings or upstands.
Landing top and bottom, minimum length not less than stair width.
Handrails both sides; circular or oval preferred as shown on next page.
Door openings onto landings to be avoided.
Full risers, no gaps.
Nosings with prominent colour 55 mm wide on tread and riser.
Nosing projections avoided, maximum of 25 mm if necessary.
Maximum 12 risers; exceptionally in certain small premises 16.
Non-slip surface to treads and landing.

Measurement of the going (AD K) ~

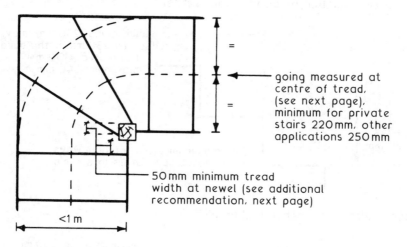

going measured at centre of tread, (see next page), minimum for private stairs 220 mm, other applications 250 mm

50 mm minimum tread width at newel (see additional recommendation, next page)

<1 m

For stair widths greater than 1 m, the going is measured at 270 mm from each side of the stair.

Additional requirements:

Going of tapered treads not less than the going of parallel treads in the same stair.
Curved landing lengths measured on the stair centre line.
Twice the rise plus the going, 550 to 700 mm.
Uniform going for consecutive tapered treads.
Other going and rise limitations as shown on the previous page and page 686.

Alternative guidance that provides reasonable safety for use in dwellings according to the requirements of Approved Document K is published in BS 585 – 1: Specification for stairs with closed risers for domestic use, including straight and winder flights and quarter or half landings.

Handrail profile ~

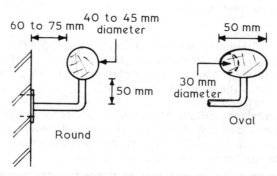

60 to 75 mm   40 to 45 mm diameter

50 mm

50 mm

30 mm diameter

Round

Oval

Basic requirements ~

- Lines of all risers to meet at one point, ie. the geometric centre.
- Centre going to be uniform and not less than the going of adjacent straight flights.
- Centre going ≥ going of adjacent parallel tread.
- Centre going ≤ 700 – (2 × rise) in mm.
- Going at newel post, not less than 75mm.
- Taper angle of treads to be uniform.
- Clear width at least 770mm.

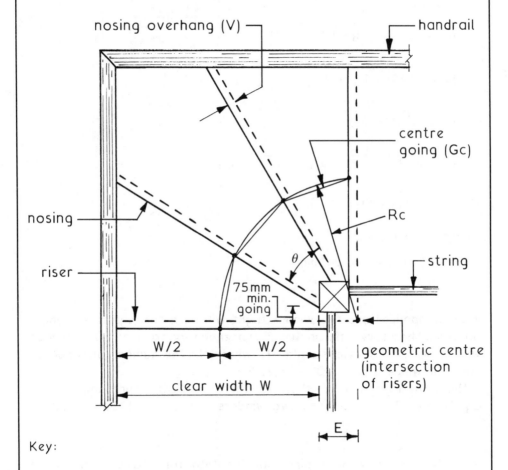

nosing overhang (V)

handrail

centre going (Gc)

nosing

Rc

riser

θ

string

75mm min. going

W/2   W/2

geometric centre (intersection of risers)

clear width W

E

Key:

Gc = centre going
Rc = radius of arc given by equation (see next page)
 θ = angle subtended by each tread
 W = clear width
 E = risers meeting point to edge of clear width
 V = nosing overhang

To check whether the design for winder flights will comply with recommended dimensions, measurements can be taken from scale drawings or calculations may be applied.

Calculations ~

$Gc = 2Rc \, (\sin \, \theta \div 2)$

where, $Rc = \sqrt{([W \div 2] + E)^2 + (V)^2}$

Eg. W = 770 mm
    E = 150 mm
    V = 16 mm

$$Rc = \sqrt{([770 \div 2] + 150)^2 + (16)^2}$$

$$Rc = \sqrt{(385 + 150)^2 + 256}$$

$$Rc = \sqrt{(535)^2 + 256}$$

$$Rc = \sqrt{286225 + 256} = \sqrt{286481} = 535 \text{ mm}$$

$Gc = 2 \times 535 \, (\sin \, 30° \div 2)$

where, $\sin 30° = 0.5$

$Gc = 1070 \times (0.5 = 2) = 267 \text{ mm}$

In most applications the winder flight will turn through a 90° angle on plan. Therefore, the angle θ subtended by each of three treads will be 30°, ie. $\theta =$ turn angle $\div N$, where N = the number of winders. For four winders, $\theta = 90° \div 4 = 22.5°$.
If the turn angle is other than 90° then $\theta = (180 -$ turn angle$) \div N$.
Eg. Turn angle of 120° with two winders, $\theta = (180 - 120) \div 2 = 30°$.

Refs. BS 585-1: Wood stairs. Specification for stairs with closed risers for domestic use, including straight and winder flights and quarter or half landings.
BS 5395-1: Stairs, ladders and walkways. Code of practice for the design, construction and maintenance of straight stairs and winders.

Summary recommendations of BS 5395-2 ~

Stair type[1]	Clear width[2]	Rise	Min. inner going[3]	Max. outer going[4]	Min. centre going[5]	2 × rise + going
A	600	170–220	120	350	145	480–800
B	800	170–220	120	350	190	480–800
C	800	170–220	150	350	230	480–800
D	900	150–190	150	450	250	480–800
E	1000	150–190	150	450	250	480–800

All dimensions in millimetres.
[1]See next page.
[2]Minimum clear width.
[3]270mm horizontally from inner handrail or column face if no handrail.
[4]270mm horizontally from outer handrail or string (take least value).
[5]Centre of clear width.

Clear headroom ~ measured from the pitch line (consecutive tread nosings at the geometric stair centre) vertically to any overhead obstruction. Normally 2.000m, but acceptable at 1.900m within 150mm of the centre column.

Landing ~ Minimum angle subtended at the stair centre is 60°. Intermediate landings, minimum angle is 45° or a plan area ≥ two treads (take greater area).

Loading guide ~ (No. of treads × 0.2) + (1.5 for the landing)
Eg. 14 treads:
= (14 × 0.2) + (1.5) = 4.3 kN.

Refs.
BS 5395-2: Stairs, ladders and walkways. Code of practice for the design of helical and spiral stairs.
Building Regulations, A.D. K: Protection from falling, collision and impact.

BS 5395-2, Stair types:

Category A ~
A small private stair for use by a limited number of people who are generally familiar with the stair. For example, an internal stair in a dwelling serving one room not being a living room or a kitchen. Also, an access stair to a small room or equipment room in an office, shop or factory not for public or general use. Subject to the provisions of A.D. B, possibly a fire escape for a small number of people. Typically, 1.300 to 1.800 m overall diameter.

Category B ~
Similar to category A, but providing the main access to the upper floor of a private dwelling. Typically, 1.800 to 2.250 m overall diameter.

Category C ~
A small semi-public stair for use by a limited number of people, some of whom may be unfamiliar with the stair. Examples include a stair in a factory, office or shop and a common stair serving more than one dwelling. Typically, 2.000 to 2.250 m overall diameter.

Category D ~
Similar to category C, but for use by larger numbers of people. Typically, 2.000 to 2.500 m overall diameter.

Category E ~
A public stair intended to be used by a large number of people at one time. For example in a place of public assembly. Typically, 2.500 to 3.500 m overall diameter.

Note 1: With regard to means of escape in event of a fire, minimum widths given may be insufficient. All applications to satisfy the requirements of Building Regulations, Approved Documents B1 and B2: Fire safety. This has particular reference to stair clear width relative to the number of persons likely to use the stair and protection of the stair well.

Note 2: In addition to an outer handrail, an inner handrail should be provided for categories C, D and E.

Functions ~ the main functions of any door are to:

1. Provide a means of access and egress.
2. Maintain continuity of wall function when closed.
3. Provide a degree of privacy and security.

Choice of door type can be determined by:-

1. Position - whether internal or external.
2. Properties required - fire resistant, glazed to provide for borrowed light or vision through, etc.
3. Appearance - flush or panelled, painted or polished, etc.

Door Schedules ~ these can be prepared in the same manner and for the same purpose as that given for windows on page 377.

Internal Doors ~ these are usually lightweight and can be fixed to a lining, if heavy doors are specified these can be hung to frames in a similar manner to external doors. An alternative method is to use door sets which are usually storey height and supplied with prehung doors.

Typical door Lining Details ~

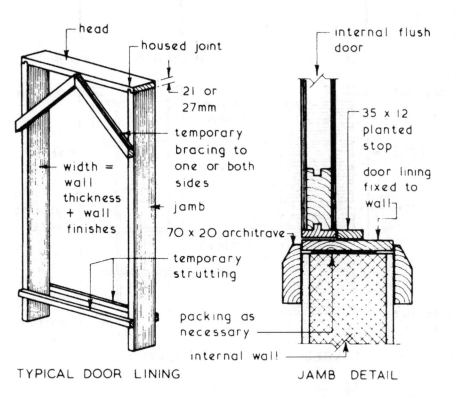

TYPICAL DOOR LINING   JAMB DETAIL

Internal Doors ~ these are similar in construction to the external doors but are usually thinner and therefore lighter in weight.

Typical Examples ~

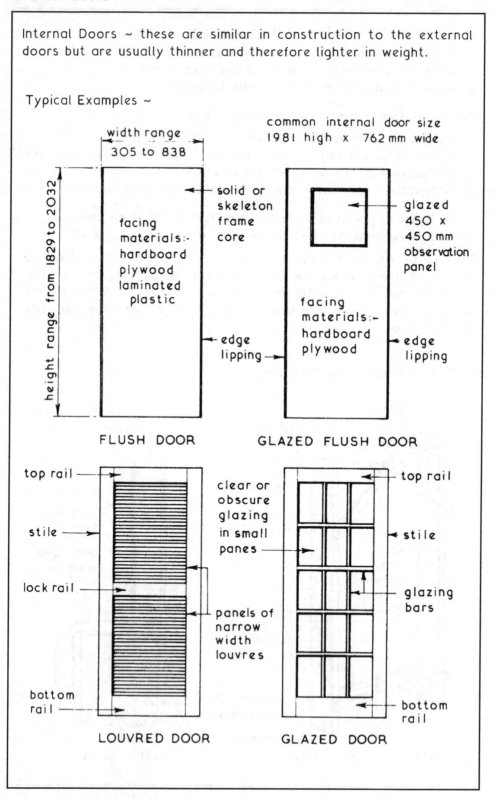

FLUSH DOOR   GLAZED FLUSH DOOR

LOUVRED DOOR   GLAZED DOOR

Internal Door Frames and linings ~ these are similar in construction to external door frames but usually have planted door stops and do not have a sill. The frames are sized to be built in conjunction with various partition thicknesses and surface finishes. Linings with planted stops ae usually employed for lightweight domestic doors.

Typical Examples ~

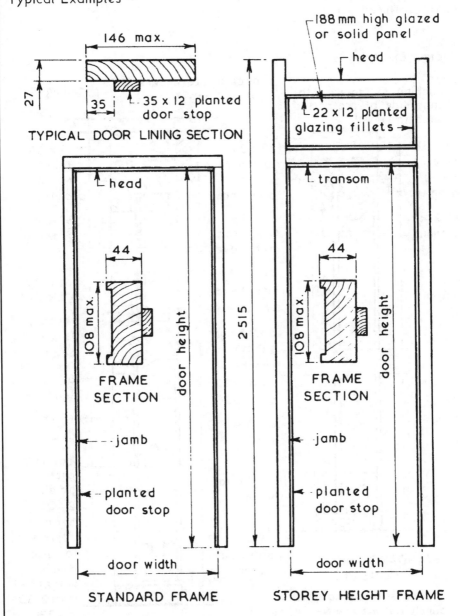

146 max.

27

35

35 x 12 planted door stop

TYPICAL DOOR LINING SECTION

188 mm high glazed or solid panel

head

22 x 12 planted glazing fillets

head

44

108 max.

FRAME SECTION

door height

2 515

jamb

planted door stop

door width

STANDARD FRAME

transom

44

108 max.

FRAME SECTION

door height

jamb

planted door stop

door width

STOREY HEIGHT FRAME

Ref. BS 4787: Internal and external wood doorsets, door leaves and frames. Specification for dimensional requirements.

Doorsets ~ these are factory produced fully assembled prehung doors which are supplied complete with frame, architraves and ironmongery except for door furniture. The doors may be hung to the frames using pin butts for easy door removal. Prehung door sets are available in standard and storey height versions and are suitable for all internal door applications with normal wall and partition thicknesses.

Typical Examples ~

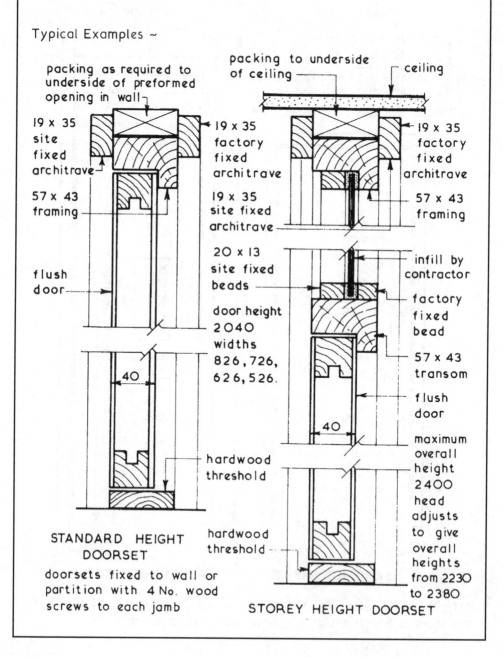

packing as required to underside of preformed opening in wall

19 x 35 site fixed architrave

57 x 43 framing

flush door

40

STANDARD HEIGHT DOORSET

doorsets fixed to wall or partition with 4 No. wood screws to each jamb

packing to underside of ceiling

ceiling

19 x 35 factory fixed architrave

19 x 35 site fixed architrave

20 x 13 site fixed beads

door height 2040 widths 826, 726, 626, 526.

hardwood threshold

19 x 35 factory fixed architrave

57 x 43 framing

infill by contractor

factory fixed bead

57 x 43 transom

flush door

40

maximum overall height 2400 head adjusts to give overall heights from 2230 to 2380

hardwood threshold

STOREY HEIGHT DOORSET

Fire doorset ~ a "complete unit consisting of a door frame and a door leaf or leaves, supplied with all essential parts from a single source". The difference between a doorset and a fire doorset is the latter is endorsed with a fire certificate for the complete unit. When supplied as a collection of parts for site assembly, this is known as a door kit.

Fire door assembly ~ a "complete assembly as installed, including door frame and one or more leaves, together with its essential hardware [ironmongery] supplied from separate sources". Provided the components to an assembly satisfy the Building Regulations – Approved Document B, fire safety requirements and standards for certification and compatibility, then a fire door assembly is an acceptable alternative to a doorset.

Fire doorsets are usually more expensive than fire door assemblies, but assemblies permit more flexibility in choice of components. Site fixing time will be longer for assemblies.

(Quotes from BS EN 12519: Windows and pedestrian doors. Terminology.)

Fire door ~ a fire door is not just the door leaf. A fire door includes the frame, ironmongery, glazing, intumescent core and smoke seal. To comply with European market requirements, ironmongery should be CE marked (see page 64). A fire door should also be marked accordingly on the top or hinge side. The label type shown below, reproduced with kind permission of the British Woodworking Federation is acceptable.

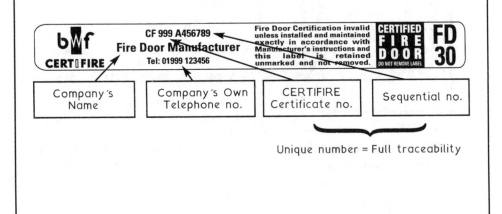

30 Minute Flush Fire Doors ~ these are usually based on the recommendations given in BS 8214. A wide variety of door constructions are available from various manufacturers but they all have to be fitted to a similar frame for testing as a doorset or assembly, including ironmongery.

A door's resistance to fire is measured by:-

1. Insulation – resistance to thermal transmittance, see BS 476-20 & 22: Fire tests on building materials and structures.

2. Integrity – resistance in minutes to the penetration of flame and hot gases under simulated fire conditions.

Typical Details ~

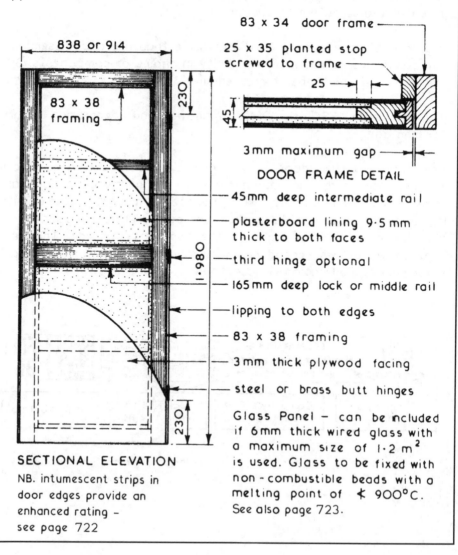

838 or 914

83 x 38 framing

230

1·980

230

SECTIONAL ELEVATION

NB. intumescent strips in door edges provide an enhanced rating –
see page 722

83 x 34 door frame

25 x 35 planted stop screwed to frame

25

45

3mm maximum gap

DOOR FRAME DETAIL

45mm deep intermediate rail

plasterboard lining 9·5mm thick to both faces

third hinge optional

165mm deep lock or middle rail

lipping to both edges

83 x 38 framing

3mm thick plywood facing

steel or brass butt hinges

Glass Panel – can be included if 6mm thick wired glass with a maximum size of 1·2 m^2 is used. Glass to be fixed with non-combustible beads with a melting point of ≮ 900°C. See also page 723.

60 Minute Flush Fire Door ~ like the 30 minute flush fire door shown on page 720 these doors are based on the recommendations given in BS 8214 which covers both door and frame. A wide variety of fire resistant door constructions are available from various manufacturers with most classified as having both insulation and integrity ratings of 60 minutes.

Typical Details ~

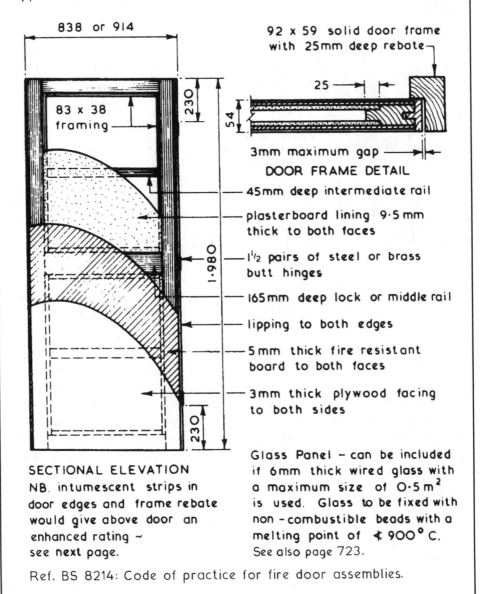

838 or 914

83 x 38 framing

230

1·980

230

92 x 59 solid door frame with 25mm deep rebate

25

54

3mm maximum gap

DOOR FRAME DETAIL

45mm deep intermediate rail

plasterboard lining 9·5mm thick to both faces

1½ pairs of steel or brass butt hinges

165mm deep lock or middle rail

lipping to both edges

5mm thick fire resistant board to both faces

3mm thick plywood facing to both sides

SECTIONAL ELEVATION
NB. intumescent strips in door edges and frame rebate would give above door an enhanced rating ~ see next page.

Glass Panel ~ can be included if 6mm thick wired glass with a maximum size of O·5 m² is used. Glass to be fixed with non-combustible beads with a melting point of ≮ 900° C. See also page 723.

Ref. BS 8214: Code of practice for fire door assemblies.

Fire and Smoke Resistance ~ Doors can be assessed for both integrity and smoke resistance. They are coded accordingly, for example FD30 or FD30s. FD indicates a fire door and 30 the integrity time in minutes. The letter 's' denotes that the door or frame contains a facility to resist the passage of smoke.

Manufacturers produce doors of standard ratings – 30, 60 and 90 minutes, with higher ratings available to order. A colour coded plug inserted in the door edge corresponds to the fire rating. See BS 8214, Table 1 for details.

Intumescent Fire and Smoke Seals ~

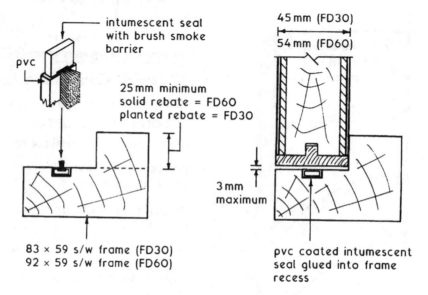

intumescent seal with brush smoke barrier

pvc

25 mm minimum
solid rebate = FD60
planted rebate = FD30

3 mm maximum

83 × 59 s/w frame (FD30)
92 × 59 s/w frame (FD60)

45 mm (FD30)
54 mm (FD60)

pvc coated intumescent seal glued into frame recess

The intumescent core may be fitted to the door edge or the frame. In practice, most joinery manufacturers leave a recess in the frame where the seal is secured with rubber based or PVA adhesive. At temperatures of about 150°C, the core expands to create a seal around the door edge. This remains throughout the fire resistance period whilst the door can still be opened for escape and access purposes. The smoke seal will also function as an effective draught seal.

Further references:

BS EN 1634-1: Fire resistance and smoke control tests for door, shutter and openable window assemblies and elements of building hardware.

BS EN 13501: Fire classification of construction products and building elements.

Apertures will reduce the potential fire resistance if not appropriately filled. Suitable material should have the same standard of fire performance as the door into which it is fitted.

Fire rated glass types ~

• Embedded Georgian wired glass
• Composite glass containing borosilicates and ceramics
• Tempered and toughened glass
• Glass laminated with reactive fire resisting interlayers

Installation ~ Hardwood beads and intumescent seals. Compatibility of glass type and sealing product is essential, therefore manufacturers details must be consulted.

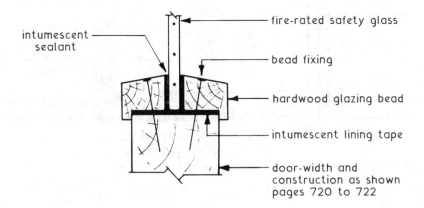

intumescent sealant

fire-rated safety glass

bead fixing

hardwood glazing bead

intumescent lining tape

door-width and construction as shown pages 720 to 722

Intumescent products ~

• Sealants and mastics 'gun' applied
• Adhesive glazing strip or tape
• Preformed moulded channel

Note: Calcium silicate preformed channel is also available. Woven ceramic fire-glazing tape/ribbon is produced specifically for use in metal frames.

Building Regulations references:

AD M, Section 2.13, visibility zones/panels between 500 and 1500 mm above floor finish.
AD N, Section 1.6, aperture size (see page 387).

Typical Details ~

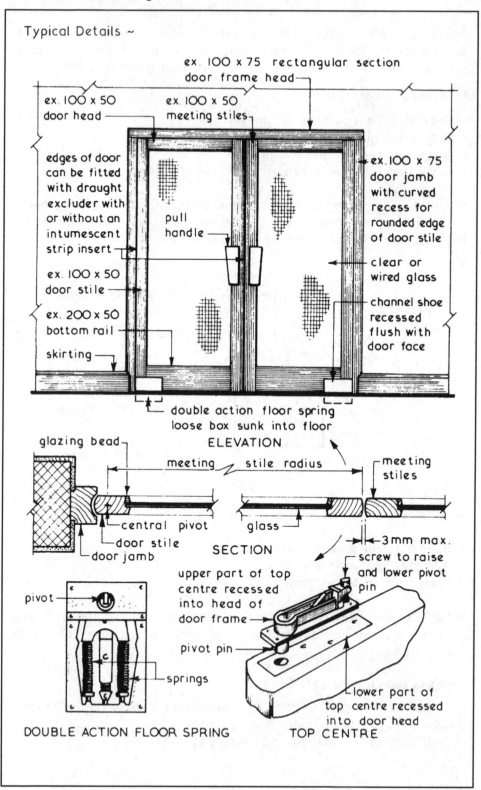

ex. 100 x 75 rectangular section door frame head

ex. 100 x 50 door head

ex. 100 x 50 meeting stiles

edges of door can be fitted with draught excluder with or without an intumescent strip insert

pull handle

ex. 100 x 75 door jamb with curved recess for rounded edge of door stile

ex. 100 x 50 door stile

clear or wired glass

ex. 200 x 50 bottom rail

channel shoe recessed flush with door face

skirting

double action floor spring loose box sunk into floor

ELEVATION

glazing bead

meeting   stile radius

meeting stiles

central pivot

door stile

door jamb

glass

3 mm max.

SECTION

screw to raise and lower pivot pin

upper part of top centre recessed into head of door frame

pivot pin

pivot

springs

DOUBLE ACTION FLOOR SPRING

lower part of top centre recessed into door head

TOP CENTRE

Plasterboard ~ a rigid board composed of gypsum sandwiched between durable lining paper outer facings. For ceiling applications, the following types can be used:

Baseboard -
1220 × 900 × 9.5mm thick for joist centres up to 400mm.
1220 × 600 × 12.5mm thick for joist centres up to 600mm.

Baseboard has square edges and can be plaster skim finished. Joints are reinforced with self-adhesive 50mm min. width glass fibre mesh scrim tape or the board manufacturer's recommended paper tape. These boards are also made with a metallised polyester foil backing for vapour check applications. The foil is to prevent any moisture produced in potentially damp situations such as a bathroom or in warm roof construction from affecting loft insulation and timber. Joints should be sealed with an adhesive metallised tape.

Wallboard -
9.5, 12.5 and 15mm thicknesses, 900 and 1200mm widths and lengths of 1800 and 2400mm. Longer boards are produced in the two greater thicknesses and a vapour check variation is available. Edges are either tapered for taped and filled joints for dry lining or square for skimmed plaster or textured finishes.

Plasterboards should be fixed breaking joint to the underside of floor or ceiling joists with zinc plated (galvanised) nails or dry-wall screws at 150mm max. spacing. The junction at ceiling to wall is reinforced with glass fibre mesh scrim tape or a preformed plaster moulding.

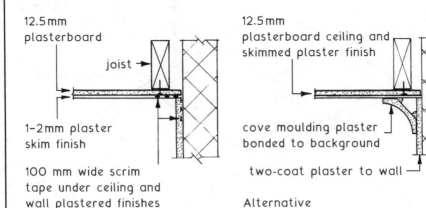

12.5mm
plasterboard

joist →

1-2mm plaster
skim finish

100 mm wide scrim
tape under ceiling and
wall plastered finishes

12.5mm
plasterboard ceiling and
skimmed plaster finish

cove moulding plaster
bonded to background

two-coat plaster to wall

Alternative

Suspended Ceilings ~ these can be defined as ceilings which are fixed to a framework suspended from main structure thus forming a void between the two components. The basic functional requirements of suspended ceilings are:-

1. They should be easy to construct, repair, maintain and clean.
2. So designed that an adequate means of access is provided to the void space for the maintenance of the suspension system, concealed services and/or light fittings.
3. Provide any required sound and/or thermal insulation.
4. Provide any required acoustic control in terms of absorption and reverberation.
5. Provide if required structural fire protection to structural steel beams supporting a concrete floor and contain fire stop cavity barriers within the void at defined intervals.
6. Conform with the minimum requirements set out in the Building Regulations governing the restriction of spread of flame over surfaces of ceilings and the exemptions permitting the use of certain plastic materials.
7. Flexural design strength in varying humidity and temperature.
8. Resistance to impact.
9. Designed on a planning module, preferably a 300 mm dimensional coordinated system.

Typical Suspended Ceiling Grid Framework Layout ~

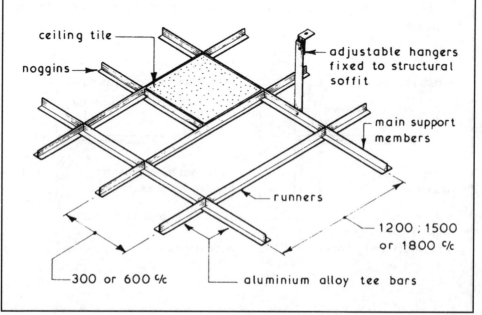

ceiling tile

noggins

adjustable hangers fixed to structural soffit

main support members

runners

1200 ; 1500 or 1800 c/c

300 or 600 c/c

aluminium alloy tee bars

Classification of Suspended Ceiling ~ there is no standard method of classification since some are classified by their function such as illuminated and acoustic suspended ceilings, others are classified by the materials used and classification by method of construction is also very popular. The latter method is simple since most suspended ceiling types can be placed in one of three groups:-

1. Jointless suspended ceilings.
2. Panelled suspended ceilings – see page 728.
3. Decorative and open suspended ceilings – see page 729.

Jointless Suspended Ceilings ~ these forms of suspended ceilings provide a continuous and jointless surface with the internal appearance of a conventional ceiling. They may be selected to fulfil fire resistance requirements or to provide a robust form of suspended ceiling. The two common ways of construction are a plasterboard or expanded metal lathing soffit with hand applied plaster finish or a sprayed applied rendering with a cement base.

Typical Details ~

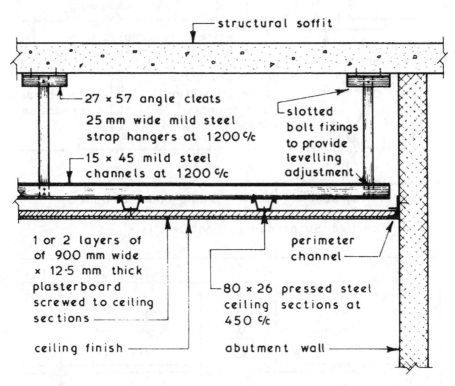

structural soffit

27 × 57 angle cleats

25 mm wide mild steel strap hangers at 1200 c/c

15 × 45 mild steel channels at 1200 c/c

slotted bolt fixings to provide levelling adjustment

1 or 2 layers of of 900 mm wide × 12·5 mm thick plasterboard screwed to ceiling sections

perimeter channel

80 × 26 pressed steel ceiling sections at 450 c/c

ceiling finish

abutment wall

See also: BS EN 13964: Suspended ceilings. Requirements and test methods.

Panelled Suspended Ceilings ~ these are the most popular form of suspended ceiling consisting of a suspended grid framework to which the ceiling covering is attached. The covering can be of a tile, tray, board or strip format in a wide variety of materials with an exposed or concealed supporting framework. Services such as luminaries can usually be incorporated within the system. Generally panelled systems are easy to assemble and install using a water level or laser beam for initial and final levelling. Provision for maintenance access can be easily incorporated into most systems and layouts.

Typical Support Details ~

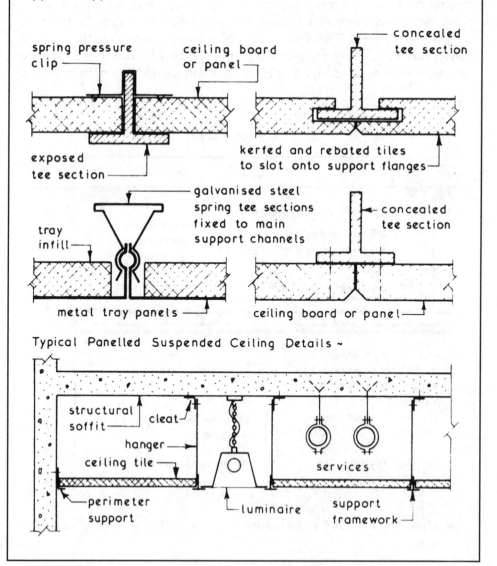

Typical Panelled Suspended Ceiling Details ~

Decorative and Open Suspended Ceilings ~ these ceilings usually consist of an openwork grid or suspended shapes onto which the lights fixed at, above or below ceiling level can be trained thus creating a decorative and illuminated effect. Many of these ceilings are purpose designed and built as opposed to the proprietary systems associated with jointless and panelled suspended ceilings.

Typical Examples ~

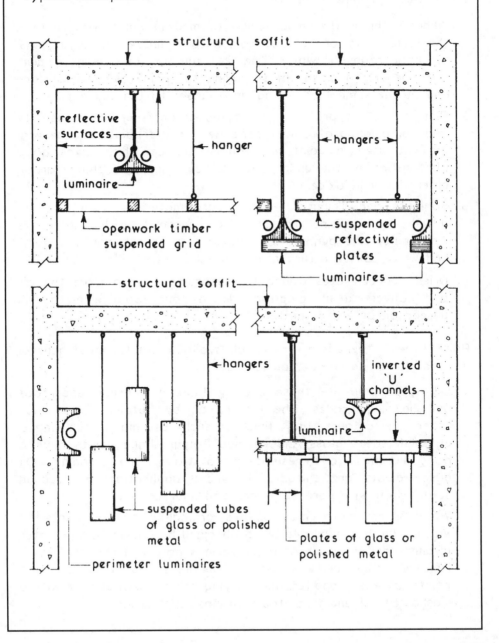

729

Functions ~ the main functions of paint are to provide:-

1. An economic method of surface protection to building materials and components.
2. An economic method of surface decoration to building materials and components.

Composition ~ the actual composition of any paint can be complex but the basic components are:-

1. Binder ~ this is the liquid vehicle or medium which dries to form the surface film and can be composed of linseed oil, drying oils, synthetic resins or water. The first function of a paint medium is to provide a means of spreading the paint over the surface and at the same time acting as a binder to the pigment.
2. Pigment ~ this provides the body, colour, durability, opacity and corrosion protection properties of the paint. The general pigment used in paint is titanium dioxide which gives good obliteration of the undercoats. Other pigments include carbon black and iron oxide.
3. Solvents and Thinners ~ these are materials which can be added to a paint to alter its viscosity. This increases workability and penetration. Water is used for emulsion paint and white spirit or turpentine for oil paint.
4. Drier ~ accelerates drying by absorbing oxygen from the air and converting by oxidation to a solid. Soluble metals in linseed oil or white spirit.

Paint Types - there is a wide range available but for most general uses the following can be considered:-

1. Oil Based paints - these are available in priming, undercoat and finishing grades. The latter can be obtained in a wide range of colours and finishes such as matt, semi-matt, eggshell, satin, gloss and enamel. Polyurethane paints have a good hardness and resistance to water and cleaning. Oil based paints are suitable for most applications if used in conjunction with correct primer and undercoat.
2. Water Based Paints - most of these are called emulsion paints the various finishes available being obtained by adding to the water medium additives such as alkyd resin & polyvinyl acetate (PVA). Finishes include matt, eggshell, semi-gloss and gloss. Emulsion paints are easily applied, quick drying and can be obtained with a washable finish and are suitable for most applications.

Supply ~ paint is usually supplied in metal containers ranging from 250 millilitres to 5 litres capacity to the colour ranges recommended in BS 381C (colours for specific purposes) and BS 4800 (paint colours for building purposes).

Application ~ paint can be applied to almost any surface providing the surface preparation and sequence of paint coats are suitable. The manufacturers specification and/or the recommendations of BS 6150 (painting of buildings) should be followed. Preparation of the surface to receive the paint is of the utmost importance since poor preparation is one of the chief causes of paint failure. The preperation consists basically of removing all dirt, grease, dust and ensuring that the surface will provide an adequate key for the paint which is to be applied. In new work the basic build-up of paint coats consists of:-

1. Priming Coats — these are used on unpainted surfaces to obtain the necessary adhesion and to inhibit corrosion of ferrous metals. New timber should have the knots treated with a solution of shellac or other alcohol based resin called knotting prior to the application of the primer.
2. Undercoats — these are used on top of the primer after any defects have been made good with a suitable stopper or filler. The primary function of an undercoat is to give the opacity and build-up necessary for the application of the finishing coat(s).
3. Finish — applied directly over the undercoating in one or more coats to impart the required colour and finish.

Paint can applied by:-

1. Brush — the correct type, size and quality of brush such as those recommended in BS 2992 (painters and decorators brushes) needs to be selected and used. To achieve a first class finish by means of brush application requires a high degree of skill.
2. Spray — as with brush application a high degree of skill is required to achieve a good finish. Generally compressed air sprays or airless sprays are used for building works.
3. Roller — simple and inexpensive method of quickly and cleanly applying a wide range of paints to flat and textured surfaces. Roller heads vary in size from 50 to 450 mm wide with various covers such as sheepskin, synthetic pile fibres, mohair and foamed polystyrene. All paint applicators must be thoroughly cleaned after use.

Painting ~ the main objectives of applying coats of paint to a surface are preservation, protection and decoration to give a finish which is easy to clean and maintain. To achieve these objectives the surface preparation and paint application must be adequate. The preparation of new and previously painted surfaces should ensure that prior to painting the surface is smooth, clean, dry and stable.

Basic Surface Preparation Techniques ~

Timber – to ensure a good adhesion of the paint film all timber should have a moisture content of less than 18%. The timber surface should be prepared using an abrasive paper to produce a smooth surface brushed and wiped free of dust and any grease removed with a suitable spirit. Careful treatment of knots is essential either by sealing with two coats of knotting or in extreme cases cutting out the knot and replacing with sound timber. The stopping and filling of cracks and fixing holes with putty or an appropriate filler should be carried out after the application of the priming coat. Each coat of paint must be allowed to dry hard and be rubbed down with a fine abrasive paper before applying the next coat. On previously painted surfaces if the paint is in a reasonable condition the surface will only require cleaning and rubbing down before repainting, when the paint is in a poor condition it will be necessary to remove completely the layers of paint and then prepare the surface as described above for new timber.

Building Boards – most of these boards require no special preparation except for the application of a sealer as specified by the manufacturer.

Iron and Steel – good preparation is the key to painting iron and steel successfully and this will include removing all rust, mill scale, oil, grease and wax. This can be achieved by wire brushing, using mechanical means such as shot blasting, flame cleaning and chemical processes and any of these processes are often carried out in the steel fabrication works prior to shop applied priming.

Plaster – the essential requirement of the preparation is to ensure that the plaster surface is perfectly dry, smooth and free of defects before applying any coats of paint especially when using gloss paints. Plaster which contains lime can be alkaline and such surfaces should be treated with an alkali resistant primer when the surface is dry before applying the final coats of paint.

Paint Defects ~ these may be due to poor or incorrect preparation of the surface, poor application of the paint and/or chemical reactions. The general remedy is to remove all the affected paint and carry out the correct preparation of the surface before applying in the correct manner new coats of paint. Most paint defects are visual and therefore an accurate diagnosis of the cause must be established before any remedial treatment is undertaken.

Typical Paint Defects ~

1. Bleeding – staining and disruption of the paint surface by chemical action, usually caused by applying an incorrect paint over another. Remedy is to remove affected paint surface and repaint with correct type of overcoat paint.

2. Blistering – usually caused by poor presentation allowing resin or moisture to be entrapped, the subsequent expansion causing the defect. Remedy is to remove all the coats of paint and ensure that the surface is dry before repainting.

3. Blooming – mistiness usually on high gloss or varnished surfaces due to the presence of moisture during application. It can be avoided by not painting under these conditions. Remedy is to remove affected paint and repaint.

4. Chalking – powdering of the paint surface due to natural ageing or the use of poor quality paint. Remedy is to remove paint if necessary, prepare surface and repaint.

5. Cracking and Crazing – usually due to unequal elasticity of successive coats of paint. Remedy is to remove affected paint and repaint with compatible coats of paint.

6. Flaking and Peeling – can be due to poor adhesion, presence of moisture, painting over unclean areas or poor preparation. Remedy is to remove defective paint, prepare surface and repaint.

7. Grinning – due to poor opacity of paint film allowing paint coat below or background to show through, could be the result of poor application; incorrect thinning or the use of the wrong colour. Remedy is to apply further coats of paint to obtain a satisfactory surface.

8. Saponification – formation of soap from alkali present in or on surface painted. The paint is ultimately destroyed and a brown liquid appears on the surface. Remedy is to remove the paint films and seal the alkaline surface before repainting.

Joinery Production ~ this can vary from the flow production where one product such as flush doors is being made usually with the aid of purpose designed and built machines, to batch production where a limited number of similar items are being made with the aid of conventional woodworking machines. Purpose made joinery is very often largely hand made with a limited use of machines and is considered when special and/or high class joinery components are required.

Woodworking Machines ~ except for the portable electric tools such as drills, routers, jigsaws and sanders most woodworking machines need to be fixed to a solid base and connected to an extractor system to extract and collect the sawdust and chippings produced by the machines.

Saws - basically three formats are available, namely the circular, cross cut and band saws. Circular are general purpose saws and usually have tungsten carbide tipped teeth with feed rates of up to 60·000 per minute. Cross cut saws usually have a long bench to support the timber, the saw being mounted on a radial arm enabling the circular saw to be drawn across the timber to be cut. Band saws consist of an endless thin band or blade with saw teeth and a table on which to support the timber and are generally used for curved work.

Planers - most of these machines are combined planers and thicknessers, the timber being passed over the table surface for planning and the table or bed for thicknessing. The planer has a guide fence which can be tilted for angle planning and usually the rear bed can be lowered for rebating operations. The same rotating cutter block is used for all operations. Planing speeds are dependent upon the operator since it is a hand fed operation whereas thicknessing is mechanically fed with a feed speed range of 6·000 to 20·000 per minute. Maximum planing depth is usually 10 mm per passing.

Morticing Machines - these are used to cut mortices up to 25 mm wide and can be either a chisel or chain morticer. The former consists of a hollow chisel containing a bit or auger whereas the latter has an endless chain cutter.

Tenoning Machines - these machines with their rotary cutter blocks can be set to form tenon and scribe. In most cases they can also be set for trenching, grooving and cross cutting.

Spindle Moulder - this machine has a horizontally rotating cutter block into which standard or purpose made cutters are fixed to reproduce a moulding on timber passed across the cutter.

Purpose Made Joinery ~ joinery items in the form of doors, windows, stairs and cupboard fitments can be purchased as stock items from manufacturers. There is also a need for purpose made joinery to fulfil client/designer/user requirement to suit a specific need, to fit into a non-standard space, as a specific decor requirement or to complement a particular internal environment. These purpose made joinery items can range from the simple to the complex which require high degrees of workshop and site skills.

Typical Purpose Made Counter Details ~

ELEVATION

2·700

900

- block board counter top with veneer, leather or similar finish
- veneer band
- timber band
- veneer faced panels
- kicking rail

SECTION

hardwood edging

920   counter top   hardwood edging

900   800   100

50   50

- drawer with runners housed in sides
- frame with projecting hardwood band
- blockboard shelf
- veneer faced blockboard front panels
- swing or sliding doors could be fitted to cupboards
- blockboard base
- kicking rail with durable facing in a contrasting colour (eg black laminate)

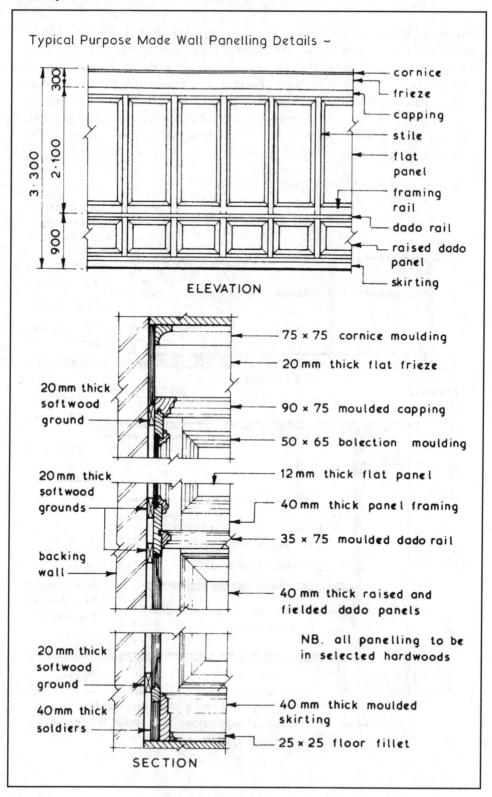

Typical Purpose Made Wall Panelling Details ~

ELEVATION

- cornice
- frieze
- capping
- stile
- flat panel
- framing rail
- dado rail
- raised dado panel
- skirting

3·300
300
2·100
900

SECTION

- 75 × 75 cornice moulding
- 20 mm thick flat frieze
- 90 × 75 moulded capping
- 50 × 65 bolection moulding
- 12 mm thick flat panel
- 40 mm thick panel framing
- 35 × 75 moulded dado rail
- 40 mm thick raised and fielded dado panels

NB. all panelling to be in selected hardwoods

- 40 mm thick moulded skirting
- 25 × 25 floor fillet

20 mm thick softwood ground

20 mm thick softwood grounds

backing wall

20 mm thick softwood ground

40 mm thick soldiers

Joinery Timbers ~ both hardwoods and softwoods can be used for joinery works. Softwoods can be selected for their stability, durability and/or workability if the finish is to be paint but if it is left in its natural colour with a sealing coat the grain texture and appearance should be taken into consideration. Hardwoods are usually left in their natural colour and treated with a protective clear sealer or polish therefore texture, colour and grain pattern are important when selecting hardwoods for high class joinery work.

Typical Softwoods Suitable for Joinery Work ~

1. Douglas Fir – sometimes referred to as Columbian Pine or Oregon Pine. It is available in long lengths and has a straight grain. Colour is reddish brown to pink. Suitable for general and high class joinery. Approximate density 530 kg/m^3.

2. Redwood – also known as Scots Pine, Red Pine, Red Deal and Yellow Deal. It is a widely used softwood for general joinery work having good durability a straight grain and is reddish brown to straw in colour. Approximate density 430 kg/m^3.

3. European Spruce – similar to redwood but with a lower durability. It is pale yellow to pinkish white in colour and is used mainly for basic framing work and simple internal joinery. Approximate density 650 kg/m^3.

4. Sitka Spruce – originates from Alaska, Western Canada and Northwest USA. The long, white strong fibres provide a timber quality for use in board or plywood panels. Approximate density 450 kg/m^3.

5. Pitch Pine – durable softwood suitable for general joinery work. It is light red to reddish yellow in colour and tends to have large knots which in some cases can be used as a decorative effect. Approximate density 650 kg/m^3.

6. Parana Pine – moderately durable straight grained timber available in a good range of sizes. Suitable for general joinery work especially timber stairs. Light to dark brown in colour with the occasional pink stripe. Approximate density 560 kg/m^3.

7. Western Hemlock – durable softwood suitable for interior joinery work such as panelling. Light yellow to reddish brown in colour. Approximate density 500 kg/m^3.

8. Western Red Cedar – originates from British Columbia and Western USA. A straight grained timber suitable for flush doors and panel work. Approximate density 380 kg/m^3.

Typical Hardwoods Suitable for Joinery Works ~

1. Beech – hard close grained timber with some silver grain in the predominately reddish yellow to light brown colour. Suitable for all internal joinery. Approximately density 700 kg/m^3.

2. Iroko – hard durable hardwood with a figured grain and is usually golden brown in colour. Suitable for all forms of good class joinery. Approximate density 660 kg/m^3.

3. Mahogany (African) – interlocking grained hardwood with good durability. It has an attractive light brown to deep red colour and is suitable for panelling and all high class joinery work. Approximate density 560 kg/m^3.

4. Mahogany (Honduras) – durable hardwood usually straight grained but can have a mottled or swirl pattern. It is light red to pale reddish brown in colour and is suitable for all good class joinery work. Approximate density 530 kg/m^3.

5. Mahogany (South American) – a well figured, stable and durable hardwood with a deep red or brown colour which is suitable for all high class joinery particularly where a high polish is required. Approximate density 550 kg/m^3.

6. Oak (English) – very durable hardwood with a wide variety of grain patterns. It is usually a light yellow brown to a warm brown in colour and is suitable for all forms of joinery but should not be used in conjunction with ferrous metals due to the risk of staining caused by an interaction of the two materials. (The gallic acid in oak causes corrosion in ferrous metals.) Approximate density 720 kg/m^3.

7. Sapele – close texture timber of good durability, dark reddish brown in colour with a varied grain pattern. It is suitable for most internal joinery work especially where a polished finish is required. Approximate density 640 kg/m^3.

8. Teak – very strong and durable timber but hard to work. It is light golden brown to dark golden yellow in colour which darkens with age and is suitable for high class joinery work and laboratory fittings. Approximate density 650 kg/m^3.

9. Jarrah (Western Australia) – hard, dense, straight grained timber. Dull red colour, suited to floor and stair construction subjected to heavy wear. Approximate density 820 kg/m^3.

Composite Boards ~ are factory manufactured, performed sheets with a wide range of properties and applications. The most common size is 2440 × 1220 mm or 2400 × 1200 mm in thicknesses from 3 to 50 mm.

1. Plywood (BS EN 636) – produced in a range of laminated thicknesses from 3 to 25 mm, with the grain of each layer normally at right angles to that adjacent. 3,7,9 or 11 plies make up the overall thickness and inner layers may have lower strength and different dimensions to those in the outer layers. Adhesives vary considerably from natural vegetable and animal glues to synthetics such as urea, melamine, phenol and resorcinol formaldehydes. Quality of laminates and type of adhesive determine application. Surface finishes include plastics, decorative hardwood veneers, metals, rubber and mineral aggregates.

2. Block and Stripboards (BS EN 12871) – range from 12 to 43 mm thickness, made up from a solid core of glued softwood strips with a surface enhancing veneer. Appropriate for dense panelling and doors.

   Battenboard – strips over 30 mm wide (unsuitable for joinery).
   Blockboard – strips up to 25 mm wide.
   Laminboard – strips up to 7 mm wide.

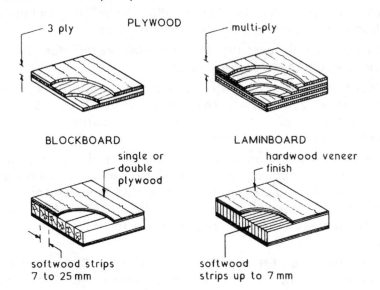

PLYWOOD

3 ply   multi-ply

BLOCKBOARD   LAMINBOARD

single or double plywood   hardwood veneer finish

softwood strips 7 to 25 mm   softwood strips up to 7 mm

3. Compressed Strawboard (BS 4046) – produced by compacting straw under heat and pressure, and edge binding with paper. Used as panels with direct decoration or as partitioning with framed support. Also, for insulated roof decking with 58 mm slabs spanning 600 mm joist spacing.

4. Particle Board

Chipboard (BS EN 319) – bonded waste wood or chip particles in thicknesses from 6 to 50 mm, popularly used for floors in 18 and 22 mm at 450 and 600 mm maximum joist spacing, respectively. Sheets are produced by heat pressing the particles in thermosetting resins.

Wood Cement Board – approximately 25% wood particles mixed with water and cement, to produce a heavy and dense board often preferred to plasterboard and fibre cement for fire cladding. Often 3 layer boards, from 6 to 40 mm in thickness.

Oriented Strand Board (BS EN 300) – composed of wafer thin strands of wood, approximately 80 mm long × 25 m wide, resin bonded and directionally oriented before superimposed by further layers. Each layer is at right angles to adjacent layers, similar to the structure of plywood. A popular alternative for wall panels, floors and other chipboard and plywood applications, they are produced in a range of thicknesses from 6 to 25 mm.

5. Fibreboards (BS EN 622-4) – basically wood in composition, reduced to a pulp and pressed to achieve 3 categories:

Hardboard density at least 800 kg/m^3 in thicknesses from 3·2 to 8 mm. Provides an excellent base for coatings and laminated finishes.

Mediumboard (low density) 350 to 560 kg/m^3 for pinboards and wall linings in thicknesses of 6·4,9, and 12·7 mm.

Mediumboard (high density) 560 to 800 kg/m^3 for linings and partitions in thicknesses of 9 and 12 mm.

Softboard, otherwise known as insulating board with density usually below 250 kg/m^3. Thicknesses from 9 to 25 mm, often found impregnated with bitumen in existing flat roofing applications. Ideal as pinboard.

Medium Density Fibreboard, differs from other fibreboards with the addition of resin bonding agent. These boards have a very smooth surface, ideal for painting and are available moulded for a variety of joinery applications. Density exceeds 600 kg/m^3 and common board thicknesses are 9, 12, 18 and 25 mm for internal and external applications.

6. Woodwool (BS EN 13168) – units of 600 mm width are available in 50, 75 and 100 mm thicknesses. They comprise long wood shavings coated with a cement slurry, compressed to leave a high proportion of voids. These voids provide good thermal insulation and sound absorption. The perforated surface is an ideal key for direct plastering and they are frequently specified as permanent formwork.

Plastics ~ the term plastic can be applied to any group of substances based on synthetic or modified natural polymers which during manufacture are moulded by heat and/or pressure into the required form. Plastics can be classified by their overall grouping such as polyvinyl chloride (PVC) or they can be classified as thermoplastic or thermosetting. The former soften on heating whereas the latter are formed into permanent non-softening materials. The range of plastics available give the designer and builder a group of materials which are strong, reasonably durable, easy to fit and maintain and since most are mass produced of relative low cost.

Typical Applications of Plastics in Buildings ~

Application	Plastics Used
Rainwater goods	unplasticised PVC (uPVC or PVC-U).
Soil, waste, water and gas pipes and fittings	uPVC; polyethylene (PE); acrylonitrile butadiene styrene (ABS), polypropylene (PP).
Hot and cold water pipes	chlorinated PVC; ABS; polypropylene; polyethylene; PVC (not for hot water).
Bathroom and kitchen fittings	glass fibre reinforced polyester (GRP); acrylic resins.
Cold water cisterns	polypropylene; polystyrene; polyethylene.
Rooflights and sheets	GRP; acrylic resins; uPVC.
DPC's and membranes, vapour control layers	low density polyethylene (LDPE); PVC film; polypropylene.
Doors and windows	GRP; uPVC.
Electrical conduit and fittings	plasticised PVC; uPVC; phenolic resins.
Thermal insulation	generally cellular plastics such as expanded polystyrene bead and boards; expanded PVC; foamed polyurethane; foamed phenol formaldehyde; foamed urea formaldehyde.
Floor finishes	plasticised PVC tiles and sheets; resin based floor paints; uPVC.
Wall claddings and internal linings	unplasticised PVC; polyvinyl fluoride film laminate; melamine resins; expanded polystyrene tiles & sheets.

741

Uses ~ to weather- and leak-proof junctions and abutments between separate elements and components that may be subject to differential movement. Also to gap fill where irregularities occur.

Properties ~

• thermal movement to facilitate expansion and contraction
• strength to resist wind and other non-structural loading
• ability to accommodate tolerance variations
• stability without loss of shape
• colour fast and non-staining to adjacent finishes
• weather resistant

Maintenance ~ of limited life, perhaps 10 to 25 years depending on composition, application and use. Future accessibility is important for ease of removal and replacement.

Mastics ~ generally regarded as non-setting gap fillers applied in a plastic state. Characterised by a hard surface skin over a plastic core that remains pliable for several years. Based on a viscous material such as bitumen, polyisobutylene or butyl rubber. Applications include bitumen treatment to rigid road construction joints (page 135) and linseed oil putty glazing (page 380). In older construction, a putty based joint may also be found between WC pan spigot outlet and cast iron socket. In this situation the putty was mixed with red lead pigments (oxides of lead), a material now considered a hazardous poison, therefore protective care must be taken when handling an old installation of this type. Modern push-fit plastic joints are much simpler, safer to use and easier to apply.

Sealants ~ applied in a plastic state by hand, knife, disposable cartridge gun, pouring or tape strip to convert by chemical reaction with the atmosphere (1 part) or with a vulcanising additive (2 part) into an elastomer or synthetic rubber. An elastomer is generally defined as a natural or synthetic material with a high strain capacity or elastic recovery, ie. it can be stretched to twice its length before returning to its original length.

Formed of polysulphide rubber, polyurethane, silicone or some butyl rubbers.

Applications ~
- Polysulphide ~ façades, glazing, fire protection, roads and paving joints. High modulus or hardness but not completely elastic.
- Polyurethane ~ general uses, façades and civil engineering. Highly elastic and resilient to abrasion and indentation, moderate resistance to ultra-violet light and chemicals.
- Silicone ~ general uses, façades, glazing, sanitary, fire protection and civil engineering. Mainly one part but set quickly relative to others in this category. Highly elastic and available as high (hard) or low (soft) modulus.

Two part ~ polysulphide and polyurethane based sealants are often used with a curing or vulcanising additive to form a synthetic rubber on setting. After the two parts are mixed the resulting sealant remains workable for up to about 4 hours. It remains plastic for a few days and during this time cannot take any significant loading. Thereafter it has exceptional resistance to compression and shear.

One part ~ otherwise known as room temperature vulcanising (RTV) types that are usually of a polysulphide, polyurethane or silicone base. Polysulphide and polyurethane cure slowly and convert to a synthetic rubber or elastomer sealant by chemical reaction with moisture in the atmosphere. Generally of less movement and loading resistance to two part sealants, but are frequently used in non-structural situations such as sealing around door and window frames, bathroom and kitchen fitments.

Other sealants ~
- one part acrylic (water based) RTV. Flexible but with limited elasticity. Internal uses such as sealing around door and window frames, fire protection and internal glazing.
- silane modified polymer in one part RTV or two parts. Highly elastic and can be used for general applications as well as for façades and civil engineering situations.

Sealants–Classification and Specification

Prior to 2003, several separate British Standards existed to provide use and application guidance for a range of sealant products. As independent publications these are now largely superseded, their content rationalised and incorporated into the current standard, BS EN ISO 11600: Building construction. Jointing products. Classification and requirements for sealants.

This International Standard covers, materials application to jointing, classification of materials, quality grading and performance testing. This enables specific definition of a sealant's requirements in terms of end use without having to understand the chemical properties of the various sealant types. Typical criteria are, movement potential, elasticity and hardness when related to particular substrate surfaces such as aluminium, glass or masonry.

Grading summary ~

BS EN ISO 11600 G, for use in glazing.
BS EN ISO 11600 F, for façade and similar applications such as movement joints.

Other suffixes or sub-classes ~

E = elastic sealant, ie. high elastic recovery or elastomeric.
P = plastic sealant, ie. low elastic recovery.
HM = high modulus, indicates hardness.*
LM = low modulus, indicates softness.*
*By definition HM and LM are high movement (20-25%) types of elastic sealants, therefore the suffix E is not shown with these.

Associated standards ~
BS EN 26927: Building construction. Jointing products. Sealants. Vocabulary.
BS 6213: Selection of construction sealants. Guide.
BS 6093: Design of joints and jointing in building construction. Guide.

8 DOMESTIC SERVICES

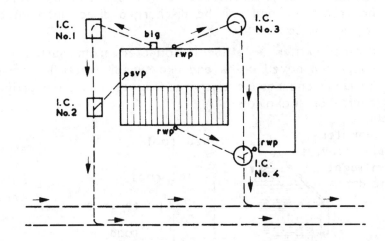

DRAINAGE EFFLUENTS

SUBSOIL DRAINAGE

SURFACE WATER REMOVAL

ROAD DRAINAGE

RAINWATER INSTALLATIONS

DRAINAGE SYSTEMS

DRAINAGE PIPE SIZES AND GRADIENTS

WATER SUPPLY

COLD WATER INSTALLATIONS

HOT WATER INSTALLATIONS

CISTERNS AND CYLINDERS

SANITARY FITTINGS

SINGLE AND VENTILATED STACK SYSTEMS

DOMESTIC HOT WATER HEATING SYSTEMS

ELECTRICAL SUPPLY AND INSTALLATION

GAS SUPPLY AND GAS FIRES

SERVICES FIRE STOPS AND SEALS

OPEN FIREPLACES AND FLUES

COMMUNICATIONS INSTALLATIONS

Effluent ~ can be defined as that which flows out. In building drainage terms there are three main forms of effluent:-

1. Subsoil Water ~ water collected by means of special drains from the earth primarily to lower the water table level in the subsoil. It is considered to be clean and therefore requires no treatment and can be discharged direct into an approved water course.

2. Surface water ~ effluent collected from surfaces such as roofs and paved areas and like subsoil water is considered to be clean and can be discharged direct into an approved water course or soakaway

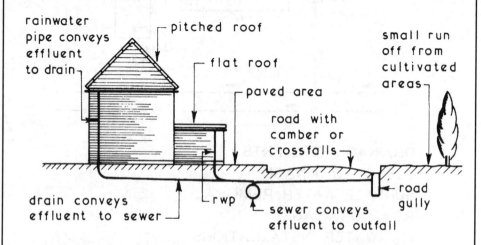

rainwater pipe conveys effluent to drain

pitched roof

flat roof

paved area

road with camber or crossfalls

small run off from cultivated areas

road gully

drain conveys effluent to sewer

rwp

sewer conveys effluent to outfall

3. Foul or Soil Water ~ effluent contaminated by domestic or trade waste and will require treatment to render it clean before it can be discharged into an approved water course.

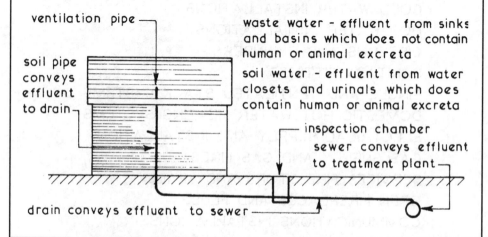

ventilation pipe

soil pipe conveys effluent to drain

waste water - effluent from sinks and basins which does not contain human or animal excreta

soil water - effluent from water closets and urinals which does contain human or animal excreta

inspection chamber

sewer conveys effluent to treatment plant

drain conveys effluent to sewer

Subsoil Drainage ~ Building Regulation C2 requires that subsoil drainage shall be provided if it is needed to avoid:-

a) the passage of ground moisture into the interior of the building or
b) damage to the fabric of the building.

Subsoil drainage can also be used to improve the stability of the ground, lower the humidity of the site and enhance its horticultural properties. Subsoil drains consist of porous or perforated pipes laid dry jointed in a rubble filled trench. Porous pipes allow the subsoil water to pass through the body of the pipe whereas perforated pipes which have a series of holes in the lower half allow the subsoil water to rise into the pipe. This form of ground water control is only economic up to a depth of 1·500, if the water table needs to be lowered to a greater depth other methods of ground water control should be considered (see pages 307 to 311).

The water collected by a subsoil drainage system has to be conveyed to a suitable outfall such as a river, lake or surface water drain or sewer. In all cases permission to discharge the subsoil water will be required from the authority or owner and in the case of streams, rivers and lakes, bank protection at the outfall may be required to prevent erosion (see page 748).

Typical Subsoil Drain Details ~

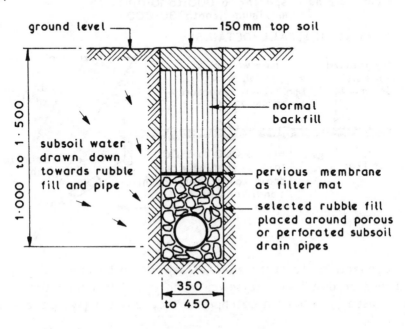

747

Subsoil Drainage Systems ~ the lay out of subsoil drains will depend on whether it is necessary to drain the whole site or if it is only the substructure of the building which needs to be protected. The latter is carried out by installing a cut off drain around the substructure to intercept the flow of water and divert it away from the site of the building. Junctions in a subsoil drainage system can be made using standard fittings or by placing the end of the branch drain onto the crown of the main drain.

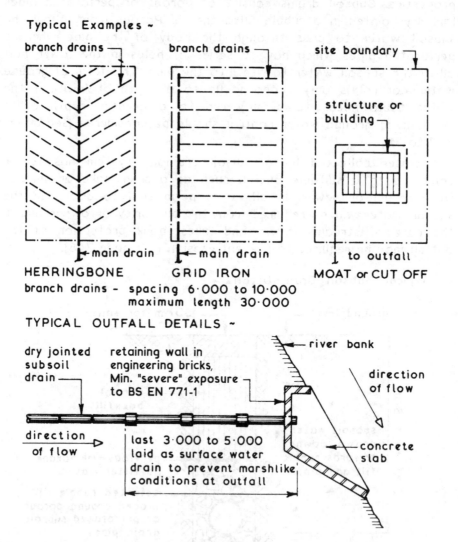

Typical Examples ~

branch drains   branch drains   site boundary

structure or building

main drain   main drain   to outfall

HERRINGBONE   GRID IRON   MOAT or CUT OFF

branch drains - spacing 6·000 to 10·000
maximum length 30·000

TYPICAL OUTFALL DETAILS ~

dry jointed subsoil drain

retaining wall in engineering bricks, Min. "severe" exposure to BS EN 771-1

river bank

direction of flow

direction of flow

last 3·000 to 5·000 laid as surface water drain to prevent marshlike conditions at outfall

concrete slab

NB. connections to surface water sewer can be made at inspection chamber or direct to the sewer using a saddle connector - it may be necessary to have a catchpit to trap any silt (see page 752).

General Principles ~ a roof must be designed with a suitable fall towards the surface water collection channel or gutter which in turn is connected to vertical rainwater pipes which convey the collected discharge to the drainage system. The fall of the roof will be determined by the chosen roof covering or the chosen pitch will limit the range of coverings which can be selected.

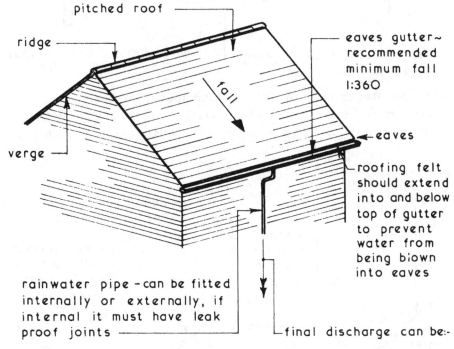

pitched roof

ridge

fall

verge

eaves gutter~ recommended minimum fall 1:360

eaves

roofing felt should extend into and below top of gutter to prevent water from being blown into eaves

rainwater pipe -can be fitted internally or externally, if internal it must have leak proof joints

final discharge can be:-

rainwater pipes and gullies must be arranged so as not to cause dampness or damage to any part of the building

Minimum Roof Pitches ~

Slates – depends on width from 25°
Hand made plain tiles – 45°
Machine made plain tiles – 35°
Single lap and interlocking tiles- depends on type from 12½°
Thatch – 45°
Timber shingles – 14°

1. Direct connection to a drain discharging into a soakaway

2. Direct connection to a drain discharging into a surface water sewer

3. Indirect connection to a drain by means of a trapped gully if drain discharges into a combined sewer

See page 755 for details

749

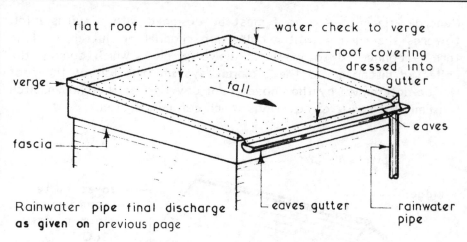

flat roof
water check to verge
roof covering dressed into gutter
verge
fascia
fall
eaves
eaves gutter
rainwater pipe

Rainwater pipe final discharge **as given on** previous page

Minimum Recommended Falls for Various Finishes ~
Aluminium – 1:60   Lead – 1:120   Copper – 1:60
Built-up roofing felts – 1:60   Mastic asphalt – 1:80

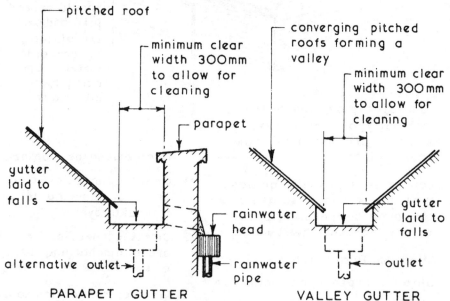

pitched roof

minimum clear width 300mm to allow for cleaning

parapet

gutter laid to falls

rainwater head

alternative outlet

rainwater pipe

converging pitched roofs forming a valley

minimum clear width 300mm to allow for cleaning

gutter laid to falls

outlet

PARAPET GUTTER

gutter formed to discharge into internal rainwater pipes or to external rainwater pipes via outlets through the parapet

VALLEY GUTTER

gutter formed to discharge into internal rainwater pipes or to external rainwater pipes sited at the gable ends

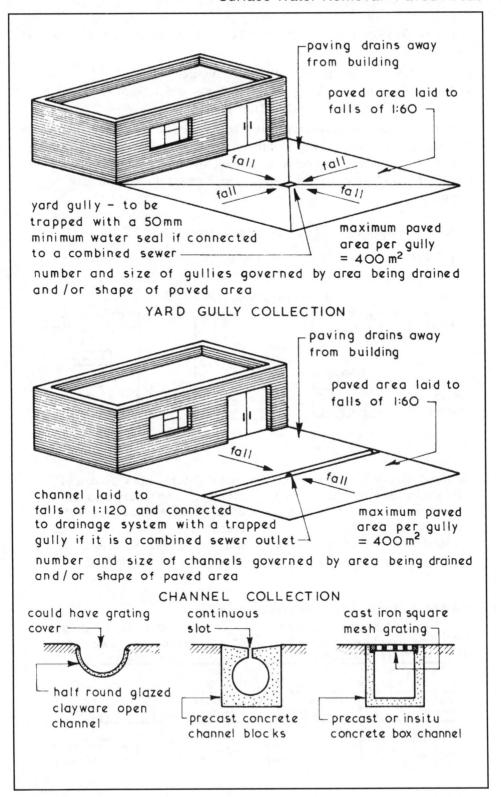

paving drains away from building

paved area laid to falls of 1:60

fall

fall

fall

fall

yard gully – to be trapped with a 50mm minimum water seal if connected to a combined sewer

maximum paved area per gully = 400 m²

number and size of gullies governed by area being drained and/or shape of paved area

YARD GULLY COLLECTION

paving drains away from building

paved area laid to falls of 1:60

fall

fall

channel laid to falls of 1:120 and connected to drainage system with a trapped gully if it is a combined sewer outlet

maximum paved area per gully = 400 m²

number and size of channels governed by area being drained and/or shape of paved area

CHANNEL COLLECTION

could have grating cover

half round glazed clayware open channel

continuous slot

precast concrete channel blocks

cast iron square mesh grating

precast or insitu concrete box channel

Highway Drainage ~ the stability of a highway or road relies on two factors –

1. Strength and durability of upper surface
2. Strength and durability of subgrade which is the subsoil on which the highway construction is laid.

The above can be adversely affected by water therefore it may be necessary to install two drainage systems. One system (subsoil drainage) to reduce the flow of subsoil water through the subgrade under the highway construction and a system of surface water drainage.

Typical Highway Subsoil Drainage Methods ~

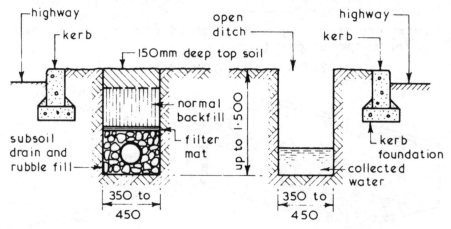

Subsoil Drain - acts as a cut off drain and can be formed using perforated or porous drain pipes. If filled with rubble only it is usually called a French or rubble drain.

Open Ditch - acts as a cut off drain and could also be used to collect surface water discharged from a rural road where there is no raised kerb or surface water drains.

Surface Water Drainage Systems ~

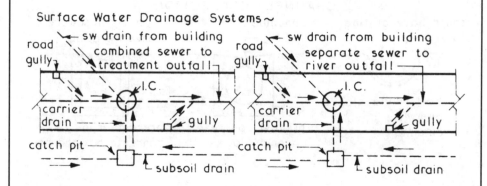

Road Drainage ~ this consists of laying the paved area or road to a suitable crossfall or gradient to direct the run-off of surface water towards the drainage channel or gutter. This is usually bounded by a kerb which helps to convey the water to the road gullies which are connected to a surface water sewer. For drains or sewers under 900 mm internal diameter inspection chambers will be required as set out in the Building Regulations. The actual spacing of road gullies is usually determined by the local highway authority based upon the carriageway gradient and the area to be drained into one road gully. Alternatively the following formula could be used:-

$$D = \frac{280\sqrt{S}}{W}$$ where D = gully spacing
S = carriageway gradient (per cent)
W = width of carriageway in metres

∴ If S = 1:60 = 1·66% and W = 4·500

$$D = \frac{280\sqrt{1·66}}{4·500} = \text{say } 80·000$$

Typical Road Gully Detail ~

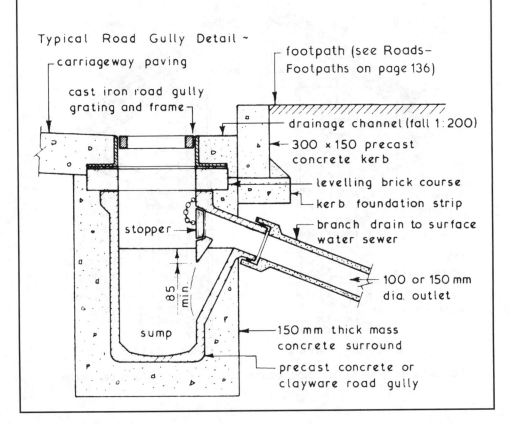

carriageway paving

cast iron road gully grating and frame

footpath (see Roads–Footpaths on page 136)

drainage channel (fall 1:200)

300 × 150 precast concrete kerb

levelling brick course

kerb foundation strip

branch drain to surface water sewer

stopper

100 or 150 mm dia. outlet

85 min.

sump

150 mm thick mass concrete surround

precast concrete or clayware road gully

753

Materials ~ the traditional material for domestic eaves gutters and rainwater pipes is cast iron but uPVC systems are very often specified today because of their simple installation and low maintenance costs. Other materials which could be considered are aluminium alloy, galvanised steel and stainless steel, but whatever material is chosen it must be of adequate size, strength and durability.

Typical Eaves Details ~

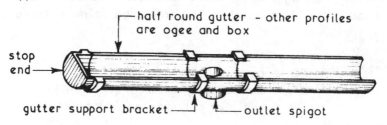

half round gutter - other profiles are ogee and box

stop end

gutter support bracket —— —— outlet spigot

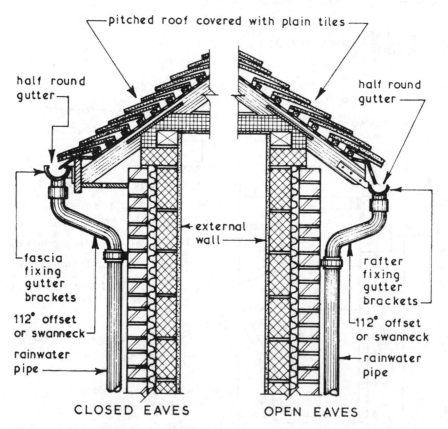

pitched roof covered with plain tiles

half round gutter

half round gutter

external wall

fascia fixing gutter brackets

rafter fixing gutter brackets

112° offset or swanneck

112° offset or swanneck

rainwater pipe

rainwater pipe

CLOSED EAVES   OPEN EAVES

For details of rainwater pipe connection to drainage see next page

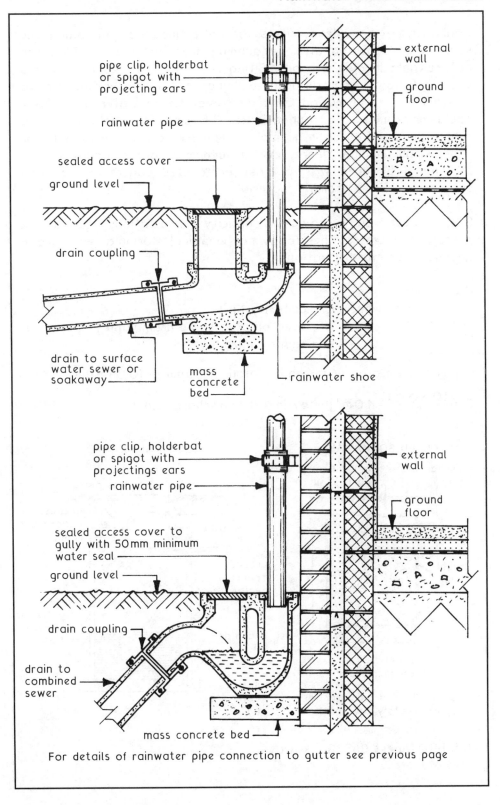

pipe clip, holderbat or spigot with projecting ears

rainwater pipe

sealed access cover

ground level

drain coupling

drain to surface water sewer or soakaway

mass concrete bed

external wall

ground floor

rainwater shoe

pipe clip, holderbat or spigot with projectings ears

rainwater pipe

sealed access cover to gully with 50mm minimum water seal

ground level

drain coupling

drain to combined sewer

mass concrete bed

external wall

ground floor

For details of rainwater pipe connection to gutter see previous page

Soakaways ~ provide a means for collecting and controlling the seapage of rainwater into surrounding granular subsoils. They are not suitable in clay subsoils. Siting is on land at least level and preferably lower than adjacent buildings and no closer than 5m to a building. Concentration of a large volume of water any closer could undermine the foundations. The simplest soakaway is a rubble filled pit, which is normally adequate to serve a dwelling or other small building. Where several buildings share a soakaway, the pit should be lined with precast perforated concrete rings and surrounded in free-draining material.

BRE Digest 365 provides capacity calculations based on percolation tests. The following empirical formula will prove adequate for most situations:-

$$C = \frac{AR}{3}$$

where: C = capacity (m³)
A = area on plan to be drained (m²)
R = rainfall (m/h)

e.g. roof plan area 60 m² and rainfall of 50 mm/h (0·05 m/h)

$$C = \frac{60 \times 0·05}{3} = 1·0 \, m^3 \text{ (below invert of discharge pipe)}$$

FILLED SOAKAWAY	HOLLOW SOAKAWAY

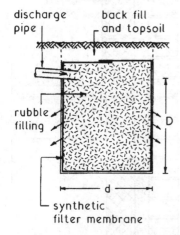

discharge pipe
back fill and topsoil
rubble filling
D
d
synthetic filter membrane

Depth (D) and diameter (d) approximately the same

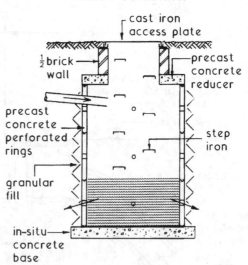

cast iron access plate
½ brick wall
precast concrete reducer
precast concrete perforated rings
step iron
granular fill
in-situ concrete base

Ref. BRE Digest 365: Soakaways.

Drains ~ these can be defined as a means of conveying surface water or foul water below ground level.

Sewers ~ these have the same functions as drains but collect the discharge from a number of drains and convey it to the final outfall. They can be a private or public sewer depending on who is responsible for the maintenance.

Basic Principles ~ to provide a drainage system which is simple efficient and economic by laying the drains to a gradient which will render them self cleansing and will convey the effluent to a sewer without danger to health or giving nuisance. To provide a drainage system which will comply with the minimum requirements given in Part H of the Building Regulations.

Typical Basic Requirements ~

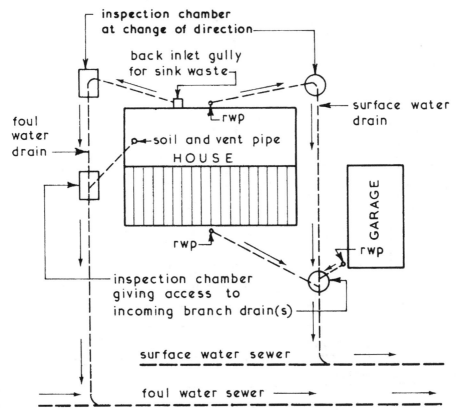

All junctions should be oblique and in direction of flow

There must be an access point at a junction unless each run can be cleared from another access point.

Separate System ~ the most common drainage system in use where the surface water discharge is conveyed in separate drains and sewers to that of foul water discharges and therefore receives no treatment before the final outfall.

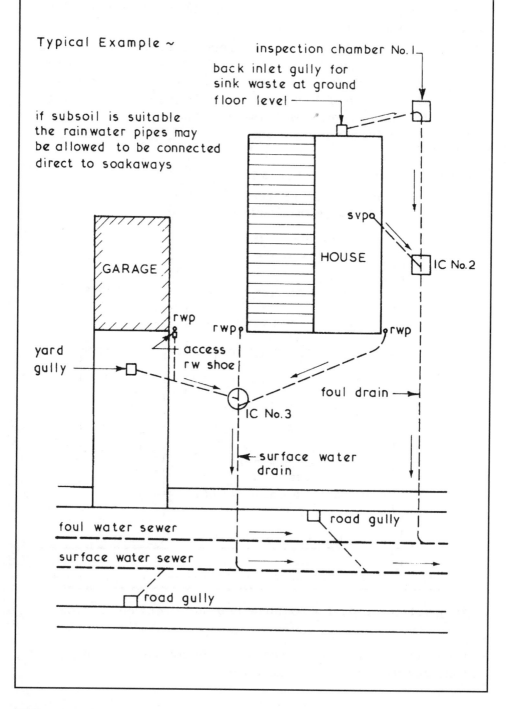

Typical Example ~

inspection chamber No. 1

back inlet gully for sink waste at ground floor level

if subsoil is suitable the rainwater pipes may be allowed to be connected direct to soakaways

svp

HOUSE

IC No. 2

GARAGE

rwp

rwp

yard gully

access rw shoe

rwp

foul drain →

IC No. 3

← surface water drain

road gully

foul water sewer

surface water sewer

road gully

Combined System ~ this is the simplest and least expensive system to design and install but since all forms of discharge are conveyed in the same sewer the whole effluent must be treated unless a sea outfall is used to discharge the untreated effluent.

Typical Example ~

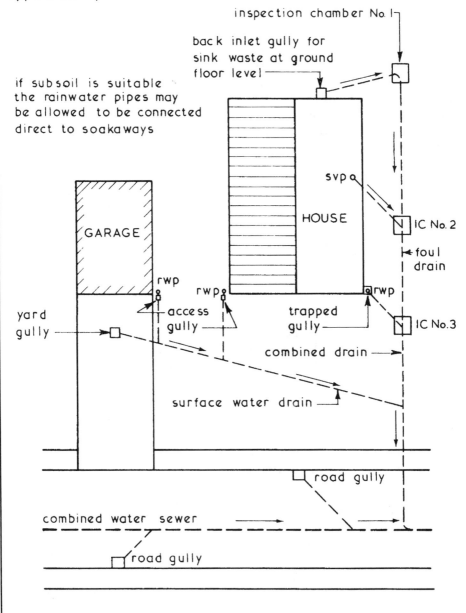

if subsoil is suitable the rainwater pipes may be allowed to be connected direct to soakaways

Ref. BS EN 752: Drain and sewer systems outside buildings.

Partially Separate System ~ a compromise system ~ there are two drains, one to convey only surface water and a combined drain to convey the total foul discharge and a proportion of the surface water.

Typical Example ~

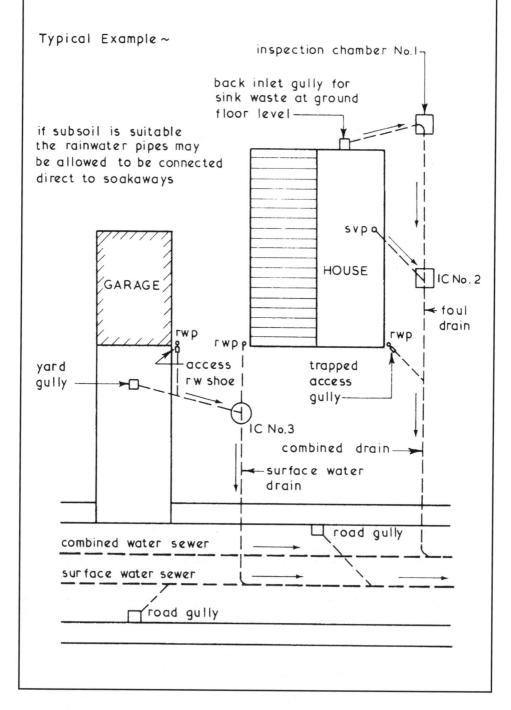

inspection chamber No.1

back inlet gully for sink waste at ground floor level

if subsoil is suitable the rainwater pipes may be allowed to be connected direct to soakaways

svp

HOUSE

IC No. 2

foul drain

GARAGE

rwp

rwp

yard gully

access rw shoe

trapped access gully

rwp

IC No.3

combined drain

surface water drain

road gully

combined water sewer

surface water sewer

road gully

Inspection Chambers ~ these provide a means of access to drainage systems where the depth to invert level does not exceed 1·000.

Manholes ~ these are also a means of access to the drains and sewers, and are so called if the depth to invert level exceeds 1·000.

These means of access should be positioned in accordance with the requirements of part H of the Building Regulations. In domestic work inspection chambers can be of brick, precast concrete or preformed in plastic for use with patent drainage systems. The size of an inspection chamber depends on the depth to invert level, drain diameter and number of branch drains to be accommodated within the chamber. Ref. BS EN 752: Drain and sewer systems outside buildings.

Typical Details ~

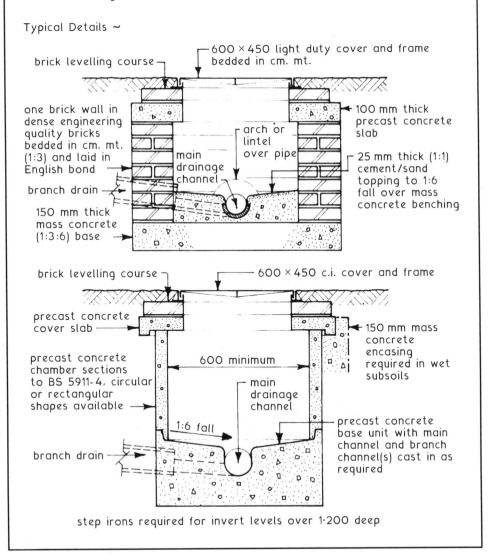

step irons required for invert levels over 1·200 deep

Plastic Inspection Chambers ~ the raising piece can be sawn horizontally with a carpenter's saw to suit depth requirements with the cover and frame fitted at surface level. Bedding may be a 100 mm prepared shingle base or 150 mm wet concrete to ensure a uniform support.

The unit may need weighting to retain it in place in areas of high water table, until backfilled with granular material. Under roads a peripheral concrete collar is applied to the top of the chamber in addition to the 150 mm thickness of concrete surrounding the inspection chamber.

Typical Example ~

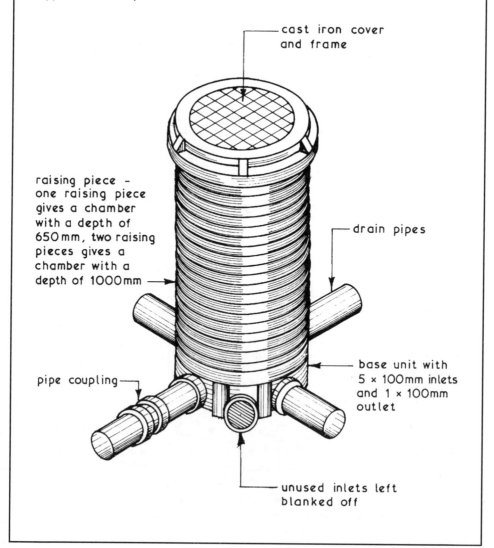

cast iron cover and frame

raising piece - one raising piece gives a chamber with a depth of 650 mm, two raising pieces gives a chamber with a depth of 1000 mm

drain pipes

pipe coupling

base unit with 5 × 100 mm inlets and 1 × 100 mm outlet

unused inlets left blanked off

Means Of Access – provision is required for maintenance and inspection of drainage systems. This should occur at:

* the head (highest part) or close to it
* a change in horizontal direction
* a change in vertical direction (gradient)
* a change in pipe diameter
* a junction, unless the junction can be rodded through from an access point
* long straight runs (see table)

Maximum spacing of drain access points (m)

To:	Small access fitting	Large access fitting	Junction	Inspection chamber	Manhole
From:					
Drain head	12	12		22	45
Rodding eye	22	22	22	45	45
Small access fitting			12	22	22
Large access fitting			22	45	45
Inspection chamber	22	45	22	45	45
Manhole				45	90

* Small access fitting is 150mm dia. or 150mm × 100mm. Large access fitting is 225mm × 100mm.

Rodding Eyes and Shallow Access Chambers – these may be used at the higher parts of drainage systems where the volume of excavation and cost of an inspection chamber or manhole would be unnecessary. SACs have the advantage of providing access in both directions. Covers to all drain openings should be secured to deter unauthorised access.

Ref. Building Regulations, Approved Document H1: Foul Water Drainage.

Excavations ~ drains are laid in trenches which are set out, excavated and supported in a similar manner to foundation trenches except for the base of the trench which is cut to the required gradient or fall.

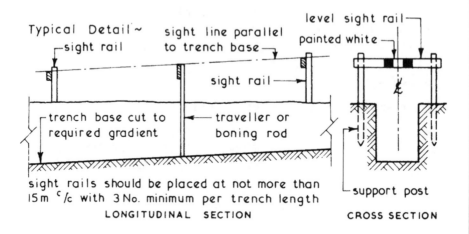

Typical Detail ~

sight rail

sight line parallel to trench base

level sight rail

painted white

sight rail

trench base cut to required gradient

traveller or boning rod

support post

sight rails should be placed at not more than 15m ᶜ/c with 3 No. minimum per trench length

LONGITUDINAL SECTION

CROSS SECTION

Joints ~ these must be watertight under all working and movement conditions and this can be achieved by using rigid and flexible joints in conjuntion with the appropriate bedding.

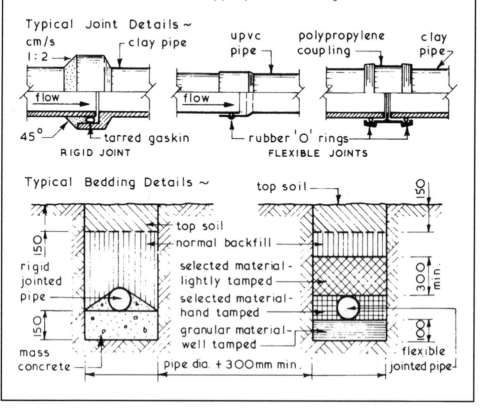

Typical Joint Details ~

cm/s 1:2

clay pipe

upvc pipe

polypropylene coupling

clay pipe

flow

flow

45°

tarred gaskin

rubber 'O' rings

RIGID JOINT

FLEXIBLE JOINTS

Typical Bedding Details ~

top soil

150

150

rigid jointed pipe

150

mass concrete

top soil

normal backfill

selected material - lightly tamped

selected material - hand tamped

granular material - well tamped

pipe dia. + 300mm min.

300 min.

100

flexible jointed pipe

Watertightness ~ must be ensured to prevent water seapage and erosion of the subsoil. Also, in the interests of public health, foul water should not escape untreated. The Building Regulaions, Approved Document H1: Section 2 specifies either an air or water test to determine soundness of installation.

AIR TEST ~ equipment: manometer and accessories (see page 783) 2 drain stoppers, one with tube attachment

Application ~

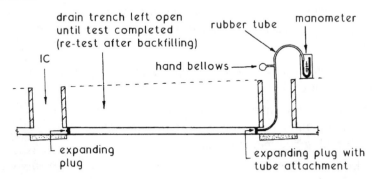

drain trench left open until test completed (re-test after backfilling)

IC

rubber tube   manometer

hand bellows

expanding plug

expanding plug with tube attachment

Test ~ 100mm water gauge to fall no more than 25mm in 5mins. Or, 50mm w.g. to fall no more than 12mm in 5mins.

WATER TEST ~ equipment : Drain stopper
Test bend
Extension pipe

Application ~

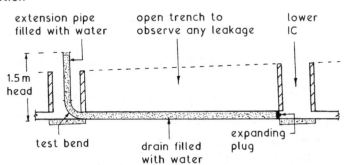

extension pipe filled with water

open trench to observe any leakage

lower IC

1.5m head

test bend   drain filled with water   expanding plug

Test ~ 1·5m head of water to stand for 2 hours and then topped up. Leakage over the next 30 minutes should be minimal, i.e.

100mm pipe – 0·05 litres per metre, which equates to a drop of 6·4mm/m in the extension pipe, and

150mm pipe – 0·08 litres per metre, which equates to a drop of 4·5mm/m in the extension pipe.

Drainage Pipes ~ sizes for normal domestic foul water applications:-

<20 dwellings   = 100 mm diameter
20–150 dwellings = 150 mm diameter

Exceptions: 75 mm diameter for waste or rainwater only (no WCs)
            150 mm diameter minimum for a public sewer

Other situations can be assessed by summating the Discharge Units from appliances and converting these to an appropriate diameter stack and drain, see BS EN 12056-2 (stack) and BS EN 752 (drain). Gradient will also affect pipe capacity and when combined with discharge calculations, provides the basis for complex hydraulic theories.

The simplest correlation of pipe size and fall, is represented in Maguire's rule:-

4" (100 mm) pipe, minimum gradient 1 in 40
6" (150 mm) pipe, minimum gradient 1 in 60
9" (225 mm) pipe, minimum gradient 1 in 90

The Building Regulations, approved Document H1 provides more scope and relates to foul water drains running at 0·75 proportional depth. See Diagram 9 and Table 6 in Section 2 of the Approved Document.

Other situations outside of design tables and empirical practice can be calculated.

eg. A 150 mm diameter pipe flowing 0·5 proportional depth.

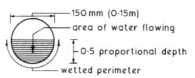

Applying the Chezy formula for gradient calculations:-
$v = c\sqrt{m \times i}$

where: v = velocity of flow, (min for self cleansing = 0·8 m/s)

$\qquad$ c = Chezy coefficient (58)

$\qquad$ m = hydraulic mean depth or;
$\qquad\qquad \dfrac{\text{area of water flowing}}{\text{wetted perimeter}}$ for 0·5 p.d. = diam/4

$\qquad$ i = inclination or gradient as a fraction 1/x

Selecting a velocity of 1 m/s as a margin of safety over the minimum:-

$1 = 58\sqrt{0.15/4 \times i}$
i = 0·0079 where i = 1/x
So, x = 1/0·0079 = 126, i.e. a minimum gradient of 1 in 126

Water supply ~ an adequate supply of cold water of drinking quality should be provided to every residential building and a drinking water tap installed within the building. The installation should be designed to prevent waste, undue consumption, misuse, contamination of general supply, be protected against corrosion and frost damage and be accessible for maintenance activities. The intake of a cold water supply to a building is owned jointly by the water authority and the consumer who therefore have joint maintenance responsibilities.

Typical Water Supply Arrangement ~

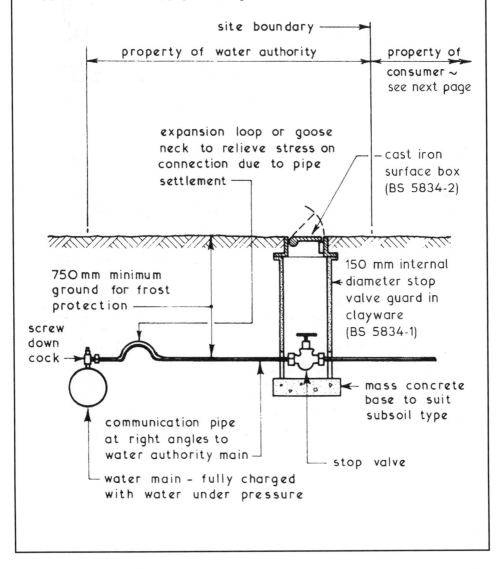

site boundary

property of water authority

property of consumer ~ see next page

expansion loop or goose neck to relieve stress on connection due to pipe settlement

cast iron surface box (BS 5834-2)

750 mm minimum ground for frost protection

150 mm internal diameter stop valve guard in clayware (BS 5834-1)

screw down cock

communication pipe at right angles to water authority main

mass concrete base to suit subsoil type

stop valve

water main - fully charged with water under pressure

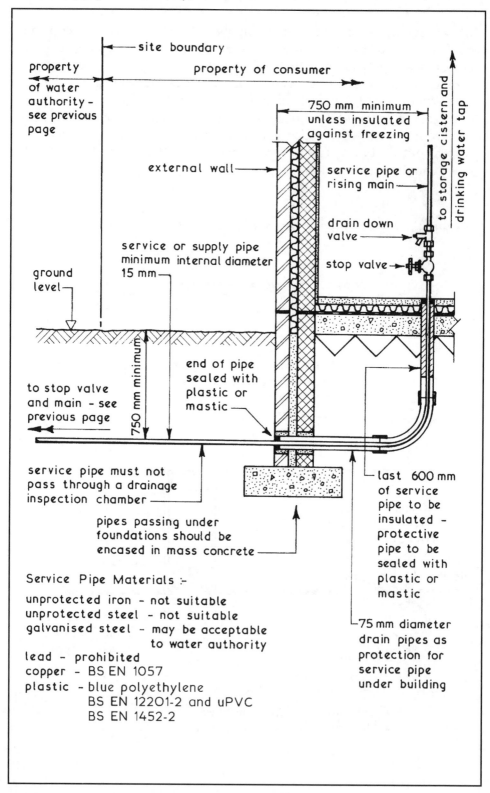

site boundary

property of water authority - see previous page

property of consumer

750 mm minimum unless insulated against freezing

external wall

service pipe or rising main

drain down valve

stop valve

service or supply pipe minimum internal diameter 15 mm

to storage cistern and drinking water tap

ground level

750 mm minimum

to stop valve and main - see previous page

end of pipe sealed with plastic or mastic

service pipe must not pass through a drainage inspection chamber

last 600 mm of service pipe to be insulated - protective pipe to be sealed with plastic or mastic

pipes passing under foundations should be encased in mass concrete

75 mm diameter drain pipes as protection for service pipe under building

Service Pipe Materials :-

unprotected iron - not suitable
unprotected steel - not suitable
galvanised steel - may be acceptable
                  to water authority
lead - prohibited
copper - BS EN 1057
plastic - blue polyethylene
          BS EN 12201-2 and uPVC
          BS EN 1452-2

General ~ when planning or designing any water installation the basic physical laws must be considered:-

1. Water is subject to the force of gravity and will find its own level.
2. To overcome friction within the conveying pipes water which is stored prior to distribution will require to be under pressure and this is normally achieved by storing the water at a level above the level of the outlets. The vertical distance between these levels is usually called the head.
3. Water becomes less dense as its temperature is raised, therefore warm water will always displace colder water whether in a closed or open circuit.

Direct Cold Water Systems ~ the cold water is supplied to the outlets at mains pressure; the only storage requirements is a small capacity cistern to feed the hot water storage tank. These systems are suitable for districts which have high level reservoirs with a good supply and pressure. The main advantage is that drinking water is available from all cold water outlets, disadvantages include lack of reserve in case of supply cut off, risk of back syphonage due to negative mains pressure and a risk of reduced pressure during peak demand periods.

Typical Direct Cold Water System ~

NB. all pipe sizes given are outside diameters for copper tube

Indirect Systems ~ Cold water is supplied to all outlets from a cold water storage cistern except for the cold water supply to the sink(s) where the drinking water tap is connected directly to incoming supply from the main. This system requires more pipework than the direct system but it reduces the risk of back syphonage and provides a reserve of water should the mains supply fail or be cut off. The local water authority will stipulate the system to be used in their area.

Typical Indirect Cold Water System ~

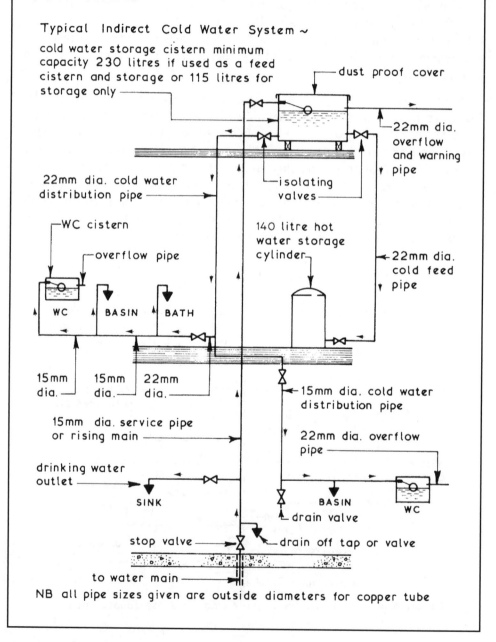

cold water storage cistern minimum capacity 230 litres if used as a feed cistern and storage or 115 litres for storage only

dust proof cover

22mm dia. overflow and warning pipe

22mm dia. cold water distribution pipe

isolating valves

WC cistern

overflow pipe

140 litre hot water storage cylinder

22mm dia. cold feed pipe

WC   BASIN   BATH

15mm dia.   15mm dia.   22mm dia.

15mm dia. cold water distribution pipe

15mm dia. service pipe or rising main

22mm dia. overflow pipe

drinking water outlet

SINK   BASIN   WC

drain valve

stop valve   drain off tap or valve

to water main

NB all pipe sizes given are outside diameters for copper tube

Direct System ~ this is the simplest and least expensive system of hot water installation. The water is heated in the boiler and the hot water rises by convection to the hot water storage tank or cylinder to be replaced by the cooler water from the bottom of the storage vessel. Hot water drawn from storage is replaced with cold water from the cold water storage cistern. Direct systems are suitable for soft water areas and for installations which are not supplying a central heating circuit.

Typical Direct Hot Water System ~

cold water storage cistern minimum capacity 230 litres — overflow

isolating valve

15 mm dia. service pipe or rising main

140 litre hot water storage cylinder

22 mm dia. cold feed pipe

22 mm dia. open vent or expansion pipe to release air and relieve pressure

450 min.

15 mm dia.

BASIN   BATH

22 mm dia. hot water supply pipe

28 mm dia. primary flow pipe

28 mm dia. primary return pipe

possible pumped secondary return pipe

15 mm dia. hot water supply pipe

in hard water areas primary circuit pipes could be 35 mm diameter

15 mm dia.   SINK

boiler   BASIN

drain valve

safety valve

NB all pipe sizes given are outside diameters for copper tube

Indirect System ~ this is a more complex system than the direct system but it does overcome the problem of furring which can occur in direct hot water systems. This method is therefore suitable for hard water areas and in all systems where a central heating circuit is to be part of the hot water installation. Basically the pipe layouts of the two systems are similar but in the indirect system a separate small capacity feed cistern is required to charge and top up the primary circuit. In this system the hot water storage tank or cylinder is in fact a heat exchanger – see page 776.

Typical Indirect Hot Water System ~

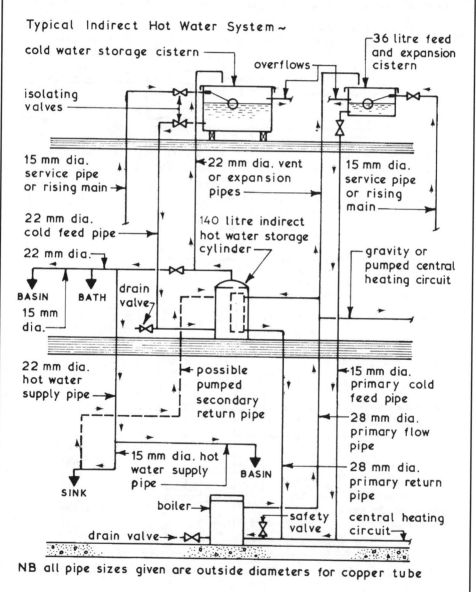

NB all pipe sizes given are outside diameters for copper tube

Mains Fed Indirect System ~ now widely used as an alternative to conventional systems. It eliminates the need for cold water storage and saves considerably on installation time. This system is established in Europe and the USA, but only acceptable in the UK at the local water authority's discretion. It complements electric heating systems, where a boiler is not required. An expansion vessel replaces the standard vent and expansion pipe and may be integrated with the hot water storage cylinder. It contains a neoprene diaphragm to separate water from air, the air providing a 'cushion' for the expansion of hot water. Air loss can be replenished by foot pump as required.

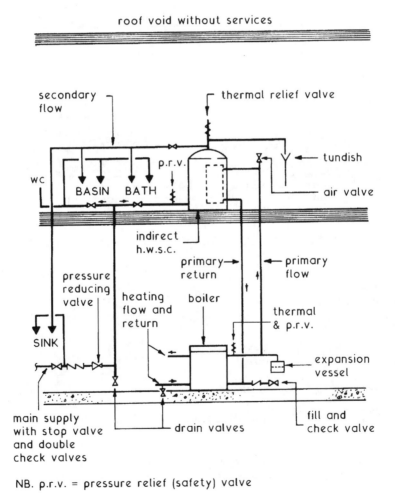

NB. p.r.v. = pressure relief (safety) valve

Flow Controls ~ these are valves inserted into a water installation to control the water flow along the pipes or to isolate a branch circuit or to control the draw-off of water from the system.

Typical Examples ~

wheel head

Spindle

packing gland

wedge shaped gate

GATE VALVE
low pressure cistern supply

crutch head

spindle

packing gland

loose jumper

flow

STOP VALVE
high pressure mains supply

seating

piston

cap

back nut

lock nut

outlet

float arm

PORTSMOUTH FLOATVALVE

nylon seating

top outlet

back nut

lock nut

float arm

DIAPHRAGM FLOATVALVE

capstan head

spindle

packing gland

easy clean cover

jumper

bib outlet

BIB TAP
horizontal inlet - used over sinks and for hose pipe outlets

capstan head

spindle

packing gland

easy clean cover

jumper

outlet

back nut

PILLAR TAP
vertical inlet - used in conjunction with fittings

Cisterns ~ these are fixed containers used for storing water at atmospheric pressure. The inflow of water is controlled by a floatvalve which is adjusted to shut off the water supply when it has reached the designed level within the cistern. The capacity of the cistern depends on the draw off demand and whether the cistern feeds both hot and cold water systems. Domestic cold water cisterns should be placed at least 750mm away from an external wall or roof surface and in such a position that it can be inspected, cleaned and maintained. A minimum clear space of 300mm is required over the cistern for floatvalve maintenance. An overflow or warning pipe of not less than 22mm diameter must be fitted to fall away to discharge in a conspicuous position. All draw off pipes must be fitted with a gate valve positioned as near to the cistern as possible.

Cisterns are available in a variety of sizes and materials such as galvanised mild steel (BS 417-2), moulded plastic (BS 4213) and reinforced plastic (BS EN 13121 and 13923). If the cistern and its associated pipework are to be housed in a cold area such as a roof they should be insulated against freezing.

Typical Details ~

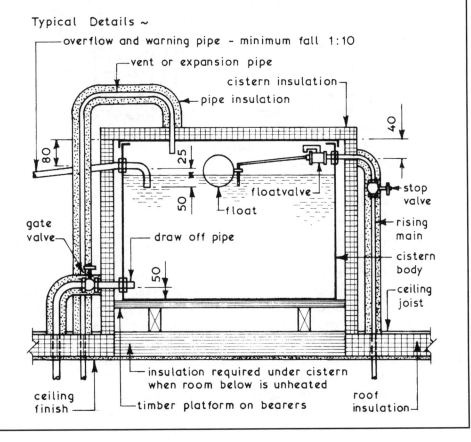

775

Indirect Hot Water Cylinders ~ these cylinders are a form of heat exchanger where the primary circuit of hot water from the boiler flows through a coil or annulus within the storage vessel and transfers the heat to the water stored within. An alternative hot water cylinder for small installations is the single feed or 'Primatic' cylinder which is self venting and relies on two air locks to separate the primary water from the secondary water. This form of cylinder is connected to pipework in the same manner as for a direct system (see page 771) and therefore gives savings in both pipework and fittings. Indirect cylinders usually conform to the recommendations of BS 417-2 (galvanised mild steel) or BS 1566-1 (copper). Primatic or single feed cylinders to BS 1566-2 (copper).

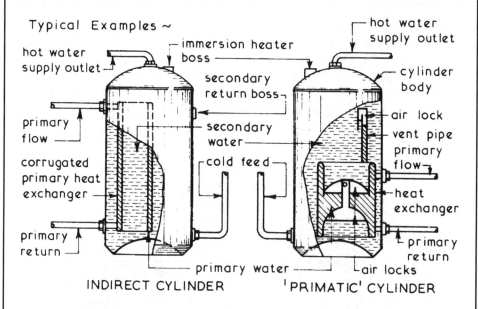

Typical Examples ~

INDIRECT CYLINDER   'PRIMATIC' CYLINDER

Primatic Cylinders ~

1. Cylinder is filled in the normal way and the primary system is filled via the heat exchanger, as the initial filling continues air locks are formed in the upper and lower chambers of the heat exchanger and in the vent pipe.

2. The two air locks in the heat exchanger are permanently maintained and are self-recuperating in operation. These air locks isolate the primary water from the secondary water almost as effectively as a mechanical barrier.

3. The expansion volume of total primary water at a flow temperature of 82°C is approximately 1/25 and is accommodated in the upper expansion chamber by displacing air into the lower chamber; upon contraction reverse occurs.

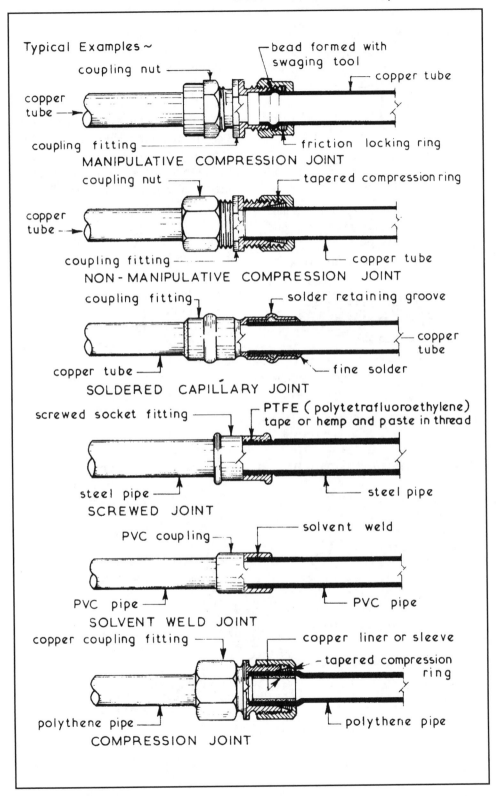

Typical Examples ~

MANIPULATIVE COMPRESSION JOINT
- coupling nut
- copper tube
- bead formed with swaging tool
- copper tube
- coupling fitting
- friction locking ring

NON-MANIPULATIVE COMPRESSION JOINT
- coupling nut
- tapered compression ring
- copper tube
- coupling fitting
- copper tube

SOLDERED CAPILLARY JOINT
- coupling fitting
- solder retaining groove
- copper tube
- copper tube
- fine solder

SCREWED JOINT
- screwed socket fitting
- PTFE (polytetrafluoroethylene) tape or hemp and paste in thread
- steel pipe
- steel pipe

SOLVENT WELD JOINT
- PVC coupling
- solvent weld
- PVC pipe
- PVC pipe

COMPRESSION JOINT
- copper coupling fitting
- copper liner or sleeve
- tapered compression ring
- polythene pipe
- polythene pipe

777

Typical Examples ~

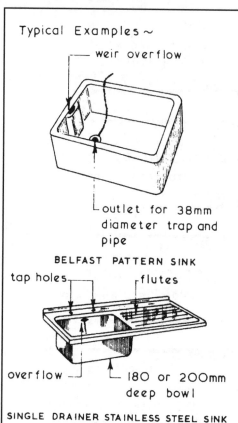

weir overflow

outlet for 38mm diameter trap and pipe

BELFAST PATTERN SINK

tap holes — flutes

overflow — 180 or 200mm deep bowl

SINGLE DRAINER STAINLESS STEEL SINK

Fireclay Sinks (BS 1206) – these are white glazed sinks and are available in a wide range of sizes from 460 × 380 × 200 deep up to 1220 × 610 × 305 deep and can be obtained with an integral drainer. They should be fixed at a height between 850 and 920 mm and supported by legs, cantilever brackets or dwarf brick walls.

Metal Sinks (BS EN 13310) – these can be made of enamelled pressed steel or stainless steel with single or double drainers in sizes ranging from 1070 × 460 to 1600 × 530 supported on a cantilever brackets or sink cupboards.

Ceramic Wash Basins (BS 1188)

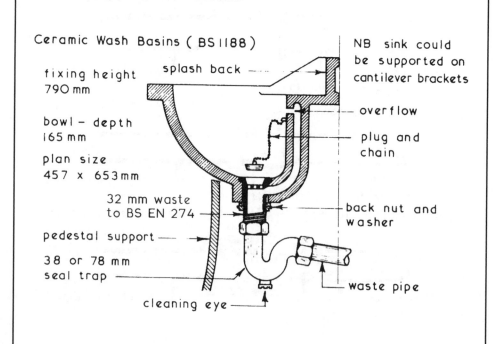

fixing height 790 mm

splash back

NB sink could be supported on cantilever brackets

overflow

bowl – depth 165 mm

plug and chain

plan size 457 x 653mm

32 mm waste to BS EN 274

back nut and washer

pedestal support

38 or 78 mm seal trap

waste pipe

cleaning eye

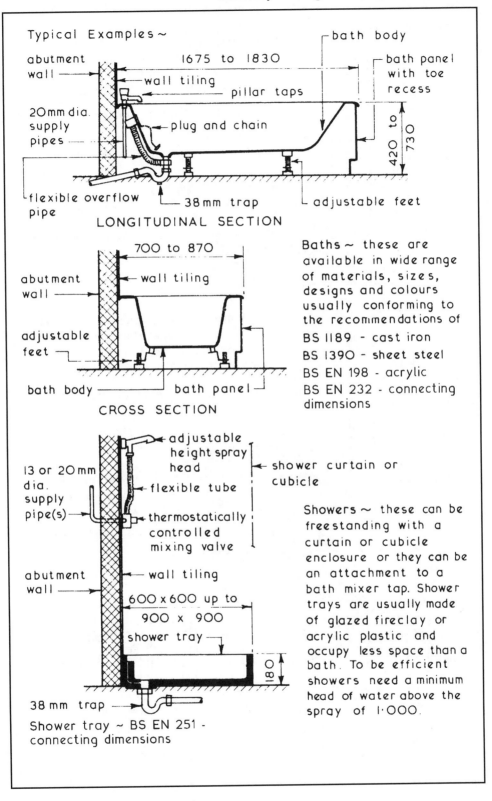

Typical Examples ~

abutment wall

1675 to 1830

wall tiling

pillar taps

bath body

bath panel with toe recess

20mm dia. supply pipes

plug and chain

420 to 730

flexible overflow pipe

38mm trap

adjustable feet

LONGITUDINAL SECTION

700 to 870

abutment wall

wall tiling

adjustable feet

bath body

bath panel

CROSS SECTION

Baths ~ these are available in wide range of materials, sizes, designs and colours usually conforming to the recommendations of

BS 1189 - cast iron
BS 1390 - sheet steel
BS EN 198 - acrylic
BS EN 232 - connecting dimensions

adjustable height spray head

shower curtain or cubicle

13 or 20mm dia. supply pipe(s)

flexible tube

thermostatically controlled mixing valve

abutment wall

wall tiling

600 x 600 up to 900 x 900

shower tray

180

38 mm trap

Showers ~ these can be freestanding with a curtain or cubicle enclosure or they can be an attachment to a bath mixer tap. Shower trays are usually made of glazed fireclay or acrylic plastic and occupy less space than a bath. To be efficient showers need a minimum head of water above the spray of 1·000.

Shower tray ~ BS EN 251 - connecting dimensions

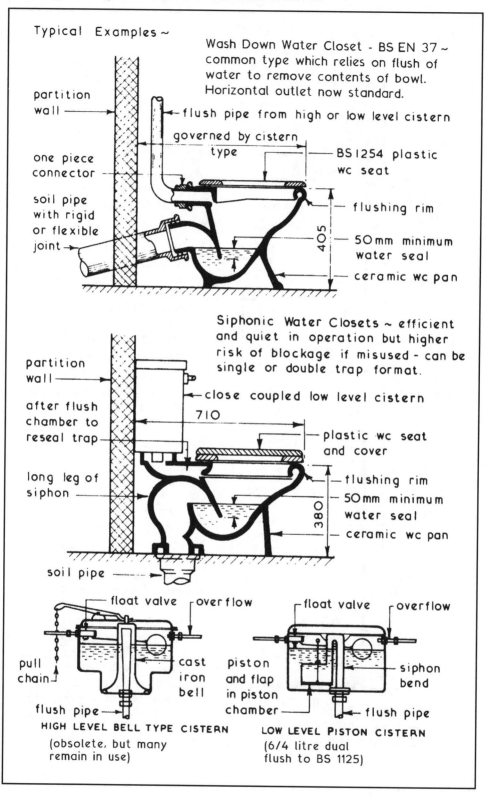

Typical Examples ~

Wash Down Water Closet - BS EN 37 ~ common type which relies on flush of water to remove contents of bowl. Horizontal outlet now standard.

partition wall

flush pipe from high or low level cistern

one piece connector

governed by cistern type

BS 1254 plastic wc seat

soil pipe with rigid or flexible joint

flushing rim

405

50 mm minimum water seal

ceramic wc pan

Siphonic Water Closets ~ efficient and quiet in operation but higher risk of blockage if misused - can be single or double trap format.

partition wall

close coupled low level cistern

after flush chamber to reseal trap

710

plastic wc seat and cover

long leg of siphon

flushing rim

50 mm minimum water seal

380

ceramic wc pan

soil pipe

float valve   overflow

float valve   overflow

pull chain

cast iron bell

piston and flap in piston chamber

siphon bend

flush pipe

flush pipe

HIGH LEVEL BELL TYPE CISTERN
(obsolete, but many remain in use)

LOW LEVEL PISTON CISTERN
(6/4 litre dual flush to BS 1125)

Single Stack System ~ method developed by the Building Research Establishment to eliminate the need for ventilating pipework to maintain the water seals in traps to sanitary fittings. The slope and distance of the branch connections must be kept within the design limitations given below. This system is only possible when the sanitary appliances are closely grouped around the discharge stack.

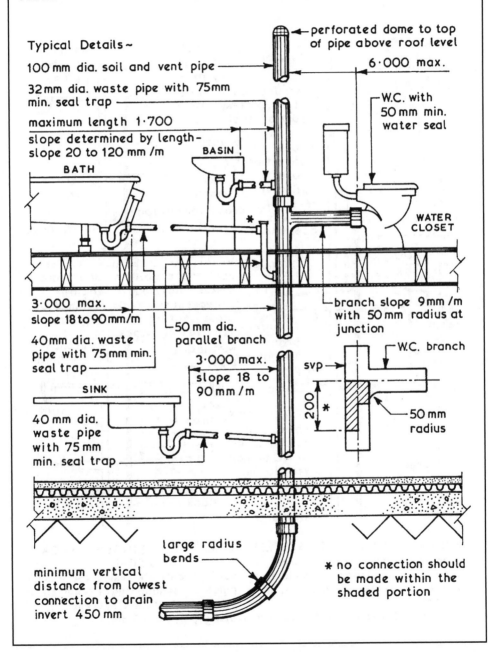

Ventilated Stack Systems ~ where the layout of sanitary appliances is such that they do not conform to the requirements for the single stack system shown on page 781 ventilating pipes will be required to maintain the water seals in the traps. Three methods are available to overcome the problem, namely a fully ventilated system, a ventilated stack system and a modified single stack system which can be applied over any number of storeys.

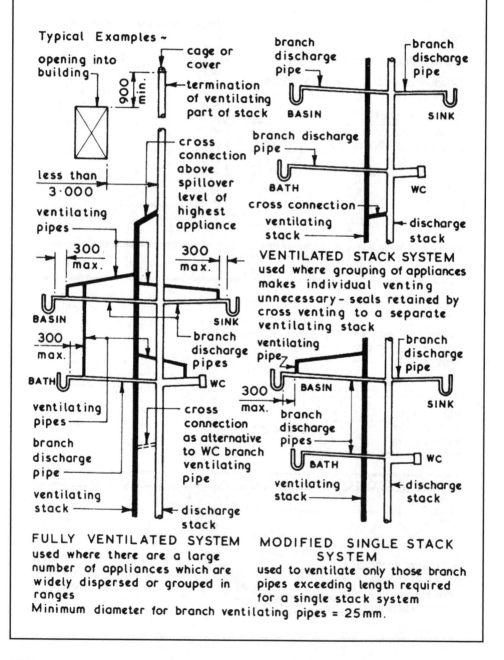

Typical Examples ~

opening into building

cage or cover

termination of ventilating part of stack

900 min.

cross connection above spillover level of highest appliance

less than 3·000

ventilating pipes

300 max.

300 max.

BASIN

SINK

300 max.

branch discharge pipes

BATH

WC

ventilating pipes

cross connection as alternative to WC branch ventilating pipe

branch discharge pipe

ventilating stack

discharge stack

FULL VENTILATED SYSTEM
used where there are a large number of appliances which are widely dispersed or grouped in ranges

branch discharge pipe

branch discharge pipe

BASIN

SINK

branch discharge pipe

BATH

WC

cross connection

ventilating stack

discharge stack

VENTILATED STACK SYSTEM
used where grouping of appliances makes individual venting unnecessary - seals retained by cross venting to a separate ventilating stack

ventilating pipe

branch discharge pipe

300 max.

BASIN

SINK

branch discharge pipes

BATH

WC

ventilating stack

discharge stack

MODIFIED SINGLE STACK SYSTEM
used to ventilate only those branch pipes exceeding length required for a single stack system

Minimum diameter for branch ventilating pipes = 25mm.

Airtightness ~ must be ensured to satisfy public health legislation. The Building Regulations, Approved Document H1: Section 1, provides minimum standards for test procedures. An air or smoke test on the stack must produce a pressure at least equal to 38 mm water gauge for not less than 3 minutes.

Application ~

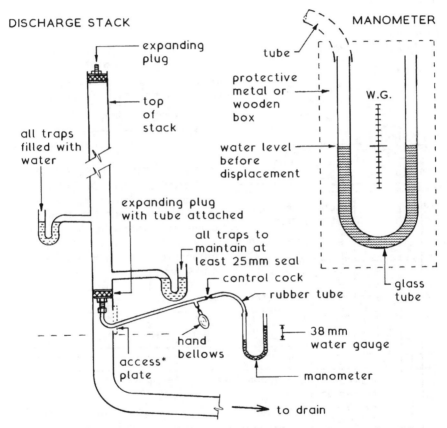

DISCHARGE STACK

expanding plug

top of stack

all traps filled with water

expanding plug with tube attached

all traps to maintain at least 25 mm seal

control cock

rubber tube

access* plate

hand bellows

38 mm water gauge

manometer

to drain

MANOMETER

tube

protective metal or wooden box

W.G.

water level before displacement

glass tube

* if access plate is not provided, top connection to first IC may be plugged and rubber tube inserted through wc pan seal.

NB. Smoke tests are rarely applied now as the equipment is quite bulky and unsuited for use with uPVC pipes. Smoke producing pellets are ideal for leakage detection, but must not come into direct contact with plastic materials.

One Pipe System ~ the hot water is circulated around the system by means of a centrifugal pump. The flow pipe temperature being about 80°C and the return pipe temperature being about 60 to 70°C. The one pipe system is simple in concept and easy to install but has the main disadvantage that the hot water passing through each heat emitter flows onto the next heat emitter or radiator, therefore the average temperature of successive radiators is reduced unless the radiators are carefully balanced or the size of the radiators at the end of the circuit are increased to compensate for the temperature drop.

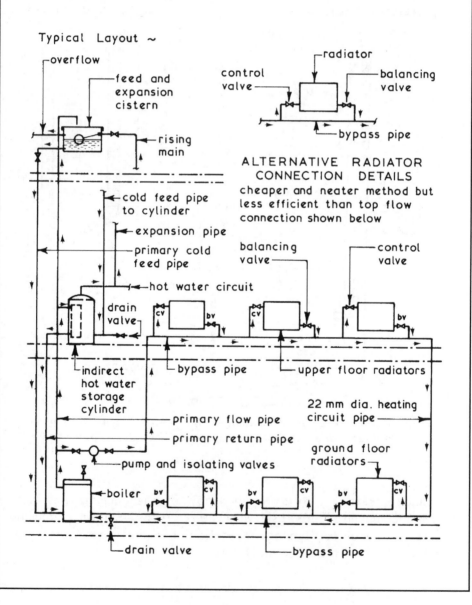

Typical Layout ~

ALTERNATIVE RADIATOR CONNECTION DETAILS
cheaper and neater method but less efficient than top flow connection shown below

Two Pipe System ~ this is a dearer but much more efficient system than the one pipe system shown on the previous page. It is easier to balance since each radiator or heat emitter receives hot water at approximately the same temperature because the hot water leaving the radiator is returned to the boiler via the return pipe without passing through another radiator.

Typical Layout ~

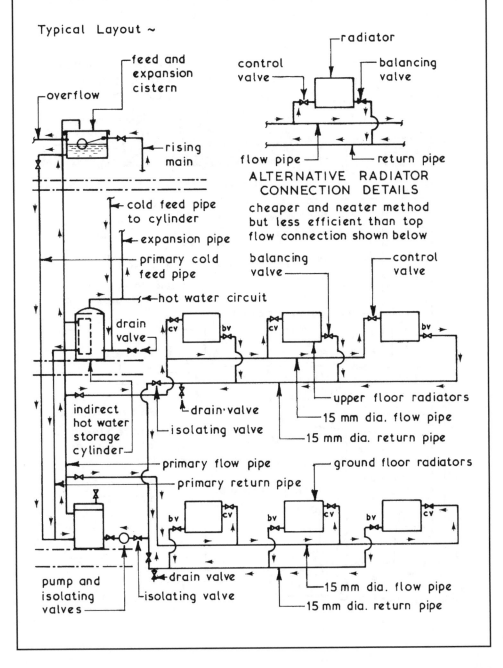

ALTERNATIVE RADIATOR CONNECTION DETAILS

cheaper and neater method but less efficient than top flow connection shown below

785

Micro Bore System ~ this system uses 6 to 12mm diameter soft copper tubing with an individual flow and return pipe to each heat emitter or radiator from a 22mm diameter manifold. The flexible and unobstrusive pipework makes this system easy to install in awkward situations but it requires a more powerful pump than that used in the traditional small bore systems. The heat emitter or radiator valves can be as used for the one or two pipe small bore systems alternatively a double entry valve can be used.

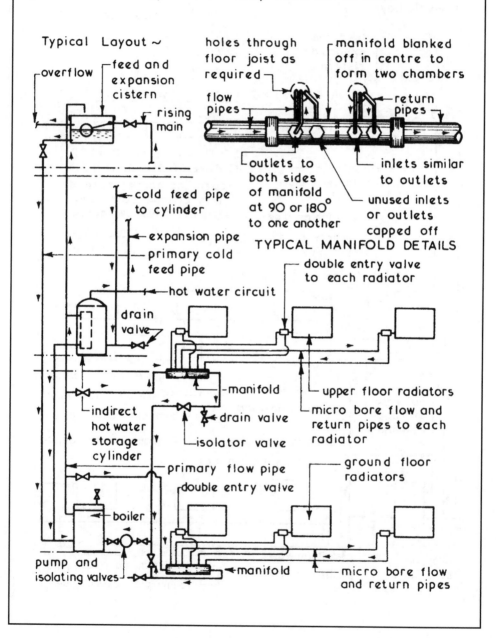

Typical Layout ~

overflow

feed and expansion cistern

rising main

holes through floor joist as required

flow pipes

manifold blanked off in centre to form two chambers

return pipes

outlets to both sides of manifold at 90 or 180° to one another

inlets similar to outlets

unused inlets or outlets capped off

TYPICAL MANIFOLD DETAILS

cold feed pipe to cylinder

expansion pipe

primary cold feed pipe

hot water circuit

drain valve

indirect hot water storage cylinder

manifold

drain valve

isolator valve

primary flow pipe

double entry valve

boiler

pump and isolating valves

manifold

double entry valve to each radiator

upper floor radiators

micro bore flow and return pipes to each radiator

ground floor radiators

micro bore flow and return pipes

Controls ~ the range of controls available to regulate the heat output and timing operations for a domestic hot water heating system is considerable, ranging from thermostatic radiator control valves to programmers and controllers.

Typical Example ~

Boiler – fitted with a thermostat to control the temperature of the hot water leaving the boiler.

Heat Emitters or Radiators – fitted with thermostatically controlled radiator valves to control flow of hot water to the radiators to keep room at desired temperature.

Programmer/Controller – this is basically a time switch which can usually be set for 24 hours, once daily or twice daily time periods and will generally give separate programme control for the hot water supply and central heating systems. The hot water cylinder and room thermostatic switches control the pump and motorised valve action.

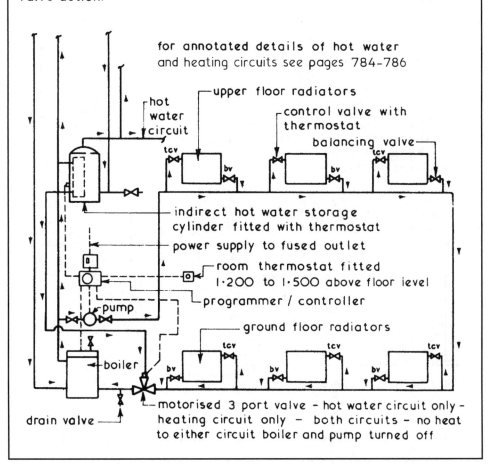

for annotated details of hot water and heating circuits see pages 784–786

upper floor radiators

control valve with thermostat

balancing valve

hot water circuit

indirect hot water storage cylinder fitted with thermostat

power supply to fused outlet

room thermostat fitted 1·200 to 1·500 above floor level

programmer / controller

pump

ground floor radiators

boiler

motorised 3 port valve – hot water circuit only – heating circuit only – both circuits – no heat to either circuit boiler and pump turned off

drain valve

Electrical Supply ~ in the UK electricity is generated mainly from gas, coal, nuclear and hydro-electricity power plants. Alternative energy generation such as wind and solar power are also viable and considered in Part 16 of the *Building Services Handbook*. Distribution is through regional companies. The electrical supply to a domestic installation is usually 230 volt single phase and is designed with the following safety objectives:-

1. Proper circuit protection to earth to avoid shocks to occupant.
2. Prevention of current leakage.
3. Prevention of outbreak of fire.

Typical Electrical Supply Intake Details ~

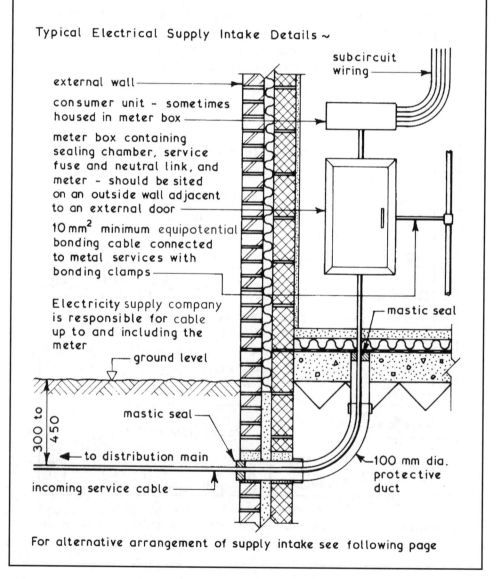

external wall

consumer unit – sometimes housed in meter box

meter box containing sealing chamber, service fuse and neutral link, and meter – should be sited on an outside wall adjacent to an external door

$10\,mm^2$ minimum equipotential bonding cable connected to metal services with bonding clamps

Electricity supply company is responsible for cable up to and including the meter

ground level

300 to 450

mastic seal

to distribution main

incoming service cable

subcircuit wiring

mastic seal

100 mm dia. protective duct

For alternative arrangement of supply intake see following page

Electrical Supply Intake ~ although the electrical supply intake can be terminated in a meter box situated within a dwelling, most supply companies prefer to use the external meter box to enable the meter to be read without the need to enter the premises.

Typical Electrical Supply Intake Details ~

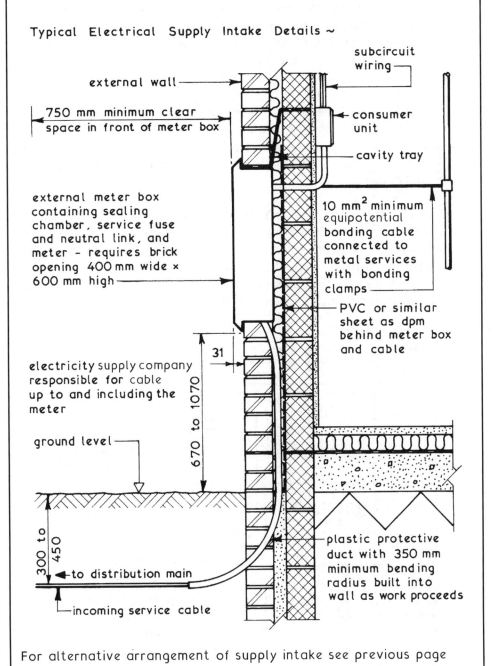

external wall

subcircuit wiring

750 mm minimum clear space in front of meter box

consumer unit

cavity tray

external meter box containing sealing chamber, service fuse and neutral link, and meter - requires brick opening 400 mm wide × 600 mm high

10 mm^2 minimum equipotential bonding cable connected to metal services with bonding clamps

PVC or similar sheet as dpm behind meter box and cable

31

electricity supply company responsible for cable up to and including the meter

670 to 1070

ground level

300 to 450

to distribution main

incoming service cable

plastic protective duct with 350 mm minimum bending radius built into wall as work proceeds

For alternative arrangement of supply intake see previous page

Entry and Intake of Electrical Service ~ the local electricity supply company is responsible for providing electricity up to and including the meter, but the consumer is responsible for safety and protection of the company's equipment. The supplier will install the service cable up to the meter position where their termination equipment is installed. This equipment may be located internally or fixed externally on a wall, the latter being preferred since it gives easy access for reading the meter - see details on the previous page.

Meter Boxes - generally the supply company's meters and termination equipment are housed in a meter box. These are available in fibreglass and plastic, ranging in size from 450mm wide × 638mm high to 585m wide × 815mm high with an overall depth of 177mm.

Consumer Control Unit - this provides a uniform, compact and effective means of efficiently controlling and distributing electrical energy within a dwelling. The control unit contains a main double pole isolating switch controlling the line and neutral conductors, called bus bars. These connect to the fuses or miniature circuit breakers protecting the final subcircuits.

Typical Layout ~

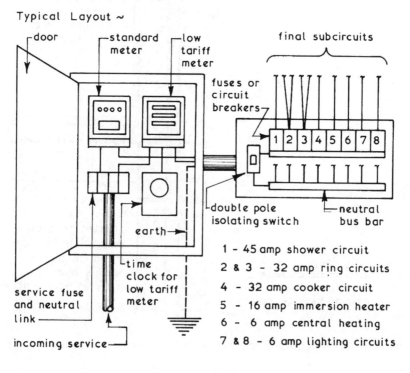

1 - 45 amp shower circuit
2 & 3 - 32 amp ring circuits
4 - 32 amp cooker circuit
5 - 16 amp immersion heater
6 - 6 amp central heating
7 & 8 - 6 amp lighting circuits

Consumer's Power Supply Control Unit – this is conveniently abbreviated to consumer unit. As described on the previous page, it contains a supply isolator switch, live, neutral and earth bars, plus a range of individual circuit over-load safety protection devices. By historical reference this unit is sometimes referred to as a fuse box, but modern variants are far more sophisticated. Over-load protection is provided by miniature circuit breakers attached to the live or phase bar. Additional protection is provided by a split load residual current device (RCD) dedicated specifically to any circuits that could be used as a supply to equipment outdoors, e.g. power sockets on a ground floor ring final circuit.

RCD – a type of electro-magnetic switch or solenoid which disconnects the electricity supply when a surge of current or earth fault occurs. See Part 11 of the *Building Services Handbook* for more detail.

Typical Split Load Consumer Unit —

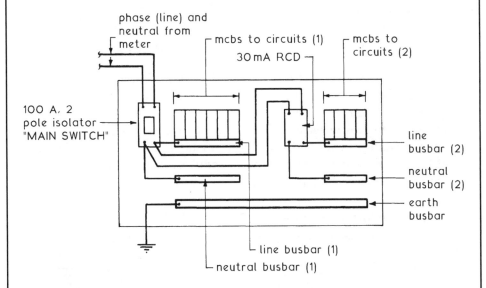

Note that with an overhead supply, the MAIN SWITCH is combined with a 100 mA RCD protecting all circuits.

Note:

Circuits (1) to fixtures, i.e. lights, cooker, immersion heater and smoke alarms.

Circuits (2) to socket outlets that could supply portable equipment outdoors.

Electric Cables ~ these are made up of copper or aluminium wires called conductors surrounded by an insulating material such as PVC or rubber.

Typical Examples ~

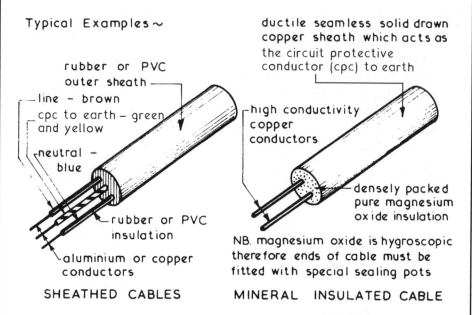

rubber or PVC outer sheath

line – brown

cpc to earth – green and yellow

neutral – blue

rubber or PVC insulation

aluminium or copper conductors

SHEATHED CABLES

ductile seamless solid drawn copper sheath which acts as the circuit protective conductor (cpc) to earth

high conductivity copper conductors

densely packed pure magnesium oxide insulation

NB. magnesium oxide is hygroscopic therefore ends of cable must be fitted with special sealing pots

MINERAL INSULATED CABLE

Conduits ~ these are steel or plastic tubes which protect the cables. Steel conduits act as a cpc to earth whereas plastic conduits will require a separate cpc drawn in. Conduits enable a system to be rewired without damage or interference to the fabric of the building. The cables used within conduits are usually insulated only, whereas in non-rewireable systems the cables have a protective outer sheath.

Typical Conduit Fittings ~

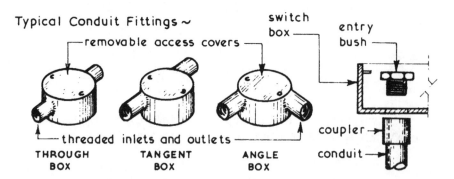

removable access covers

switch box

entry bush

threaded inlets and outlets

coupler

conduit

THROUGH BOX   **TANGENT BOX**   **ANGLE BOX**

Trunking – alternative to conduit and consists of a preformed cable carrier which is surface mounted and is fitted with a removable or 'snap on' cover which can have the dual function of protection and trim or surface finish.

Wiring systems ~ rewireable systems housed in horizontal conduits can be cast into the structural floor slab or sited within the depth of the floor screed. To ensure that such a system is rewireable, draw-in boxes must be incorporated at regular intervals and not more than two right angle boxes to be included between draw-in points. Vertical conduits can be surface mounted or housed in a chase cut in to a wall provided the depth of the chase is not more than one-third of the wall thickness. A horizontal non-rewireable system can be housed within the depth of the timber joists to a suspended floor whereas vertical cables can be surface mounted or housed in a length of conduit as described for rewireable systems.

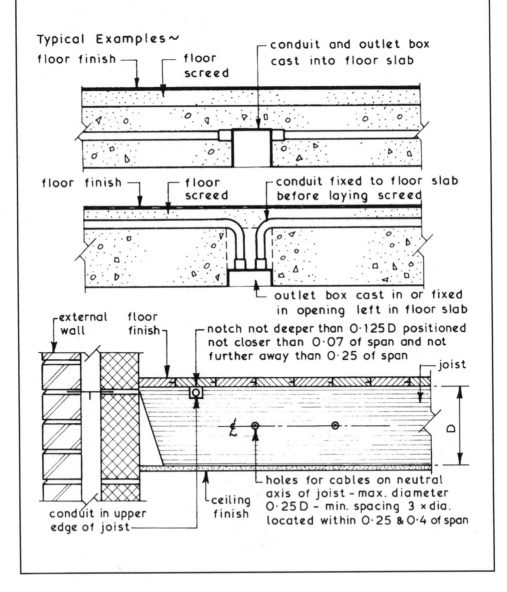

Typical Examples~

floor finish — floor screed — conduit and outlet box cast into floor slab

floor finish — floor screed — conduit fixed to floor slab before laying screed

outlet box cast in or fixed in opening left in floor slab

external wall — floor finish — notch not deeper than 0·125 D positioned not closer than 0·07 of span and not further away than 0·25 of span — joist

conduit in upper edge of joist — ceiling finish — holes for cables on neutral axis of joist – max. diameter 0·25 D – min. spacing 3 × dia. located within 0·25 & 0·4 of span

Cable Sizing ~ the size of a conductor wire can be calculated taking into account the maximum current the conductor will have to carry (which is limited by the heating effect caused by the resistance to the flow of electricity through the conductor) and the voltage drop which will occur when the current is carried. For domestic electrical installations the following minimum cable specifications are usually suitable -

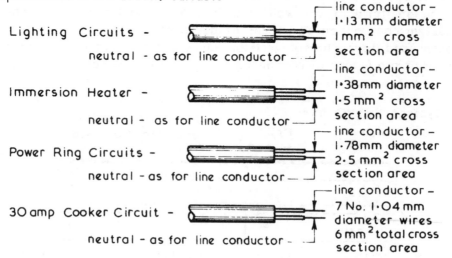

Lighting Circuits -

neutral - as for line conductor

line conductor -
1·13 mm diameter
1 mm² cross
section area

Immersion Heater -

neutral - as for line conductor

line conductor -
1·38 mm diameter
1·5 mm² cross
section area

Power Ring Circuits -

neutral - as for line conductor

line conductor -
1·78 mm diameter
2·5 mm² cross
section area

30 amp Cooker Circuit -

neutral - as for line conductor

line conductor -
7 No. 1·04 mm
diameter wires
6 mm² total cross
section area

All the above ratings are for the line and neutral conductors which will be supplemented with a circuit protective conductor as shown on page 792.

Electrical Accessories ~ for power circuits these include cooker control units and fused switch units for fixed appliances such as immersion heaters, water heaters and central heating controls.

Socket Outlets ~ these may be single or double outlets, switched or unswitched, surface or flush mounted and may be fitted with indicator lights. Recommended fixing heights are -

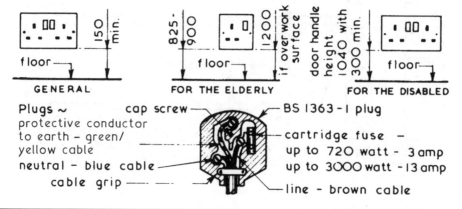

GENERAL

FOR THE ELDERLY

FOR THE DISABLED

Plugs ~
protective conductor
to earth - green/
yellow cable
neutral - blue cable
cable grip

cap screw

BS 1363-1 plug

cartridge fuse -
up to 720 watt - 3 amp
up to 3000 watt - 13 amp

line - brown cable

Power Sockets ~ in new domestic electrical installations the ring final circuit is usually employed instead of the older obsolete radial system where socket outlets are on individual fused circuits with unfused round pin plugs. Ring circuits consist of a fuse or miniature circuit breaker protected subcircuit with a 32 amp rating for a line conductor, neutral conductor and a cpc to earth looped from socket outlet to socket outlet. Metal conduit systems do not require a cpc wire providing the conduit is electrically sound and earthed. The number of socket outlets on a ring final circuit is unlimited but a separate circuit must be provided for every 100 m^2 of floor area. To conserve wiring, spur outlets can be used as long as the total number of spur outlets does not exceed the total number of outlets connected to the ring and that there are not more than two outlets per spur.

Typical Ring Final Circuit Wiring Diagram ~

795

Lighting Circuits ~ these are usually wired by the loop-in method using a line, neutral and circuit protective conductor to earth cable with a 6 amp fuse or miniature circuit breaker protection. In calculating the rating of a lighting circuit an allowance of 100 watts per outlet should be used. More than one lighting circuit should be used for each installation so that in the event of a circuit failure some lighting will be in working order.

Typical Lighting Circuit Wiring Diagram ~

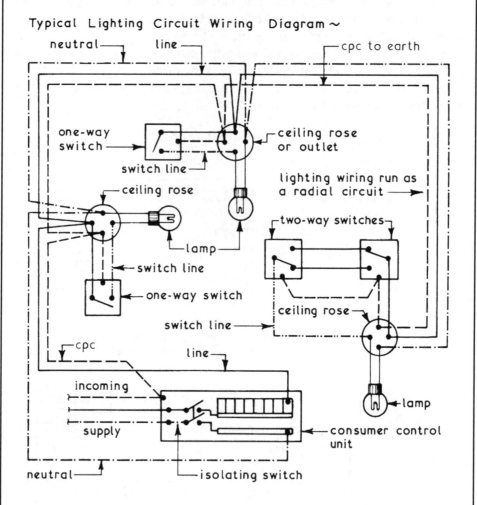

Electrical Accessories ~ for lighting circuits these consist mainly of switches and lampholders, the latter can be wall mounted, ceiling mounted or pendant in format with one or more bulb or tube holders. Switches are usually rated at 5 amps and are available in a variety of types such as double or 2 gang, dimmer and pull or pendant switches. The latter must always be used in bathrooms.

Gas Supply ~ potential consumers of mains gas may apply to their local utilities supplier for connection, e.g. Transco (Lattice Group plc). The cost is normally based on a fee per metre run. However, where the distance is considerable, the gas authority may absorb some of the cost if there is potential for more customers. The supply, appliances and installation must comply with the safety requirements made under the Gas Safety (Installation and Use) Regulations, 1998, and Part J of the Building Regulations.

Typical Gas Supply Arrangement ~

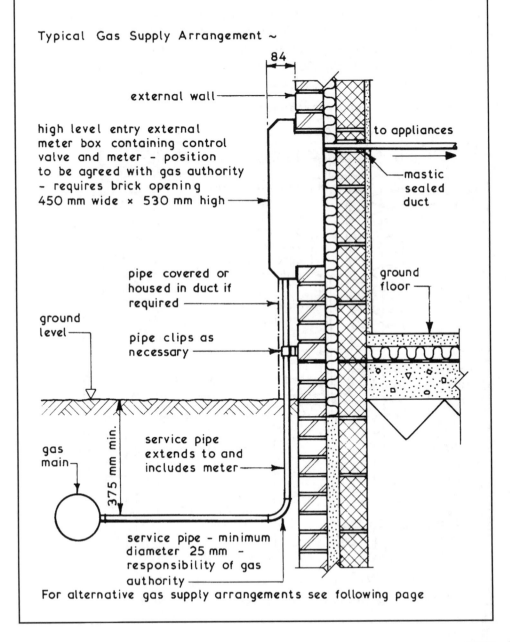

external wall

high level entry external meter box containing control valve and meter - position to be agreed with gas authority - requires brick opening 450 mm wide × 530 mm high

84

to appliances

mastic sealed duct

pipe covered or housed in duct if required

ground floor

ground level

pipe clips as necessary

gas main

375 mm min.

service pipe extends to and includes meter

service pipe - minimum diameter 25 mm - responsibility of gas authority

For alternative gas supply arrangements see following page

Gas Service Pipes ~

1. Whenever possible the service pipe should enter the building on the side nearest to the main.
2. A service pipe must not pass under the foundations of a building.
3. No service pipe must be run within a cavity but it may pass through a cavity by the shortest route.
4. Service pipes passing through a wall or solid floor must be enclosed by a sleeve or duct which is end sealed with mastic.
5. No service pipe shall be housed in an unventilated void.
6. Suitable materials for service pipes are copper (BS EN 1057) and steel (BS EN 10255). Polyethylene (BS EN 1555) is normally used underground, but not above ground.

Typical Gas Supply Arrangement ~

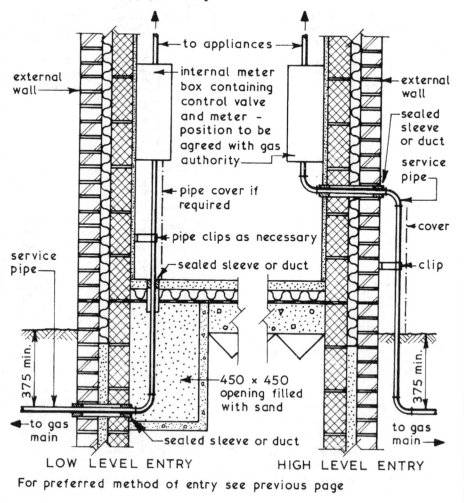

LOW LEVEL ENTRY HIGH LEVEL ENTRY

For preferred method of entry see previous page

Gas Fires ~ for domestic use these generally have a low energy rating of less than 7 kW net input and must be installed in accordance with minimum requirements set out in Part J of the Building Regulations. Most gas fires connected to a flue are designed to provide radiant and convected heating whereas the room sealed balanced flue appliances are primarily convector heaters.

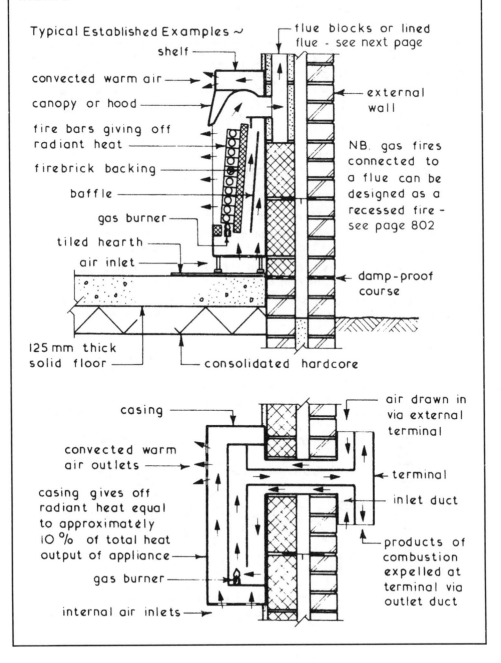

Typical Established Examples ~

flue blocks or lined flue - see next page

shelf

convected warm air

canopy or hood

external wall

fire bars giving off radiant heat

firebrick backing

baffle

gas burner

tiled hearth

air inlet

NB. gas fires connected to a flue can be designed as a recessed fire - see page 802

damp-proof course

125 mm thick solid floor

consolidated hardcore

casing

air drawn in via external terminal

convected warm air outlets

casing gives off radiant heat equal to approximately 10 % of total heat output of appliance

gas burner

internal air inlets

terminal

inlet duct

products of combustion expelled at terminal via outlet duct

Gas Fire Flues ~ these can be defined as a passage for the discharge of the products of combustion to the outside air and can be formed by means of a chimney, special flue blocks or by using a flue pipe. In all cases the type and size of the flue as recommended in Approved Document J, BS EN 1806 and BS 5440 will meet the requirements of the Building Regulations.

Typical Single Gas Fire Flues ~

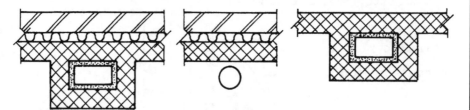

| LINED CHIMNEY ON EXTERNAL WALL | FLUE PIPE ON EXTERNAL WALL | LINED CHIMNEY ON INTERNAL WALL |

Flue Size Requirements :-

1. No dimension should be less than 63 mm.

2. Flue for a decorative appliance should have a minimum dimension measured across the axis of 175 mm.

3. Flues for gas fires - min. area = 12000 mm^2 if round, 16500 mm^2 if rectangular and having a minimum dimension of 90 mm.

4. Any other appliance should have a flue with a cross-sectional area at least equal to the outlet size of the appliance.

Flue Blocks ~

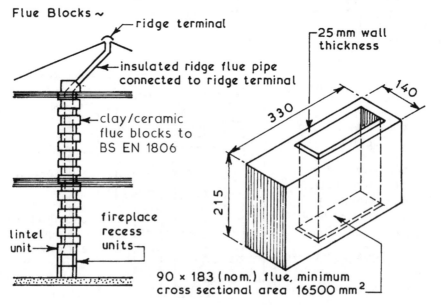

ridge terminal

insulated ridge flue pipe connected to ridge terminal

clay/ceramic flue blocks to BS EN 1806

lintel unit

fireplace recess units

25 mm wall thickness

330

140

215

90 × 183 (nom.) flue, minimum cross sectional area 16500 mm^2

Open Fireplaces ~ for domestic purposes these are a means of providing a heat source by consuming solid fuels with an output rating of under 50 kW. Room-heaters can be defined in a similar manner but these are an enclosed appliance as opposed to the open recessed fireplace.

Components ~ the complete construction required for a domestic open fireplace installation is composed of the hearth, fireplace recess, chimney, flue and terminal.

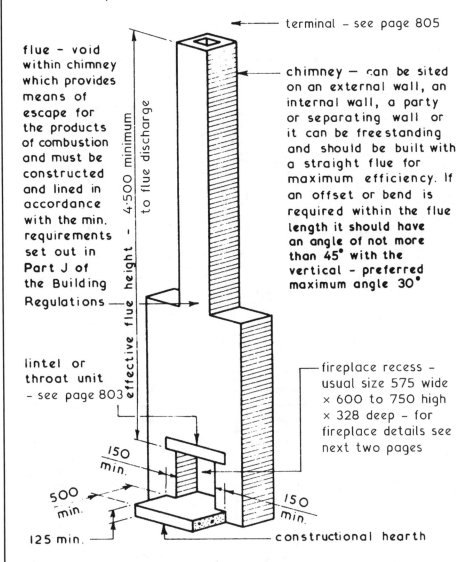

terminal – see page 805

flue – void within chimney which provides means of escape for the products of combustion and must be constructed and lined in accordance with the min. requirements set out in Part J of the Building Regulations

chimney — can be sited on an external wall, an internal wall, a party or separating wall or it can be freestanding and should be built with a straight flue for maximum efficiency. If an offset or bend is required within the flue length it should have an angle of not more than 45° with the vertical – preferred maximum angle 30°

4·500 minimum to flue discharge

effective flue height

lintel or throat unit – see page 803

fireplace recess – usual size 575 wide × 600 to 750 high × 328 deep – for fireplace details see next two pages

150 min.

500 min.

150 min.

125 min.

constructional hearth

See also BS 5854: Code of practice for flues and flue structures in buildings.

Open Fireplace Recesses ~ these must have a constructional hearth and can be constructed of bricks or blocks of concrete or burnt clay or they can be of cast in-situ concrete. All fireplace recesses must have jambs on both sides of the opening and a backing wall of a minimum thickness in accordance with its position and such jambs and backing walls must extend to the full height of the fireplace recess.

Typical Examples ~

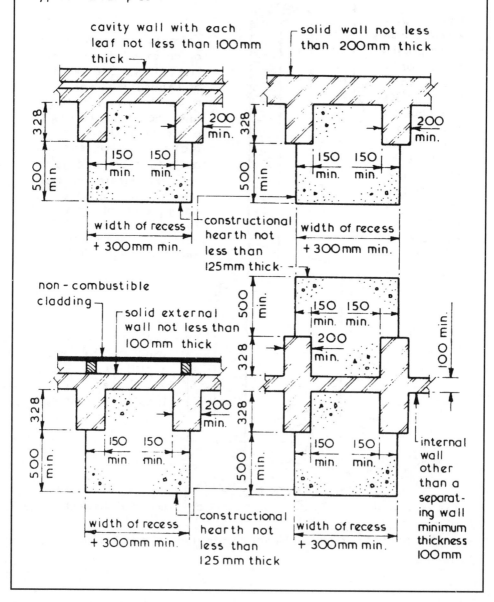

Traditional Fireplace Details ~

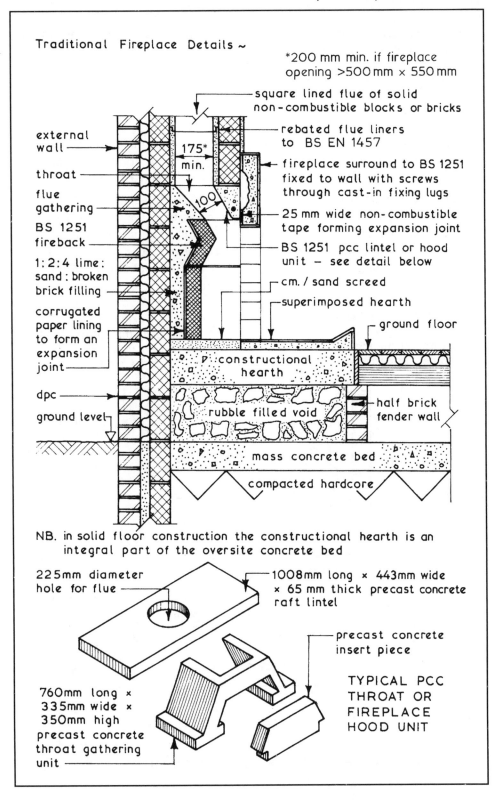

*200 mm min. if fireplace opening >500 mm × 550 mm

square lined flue of solid non-combustible blocks or bricks

rebated flue liners to BS EN 1457

external wall

175* min.

100

throat

flue gathering

BS 1251 fireback

1:2:4 lime: sand: broken brick filling

corrugated paper lining to form an expansion joint

dpc

ground level

fireplace surround to BS 1251 fixed to wall with screws through cast-in fixing lugs

25 mm wide non-combustible tape forming expansion joint

BS 1251 pcc lintel or hood unit — see detail below

cm. / sand screed

superimposed hearth

ground floor

constructional hearth

rubble filled void

half brick fender wall

mass concrete bed

compacted hardcore

NB. in solid floor construction the constructional hearth is an integral part of the oversite concrete bed

225mm diameter hole for flue

1008mm long × 443mm wide × 65 mm thick precast concrete raft lintel

precast concrete insert piece

760mm long × 335mm wide × 350mm high precast concrete throat gathering unit

TYPICAL PCC THROAT OR FIREPLACE HOOD UNIT

803

Open Fireplace Chimneys and Flues ~ the main functions of a chimney and flue are to:-

1. Induce an adequate supply of air for the combustion of the fuel being used.
2. Remove the products of combustion.

In fulfilling the above functions a chimney will also encourage a flow of ventilating air promoting constant air changes within the room which will assist in the prevention of condensation.

Approved Document J recommends that all flues should be lined with approved materials so that the minimum size of the flue so formed will be 200 mm diameter or a square section of equivalent area. Flues should also be terminated above the roof level as shown, with a significant increase where combustible roof coverings such as thatch or wood shingles are used.

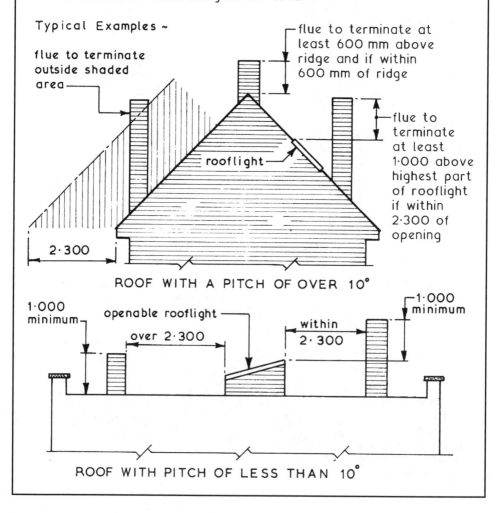

Typical Examples ~

flue to terminate outside shaded area

flue to terminate at least 600 mm above ridge and if within 600 mm of ridge

flue to terminate at least 1·000 above highest part of rooflight if within 2·300 of opening

rooflight

2·300

ROOF WITH A PITCH OF OVER 10°

1·000 minimum

openable rooflight

over 2·300

within 2·300

1·000 minimum

ROOF WITH PITCH OF LESS THAN 10°

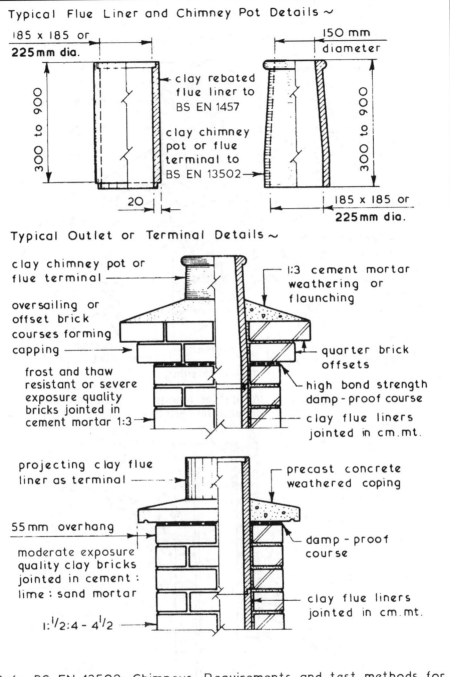

Typical Flue Liner and Chimney Pot Details ~

185 x 185 or
225mm dia.

300 to 900

20

clay rebated
flue liner to
BS EN 1457

clay chimney
pot or flue
terminal to
BS EN 13502

150 mm
diameter

300 to 900

185 x 185 or
225mm dia.

Typical Outlet or Terminal Details ~

clay chimney pot or
flue terminal

oversailing or
offset brick
courses forming
capping

frost and thaw
resistant or severe
exposure quality
bricks jointed in
cement mortar 1:3

1:3 cement mortar
weathering or
flaunching

quarter brick
offsets

high bond strength
damp-proof course

clay flue liners
jointed in cm.mt.

projecting clay flue
liner as terminal

55mm overhang

moderate exposure
quality clay bricks
jointed in cement :
lime : sand mortar

1:$\frac{1}{2}$:4 - 4$\frac{1}{2}$

precast concrete
weathered coping

damp-proof
course

clay flue liners
jointed in cm.mt.

Refs. BS EN 13502: Chimneys. Requirements and test methods for clay/ceramic flue terminals.

BS EN 1457: Chimneys. Clay/ceramic flue liners. Requirements and test methods.

BS EN 771-1: Specification for (clay) masonry units.

805

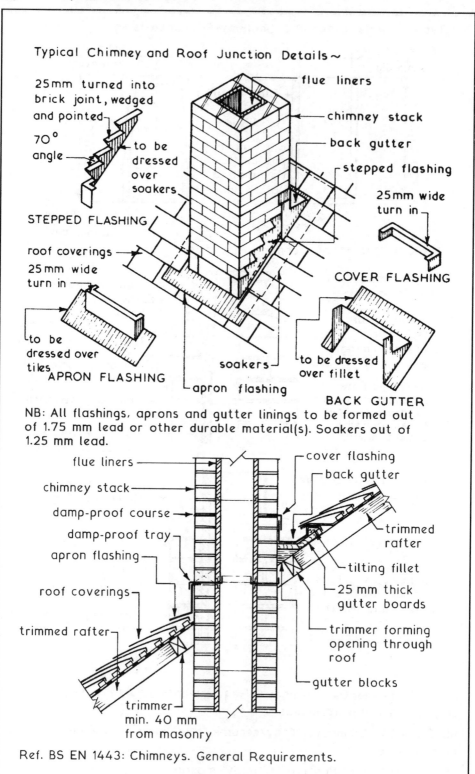

Typical Chimney and Roof Junction Details ~

25mm turned into brick joint, wedged and pointed

70° angle

to be dressed over soakers

STEPPED FLASHING

roof coverings
25mm wide turn in

to be dressed over tiles

APRON FLASHING

flue liners

chimney stack

back gutter

stepped flashing

25mm wide turn in

COVER FLASHING

soakers

apron flashing

to be dressed over fillet

BACK GUTTER

NB: All flashings, aprons and gutter linings to be formed out of 1.75 mm lead or other durable material(s). Soakers out of 1.25 mm lead.

flue liners

chimney stack

damp-proof course

damp-proof tray

apron flashing

roof coverings

trimmed rafter

trimmer min. 40 mm from masonry

cover flashing

back gutter

trimmed rafter

tilting fillet

25 mm thick gutter boards

trimmer forming opening through roof

gutter blocks

Ref. BS EN 1443: Chimneys. General Requirements.

Chimney construction –

- chimney pot or flue terminal
- flaunching cement-sand (1:3)
- $\frac{1}{4}$ brick oversailing course
- dpc
- brick masonry 1.75mm lead saddle flashing
- dpc (optional)
- dpc tray

600mm min. Max. 4·5 times least lateral dimension

highest point of intersection with the roof surface

Tray

Soaker

75mm

100mm

gauge + lap + 25mm

Typical chimney outlet

Clay bricks – Frost and thaw, severe exposure resistant quality. Min. density 1500 kg/m^3.

Calcium silicate bricks – Min. compressive strength 20.5 N/mm^2 (27.5 N/mm^2 for cappings).

Precast concrete masonry units – Min. compressive strength 15 N/mm^2.

Mortar – A relatively strong mix of cement and sand 1:3. Cement to be specified as sulphate resisting because of the presence of soluble sulphates in the flue gas condensation.

Chimney pot – The pot should be firmly bedded in at least 3 courses of brickwork to prevent it being dislodged in high winds.

Flashings and dpcs – Essential to prevent water which has permeated the chimney, penetrating into the building. The minimum specification is 1.75mm lead, 1.25mm for soakers. This should be coated both sides with a solvent-based bituminous paint to prevent the risk of corrosion when in contact with cement. The lower dpc may be in the form of a tray with edges turned up 25mm, except where it coincides with bedded flashings such as the front apron upper level. Here weep holes in the perpends will encourage water to drain. The inside of the tray is taken through a flue lining joint and turned up 25mm.

Combustion Air ~ it is a Building Regulation requirement that in the case of open fireplaces provision must be made for the introduction of combustion air in sufficient quantity to ensure the efficient operation of the open fire. Traditionally such air is taken from the volume of the room in which the open fire is situated, this can create air movements resulting in draughts. An alternative method is to construct an ash pit below the hearth level fret and introduce the air necessary for combustion via the ash by means of a duct.

Typical Established Details ~

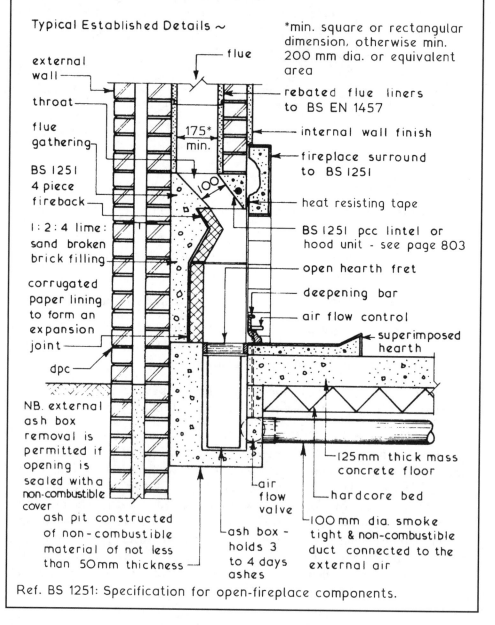

external wall

throat

flue gathering

BS 1251 4 piece fireback

1:2:4 lime: sand broken brick filling

corrugated paper lining to form an expansion joint

dpc

NB. external ash box removal is permitted if opening is sealed with a non-combustible cover

ash pit constructed of non-combustible material of not less than 50mm thickness

flue

175* min.

100

air flow valve

ash box - holds 3 to 4 days ashes

*min. square or rectangular dimension, otherwise min. 200 mm dia. or equivalent area

rebated flue liners to BS EN 1457

internal wall finish

fireplace surround to BS 1251

heat resisting tape

BS 1251 pcc lintel or hood unit - see page 803

open hearth fret

deepening bar

air flow control

superimposed hearth

125mm thick mass concrete floor

hardcore bed

100mm dia. smoke tight & non-combustible duct connected to the external air

Ref. BS 1251: Specification for open-fireplace components.

Lightweight Pumice Chimney Blocks ~ these are suitable as a flue system for solid fuels, gas and oil. The highly insulative properties provide low condensation risk, easy installation as a supplement to existing or on-going construction and suitability for use with timber frame and thatched dwellings, where fire safety is of paramount importance. Also, the natural resistance of pumice to acid and sulphurous smoke corrosion requires no further treatment or special lining. A range of manufacturer's accessories allow for internal use with lintel support over an open fire or stove, or as an external structure supported on its own foundation. Whether internal or external, the units are not bonded in, but supported on purpose made ties at a maximum of 2 metre intervals.

flue (mm)	plan size (mm)
150 dia.	390 × 390
200 dia.	440 × 440
230 dia.	470 × 470
260 square	500 × 500
260 × 150 oblong	500 × 390

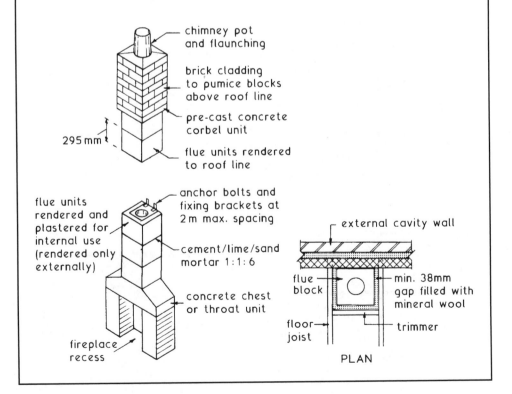

chimney pot and flaunching

brick cladding to pumice blocks above roof line

pre-cast concrete corbel unit

flue units rendered to roof line

295 mm

flue units rendered and plastered for internal use (rendered only externally)

anchor bolts and fixing brackets at 2 m max. spacing

cement/lime/sand mortar 1:1:6

concrete chest or throat unit

fireplace recess

external cavity wall

flue block

min. 38mm gap filled with mineral wool

floor joist

trimmer

PLAN

Fire Protection of Services Openings ~ penetration of compartment walls and floors (zones of restricted fire spread, e.g. flats in one building), by service pipes and conduits is very difficult to avoid. An exception is where purpose built service ducts can be accommodated. The Building Regulations, Approved Document B3: Sections 7 [Vol. 1] and 10 [Vol. 2] determines that where a pipe passes through a compartment interface, it must be provided with a proprietary seal. Seals are collars of intumescent material which expands rapidly when subjected to heat, to form a carbonaceous charring. The expansion is sufficient to compress warm plastic and successfully close a pipe void for up to 4 hours.

In some circumstances fire stopping around the pipe will be acceptable, provided the gap around the pipe and hole through the structure are filled with non-combustible material. Various materials are acceptable, including reinforced mineral fibre, cement and plasters, asbestos rope and intumescent mastics.

Pipes of low heat resistance, such as PVC, lead, aluminium alloys and fibre cement may have a protective sleeve of non-combustible material extending at least 1 m either side of the structure.

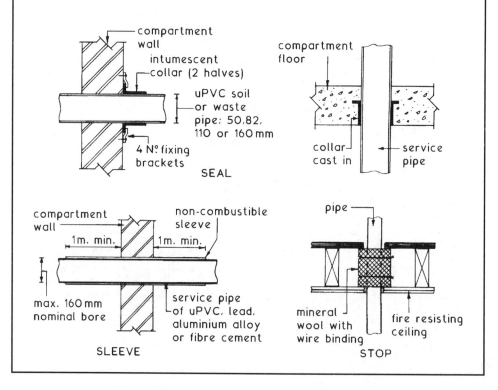

SEAL

SLEEVE

STOP

Telephone Installations ~ unlike other services such as water, gas and electricity, telephones cannot be connected to a common mains supply. Each telephone requires a pair of wires connecting it to the telephone exchange. The external supply service and connection to the lead-in socket is carried out by telecommunication engineers. Internal extensions can be installed by the site electrician.

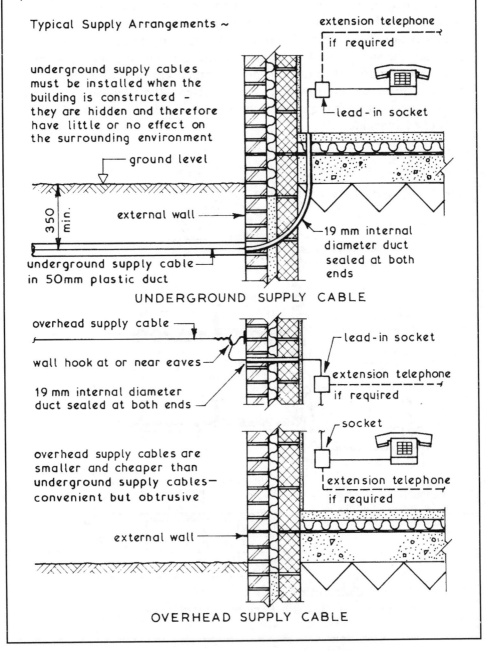

Typical Supply Arrangements ~

extension telephone
if required

underground supply cables must be installed when the building is constructed - they are hidden and therefore have little or no effect on the surrounding environment

lead-in socket

ground level

350 min.

external wall

19 mm internal diameter duct sealed at both ends

underground supply cable in 50mm plastic duct

UNDERGROUND SUPPLY CABLE

overhead supply cable

lead-in socket

wall hook at or near eaves

extension telephone
if required

19 mm internal diameter duct sealed at both ends

socket

overhead supply cables are smaller and cheaper than underground supply cables— convenient but obtrusive

extension telephone
if required

external wall

OVERHEAD SUPPLY CABLE

Electronic Installations – in addition to standard electrical and telecommunication supplies into buildings, there is a growing demand for cable TV, security cabling and broadband access to the Internet. Previous construction practice has not foreseen the need to accommodate these services from distribution networks into buildings, and retrospective installation through underground ducting is both costly and disruptive to the structure and surrounding area, particularly when repeated for each different service. Ideally there should be a common facility integral with new construction to permit simple installation of these communication services at any time. A typical installation will provide connection from a common external terminal chamber via underground ducting to a terminal distribution box within the building. Internal distribution is through service voids within the structure or attached trunking.

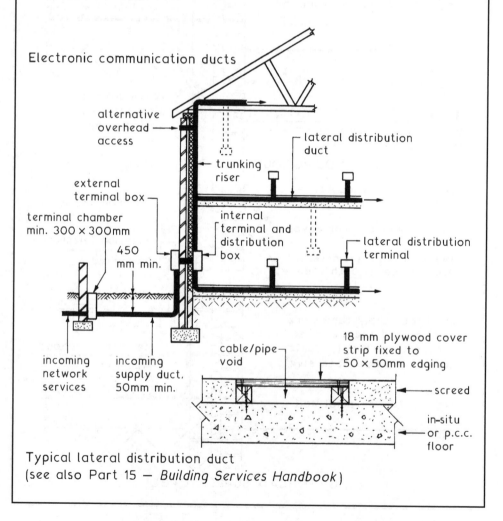

Electronic communication ducts

Typical lateral distribution duct
(see also Part 15 – *Building Services Handbook*)

INDEX

Access for disabled, 492−4
Access to sites, 88
Accommodation on sites, 89−90
Active earth pressures, 251, 263−4
Adhesives, 556
Admixtures, 272, 356
Aerated concrete floor units, 673
Aggregate samples, 108
Air lock hopper, 230
Air test, 765, 783
Air tightness, infiltration and permeability,
    477, 480, 483−4, 491, 559
Alternating tread stair, 692
Aluminium alloy infill panels, 598
Aluminium alloy windows, 382, 384
Anchor bars, 500
Anchor straps, 462, 660
Anchorages, 270, 527−9, 531, 588
Angle Piling, 300
Angledozers, 171
Angles:
    structural steel, 532−3
    surveying, 131
Angles of repose, 250, 263−4, 279, 281
Annexed buildings, 330−1
Anti-shear studs, 499
Apex hinges, 551
Approved Documents, 53−5
Approved inspector, 53, 56−8
Apron cladding panels, 608
Apron flashing, 806−7
Arches, 358−61
Asphalt tanking, 274−5
Asymmetric beam, 677
Attached piers, 325−6, 329−30
Augers, 78, 232−4
Automatic level, 129
Axial grid, 48
Axonometric projection, 24

Backacter/hoe, 33, 170, 177, 179, 268
Backgutter, 806
Backshore, 155
Balanced flue, 799
Balloon frame, 404
Balusters, 685, 688−91, 694−5, 698, 704,
    706, 708

Balustrade, 13, 698−9, 704, 706−8
Bar chart, 34, 88
Bargeboard, 565
Barrel vaults, 9, 580−2
Basement excavation, 265−8
Basement insulation, 277
Basement waterproofing, 272−6
Basements, 54, 265−77
Basic forms and types of structure, 5−10
Basic formwork, 514−20
Basic module grid, 47
Basic roof forms, 417−19
Basins, 778
Batch mixers, 199−200
Baths, 779
Bay window, 376
Beam and block floor, 650−1, 674
Beam and pot floor, 674
Beam and slab raft, 226
Beam formwork, 515−17
Beam to column connection, 524, 537−8
Beam design:
    concrete, 501−3
    steel, 544−6
    timber, 663−6
Beams, 500, 533−4, 553−4, 557−9, 666,
    676−7
Bearing piles, 228
Beech, 738
Belfast truss, 573−5
Benchmark, 71, 125, 127−30
Bending moment, 544, 549, 664
Bentonite, 231, 309−10
Bib tap, 774
Binder, 730
Binders/binding − reinforcement, 226, 500,
    504−6
Binders/binding − shoring, 155, 267
Birdcage scaffold, 148
Block bonding, 621
Block plan, 27, 42
Block walls, 621−5
Blockboard, 739
Bloom base, 225, 539
Boarded web girder, 557
Bolection moulding, 736
Bolt box, 522, 537, 539

Bolted connections, 537–8, 541
Bolts – types, 541
Bonding, bricks, 323–9
Bonnet tiles, 431, 437
Boot lintels, 362
Bore hole, 72, 80
Borrowed light, 634
Bottom shore, 155
Bowstring truss, 573, 575
Box beam, 557
Box caisson, 292
Box pile foundation, 236, 240
Braced structures, 561
Breather membrane, 409–10, 415, 430,
    434–6
Bressummer beam, 666
Brick:
    corbel, 335–6, 805, 807
    dentil, 336
    dog-tooth, 336
    infill panels, 597
    internal walls, 621
    jointing and pointing, 332
    panel walls, 594
    plinths, 334–5
    purpose made, 334
    retaining walls, 248–9, 252–3, 258–9
    specials, 333–4
    strength, 353–4
    testing, 106
    underpinning, 295–6
Brick bonding:
    attached piers, 325–6, 329
    English bond, 323, 325, 761
    Flemish bond, 326
    principles, 324
    quetta bond, 258–9
    special bonds, 258, 323–4, 327–8
    stack bond, 328
    stretcher bond, 324
    zipper bond, 323
Brickwork cladding support, 595
Bridging of dpc's, 349
British Standards, 63
Brown field, 318
Buckling factor, 507
Builders plant, 168–204
Building control, 56–8
Building Regulations, 53–61
Building Regulations:
    application, 61
    applied, 54
    approval, 57
    Approved Documents, 54
    building notice, 56, 59
    competent persons, 57

compliance, 58
control, 56
exemptions, 60
full plans, 56, 59, 61
regularisation, 59
validation, 59
Building Research Establishment, 66
Building surveyor, 28–32
Built environment, 2–4
Built-up roofing felt, 448–51
Bulk density of soil, 86
Bulking of sand, 107
Bulldozer, 171
Busbar, 790–1
Butt weld, 540

Cab crane control, 194
Cable sizing, 794
Caissons, 292–3
Calculated brickwork, 353–5
Calculation of storage space, 102
Camber arch, 359
Cantilever:
    beam, 5, 500
    foundations, 227
    retaining wall, 254
    scaffold, 150
    stairs, 697, 703
    structures, 560
Capita Group, 479
Cap tendons, 528
Carbon emissions, 32, 481–3
Carbon index, 481
Cartridge fuse, 794
Cased pile foundation, 241
Casein adhesive, 556
Casement windows:
    ironmongery, 371
    metal, 370, 384
    timber, 368–9
Castellated beam, 535
Cast-in wall ties, 597
Cast insitu diaphragm walls, 270, 309
Cast insitu pile foundation, 244
Cast-on finishes, 615
Cavity barrier/closer, 350, 405, 434, 479,
    489–90, 606, 627–8
Cavity fire stop, 627–8
Cavity tray, 222, 360–1
Cavity walls, 338–43, 486–7
Cedar, 737
Ceilings:
    plasterboard, 642, 725, 727
    suspended, 726–9
Cellular basement, 269
Cement, 284–5, 354, 356–7

Cement grouts, 311
Central heating, 784−7
Centres, 359, 361
Certificates of ownership, 42, 46
Channel floor units, 674
Channels − structural steel, 532−3
Checked rebates in walls, 365
Chemical dpc, 347−8
Chemical grouts, 311
Cheshire auger, 78
Chezy formula, 766
Chimney pot, 22, 805, 809
Chimneys, 801, 804−7
Chipboard, 653, 740
Circular bay window, 376
Circular column, 504
CI/SfB system, 68
Cisterns, 769−72, 775, 780, 784−6
Cladding panels, 595−6, 610−13
Clamp vibrator, 203
Classification of piled foundations, 228−9
Clay cutter, 230
Clear glass, 379
Client, 50
Climbing cranes, 193, 197
Climbing formwork, 256
Closed couple roof, 422
Code for Sustainable Homes, 62
Codes of Practice, 63
Cofferdams, 290−1
Coil wall ties, 255, 519−20
Cold bridging, 350, 384, 488−9, 651
Cold deck/roof, 448, 450−1
Cold water installations, 767−70
Collar roof, 422, 436
Column design, 506−7, 548
Column formwork, 517−18
Column to column connection, 523, 538
Column underpinning, 301
Columns, 504, 533−4, 537−9
Combined column foundation, 225
Combined method, 475
Combined pad foundation, 214
Combined system, 759
Combustion air, 808
Communicating information:
    bar chart, 34, 88
    elevations, 23, 26
    floor plans, 21, 23, 26, 41
    isometric projection, 24
    orthographic projection, 23
    perspective projection, 25
    sketches, 22
Compartment floor, 18, 513
Compartment wall, 18, 513, 627
Competent persons, 57

Complete excavation, 267
Completion certificate, 31
Composite beams and joists, 557−9
Composite boarding, 739−40
Composite floors, 499, 674
Composite lintels, 362
Composite panel, 598
Composite piled foundation, 241−2
Compound sections, 534, 676
Compressed strawboard, 739
Compressible joints, 594, 610−12, 624, 701−2
Concrete:
    admixtures, 272
    beam design, 501−3
    claddings, 607−13
    column design, 506−7
    floor screed, 656−7
    mixers, 199−200
    placing, 202−3
    production, 284−9
    pumps, 202
    slab design, 501−3
    stairs, 696−704
    surface defects, 616
    surface finishes, 614−15
    test cubes, 109
    testing, 110−12
Concrete production:
    designated mix, 288
    designed mix, 288
    materials, 284
    prescribed mix, 288
    site storage, 90, 285
    specification, 288
    standard mix, 288
    supply, 289
    volume batching, 286
    weight (weigh) batching, 287
Concreting, 198
Concreting plant, 199−204
Conductivity, 470−1
Conduit and fittings, 792−3
Conoid roof, 9, 582
Consolidation of soil, 87
Construction activities, 19, 34
Construction, Design and Management
    Regs., 49−50
Construction joints, 135, 273, 624, 654
Construction Products Directive, 65
Construction Regulations, 49−50
Consumer control unit, 788−91, 795−6
Contaminated subsoil, 318−19
Contiguous piling, 308
Continuous column foundations, 224
Contraction joints, 135, 624

Index

Controlling dimensions, 48
Controlling grid and lines, 47–8
Cooker circuit cable, 794
Coping stones, 248, 252–3, 341
Corbelled brickwork, 335–6, 805, 807
Core drilling, 79
Core structures, 561
Cored floor units, 674
CORGI, 479
Cor-ply beam, 557
Cornice, 736
Corrugated sheet, 564–8
Coulomb's line, 86
Coulomb's wedge theory, 264
Counter batten, 415, 430, 435–6
Couple roof, 422
Cove mouldings, 641, 725
Cover flashing, 806
CPI System of Coding, 67
Cracking in walls, 207–8, 211
Cradles, 149
Crane operation, 194
Crane rail track, 196
Crane skips, 198, 201
Cranes, 186–97
Cranked slab stairs, 696, 702
Crawler crane, 190
Crib retaining wall, 260
Crosswall construction, 400–3
Crown hinges, 551
Curtain walling, 603–6
Curved laminated timber, 555
Curved tendons, 528
Cut and fill, 70, 278
Cylinders – hot water, 776

Dado panel and rail, 736
Damp proof course, 344–50
Damp proof course materials, 345–6
Damp proof membrane, 344, 351, 647–8, 657
Datum, 125–30
Datum post, 125, 128
Dead loads, 35–6, 159, 218
Dead shoring, 153–4, 159–60
Death Watch beetle, 463
Decorative suspended ceiling, 729
Decorative roof/wall tiles, 413
Deep basements, 270–1
Deep strip foundation, 213, 224
Defects in painting, 733
Deflection, 496, 546, 557, 665
Demolition, 162–6
Demountable partitions, 634–5
Dense monolithic concrete, 272
Dentil course, 336

Density of materials, 36, 470–1
Designer, 50
Desk study, 72
Dewatering, 302–6
Diaphragm floatvalve, 774
Diaphragm walls, 270–1, 309–10, 343
Diesel hammers, 246
Dimensional coordination, 47–8
Dimensional grids, 47–8
Direct cold water system, 769
Direct hot water system, 771
Displacement pile foundation, 228, 236–44
Disturbed soil sample, 73–4, 77
Documents for construction, 20
Dog leg stair, 693–4
Dog toothed brickwork, 336
Domelights, 593
Domes, 9, 578–9
Domestic drainage, 746–66
Domestic electrical installations, 788–96
Domestic floor finishes, 652–3
Domestic ground floors, 647–51
Domestic heating systems, 784–7
Doors:
    external, 391–4
    fire, 719–23
    frames, 394, 717–22
    glazed double swing, 724
    industrial, 396–9
    internal, 391, 715–24
    ironmongery, 395
    linings, 715, 717
    performance, 391
    sliding, 396–8
    types, 392–3
Door set, 718–19
Dormer window, 419, 456–8
Double acting hammers, 246
Double action floor spring, 724
Double flying shore, 157
Double glazing, 369, 375, 381–5
Double hung sash windows, 367, 372–3
Double lap roof tiling, 430–4, 437–8
Double lap wall tiling, 410–14
Double layer grids, 578
Double rafter roof, 423
Double swing doors, 724
Douglas fir, 737
Dovetail anchor slots, 594, 597
Draft for Development, 63
Draglines, 178
Drainage:
    effluent, 746
    gradient, 763–4, 766
    paved areas, 751
    pipe sizes, 766

proportional depth, 766
rainwater, 754−6
roads, 752−3
roofs, 750
simple, 757−64
systems, 757−60
testing, 765
Drained cavities, 276
Draught proofing, 491
Drawings:
    axonometric projection, 24
    construction process, 21
    hatchings, symbols and notation, 38−41
    isometric projection, 24
    orthographic projection, 23
    perspective projection, 25
    plans and elevations, 26
    sketches, 22
Drier, 730
Drilling rigs, 230−3
Driven insitu piled foundations, 243
Drop arch, 359
Drop hammer, 230, 238, 241−3, 245
Dry linings, 639−41
Dry rot, 121−2
Dumpers, 181, 201
Dumpling, 266
Dynamic compaction, 313, 316

Earth pressures, 251, 263−4
Earthworks for roads, 132−3
Eaves:
    closure piece, 565−6
    details, 433−5, 440−1, 446, 448, 451, 485
    filler piece, 565−6
    gutter, 433, 440−1, 485, 568, 750, 754
    ventilation, 433−4, 440−1, 448, 451, 459−61
Effective height of walls, 355
Effective length of columns, 507, 548
Effective thickness of walls, 353, 355
Effluents, 746
Electrical cables, 794
Electrical installations, 792−6
Electrical site lighting, 94−6
Electricity:
    domestic supply, 788−91
    supply to sites, 97
Elevations, 23, 26
End bearing pile foundation, 228
Energy efficiency, 32, 477−8, 483−4
Energy performance certificate, 32
Energy rating, 32, 477, 480−1
Energy roof system, 567
English bond, 323, 325, 761

Environment, 2−4
Equivalent area, 506−7
Espagnolette, 375
European spruce, 737
European Standards, 64
Excavating machines, 175−9, 268
Excavations:
    basement, 265−8
    dewatering, 303, 306
    oversite, 278
    pier holes, 279
    reduced level, 278
    setting out, 125−6
    trench, 279
    temporary support, 281−3
Expansion joint, 135, 273, 581, 594−5, 610, 612−13, 624, 652
Exposed aggregate, 614−15
External asphalt tanking, 274
External doors, 391−4
External envelope, 17, 322
External escape stairs, 705
Eyebrow dormer window, 458

Facade bracing, 140
Face grid, 48
Face shovel, 176, 268
Facings to panel walls, 596
Factories Acts, 49
Fencing, 89−93, 100, 123
FENSA, 57
Festoon lighting, 96
Fibreboard, 740
Field study, 72
Fillet weld, 537−40, 676
Finger joint, 553
Finishes, 14, 614−15, 636−8, 652−3, 720−3
Fin wall, 342
Fire back, 803, 808
Fire doors, 719−23
Fire doorsets and assemblies, 719
Fire protection of masonry walls, 626
Fire protection of steelwork, 542−3, 633
Fire protection of timber floors, 667
Fire resistance, 350, 542−3, 626, 633, 667, 720
Fire resistance of steelwork, 542
Fire resistance/insulation, 667, 720
Fire resistance/integrity, 350, 667, 720
Fire resisting doors, 722
Fire stops and seals, 627−8, 810
Fireplaces and flues, 800−9
Fish plates, 538
Fixed portal frame, 549−50, 552
Flashings, 341, 438, 449, 451, 806−7

Index

Flat roofs, 417, 447–56, 461, 466
Flat sawn timber, 554
Flat slabs, 496–8, 668
Flat top girder, 569, 572
Flemish bond, 326
Flexible paving, 133
Flight auger, 78, 234
Flitch beam, 557
Float valves, 774–5
Floating floor, 651, 683–4
Floating pile foundation, 228
Floor plans, 26, 41
Floor springs, 724
Floors:
  domestic ground, 647–51
  finishes, 652–3
  fixings, 672
  flat slab, 496–8, 668
  hollow pot, 671–2
  insitu RC suspended, 496–8, 668
  large cast insitu, 654–5
  loads, 37
  precast concrete, 673–5
  ribbed, 669–70
  screeds, 647–8, 656–7
  sound insulation, 679–80, 683–4
  suspended concrete ground, 352, 650–1
  suspended timber ground, 547, 549
  suspended timber upper, 658–63, 667
  waffle, 669–70
Flue blocks, 800, 809
Flue lining, 800–1, 803, 805–6, 808
Flush bored pile foundation, 229, 231
Fly jib, 189–90
Flying shore, 153, 156–8
Folded plate construction, 8, 587
Footpaths, 136
Forklift truck, 182
Formwork:
  beams, 515–17
  columns, 517–18
  principles, 514
  slab, 498
  stairs, 700
  walls, 519–20
Foundation hinges, 551
Foundations:
  basic sizing, 216
  beds, 215
  calculated sizing, 218
  defects, 207–9
  design procedure, 216–18, 223
  functions, 205
  grillage, 225
  isolated pad, 225
  piled, 221, 230–47

raft, 222–3
short bored piled, 221, 228–30
simple reinforced, 220
stepped, 219
strip width guide, 217
subsoil movement, 207–10
types, 223–7
Four-in-one bucket, 174
Four wheeled grader, 173
Framed structures, 500, 504, 521–4, 537–9, 560
Freyssinet anchorage, 529
Friction hinge, 375
Friction piling, 228
Frieze, 736
Frodingham box pile, 240
Frodingham sheet pile, 291
Full height casting, 255
Fungicide, 122
Furniture beetle, 463

Gabion wall, 262
Gable end roof, 418–20
Gambrel roof, 419, 427
Gantry crane, 191
Gantry girder, 534
Gantry scaffolding, 152
Garage, 330–1
Garden wall bonds, 327
Gas:
  fires, 799
  flues, 799–800, 809
  service intake, 797–8
  supply, 797–8
Gas resistant membranes, 351–2, 651
Gas Safe Register, 57, 479
Gate valve, 774–5
Gauge box, 286
Geodistic dome, 578
Girders, 534, 569, 571–2
Glass and glazing, 379–88
Glass block wall, 389–90
Glass manifestation, 388
Glazed cladding, 601–2
Gluelam, 553–6
Grab bucket, 178
Graders, 173
Granular soil shell, 230
Green field, 318, 351
Green roof, 465–6
Grillage foundation, 225
Groins, 582
Ground anchors, 270, 310, 531, 588
Ground freezing, 312
Ground vibration, 313–15
Ground water, 70, 72, 80, 302

Ground water control, 303–12
Grouting-subsoil, 311
Grouting-tiles, 646
Guard rail, 141–3, 145, 147, 150–1
Gusset base, 539
Gusset plate, 552, 563, 566

Half hour fire door, 720
Hammer head junction, 132
Hand auger, 78
Hand auger holes, 73
Handrail, 685–6, 688, 690, 693–5, 698–9,
    704–6, 708, 710–11
Hardwoods, 738
Half space landing, 693–4
Health and Safety at Work, etc. Act, 49, 51
Hearths, 801–3, 808
Helical binding, 504–5
Helical stairs, 698
Helmets, 238, 242
Hemispherical dome, 579
Herringbone strutting, 659
HETAS, 479
High performance window, 369
Highway dumpers, 181
Hinge or pin joints, 548–9, 551
Hipped roof, 418–21
Hip tiles, 431, 437
Home Information Pack, 32
Hoardings, 89–93
Hoists, 183–4
Holding down bolts, 522, 537, 539, 551–2
Hollow box floor units, 673
Hollow pot floor, 671–2
Hollow steel sections, 532–3
Horizontal shore, 156–7
Horizontal sliding sash window, 374
Hot water:
    cylinders, 769–73, 776
    direct system, 771
    expansion vessel, 772
    heating systems, 784–7
    indirect system, 772
    mains fed, 773
House Longhorn beetle, 463
Hull core structures, 561
Hyperbolic paraboloid roof, 9, 583–5, 588

Immersion heater cable, 794
Imposed floor load, 37
Inclined slab stair, 696, 699
Independent scaffold, 124
Indirect cold water system, 770
Indirect cylinder, 772–3, 776
Indirect hot water system, 772

Industrial doors, 396–9
Infill panel walls, 596–9
Inspection chambers, 746, 752, 757–63–9
Insulating dpc, 350, 489
Insulation of basements, 277
Insulation – fire, 720
Insulation – sound, 679–84
Insulation – thermal, 467–91
Integrity – fire, 720, 722
Interest on capital outlay costing, 169
Internal asphalt tanking, 275
Internal doors, 391, 716–23
Internal drop hammer, 230, 241, 243
Internal elements, 618
Internal environment, 2
Internal partitions, 629–35
Internal walls:
    block, 622–9, 680–1
    brick, 621, 626–9, 680–1
    functions, 619
    plaster finish, 636–8
    plasterboard lining, 632–4, 639–44
    types, 620
International Standards, 64
Intumescent collar, 810
Intumescent strips/seals, 720–3
Inverted warm deck, 450
Iroko, 738
Ironmongery, 371, 375, 395, 719
Isolated foundations, 214, 225
Isometric projection, 24

Jack pile underpinning, 297
Jambs, 344, 350, 363, 365, 488–9, 491
Jarrah, 738
Jet grouting, 311, 313, 317
Jetted sumps, 304
Jetted well points, 305–6
Joinery production, 734–8
Joinery timbers, 737–8
Jointing and pointing, 332
Jointless suspended ceiling, 727
Joints:
    basements, 273
    blockwork, 621–3
    cladding, 610–13
    drainage, 764
    laminated timber, 553
    portal frame, 549–52
    roads, 135
Joists:
    timber, 559, 649, 658–67
    steel, 533–4
Joist sizing – timber, 658, 663–5
Joist sizing – steel, 544–6

Index

Kerbs, 137
Kelly bar, 231—4
Kentledge, 149, 247, 301
Kitemark, 63

Laboratory analysis of soils, 72, 81—2,
    85—7
Ladders, 141—2, 145
Laminated timber, 553—6
Land caissons, 292
Land reclamation, 318
Landings, 693—706
Lantern lights, 592
Large diameter piled foundations, 229, 233
Larssen sheet piling, 291
Latent defects, 28
Lateral restraint, 462, 658—61
Lateral support — basements, 271
Lateral support — walls, 425, 462, 658,
    661—2
Lattice beam, 426, 536, 557, 569, 571—2
Lattice girder, 536
Lattice jib crane, 189
Lattice truss, 426, 559, 569—75
Leader — piling rig, 238
Lean-to roof, 418, 422, 461
Lens light, 592
Levelling, 129—31
Lift casting, 256
Lighting:
    cable, 794
    circuits, 794, 796
    sites, 94—6
Lightweight decking, 577
Lightweight infill panels, 596, 598—9
Limiting U values, 478, 483—4
Lintels, 358, 362, 365, 621, 625
Litzka beam, 535
Load-bearing concrete panels, 607
Load-bearing partitions, 618, 620—2, 629
Local authority building control, 57—8
Locating services, 124
Loft hatch, 491
Long span roofs, 569—78
Loop ties, 257
Lorries, 180
Lorry mounted cranes, 188—9
Loss Prevention Certification Board, 66
Low emissivity glass, 383, 385—6
Luffing jib, 192
Lyctus Powder Post beetle, 463

Maguire's rule, 766
Mahogany, 738
Main beams, 7, 16, 538

Mandrel, 242
Manifestation of glass, 388
Manometer, 765, 783
Mansard roof, 419, 427
Masonry cement, 356
Masonry partitions, 629
Mass concrete retaining wall, 253
Mass retaining walls, 252—3
Mast cranes, 192
Mastic asphalt tanking, 274—5
Mastic asphalt to flat roofs, 451
Mastic sealant, 742
Materials:
    conductivity, 470—1
    density, 36, 470—1
    hoist, 183
    storage, 89—91, 100—5, 284—5
    testing, 106—19
    weights, 35—6
Mathematical tiles, 415
Mattress retaining wall, 262
MD Insurance, 28, 58
Mechanical auger, 78
Meeting rails, 372—3
Meeting stiles, 374, 724
Metal casement windows, 370
Metal section decking, 499, 677
Metal stairs, 705—7
Meter box, 788—90, 797—8
Methane, 351
Method statement, 33
Micro-bore heating, 786
Middle shore, 155
Middle third rule, 249
Mineral insulating cable, 792
Miniature circuit breaker, 790—1
Mixing concrete, 198—200, 286—8
Mobile cranes, 186—90
Mobile scaffold, 145
Modular coordination, 47—8
Modular ratio, 506
Mohr's circle, 86
Moment of resistance, 160, 544
Monitor roof, 569, 571
Monogroup cable, 526
Monolithic caissons, 292
Monopitch roof, 418, 461
Monostrand anchorage, 529
Monostrand cable, 526
Mortar mixes and strength, 354, 356—7
Morticing machine, 734
Movement joint, 135, 581, 595, 610—13,
    624, 701
Mud-rotary drilling, 79
Multi-purpose excavators, 175, 179
Multi-span portal frames, 550, 552

Multi-stage wellpoints, 306
Multi-storey structures, 560−1

National Building Specification, 67
Needle and pile underpinning, 298
Needle design, 159−60
Needles, 153−60
Newel post, 685, 688, 690−1, 693−5, 710−11
NHBC, 28, 58
Non-load-bearing partitions, 618, 620, 629−34
Non-residential buildings, 330−1
Northlight barrel vault, 582
Northlight roofs, 569−70

Oak, 738
Oedometer, 87
OFTEC, 479
Oil based paint, 730
One hour fire door, 721
One pipe heating, 784
Open caissons, 292
Open excavations, 265
Open fireplaces, 801−9
Open newel stairs, 693
Open riser stairs, 690−2, 695, 703
Open suspended ceiling, 729
Open web beam, 535
Openings in walls:
    arches, 358−61
    heads, 362, 365
    jambs, 363, 365
    sills, 364−5
    support, 358
    threshold, 365, 492
Oriel window, 376
Oriented strand board, 740
Orthographic projection, 23
Out-of-service crane condition, 193
Output and cycle times, 170
Overcladding, 596, 600
Overhead and forklift trucks, 182

Pad foundation, 214, 225, 504, 537, 550, 552
Pad foundation design, 216
Pad template, 127
Paint defects, 733
Paints and painting, 730−3
Panelled suspended ceiling, 727−8
Parallel lay cable, 526
Parallel strand beam, 558
Parana pine, 737
Parapet wall, 341

Partially preformed pile, 242
Particle board, 653, 740
Partitions:
    demountable, 629, 634−5
    load-bearing, 618, 620−2, 629
    metal stud, 632−3
    non-load-bearing, 618, 620, 629−35
    timber stud, 629−31
Party wall, 627−8
Passenger hoist, 184
Passenger vehicles, 180
Passive earth pressures, 251
Patent glazing, 570−1, 590−1
Patent scaffolding, 147
Paved areas − drainage, 751
Paving, 132−8
Paving flags, 138
Pedestal floor, 678
Peg tile, 432
Pendentive dome, 9, 579
Penetration test, 83, 111
Percussion bored piling, 229−30
Perforated beam, 535
Performance requirements:
    doors, 391
    roofs, 416
    windows, 366
Perimeter trench excavation, 266
Permanent exclusion of water, 302, 307−11
Permanent formwork, 499, 677
Permitted development, 42
Perspective drawing, 25
Phenol formaldehyde, 556, 741
Pigment, 730
Pile:
    angle, 300
    beams, 247
    caps, 232, 247
    classification, 228
    root, 300
    testing, 247
    types, 229, 236
Piled basements, 269
Piled foundations, 228−47
Piling:
    contiguous, 308
    hammers, 245−6
    helmets, 238, 242
    rigs, 238, 241−2
    secant, 308
    steel sheet, 290−1
Pillar tap, 774
Pin or hinge joint, 548−9, 551
Pitch pine, 737
Pitched roofs − domestic, 416, 418−46
Pitched roofs − industrial, 562−70

Pitched trusses, 424−5, 562−3, 569−70
Pivot window, 367, 375
Placing concrete, 198
Plain tiles and tiling, 410−14, 430−4, 437−8
Planer, 734
Plank and pot floor, 674
Planning application, 42−6
Planning grid, 47
Plant:
  bulldozer, 171
  concreting, 198−204
  considerations, 89, 168
  costing, 169
  cranes, 186−97
  dumpers, 181, 201, 280
  excavators, 175−9, 280
  forklift trucks, 182
  graders, 173
  hoists, 183−4
  scrapers, 172
  skimmers, 175
  tractor shovel, 174
  transport vehicles, 180−1
Plaster cove, 641, 725
Plaster finish, 637−8, 642
Plasterboard, 470, 639−44
Plasterboard ceiling, 433, 440−1, 476, 485, 683−4, 725
Plasterboard dry lining, 639−41, 643−4
Plasters, 636
Plasticiser, 356
Platform floor, 684
Platform frame, 404
Plinths, 334−5
Plug wiring, 794
Plywood, 739
Pneumatic caisson, 293
Poker vibrator, 203
Poling boards, 282−3
Polyurethane paint, 730
Portal frames, 549−52, 555
Portsmouth float valve, 774
Post-tensioned retaining wall, 259
Post-tensioning, 259, 528−9
Power circuit, 794−1
Power float, 204, 654−5
Precast concrete:
  diaphragm wall, 310
  floors, 650−1, 673−7
  frames, 521−4
  portal frames, 550−1
  stairs, 701−4
Preformed concrete pile, 238−9
Preservative treatment, 122, 463−4, 556
Pressed steel lintels, 362, 365

Pressure bulbs, 75−6
Prestressed concrete, 270, 362, 525−31
Pretensioning, 527
Primary elements, 12
Primatic cylinder, 776
Principal contractor, 50
Private sector building control, 57
Profile boards, 126−7
Profiled surface, 138, 493, 614
Program of work, 34
Programmer/controller, 787
Project co-ordinator, 50
Prop design, 160−1, 631
Proportional area method, 474, 476
Propped structures, 5, 560
Protection orders, 123
Public and general use stairs, 709
Public utility services, 124
Published Document, 63
Pump sizing, 303
Purlin fixings, 565−8, 570
Purlin roof, 421, 423
Purpose designed excavators, 175
Purpose made joinery, 735−6
Putlog scaffold, 141
Putty, 380
Pynford stool underpinning, 299

Quarter sawn timber, 554
Quarter space landing, 693
Quetta bond, 258−9

Radiators, 784−7
Radius of gyration, 547
Radon, 351
Raft basements, 269
Raft foundations, 222−3, 226
Rail tracks for cranes, 196
Rainscreen cladding, 596, 600
Rainwater drainage, 749−51, 754−6
Raised access floor, 678
Raking shore, 153, 155, 158
Raking struts, 156−7, 267
Rankine's formula, 263
Rat trap bond, 327
Ready mixed concrete, 289
Ready mixed concrete truck, 201, 289
Ready mixed mortar, 357
Redwood, 737
Reinforced concrete:
  beams, 500−2
  column design, 506−7
  columns, 504−7
  floors, 499, 668−72
  formwork, 498, 514−18, 700

foundations, 220–2, 224–6
lintels, 362
pile caps and beams, 232, 247
raft foundation, 222, 226
reinforcement, 497–513
retaining walls, 254
slabs, 496–9, 501–3
stairs, 696–700
strip foundations, 220, 224
Reinforced masonry, 258–9, 328, 625
Reinforcement:
    bar coding and schedules, 508–9
    concrete cover, 512–13
    grip length, 501, 503
    materials, 510–11
    spacers, 512
    spacing, 505
    types of bar, 510
    types of mesh, 511
Remedial dpc, 347–8
Remote crane control, 194
Rendering, 408–9
Rendhex box pile, 240
Replacement piling, 228–34
Residual current device, 791
Resin grout, 311
Resorcinol formaldehyde, 556, 739
Restraint straps, 462, 658–61
Retaining walls, 248–64
Retaining walls – design, 263–4
Retro-ties and studs, 661
Reversing drum mixer, 200
Ribbed floor, 669–70, 684
Rider, 155
Ridge detail, 433–4, 440–1, 446, 566, 568, 570
Ridge piece, 565–6, 568, 570
Ridge roll, 446
Ridge tiles, 433–4, 437, 440–1
Ridge ventilation, 434, 459–61
Rigid pavings, 134–5
Rigid portal frames, 549–50, 552
Ring final circuit wiring, 795
Rising damp, 346–9
Roads:
    construction, 132–5
    drainage, 752–3
    earthworks, 133
    edgings, 136–7
    footpaths, 136
    forms, 134
    gullies, 751–3
    joints, 135
    kerbs, 137
    landscaping, 139
    pavings, 138
    services, 139
    setting out, 132
    signs, 139
Robust Details Ltd., 58
Rolled steel joist, 533
Roll over crane skip, 198
Roller shutters, 396, 399
Roofs and roof covering:
    basic forms, 417–19
    built-up felt, 448–9, 450–1
    flat top girder, 569, 572
    garden/garden, 465–6
    lead sheet, 452–6
    long span, 569–75
    mastic asphalt, 451
    monitor, 569, 571
    northlight, 569–70
    performance, 416
    purlin, 421, 423
    sheet coverings, 564–8
    shells, 579–86
    slating, 441–5
    space deck, 577
    space frame, 578
    sprocket eaves, 429
    surface water removal, 746, 749–60
    thatching, 446
    thermal insulation, 433–6, 440–1, 448–51, 459–61, 466, 476, 478–9, 484–5, 565–8, 570
    tiling, 430–4, 437–40
    timber flat, 417, 448–56
    timber pitched, 418–29
    trussed rafter, 425
    trusses, 424, 563, 569–70
    underlay, 430, 433–8, 440–1, 444–5, 453–4
    valley, 428
    ventilation, 459–61
Rooflights, 419, 569, 571, 591–3
Room sealed appliance, 799
Root pile, 300
Rotary bored piling, 229, 232–4
Rotational dome, 9, 579
Rubble chutes and skips, 185
Runners, 283

Saddle vault, 583
Safe bearing, 216–18, 666
Safety signs, 51–2, 139
Sampling shells, 78
Sand bulking test, 107
Sand compaction, 315
Sand pugging, 684
Sanitary fittings:
    basin, 778

Index

Sanitary fittings (*continued*)
  bath, 779
  discharge systems, 781–2
  shower, 779
  sink, 778
  wc pan, 780
Sanitation systems, 781–3
Sanitation system testing, 783
Sapele, 738
Sarking felt, 430
Sash weights, 372
Saws, 734
Scaffolding:
  birdcage, 148
  boards, 141–3
  cantilever, 150
  component parts, 140, 146
  fittings, 146
  gantry, 152
  independent, 142
  ladders, 141–2, 145
  mobile, 145
  patent, 147
  putlog, 141
  slung, 148
  suspended, 149
  truss-out, 151
  tying-in, 144
  working platform, 143
Scarf joint, 553
Scrapers, 172
Screed, 647–8, 657–8
Sealants, 742–4
Secant piling, 308
Secondary beams, 7, 16, 538
Secondary elements, 13
Secondary glazing, 382
Section factor, 543
Section/elastic modulus, 160, 544
SEDBUK, 478
Segmental arch, 359
Self propelled crane, 187
Self supporting static crane, 193–4
Separate system, 758, 760
Separating wall, 18, 627
Services fire stops and seals, 810
Setting out:
  angles, 127, 131
  bases, 127
  basic outline, 125
  drainage, 764
  grids, 127
  levelling, 129–30
  reduced levels, 128
  roads, 132
  theodolite, 127, 131

trenches, 126
Shear, 499–500, 545, 666
Shear bars, 500
Shear box, 87
Shear leg rig, 78–9, 230
Shear plate connector, 576
Shear strength of soils, 85–7
Shear strength of timber, 666
Shear wall structures, 561
Sheathed cables, 792
Sheet coverings, 564–8
Shell roofs, 579–86
Shoring, 153–60
Short bored pile foundation, 221
Shower, 779
Shutters, 399
Sight rails, 127–8, 132, 764
Sills, 364–5, 368–70, 372–6, 384, 386
Silt test for sand, 108
Simple drainage:
  bedding, 764
  inspection chambers, 757–63
  jointing, 764
  roads, 752–3
  setting out, 764
  systems, 757–60
Simply supported RC slabs, 496–8
Single acting hammer, 245
Single barrel vault, 580
Single flying shore, 153, 156
Single lap tiles and tiling, 439–40
Single span portal frames, 549–52, 555
Single stack drainage, 781
Sink, 778
Site:
  construction activities, 19
  electrical supply, 97
  health and welfare, 99
  investigations, 71–2, 74, 313
  layout, 88–90
  lighting, 94–6
  materials testing, 106–19
  offices, 89–90, 98, 103
  plan, 21, 27, 42
  preparation, 318–19
  security, 89, 91
  setting out, 125–32
  soil investigations, 72–9
  storage, 100–5, 284–5
  survey, 70, 72, 318
Sitka spruce, 737
Six wheeled grader, 173
Sixty minute fire door, 721
Sketches, 22
Skimmer, 175
Skips, 185

Slab base, 539
Slates and slating, 441−5
Slenderness ratio, 160−1, 355, 547
Sliding doors, 396−8
Sliding sash windows, 372−4
Slump test, 109
Slung scaffold, 148
Small diameter piled foundation, 221, 229−32
Smoke seal, 722
Smoke test, 783
Snow loading, 218, 544
Soakaway, 756
Soakers, 411−12, 438, 444, 806
Socket outlets, 794−5
Softwoods, 737
Soil:
    assessment, 81−7
    classification, 81−3, 217
    investigation, 70−9, 313
    nailing, 261
    particle size, 81
    samples, 73
    stabilization and improvement, 313−17
    testing, 81−7
    washing, 319
Solid block walls, 337, 622, 624−6, 680−2
Solid brick walls, 323−31, 626−7
Solid slab raft, 226
Solvent, 730
Sound insulating:
    floors, 680, 683−4
    walls, 680−2
Sound insulation, 679−84
Sound reduction, 381, 679
Space deck, 10, 577
Space frame, 10, 578
Spindle moulder, 734
Spine beam stairs, 703
Spiral stair, 695, 698, 704, 706, 713−14
Splice joint, 552
Splicing collar, 238−9
Split barrel sampler, 83
Split load consumer unit, 791
Split ring connector, 576
Spruce, 737
Stack bond, 328
Stairs:
    alternating tread, 692
    balusters, 685, 688−91, 694−5, 698, 704, 706, 708
    balustrade, 13, 698−9, 704, 706−8
    formwork, 700
    handrail, 685−6, 688, 690, 693−5, 698−9, 704−6, 708, 710−11
    insitu RC, 696−700

metal, 705−7
open riser, 690−2, 695, 703
precast concrete, 701−4
public, 709
timber, 685−95
Standard Assessment Procedure, 32, 477, 481
Standard dumper, 181, 201
Standing crane skip, 198
Star rating, 62
Steel:
    beam, 533−5, 538, 676−7
    beam design, 544−6
    bolt connections, 541
    cold rolled sections, 532
    column, 533−4
    column design, 547
    compound sections, 534, 676
    fire protection, 542−3
    frame construction, 406−7
    gantry girder, 534
    grillage, 225
    hot rolled sections, 533
    lattice beams, 536
    portal frames, 552
    roof trusses, 562−3, 566, 569−71
    screw pile, 240
    sheet piling, 290−1
    standard sections, 532−3
    string stairs, 707
    tube pile, 241
    web lattice joist, 559
    welding, 539−40
Stepped barrel vault, 582
Stepped flashing, 438, 807
Stepped foundation, 219
Stock holding policy, 103
Stop valve, 767−70, 774
Storage of materials, 90, 100−5, 284−5
Storey height cladding, 613
Straight flight stairs, 685−92, 701
Straight line costing, 169
Straight mast forklift truck, 182
Strand, 526
Strawboard, 739
Stress reduction in walls, 355
Stretcher bond, 258, 324
String beam stair, 697
Strip foundations, 213, 216−19
Stripboard, 739
Structural glazing, 601−2
Structural grid, 47
Structure:
    basic forms, 7−10
    basic types, 5−6

Structure (*continued*)
  functions, 15−16
  protection orders, 123
Strut design, 161, 631
Strutting of floors, 659
Stud partitions, 629−33
Subsoil:
  drainage, 747−8, 752
  movements, 207−11
  water, 70, 72, 80, 302
Substructure, 11
Sump pumping, 303
Supply and storage of concrete, 198
Supported static tower crane, 193, 195
Surcharging of soil, 313
Surface water, 302, 746, 749−53
Surface water removal, 749−51
Survey:
  defects, 28−32, 211
  desk, 72
  field, 72
  site, 70−2, 74, 313, 318
  structure, 28−32
  thermographic, 484
  walk-over, 72
Suspended ceilings, 726−9
Suspended scaffold, 149
Suspended structures, 560
Suspended timber ground floors, 647, 649
Suspended timber upper floors, 658−63, 667
Sustainability, 62
Sustainable demolition, 166
Swivel skip dumper, 181

Tactile pavings, 138
Tamping board vibrator, 203
Tapered treads, 695, 698, 704, 706, 710−12
Teak, 738
Telephone installation, 811−12
Telescopic boom forklift truck, 182
Telescopic crane, 188
Temporary bench mark, 125, 127−9
Temporary exclusion of water, 303−6
Temporary services, 89
Tendons, 261, 526
Tenoning machine, 734
Test cubes, 109
Testing of materials, 106−19
Textured surfaces, 614
Thatched roof, 446
Theodolite, 127, 131
Thermal break, 384
Thermal bridging, 479, 488−90
Thermal conductivity, 468, 470−4, 476
Thermal insulation, 467−91
Thermal resistance, 467−9, 473−6

Thin grouted membranes, 307
Thinners, 730
Thirty minute fire door, 720
Three axle scraper, 172
Three centre arch, 359
Three pin portal frame, 549
Tile hanging/cladding, 410−15
Tilting drum mixer, 199−200
Tilting level, 129
Timber:
  beam design, 663−6
  beam notching, 666
  casement windows, 368−9
  connectors, 575−6
  doors, 391−4, 397−8, 715−6, 718, 720−3
  flat roofs, 417, 447−56
  frame construction, 404−5
  girders, 572
  grade markings, 113
  hardwoods, 738
  joinery production, 734−8
  pile foundation, 236−7
  pitched roofs, 416, 418−30, 433−8, 440−1, 462
  preservation, 122, 463−4
  sizes and finishes, 120
  softwoods, 737
  stairs, 685−95
  storage, 105
  strength classes/properties, 114−15
  stud partition, 629−31
  trestle, 241
  truss, 424−5, 573−6
  visual strength grading/characteristics, 116−19
Timbering trenches, 282−3
Tooled surface, 614
Toothed plate connector, 424, 575−6
Top shore, 155, 158
Towed scraper, 172
Tower cranes, 90, 186, 193−6
Tower scaffold, 145
Track mounted crane, 190
Tractor shovel, 174
TRADA, 55, 113
Traditional strip foundation, 213, 224
Traditional underpinning, 294−6
Translational dome, 9, 579
Translucent glass, 379
Transport vehicles, 180−1
Transporting concrete, 198, 201
Traveller, 126−8, 764
Travelling tower crane, 193, 196
Tree protection, 123, 210
Trees − foundation damage, 208−9

Tremie pipe, 231, 309
Trench fill foundation, 213, 224
Trench setting out, 126
Trench sheeting, 283, 290−1
Trial pits, 73−4, 77
Triangle of forces, 249, 264
Triangular chart, 82
Triaxial compression test, 86
Triple glazing, 385−6
Tripod rig, 78−9, 230
Truss out scaffold, 151
Trussed purlin, 426, 557
Trussed rafter roof, 425
Tubular scaffolding, 140−52
Two axle scraper, 172
Two pin portal frame, 549
Two pipe heating, 785
Tying-in scaffolding, 144

U value calculations, 467−76
U value limits, 478, 483−4
U value objectives, 478−9, 483−4
U value targets, 478, 484
U value weighted average, 478−9, 484
U values, 381, 383, 467−76, 478−80,
    483−4, 486−7, 567
Unconfined compression test, 85
Undercoat, 731
Underlay, 430, 433−8, 440−1, 444−5,
    453−4
Underpinning, 294−301
Under-reaming, 233
Undersill cladding panels, 608
Undisturbed soil sample, 73−4, 77, 85−6
Universal beam, 533
Universal bearing pile, 240
Universal column, 533
Universal excavator, 175
Urea formaldehyde, 556, 741

Vacuum dewatering, 655
Valley beam, 570
Valley gutter, 440, 568, 570
Valley tiles, 431, 437
Valves and taps, 774
Vane test, 84
Vans, 180
Vapour check plasterboard, 433, 435−6,
    440, 448, 451, 459, 642
Vapour control layer, 405, 407, 433, 435−6,
    448−51, 466, 567−8
Vapour permeable underlay, 409−10, 415,
    430, 433
Vaults, 580−2
Ventilated stack discharge system, 782

Ventilation of roof space, 430, 433−6,
    440−1, 448, 451, 459−61
Ventilation spacer, 433−4, 440−1, 485
Verge details, 418−20, 438, 449
Vertical casting, 615
Vertical laminations, 554
Vertical tile hanging, 410−15
Vestibule frame, 387
Vibrating tamping beam, 203
Vibration of soil, 313−15
Vibro cast insitu piling, 244
Volume batching, 286

Walings, 282−3
Walk-over syrvey, 72
Wall hooks, 155, 157−8
Wall plates, 155−8, 422−9, 433−5, 440−1,
    446, 448, 451, 462, 485, 649
Wall profiles, 623
Wallboard, 641−2
Walls:
  cavity, 338−43, 486−7
  curtain, 603−6
  design strength, 353−4
  diaphragm, 270−1, 309−10, 343
  fin, 342
  formwork, 255−7, 519−20
  glass block, 389−90
  infill panel, 596−9
  internal, 619−29, 680−2
  openings, 358−65
  panelling, 736
  remedial work, 661
  retaining, 248−64
  rising dampness, 346−9
  slenderness ratio, 160, 355
  solid block, 337, 622, 624−6, 680−2
  solid brick, 323−31, 680−2
  sound insulation, 680−2
  thermal insulation, 468−71, 474−5,
      478−9, 484, 486−7
  thickness, 330−1, 353, 355
  tiling, 410−15, 645−6
  underpinning, 294−300
  waterproofing basements, 272−6
Warm deck/roof, 436, 448−50
Wash boring, 79
Water bar, 273−7, 394
Water based paint, 730
Water/cement ratio, 272, 287
Water closet pan, 780
Water installations:
  cisterns, 769−72, 775, 780, 784−6
  cold, 767−70
  hot, 770−2
  pipe joints, 777

Water installations (*continued*)
  supply, 767—8
  valves and taps, 774
Water jetting, 304—6, 315
Water table, 70, 72, 80, 302
Water test, 765
Waterproofing basements, 272—6
Weatherboarding, 410
Web cleats, 537—8
Weep holes, 248—54, 362, 365
Weight batching, 287
Weights of materials, 35—6
Welded connections, 537—40
Wellpoint systems, 305—6
Western hemlock, 737
Western red cedar, 737
Wet rot, 121—2
Winchester cutting, 414
Windows:
  aluminium casement, 384
  bay, 376
  double glazing, 369, 375, 381—5
  energy rating, 480
  glass and glazing, 379—88
  high performance, 369
  ironmongery, 371
  metal casement, 370
  notation, 378
  oriel, 376
  performance requirements, 366
  pivot, 367, 375
  schedules, 377
  secondary glazing, 382
  sliding sash, 372—4
  timber casement, 368—9
  triple glazing, 383, 385—6
  types, 367
  uPVC, 386
Windsor probe test, 111
Wired glass, 379, 723
Wiring systems, 795—6
Wood cement board, 740
Wood rot, 121—2
Wood wool, 740
Woodworm, 463—4
Woodworking machinery, 734
Working platform, 143
Wrought iron dogs, 154, 156—7

Yokes, 518

Zed beam, 532, 565, 567—8
Zero carbon home, 62
Zipper bond, 323
Zones, 48